Exploring the World Ocean

W. Sean Chamberlin
Fullerton College

Tommy D. Dickey
University of California –
Santa Barbara

Foreword by
George Philander
Princeton University

 **Higher Education**

Boston   Burr Ridge, IL   Dubuque, IA   New York   San Francisco   St. Louis
Bangkok   Bogotá   Caracas   Kuala Lumpur   Lisbon   London   Madrid   Mexico City
Milan   Montreal   New Delhi   Santiago   Seoul   Singapore   Sydney   Taipei   Toronto

Higher Education

EXPLORING THE WORLD OCEAN

Published by McGraw-Hill, a business unit of The McGraw-Hill Companies, Inc., 1221 Avenue of the Americas, New York, NY 10020. Copyright 2008 by The McGraw-Hill Companies, Inc. All rights reserved. No part of this publication may be reproduced or distributed in any form or by any means, or stored in a database or retrieval system, without the prior written consent of The McGraw-Hill Companies, Inc., including, but not limited to, in any network or other electronic storage or transmission, or broadcast for distance learning.

Some ancillaries, including electronic and print components, may not be available to customers outside the United States.

This book is printed on acid-free paper.

1 2 3 4 5 6 7 8 9 0 DOW/DOW 0 9 8 7 6

ISBN 978–0–07–301654–2
MHID 0–07–301654–3

Publisher: *Thomas D. Timp*
Executive Editor: *Margaret J. Kemp*
Developmental Editor: *Debra A. Henricks*
Sr. Editorial Assistant: *Tamara R. Ben*
Marketing Manager: *Todd Turner*
Sr. Marketing Assistant: *Traci A. Andre*
Lead Project Manager: *Joyce M. Berendes*
Lead Production Supervisor: *Sandy Ludovissy*
Lead Media Project Manager: *Judi David*
Media Producer: *Daniel M. Wallace*
Designer: *Rick D. Noel*
Cover/Interior Designer: *Rokusek Design*
(USE) Cover Image: *©Corbis, Divers on Underwater Propulsion Vehicles, Stephen Frink*
Senior Photo Research Coordinator: *John C. Leland*
Compositor: *Electronic Publishing Services Inc., NYC*
Typeface: *10/12 Adobe Caslon*
Printer: *R.R. Donnelley Willard, OH*

The credits section for this book begins on page 369 and is considered an extension of the copyright page.

Library of Congress Cataloging-in-Publication Data

Chamberlin, W. Sean.
    Exploring the world ocean / W. Sean Chamberlin, Tommy D. Dickey.
       p.cm.
    Includes index.
    ISBN 978–0–07–301654–2 - - - ISBN 0–07–301654–3 (alk. paper)
    1. Oceanography–Textbooks. I. Dickey, Tommy D. (Tommy De), 1945– II. Title.

GC11.2C447 2008
551.46–dc22

                                2006047224

www.mhhe.com

To Shae, Mom, Ariel and Leia.
Sean Chamberlin

To Todd, David, and Duke.
Tommy Dickey

And to our students…

Brief Contents

Table of Contents

From Microbes to Mammals   239

The Foundations of Evolutionary Theory   240

The Tree of Live   242

    Classification and Systematics   242
    Molecular Approaches to Classification   242
    Life in Three Domains   243

Ocean Biodiversity   246

    The Benthos   246
    The Plankton   247
    The Nekton   249

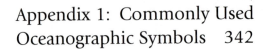

Chapter 16 Future Explorations   336

Foreword

The ocean is of great cultural importance to humans. Recently, oceanographers started making their contributions to that culture by exploring the world beneath the ocean surface, a world with a life of its own. The conditions where we live, on terra firma, are profoundly affected by conditions in the ocean, even though it is a mere film of salty water, some 4 km deep, on a sphere whose diameter exceeds 12,000 km. (If Earth were the size of an apple, the ocean would correspond to the peel of the apple.) We have only started to explore that film, so it still retains much of the "sweet mystery" Herman Melville mentioned more than a century ago. That is why oceanography is an exciting and rapidly-developing branch of science. This book is an excellent introduction to this important new field.

There is, one knows not what sweet mystery about this sea, whose gently awful stirrings seem to speak of some hidden soul beneath.

–*Moby Dick*, Herman Melville

Originally, the oceans were studied mainly for the practical purpose of making journeys across the oceans shorter and safer. Studies of the ocean for its own sake—of conditions far below the surface for example—started when the *HMS Challenger* circumnavigated the globe between 1872 and 1876. Numerous similar expeditions, sponsored by various European colonial powers, followed immediately afterwards. From the sum of all the measurements made on the various expeditions, oceanographers inferred the major oceanic currents. The results were often published in atlases because, beneath its restless surface, the ocean was assumed to be unchanging. Only after World War II did scientists gain access to data that implied that not only the ocean, but the ocean floor, is in motion.

Data concerning unexpected ridges and trenches on the ocean floor brought geologists to the remarkable conclusion that terra firma is anything but firm, that the continents drift! This dramatic result beautifully unified seemingly unrelated phenomena that include the shapes of continents, the locations of volcanoes, and the occurrence of earthquakes. It furthermore permitted reconstruction of the habitats of earlier life-forms such as dinosaurs. Plate tectonics amounted to a revolution, not only in the earth sciences, but in all of science.

The launching by the Soviet Union of the first artificial satellite, Sputnik, in 1957, heightened Cold War tensions, but proved a bonanza to science in general, oceanography in particular. (This happened because of the military importance of the oceans. Submarines are difficult to detect because seawater conducts electricity, thus ruling out the use of tools such as radar.) The significant increase in the number of oceanographers, and in the resources available to them, permitted the traditional mode of operation—isolated cruises to unexplored regions—to be complemented with large, coordinated programs involving numerous investigators from different institutions to study phenomena such as coastal upwelling over prolonged periods. The results showed that, below its restless surface, the ocean is anything but unchanging; the ocean is as changeable as the weather. The oceanic counterpart to oscillations of the atmospheric jet stream are meanders of the Gulf Stream and Kuroshio Currents. Below these swift, wind-driven currents of the upper ocean, is the deep, slow thermohaline circulation. Advances in ocean chemistry permitted detailed descriptions of its various components, and led to the realization that the ocean has a conveyor belt that transports large amounts of heat to the higher latitudes of the northern Atlantic Ocean.

The least explored aspect of the ocean is its great variety of life-forms. Particularly important are the microscopic bacteria. For example, only in the 1990s did scientists discover the bacteria that are responsible for much of the photosynthesis on Earth, and hence for much of the oxygen in the air. They are not distributed uniformly throughout the ocean but prefer certain climatic zones. In the same way that the continents have a great diversity of climatic zones that favor a diversity of life-forms, so have the oceans. The critical factors that determine the character of a zone are temperature and rainfall on land, temperature and nutrient concentration in the oceans. Hence, an understanding of biological oceanography requires knowledge of the chemistry and physics of the ocean. The physical conditions, in turn, depend on the atmospheric conditions above the ocean—the winds, for example. The winds, however, depend on oceanic conditions, specifically the sea surface temperature patterns. This means that we inhabit a planet where conditions in the oceans, on land, and in the atmosphere are all interconnected. Now that we have become the stewards of this remarkable planet, we need to familiarize ourselves with how it works. Learning about the ocean is an appropriate way to start. I congratulate Sean Chamberlin and Tommy Dickey on their excellent book that introduces readers to the fascinating and important "sweet mysteries" of the ocean.

S. George Philander

Professor George Philander received his Ph.D. from Harvard University and presently serves as the Knox Taylor Professor of Geosciences and Director of the Program in Atmospheric and Oceanic Sciences at Princeton University. In addition to numerous scientific publications, he has authored three timely and important best-selling books on oceanography, atmospheric sciences, and climate: *El Nino, La Nina, and the Southern Oscillation* (1990), *Is the Temperature Rising?* (1998), and *Our Affair with El Nino* (2004). Professor Philander has been internationally recognized for his exceptional ability to write clearly and imaginatively to make scientifically challenging concepts understandable to the general public. Among the many awards bestowed upon Professor Philander are the NOAA Environmental Research Laboratories Distinguished Authorship Award (1979 and 1983), the Sverdrup Gold Medal of the American Meteorological Society (1985), and the Department of Commerce Gold Medal (1985). Professor Philander is a Fellow of the American Meteorological Society and the American Geophysical Union and a member of the U.S. National Academy of Sciences.

Preface

Exploring the World Ocean was written explicitly to inform and educate students about the world ocean and the ongoing efforts of 21st century oceanographers to understand it. Students are a vital part of this effort. An ocean-literate public is an engaged and responsible public. Humans use the ocean for recreation, commerce, transportation, medicine, national security, mineral resources, energy supplies, simple enjoyment, and much more. Ocean-educated students can articulate the consequences of these activities and weigh their merits against concerns for human health and welfare, marine life conservation and biodiversity, and coastal preservation and management, to name a few. Ocean-educated students help oceanographers to promote an open-minded, balanced, and fair debate on ocean issues. Together, *oceanographers and the ocean-literate public share responsibility for communicating the importance of the world ocean to our daily lives.*

A Contemporary Approach

Exploring the World Ocean embodies our belief that oceanography is best taught as an earth systems science, one that acknowledges the connections and interactions between the atmosphere, Earth, and ocean. Although we have arranged the book in a traditional, chapter-by-chapter, single-subject format, chapters and topics are integrated to the greatest degree possible to build upon the theme that Earth has one world ocean and that it is studied as a system. For example, a complete understanding of the biological productivity of the ocean includes topics in ocean chemistry, ocean physics, ocean-atmosphere interactions, ocean sediments, and ocean biology. To facilitate that understanding, we revisit topics from time-to-time to build a comprehensive and integrated picture of the world ocean. Our goal is for students to understand that *the world ocean is a single, interdependent system of interacting geological, physical, chemical, and biological processes.*

Key Themes

Remarkable new discoveries made possible by the inventions of ever more sophisticated technology have revolutionized our understanding of the world ocean and how it works. *Exploring the World Ocean* aims to provide students with the sense of adventure experienced by oceanographers as they embark on missions to observe and study the world ocean. We purposefully emphasize the role of technology and the spirit of human endeavor in the historical and modern-day practice of ocean science. Each chapter begins with a brief history of the science and describes the mile-stones in technologies and conceptual revelations that contributed to the development of a particular scientific idea from past to present. Truly, the 21st century oceanographer stands on the shoulders of the men and women who came before. Yet contributions to oceanography come from more than just scientists. Perhaps more than any other scientific endeavor except perhaps manned space exploration, oceanography relies on teams of scientists, engineers, marine technicians, ship personnel, and computer specialists working together to discover new knowledge. The *esprit de corps* of oceanography results from the shared experience of men and women working toward a common goal. Success in oceanography depends on *the mutual cooperation of teams of researchers and support personnel engaged in the application of technology to observing and understanding the world ocean.*

Investigative Science

This text is designed to meet the needs of 21st century oceanography students and instructors. It can be used in a traditional, content-rich, format or applied to active and inquiry-based exploration of fundamental questions and current topics. Throughout the text, we emphasize the concept of inquiry as a means for helping students to become ocean literate. This approach helps students to understand that science is a cumulative and tentative process, subject to careful scrutiny by peers, and subject to new evidence and information that may overturn currently held ideas. The presentation of oceanographic knowledge in this way illustrates the concept of *oceanography as an ongoing inquiry into the nature of the world ocean.*

- Each chapter begins with **Questions to Consider**, a set of questions that may be used as an introduction to a topic, or a starting point for inquiry-based studies. These questions are intended to stimulate curiosity and promote thinking like an oceanographer.
- **Chapter outlines** provide a convenient guide to the topics covered in the chapter.
- **Historical content and environmental issues** are woven throughout the text in the context of chapter-specific subject matter.
- **More than 400 full-color photographs, maps, and illustrations** have been carefully selected and designed to complement and reinforce the text.
- **Spotlight** boxed readings highlight cutting-edge oceanographic research and technology through discussions of ocean-related environmental and societal issues, as well as controversies and areas of research where oceanographers disagree. These boxes provide added interest and demonstrate the fact that science is a works-in-progress.

- **You Might Wonder** offers general interest questions and answers that provide interesting tidbits, strange and amazing facts, common misconceptions, and miscellaneous topics of interest to students.
- **Key Concepts** outlines the main points of the chapter in bulleted format for quick review.
- **Terms to Remember** lists the most important vocabulary words from the chapter along with page numbers for further review.
- Each chapter ends with **Critical Thinking** questions which are designed to help students assess their understanding of chapter material. These questions also provide an excellent starting point for deeper explorations of a topic in classroom or online discussions.
- **Exploration Activities** are inquiry-based activities that can be used to supplement or substitute for a traditional course of lecture and laboratory studies.
- A **fold-out map** at the end of the book provides a quick reference to the oceans, seas, bays, gulfs, straights, channels, and waters that form the world ocean.

Organization

Exploring the World Ocean generally follows the order of topics presented in traditional oceanography textbooks. However, we have reduced, merged, and expanded many topics to offer you contextual as well as pedagogical improvements over existing texts. Rather than devote separate chapters to oceanographic history and environmental issues, these topics are embedded within each chapter. By featuring a brief history of the science at the start of each chapter, students can trace the historical development of a particular field of study and better see science as a progression of ideas and discoveries that builds on previous work. In a similar light, environmental issues are addressed within the context of the topic being discussed so that students can witness the application of a particular field of study to a specific human problem. We have deliberately chosen not to include a chapter on ocean resources. Rather, we discuss practical uses of the ocean where appropriate throughout the text and feature expanded coverage of this topic on our website for those instructors who wish to devote more time to this subject. In some instances, we have juxtaposed topics that often appear in different chapters. For example, waves and beaches have been combined in a single chapter to focus student attention on the processes that shape beaches and generate interesting wave-related phenomena, such as extreme surf and tsunami. In other chapters, we have expanded coverage of a topic, especially those that pertain to climate change. Nearly every chapter addresses climate change in some fashion. We have also chosen to devote greater attention to overfishing in our chapter on food webs because of the severity of this problem in modern times. Our final chapter specifically addresses coastal ecosystems and human impacts in contrast to the traditional description of coastal landforms and morphology. Some instructors may wish to begin a course of study with this chapter and use its emphasis to provide an environmental context for studies of the geology, physics, chemistry, and biology of the world ocean. For those top-

ics we were unable to include or expand upon in the printed text, we offer extensive additional content accessible from our website: **www.mhhe.com/chamberlin1e**.

Acknowledgements

Our sincere appreciation to George Philander, the National Aeronautics and Space Administration, the National Oceanic and Atmospheric Administration, the National Oceanographic Partnership Program, the National Science Foundation, the Office of Naval Research, and our students and colleagues from around the globe.

We would like to acknowledge fruitful discussions with: Richard Lozinsky, Karl Banse, Richard Behl, Will Berelson, Annie Bianchino, Robert Busby, Maureen Conte, John Cooper, Bruce Cordell, John Cullen, Geoff Davies, Rick Dickert, William Dickinson, Michael Drake, Robert Ellis, Bruce Frost, Donn Gorsline, Peggy Hamner, Warren Hamilton, Carolyn Heath, Pete Jumars, Dennis Kelley, John Knauss, Andrew Knoll, Pat Kremer, Charles Leavell, Michael Lizotte, Donal Manahan, John Marra, Dean McManus, Tom Morris, John Morrow, Leslie Newman, Mary Jane Perry, James Price, David Pugh, Francisco Rey, John Rogers, Virginia Roundy, Egil Sakshaug, Allan Schoenherr, Donald Simanek, Lynn Talley, Albert Theberge, Jr., Libe Washburn, Marc Willis, and Cal Young. We would also like to acknowledge invaluable discussions with participants in the Faculty Institutes for Reforming Science Teaching and the EarthEd Advisory Group.

We express our gratitude to the following persons who provided helpful source materials:

Marcel Babin, Richard Barber, Art Berman, Wouter Bleeker, Dee Breger, David Caron, Peter Flood, Gillian Foulger, Chris Gallienne, Steven Graham, Chris Goldfinger, Alastair Grant, Dennis Hartmann, Bill Haxby, Greg Huglin, Pete Klimley, Sebastian Krastel, Walter Lenz, Kerry Marchinko, George McManus, Richard Murphy, Don Reed, Phil Richardson, Kevin Righter, Gary Robbins, David Sandwell, James Sprinkle, Herwig Stibor, Ted Strub, Paul Taylor, Regina Wetzer, and Jin Xiong.

We wish to recognize the following reviewers who offered valuable suggestions and constructive criticism during the development process. Their comments helped shape this edition.

Patty Anderson, *Scripps Institution of Oceanography*
Kevin R. Arrigo, *Stanford University*
Emily Fast Christensen, *Missouri Baptist University*
Brent Dugolinsky, *State University College–Oneonta*
Henrietta Edmonds, *University of Texas–Austin*
Dan V. Ferandez, *Anne Arundel Community College*
Anne Gardulski, *Tufts University*
Eric A. Harms, *Brevard Community College*
Christi Hill, *Grossmont College*
Jerry D. Horne, *San Bernardino Valley College*
Matthew Horrigan, *San Francisco State University*
Martha House, *Pasadena City College*
Richard O. Hughes III, *Crafton Hills College*

William M. Landing, *Florida State University*
Jennifer C. Latimer, *Indiana State University*
John Leland, *Glendale Community College*
Michael E. Lyle, *Tidewater Community College*
Jennifer McCabe, *California State University–Sacramento*
Nancy A. Penncavage, *Suffolk Community College–Grant Campus*
James L. Pinckney, *University of South Carolina*
Elizabeth Simmons, *Metropolitan State College, Denver*
Reed A. Schwimmer, *Rider University*
Martha R. Scott, *Texas A & M University*
Gregory B. Smith, *Edison College*
Carrie E. Sweitzer, *Kent State University*
Douglas F. Williams, *Marine Science Program–University of South Carolina*
Jorge F. Willemsen, *University of Miami*
Donna L. Witter, *Kent State University*
Andrew Zimmerman, *University of Florida*

Finally, a special note of thanks for the assistance of: Marge Kemp, Executive Editor; Debra Henricks, Developmental Editor; Joyce Berendes, Lead Production Manager; Dean Richmond, Freelance Developmental Copyeditor; and the entire McGraw-Hill Higher Education team.

W. Sean Chamberlin
Tommy D. Dickey

Supplements

For the Student:

The **Online Learning Center** for *Exploring the World Ocean* is a great place to review chapter material and enhance your study routine. Visit **www.mhhe.com/chamberlin1e** for access to the following online study tools:

- Chapter quizzing
- Web links
- Key term flashcards
- Scripps video clips
- And other activities that support inquiry-based learning

For the Instructor:

The *Exploring the World Ocean* **Online Learning Center (www.mhhe.com/chamberlin1e)** offers a wealth of teaching and learning aids for instructors and students. Instructors will appreciate:

- A password-protected Instructor's Manual
- Access to the new online **Presentation Center** including all of the illustrations, photographs, and tables from the text in convenient jpeg format
- Key references and suggested readings
- Access to supplemental marine biology material
- Online activities for students such as chapter quizzing, web links, key term flash cards, Scripps video clips, and other activities that support inquiry-based learning
- A link to author Sean Chamberlin's website (http://exploreworldocean.com) for additional marine biology material. Simply click on "Marine Life" and register as a new user to access the information.

Simply visit **www.mhhe.com/chamberlin1e.**

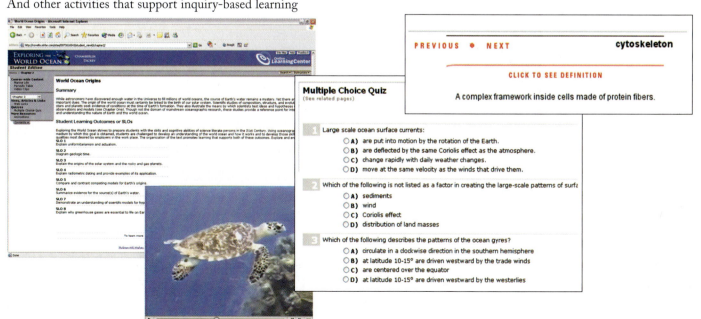

Find digital assets quickly and easily with the new Presentation Center!

NEW! Presentation Center
www.mhhe.com/chamberlin1e

Build instructional materials wherever, whenever, and however you want!

Accessed through the Online Learning Center, the **Presentation Center** is an online digital library containing assets such as art, photos, tables, PowerPoints®, video clips, and other media types that can be used to create customized lectures, visually enhanced tests and quizzes, compelling course websites, or attractive printed support materials.

Access to your book, access to all books!

The **Presentation Center Library** includes thousands of assets from many McGraw-Hill titles. This ever-growing resource gives instructors the power to utilize assets specific to an adopted textbook as well as content from all other books in the library.

Nothing could be easier!

Presentation Center's dynamic search engine allows you to explore by discipline, course, textbook chapter, asset type, or keyword. Simply browse, select, and download the files you need to build engaging course materials. All assets are copyrighted by McGraw-Hill Higher Education but can be used by instructors for classroom purposes.

Here's how it works:

1. **Search a user-friendly navigation screen using a variety of parameters** that allow you to control the amount of information you retrieve. Search by:
   - Discipline/course area
   - Author
   - Text title/edition
   - Chapter

Unsure of where to search? Simply enter the keyword and let the search engines do the work for you. Or enter the ISBN of the book to immediately go to those files.

2. **Select from a multitude of digital assets including art, photos, tables, PowerPoints®, video clips, and other media types.**
   - Your results will be displayed for you to easily review and select items.
   - Click "add to library" below each item to add items to your library or create a download list.

3. **Export to fit your needs.**
   - Review your list.
   - Click "export/save my list" and the items listed will be saved/exported in one zip file. This zip download allows you to easily gather and organize information for a particular lecture or subsection of the course. It's that easy!

Instructor's Testing & Resource CD-ROM

This cross-platform CD-ROM includes an Instructor's Manual and test bank utilizing McGraw-Hill's EZ Test software. EZ Test is a flexible and easy-to-use electronic testing program that allows instructors to create tests in a wide variety of question types. Instructors may use the test questions provided by McGraw-Hill, add their own questions, create multiple versions of a test, and export tests for use with course management systems. Answers to the Questions to Consider and Critical Thinking questions from the text and Learning Outcomes for each chapter are also provided.

Earth and Environmental Science Digitized Videos

This exciting cross-platform DVD produced by Discovery Education offers short, three-to-five minute videos on ecology, evolution, conservation, energy, and physical sciences. Instructors can search from 50 different topics and download videos into their PowerPoint presentations to provide stunning examples of plate tectonics, solar energy, currents, waves, tides, rocks and minerals, and more! (Available to colleges and universities.) Contact your sales representative for a detailed listing of topics. ISBN 978-0-07-352541-9 (MHID 0-07-352541-3)

Classroom Performance System and Questions

The Classroom Performance System (CPS) is a wireless response system that brings interactivity into the classroom. Students use the wireless response pads (which are essentially easy-to-use remotes) to answer questions during class, providing instructors with immediate feedback on how well they understand the material. Instructors can create their own questions for use with CPS, or take advantage of the questions provided by McGraw-Hill. A text-specific set of questions, formatted for both CPS and PowerPoint, is available via download from the Instructor area of the Chamberlin/Dickey Online Learning Center.

Guided Tour

The *Exploring the World Ocean* Learning System

Designed to promote understanding, *Exploring the World Ocean* prompts students to question the ocean and its systems, and to seek answers by investigating the different tools and methods scientists use to explain the processes they observe.

Questions to Consider

Each chapter begins with *Questions to Consider*, a set of questions that may be used as an introduction to a topic or a starting point for inquiry-based studies. These questions are intended to stimulate curiosity and promote thinking like an oceanographer.

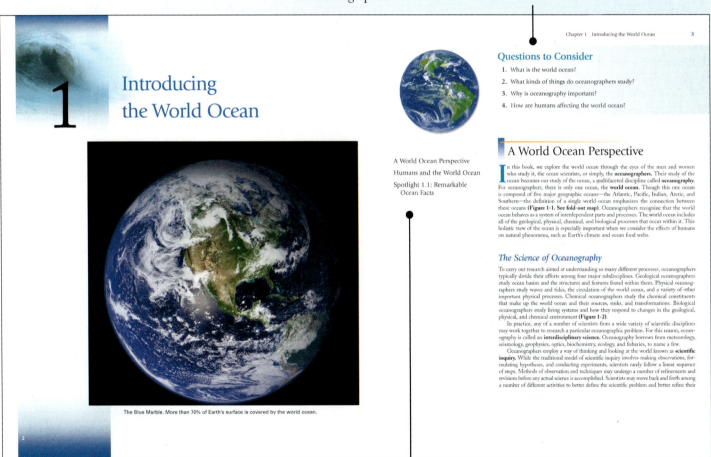

1

Introducing the World Ocean

A World Ocean Perspective

Humans and the World Ocean

Spotlight 1.1: Remarkable Ocean Facts

Questions to Consider

1. What is the world ocean?
2. What kinds of things do oceanographers study?
3. Why is oceanography important?
4. How are humans affecting the world ocean?

A World Ocean Perspective

In this book, we explore the world ocean through the eyes of the men and women who study it, the ocean scientists, or simply, the **oceanographers.** Their study of the ocean becomes our study of the ocean, a multifaceted discipline called **oceanography.** For oceanographers, there is only one ocean, the **world ocean.** Though this one ocean is composed of five major geographic oceans—the Atlantic, Pacific, Indian, Arctic, and Southern—the definition of a single world ocean emphasizes the connection between these oceans **(Figure 1-1. See fold-out map)**. Oceanographers recognize that the world ocean behaves as a system of interdependent parts and processes. The world ocean includes all of the geological, physical, chemical, and biological processes that occur within it. This holistic view of the ocean is especially important when we consider the effects of humans on natural phenomena, such as Earth's climate and ocean food webs.

The Science of Oceanography

To carry out research aimed at understanding so many different processes, oceanographers typically divide their efforts among four major subdisciplines. Geological oceanographers study ocean basins and the structures and features found within them. Physical oceanographers study waves and tides, the circulation of the world ocean, and a variety of other important physical processes. Chemical oceanographers study the chemical constituents that make up the world ocean and their sources, sinks, and transformations. Biological oceanographers study living systems and how they respond to changes in the geological, physical, and chemical environment **(Figure 1-2)**.

In practice, any of a number of scientists from a wide variety of scientific disciplines may work together to research a particular oceanographic problem. For this reason, oceanography is called an **interdisciplinary science.** Oceanography borrows from meteorology, seismology, geophysics, optics, biochemistry, ecology, and fisheries, to name a few.

Oceanographers employ a way of thinking and looking at the world known as **scientific inquiry.** While the traditional model of scientific inquiry involves making observations, formulating hypotheses, and conducting experiments, scientists rarely follow a linear sequence of steps. Methods of observation and techniques may undergo a number of refinements and revisions before any actual science is accomplished. Scientists may move back and forth among a number of different activities to better define the scientific problem and better refine their

The Blue Marble. More than 70% of Earth's surface is covered by the world ocean.

2

Chapter Outlines

Chapter outlines provide a convenient guide to the topics covered in the chapter.

Historical content and environmental issues are woven throughout the text in the context of chapter-specific subject matter.

Humans and the World Ocean

Humans have always been curious about the world ocean, especially about how it might help or harm them. In that vein, oceanography was born of a practical desire to know and use the sea. The first peoples of the world took to the ocean to obtain food, gather resources, avoid enemies, find new lands, and seek wealth. Humans depend on the world ocean for food. Seaweeds, shellfishes, fishes, and a number of other marine organisms supply food to more than a billion people daily. In addition to food resources, the world ocean supplies a considerable bounty of energy, mineral, and biological resources **(Table 1.1)**. As early as 3000 B.C., the Polynesians sailed from island to island to buy and sell goods, establishing trade and shipping. Even today, the world ocean serves as a "highway" across which people and their goods can be

Exploring the World Ocean contains more than *400 full-color maps, illustrations, and photographs.* All of the art has been carefully designed and selected to complement and reinforce the text. A number of maps have been put on the Robinson projection for added visual appeal.

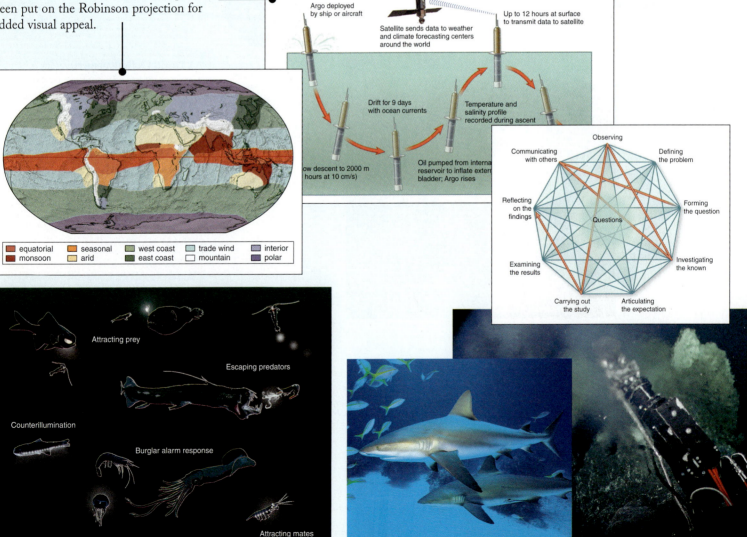

| equatorial | seasonal | west coast | trade wind | interior |
| monsoon | arid | east coast | mountain | polar |

Argo deployed by ship or aircraft

Satellite sends data to weather and climate forecasting centers around the world

Up to 12 hours at surface to transmit data to satellite

Drift for 9 days with ocean currents

Temperature and salinity profile recorded during ascent

ow descent to 2000 m hours at 10 cm/s)

Oil pumped from interna reservoir to inflate exter bladder; Argo rises

Observing

Communicating with others

Defining the problem

Reflecting on the findings

Questions

Forming the question

Examining the results

Investigating the known

Carrying out the study

Articulating the expectation

Attracting prey

Escaping predators

Counterillumination

Burglar alarm response

Attracting mates

Spotlight Boxes

Spotlight boxed readings highlight cutting-edge oceanographic research and technology through discussions of ocean-related environmental and societal issues, as well as controversies and areas of research where oceanographers disagree. These boxes provide added interest and demonstrate the fact that science is a works-in-progress.

SPOTLIGHT — 15.1

Coastal Ocean Observing Systems: Sentinels for Science and Public Safety

The importance of weather to human affairs has long been recognized. One of the primary objectives of US National Weather Service and the World Meteorological Organization is to provide information and forecasts about atmospheric conditions that may endanger human lives and cause damage to property. Weather forecasting places many demands on scientists and weather observers, who may be responsible for tracking and predicting any number of episodic and extreme phenomena, including tornadoes, hurricanes, typhoons, and winter storms. Though less appreciated, forecasting of ocean weather is equally important and even more daunting. The number of oceanographic processes and variables of interest far outnumber those of the atmosphere. The diversity of oceanic hazards and public safety concerns are considerable. Because more than 60% of the world population now resides within the coastal zone, the necessity for timely delivery of accurate oceanographic data to provide nowcasts and forecasts for industry, tourism, recreation, and public safety has become more widely recognized. To this end, a number of local, state, federal, and international public and private agencies have begun to commission and develop coastal ocean observing systems (COOS).

These systems will have many components, and virtually all portions of the world ocean will be systematically observed using a broad range of sampling platforms. Clearly, implementation of COOS is a formidable proposition, but on the positive side, it is bringing together oceanographers with a multiplicity of disciplines and interests from all over the globe. In the not-too-distant-future, the entire coastline of the United States will be equipped with land-based and *in situ* platforms supported by satellite observations to provide comprehensive temporal and spatial information on ocean temperature, salinity, currents, waves, sea level, bottom topography, phytoplankton (including harmful algal blooms), and sediments (**Figure 15a**). Coastal oceans interact with the open ocean, so efforts are under way to incorporate data from offshore ocean observatories to maximize weather and climate forecasting capability and to integrate these observations with other Earth-observing systems.

FIGURE 15a Concept of an integrated coastal ocean observing

You Might Wonder

You Might Wonder sections offer general interest questions and answers that provide interesting tidbits, strange and amazing facts, common misconceptions, and miscellaneous topics of interest to students.

YOU Might Wonder

1. What is the difference between oceanography and marine biology?
The distinctions between oceanography and marine biology are somewhat arbitrary, and the two disciplines naturally overlap. In general, oceanographers are concerned with the dynamics of ocean systems, whereas marine biologists focus on the organisms within those systems. Oceanography emphasizes the study of ocean systems and the interaction of geological, physical, chemical, and biological processes. Marine biology typically gives greater attention to the biology, ecology, and evolutionary adaptations of marine organisms.

2. What kinds of careers are available in oceanography?
Among the fields of study associated with oceanography are (in no particular order) physics, chemistry, biochemistry, geology, geophysics, geochemistry, biogeochemistry, geography, biogeography, meteorology, climatology, biology, marine biology, ecology, marine ecology, benthic ecology, systems ecology, evolutionary biology, zoology, archaeology, marine archaeology, paleontology, cultural anthropology, hydrological optics, bio-optics, hydrodynamics, hydrology, agriculture, aquaculture, mathematics, mechanical engineering, electrical engineering, aeronautical engineering, chemical engineering, architecture, marine architecture, cosmology, law, and lots more. This leaves a whole lot of opportunity for combining a field of study in which you are interested with a career in oceanography. Oceanography offers possibilities in pure and applied research, academics and industry, government and nonprofit organizations, design and engineering, technical and support roles, education and public advocacy. You are limited only by your imagination should you wish to pursue an ocean-related career.

Key Concepts sections outline the main points of the chapter in bulleted format for quick review.

Terms to Remember lists the most important vocabulary words from the chapter along with page numbers for further review.

Each chapter ends with ***Critical Thinking*** questions that are designed to help students assess their understanding of chapter material. These questions also provide an excellent starting point for deeper explorations of a topic in classroom or online discussions.

The ***Exploring the World Ocean Online Learning Center*** offers a wealth of teaching and learning aids for instructors and students.
www.mhhe.com/chamberlin1e

Key Concepts

- The world ocean encompasses all of the waters of the ocean basins.
- Oceanographers view the world ocean as a single, interdependent system.
- Oceanography is a multidisciplinary, international, team endeavor.
- Oceanography requires the knowledge and skills of an enormous number of people with various backgrounds, training, and interests.
- Oceanography serves practical and basic goals driven fundamentally by a desire to understand how the ocean works.
- Humans negatively impact the ocean on global and local scales.
- Educating yourself about the ocean will prepare you to make informed and intelligent decisions regarding the conservation and protection of ocean resources.

Terms to Remember

abrupt climate change, 6
activity model of the scientific method, 5
climate, 5

global warming, 5
greenhouse effect, 5
interdisciplinary science, 3
marine communities, 6

oceanographers, 3
oceanography, 3
scientific inquiry, 3

spatial scales, 5
temporal scales, 5
world ocean, 3

Critical Thinking

1. What is the difference between a world ocean perspective and a view of the ocean as a series of independent oceans?
2. In what ways does oceanography exemplify the activity model of the scientific method? Why is the scientific method more than a simple linear sequence of observation, hypotheses, experiments, results, and conclusions?
3. Describe another field of study that relies on teams of scientists, technicians, and support personnel working together toward a set of common scientific goals.
4. Make a list of ocean resources that you use in your daily life. Discuss the positive and negative impacts of your use of two of those resources.
5. Why is oceanography important to human society, and why is it important to you personally?

Explore Online

Visit **www.mhhe.com/chamberlin1e** for access to chapter quizzing, key term flash cards, video clips, interactive activities, and more. Further enhance your knowledge with web links to chapter-related material!

Exploration Activities are inquiry-based activities that can be used to supplement or substitute for a traditional course of lecture and laboratory studies.

Exploration Activity 1-1: Exploring Our Troubled Ocean Through Critical Thinking

Exploration Activities require you to dive deeper into ocean topics of current scientific and societal interest. Based on the principles of inquiry, they are designed to help you make connections between what you already know and what you are learning. They help you construct new understandings of scientific concepts, models, and ideas, and to apply those understandings to solving complex problems. Inquiry-based activities have many possible answers; what's important is that your statements and solutions are supported by evidence or data. The lack of a clear-cut answer often leaves students feeling very uncertain of their work. For that reason, we provide rubrics at the end of this chapter for assessing the depth and quality of your work. As you complete more activities, you will discover the way of organizing, analyzing, and synthesizing ideas that works best for you.

Question for Inquiry
How are humans impacting the world ocean?

Summary
In this chapter, we introduced the idea that human activities are having significant and harmful impacts on the world ocean. Although we explore several of these impacts in greater detail in the chapters that follow, it is instructive (and important) to gain a deeper appreciation for the magnitude of these problems. In this activity, you will become an "expert" on one of the human impacts on the world ocean presented in Table 1.2.

Learning Objectives
By the end of this activity, you should be able to:
- Distinguish between reliable and unreliable sources of scientific information
- Analyze scientific information and summarize its key evidence
- Present scientific evidence that a particular human activity is harming the ocean

Part I: Developing Analytical Tools
Review the following statement from the Pew Ocean Commission's report *America's Living Oceans: Charting a Course for Sea Change* (May 2003).

> Just as the 20th century brought us into knowledgeable contact with outer space, the 21st will almost certainly connect us more intimately to our ocean. In fact, it is imperative because—as much as we love our ocean—our ignorance has been destroying them. We love clean beaches, but what we discharge into the ocean befouls

them. We destroy the very coastal wetlands we need to buffer storms and filter fresh water. A nation of seafood lovers, we are careless about how we treat the ocean "nurseries" and brood stocks that replenish our fish supply.

Furthermore, the size of the world's human population and the extent of our technological creativity have created enormously damaging impacts on all of the oceans. We are now capable of altering the ocean's chemistry, stripping it of fish and the many other organisms which comprise its amazingly rich biodiversity, exploding and bleaching away its coral nurseries, and even reprogramming the ocean's delicate background noise.

We love our freedom to move about the ocean surface where no streets, signs, or fences impede us, yet our sense that no one owns this vast realm has allowed us to tolerate no one caring for it.

During the 20th century our nation has come to regard the air we breathe, the fresh water we drink, and the open lands as "common goods," part of our public trust. Now we must acknowledge that the oceans, too, are part of our common heritage and our common responsibility.*

In teams or individually, attempt to answer the following questions based on what you already know and have just read:
1. What is meant by the statement "our ignorance" is destroying the ocean?
2. What kinds of things do humans discharge into the ocean?
3. Why are coastal wetlands important?
4. What are the ocean's "nurseries" and why are they important?
5. How have humans altered the ocean's chemistry?
6. What is biodiversity?
7. How have humans reprogrammed the ocean's background noise?
8. Why is it difficult for humans to recognize the world ocean as a "common good"?

Review your answers and identify areas where your knowledge is uncertain or the author's meaning is unclear. What kinds of information and data would you need to understand and verify the author's statements and arguments? Brainstorm a list of missing information and unclear points and write it on a separate piece of paper.

*America's Living Oceans: Charting a Course for Sea Change (May 2003). © 2003 The Pew Charitable Trusts. Reprinted with permission.

Part II: Preliminary Research
From Table 1.2, find a particular human impact on the world ocean that interests you. Then do some preliminary research on your topic using the materials in this textbook (the index is a great place to start), references in the library (ask your librarian for help), or on our website (see the Resources section). Be sure to use verifiable sources of scientific information. If you are uncertain whether a particular source is valid, check with your instructor. During this initial phase of your research, make notes, write down any questions, and list terms or concepts that are unclear to you. In essence, apply the same techniques you used to analyze and identify the key topics and missing knowledge in Part I. You may find that the topic you picked is not what you expected, overly difficult, or lacking information. If so, then refine or change your topic. Return to your previous sources and find new ones. Be sure to note the sources of your information. Extract the scientific evidence that the impact

you have chosen is harmful. Be careful to distinguish between fact and opinion. Compare and cross-check evidence from different sources. Does it add up? You may have to do additional research if your original sources are less than reliable or your information is incomplete.

Part III: Defend Your Data!
Using your notes, write a position statement (a thesis sentence or paragraph) focused on the issue and problem you chose. You may style it after one of the sentences in the paragraph above. Review your thesis statement with a classmate or your instructor to refine it. Then write a two-page paper that presents three to four key pieces of evidence to support your thesis statement. Your arguments should be logical, self-consistent, and supportive of your thesis. Use the scoring rubric below as a guide to writing and improving your paper.

Rubric for Analyzing and Presenting Scientific Data and Information

	Exemplary	Adequate	Minimal
Quality of sources or data	Cites mostly primary scientific literature, peer-reviewed or validated web sites, other reviewed and substantiated sources or data.	Cites some reviewed and substantiated sources and some unreviewed or unvalidated sources or data.	Cites mostly unreviewed and unvalidated sources or data.
Analysis of sources or data	Offers clear presentation of facts/evidence, distinguishes facts/evidence from opinion/interpretation, provides multiple sources or lines of evidence	Presentation of facts/evidence not always clear, mixes some facts/evidence with opinions or interpretations, relies mostly on single sources of data and information.	Facts/evidence and opinions/interpretations not clearly distinguished, sources not reliable
Summary of key points and concepts	Synthesizes key points and conclusions; makes connections between concepts in multiple sources and data	Less synthesis and fewer and unclear connections between sources and data	Lists facts/evidence with little organization and interpretation
Organization and presentation	Arguments presented in a thoughtful and logical manner; evidence supports arguments; no grammatical or syntactical errors	Some arguments presented well; evidence does not always support arguments; some grammatical and syntactical errors	Poor arguments, lacking good evidence; many grammatical and syntactical errors

Instructors will appreciate:
- A password-protected Instructor's Manual
- Access to the new online PrepCenter, including all of the illustrations, photographs, and tables from the text in convenient jpeg format
- Key references and suggested readings
- A marine biology supplement

Students will find:
- Chapter quizzing
- Web links
- Key term flash cards
- Scripps video clips
- And other activities that support inquiry-based learning

A *fold-out map* at the end of the book provides a quick reference to the oceans, seas, bays, gulfs, straights, channels, and waters that form the world ocean.

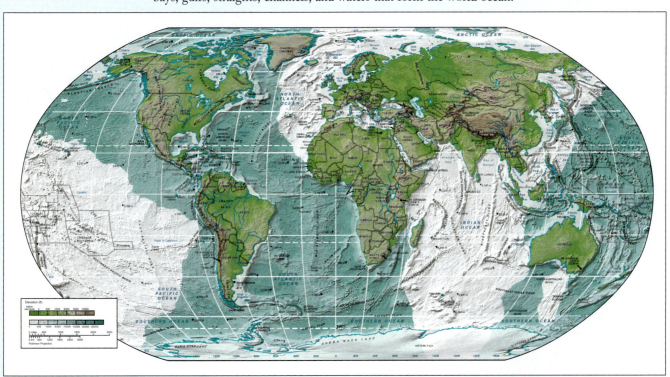

About the Authors

Professor Sean Chamberlin teaches and conducts research at Fullerton Community College in Fullerton, California. He grew up in South Florida, where he spent much of this youth diving or watching *Apollo* moon rockets soar overhead. Intrigued by astronauts working underwater and enamored of the TV series *Undersea World of Jacques Cousteau*, he decided at the age of 10 to pursue a career in ocean science. Traversing the United States to one of the few colleges that offered an undergraduate degree in oceanography, he enrolled in the University of Washington in Seattle, where he earned bachelor's degrees in Oceanography and English. Soon thereafter, he pursued graduate studies at the University of Southern California (USC) during which time he was invited to conduct research aboard Jacques Cousteau's Calypso in the South Pacific Ocean. His work on the natural fluorescence of phytoplankton earned him a PhD in Biology. It was also at USC that he met and took courses from Professor Dickey. Following two years as a postdoctoral researcher at Lamont Doherty Earth Observatory, Professor Chamberlin worked in the environmental industry. In 1996, he joined the Natural Sciences faculty at Fullerton College, where he has taught oceanography, geology, and meteorology for more than ten years. Professor Chamberlin has been especially active in the development of tools and teaching methodologies for online education; his online oceanography course attracts students from around the world. Professor Chamberlin is also active in science education research and national initiatives for the reform of science teaching, and has presented numerous presentations and workshops on these topics. When not teaching, Professor Chamberlin can be found with camera in hand somewhere along the California coast. Otherwise, he devotes time to his website, **http://exploreworldocean.com**, an excellent source of additional information about the world ocean.

Professor Tommy Dickey teaches and conducts research at the University of California at Santa Barbara (UCSB). He grew up in rural Indiana and received BS and BA degrees in physics and mathematics at Ohio University. He earned an MS degree in physics by doing night studies at Stevens Institute of Technology while serving in the US Coast Guard as an instructor of electronics and human relations.

Subsequently, he studied at Princeton University, where he received MA and PhD degrees in geophysical fluid dynamics. Following a year of postgraduate research as a Rosenstiel Fellow at the University of Miami, Professor Dickey joined the faculty of the University of Southern California, teaching there for 18 years before joining the faculty at UCSB, where he has taught for over a decade. Thus far, he has instructed approximately 10,000 students during his career. Professor Dickey's research group has participated in over 180 research cruises, and he has led five major multi-institutional research programs. He has been internationally recognized for his interdisciplinary oceanographic research and development of ocean observational systems.

Professor Dickey has served in editorial capacities for four major oceanographic journals and has authored or coauthored over 100 reviewed publications. He was elected as a Fellow of the American Geophysical Union in 2006. More information concerning Professor Dickey's past and current research may be found on the website **www.opl.ucsb.edu**.

Exploring
the
World
Ocean

1

Introducing the World Ocean

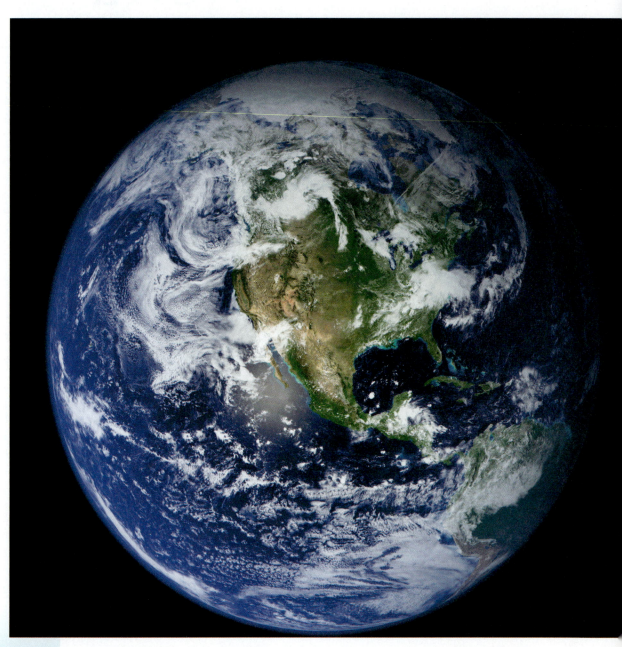

The Blue Marble. More than 70% of Earth's surface is covered by the world ocean.

Questions to Consider

1. What is the world ocean?

2. What kinds of things do oceanographers study?

3. Why is oceanography important?

4. How are humans affecting the world ocean?

A World Ocean Perspective

In this book, we explore the world ocean through the eyes of the men and women who study it, the ocean scientists, or simply, the **oceanographers.** Their study of the ocean becomes our study of the ocean, a multifaceted discipline called **oceanography.** For oceanographers, there is only one ocean, the **world ocean.** Though this one ocean is composed of five major geographic oceans—the Atlantic, Pacific, Indian, Arctic, and Southern—the definition of a single world ocean emphasizes the connection between these oceans **(Figure 1-1. See fold-out map)**. Oceanographers recognize that the world ocean behaves as a system of interdependent parts and processes. The world ocean includes all of the geological, physical, chemical, and biological processes that occur within it. This holistic view of the ocean is especially important when we consider the effects of humans on natural phenomena, such as Earth's climate and ocean food webs.

The Science of Oceanography

To carry out research aimed at understanding so many different processes, oceanographers typically divide their efforts among four major subdisciplines. Geological oceanographers study ocean basins and the structures and features found within them. Physical oceanographers study waves and tides, the circulation of the world ocean, and a variety of other important physical processes. Chemical oceanographers study the chemical constituents that make up the world ocean and their sources, sinks, and transformations. Biological oceanographers study living systems and how they respond to changes in the geological, physical, and chemical environment **(Figure 1-2)**.

In practice, any of a number of scientists from a wide variety of scientific disciplines may work together to research a particular oceanographic problem. For this reason, oceanography is called an **interdisciplinary science.** Oceanography borrows from meteorology, seismology, geophysics, optics, biochemistry, ecology, and fisheries, to name a few.

Oceanographers employ a way of thinking and looking at the world known as **scientific inquiry.** While the traditional model of scientific inquiry involves making observations, formulating hypotheses, and conducting experiments, scientists rarely follow a linear sequence of steps. Methods of observation and techniques may undergo a number of refinements and revisions before any actual science is accomplished. Scientists may move back and forth among a number of different activities to better define the scientific problem and better refine their

a.

b.

c.

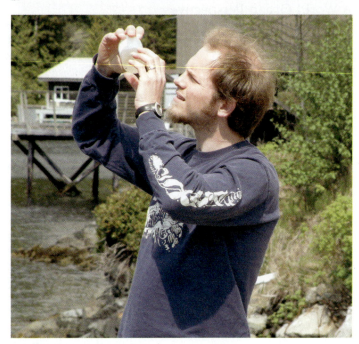

d.

■ **FIGURE 1-2** Major subdisciplines of oceanography. **(a)** Geological oceanographers strive to understand processes that create and modify ocean basins and to identify the origins of the structures and features found within them. Most recently, geological oceanographers have focused their attention on the history of Earth's climate as evidenced in sediments. Geological oceanographers also study deep-sea hydrothermal vents, regions of the seafloor that heat and enrich seawater with minerals and support a myriad of unusual life forms. Geological oceanographers represent a wide range of geologic disciplines, including geophysics, seismology, volcanology, paleoceanography, sedimentary geology, geodetics, geomorphology mineralogy, petrology, and rock mechanics. **(b)** Physical oceanographers aim to understand the circulation of the world ocean and transformations of energy over large temporal and spatial scales. The role of the world ocean in global climate has received great attention in recent years, especially toward understanding phenomena such as El Niño/La Niña and other connections between the ocean and the atmosphere. Physical oceanographers come from a number of different backgrounds, including thermodynamics, electromagnetism, optics, mechanics, fluid mechanics, geophysical fluid dynamics, and acoustics. **(c)** Chemical oceanographers study the chemical constituents that make up the world ocean and their sources, sinks, and transformations. Of particular concern to chemical oceanographers are the biogeochemical cycling of elements and their interactions with seafloor processes, sediments, marine organisms, and the atmosphere. Recently, chemical oceanographers have engaged in large-scale experiments designed to test hypotheses concerning iron limitation in the oceans and its importance in the carbon cycle. Chemical oceanographers include physical chemists, organic chemists, trace metal chemists, geochemists, biochemists, sediment chemists, and atmospheric chemists, among others. **(d)** Biological oceanographers endeavor to understand the flow of energy and materials through marine ecosystems. In particular, they seek to understand the structure and dynamics of ocean food webs and how they respond to changes in the physical and chemical environment. Newly discovered microbes and chemosynthetic organisms have revolutionized scientific understanding of energy and material flows in the world ocean. New insights have been gained from new technologies ranging from molecular probes to Earth-orbiting satellites. One important applied aspect of biological oceanography is fisheries oceanography, especially the population dynamics of commercially important fishes.

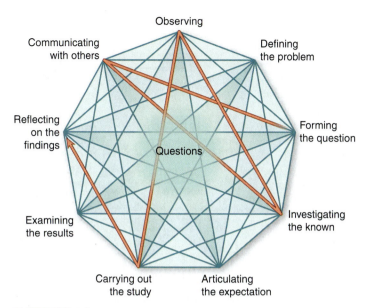

Observing

Communicating with others

Defining the problem

Reflecting on the findings

Questions

Forming the question

Examining the results

Investigating the known

Carrying out the study

Articulating the expectation

■ **FIGURE 1-3** Activity model of the scientific method. At least ten interrelated activities may be carried out during scientific research. Collectively, these activities comprise the scientific method.

ability to answer it. Science becomes a dynamic and social process as a result. This approach to scientific inquiry is called the **activity model of the scientific method (Figure 1-3)**. This model provides an ideal framework for the collaborative approach of oceanographers who may spend a great deal of time on a ship at sea, in the laboratory, or at a computer terminal acquiring, analyzing, and discussing data with students and colleagues.

The Challenges of Space and Time

Chief among the oceanographer's challenges is the ocean's variability over a tenfold range of space and time, expressed in units of **spatial scales** and **temporal scales** (see inside front cover). The microscopic environment of a tiny shrimp differs considerably from that of a blue whale, yet an intimate knowledge of both environments is required to understand how these organisms interact. So, too, the second-by-second pace of waves on a beach is a blip in time compared to the millions of years of continental drift, yet both ultimately affect the shape of our coastlines. Episodic and extreme events, such as tsunami, may occur once in a lifetime, meaning that oceanographers have to be ready at a moment's notice if they are to learn about these phenomena. To meet this challenge, oceanographers have begun to deploy a new generation of sensors above, within, and at the bottom of the world ocean. Earth-orbiting satellites, aircraft, moorings, ships, robotic vehicles, and even marine animals carry devices for acquiring knowledge of the world ocean. With these sensors and advances in genomics, geochemistry, and communications, oceanographers now have the capability to study the world ocean from molecular

to global scales. These tools have brought about a revolution as oceanographers can now better measure and understand the complex scales of ocean processes. The twenty-first century is a time of great promise and excitement for oceanographers.

Humans and the World Ocean

Humans have always been curious about the world ocean, especially about how it might help or harm them. In that vein, oceanography was born of a practical desire to know and use the sea. The first peoples of the world took to the ocean to obtain food, gather resources, avoid enemies, find new lands, and seek wealth. Humans depend on the world ocean for food. Seaweeds, shellfishes, fishes, and a number of other marine organisms supply food to more than a billion people daily. In addition to food resources, the world ocean supplies a considerable bounty of energy, mineral, and biological resources **(Table 1.1)**. As early as 3000 B.C., the Polynesians sailed from island to island to buy and sell goods, establishing trade and shipping. Even today, the world ocean serves as a "highway" across which people and their goods can be transported. In the near future, it is hoped that the world ocean will supply new forms of energy and life-sustaining medicines. Of course, the world ocean has long been used to defend a nation's borders or prosecute actions against an enemy. Deployments of ships and navies above, upon, and within the world ocean require considerable knowledge and understanding of the world ocean and its conditions. Whether out of conflict, necessity, or curiosity, humans have ventured to sea in search of new surroundings and a better life. The extraordinary migration of humans across the world ocean continues to fascinate and perplex historians. And while piracy and buccaneering once attracted men and women to seek riches on the world ocean, today, enterprising businesspeople and entertainers use the world ocean to attract tourists to cruise ships or seaside resorts and amusement parks **(Figure 1-4)**. Globally, ocean economies employ millions of people and generate trillions of dollars.

A major concern of oceanographic research in the twenty-first century is the impact of humans on Earth's **climate,** the long-term fluctuations in the mean temperature and weather conditions on our planet. Carbon dioxide, methane, and similar gases trap heat in the atmosphere and warm our planet naturally via the **greenhouse effect.** Combustion of fossil fuels over the past century, however, has caused an increase in these gases, resulting in **global warming.** While fluctuations in greenhouse gases and global warming and cooling have occurred in Earth's past, the rate of warming over the past several decades is unprecedented. Global warming affects agriculture, biodiversity, energy consumption, resource utilization, and a host of other human activities. Ocean warming causes a rise in sea level, melting of polar sea ice, shifts in species distributions, increases in storm and cyclone activity, subsidence of low-lying islands and coastlines, and increased erosion of

TABLE 1.1 Resources from the World Ocean

Energy resources

Petroleum
Natural gas
Methane hydrates
Tide power
Wind power
Wave power
Heat (thermal gradient) power

Mineral resources

Sand and gravel
Magnesium compounds
Salts
Manganese nodules
Phosphorite deposits
Metallic sulfides
Diamonds
Mud

Biological resources

Fish and shellfish
Plankton and krill
Marine algae/seaweeds
Marine mammals
Sea turtles
Medicines from natural products
Sea snake skins
Shells

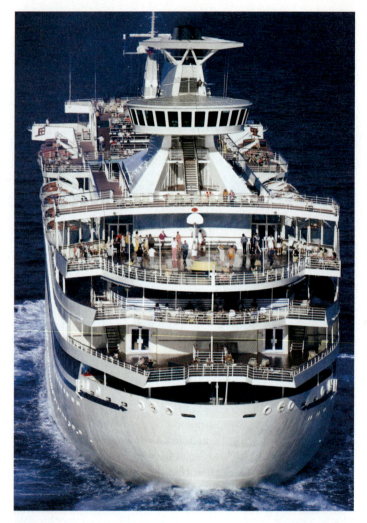

■ **FIGURE 1-4** The cruise ship industry contributed more than $30 billion to the U.S. economy in 2006.

coastal landforms. Some scientists fear continued global warming could bring **abrupt climate change,** a sudden shift in climate that plunges some parts of the globe into a deep freeze and heats other parts to intolerable levels. As George Philander puts it, global warming endangers "the *habitability* of our planet."

On a local scale, human activities directly affect **marine communities,** the plants and animals that inhabit the world ocean. In many parts of the world ocean, some species of fishes and marine animals have been reduced to near-extinction by over-fishing. Depletion of fish stocks alters marine food webs with uncertain consequences for fisheries and ecosystem health. Runoff of human-made fertilizers has created dead zones where oxygen is reduced to lethal levels for most forms of marine life. Habitat destruction and floating debris have caused declines in a number of marine animals, especially fishes, birds, and mammals. Retired Admiral James D. Watkins, of the U.S. Commission on Ocean Policy, wrote in April 2004, "Our oceans and coasts are in trouble, and we as a nation have an historic opportunity to make a positive and lasting change in the way we manage them before it is too late" **(Table 1.2)**.

TABLE 1.2   Human Impacts on the World Ocean

Problem	Cause	Effect	Remedy
Global warming	Buildup of greenhouse gases in atmosphere	Increased global temperatures; increased sea level; disruption of normal weather patterns; habitat alteration; potentially serious economic consequences	Reduce dependency on fossil fuels; burial of carbon dioxide in deep-ocean sediments; continued scientific study of ocean and atmospheric processes; public education
Ozone depletion	Past buildup of chlorofluorocarbons (CFCs)	Increased transmission of ultraviolet radiation at high latitudes; increased incidence of human skin cancer	Continued ban of CFCs; continued monitoring of ultraviolet light; public education
Habitat destruction	Coastal development; sea-level rise; trawling of the seafloor; destruction of coral reefs	Reduced coral, estuary, and mangrove habitat for juvenile fishes and migrating birds; increased eutrophication and marine pollution; destruction of deep-sea corals	Responsible development; mitigation of negative effects; setting aside trawl-free zones; studying effects of trawling on deep-sea habitat; public education
Overfishing	Overharvest of commercially important species; incidental "by-catch" of nonconsumed species	Shifts in structure and functioning of marine ecosystems; loss of biodiversity; negative economic consequences; population increases in harmful species, such as jellyfish and toxic algae	Responsible management of marine fisheries; responsible development of aquaculture; change in fishing practices; creation of marine reserves; public education
Marine pollution	Non-point source runoff of pollutants from land; aerial deposition of pollutants; marine debris; many other sources	Unhealthy conditions and water-borne illness; loss of biodiversity; biomagnification of toxic compounds; alteration of marine ecosystems	Water treatment of all runoff; reduced litter; increased street cleaning; development of alternative industrial and household compounds; public education
Eutrophication	Overabundance of nutrients from sewage and agriculture leading to overgrowth of algae and depletion of oxygen	Fish and invertebrate kills; dead zones in large regions of the world ocean	Improved sewage treatment; altered agricultural practices; growth of estuaries; public education
Oil spills	Spillage of oil during drilling, pumping, transportation, or transfer	Harm to marine life, especially birds and mammals	Better enforcement of international oil spill regulations; public education
Invasive species	Release of ballast water from ships; release of "pests"	Alteration of marine ecosystems; hypergrowth of nuisance or harmful species; possible negative economic consequences	Enforcement of international laws on the transfer of ballast water; enforcement of laws on the importation of exotic marine species; public education

Remarkable Ocean Facts

The existence of liquid water on our planet makes it unique among all known planets. The world ocean covers approximately 70.8% of our planet, a figure that emphasizes the prominence of the ocean across Earth's surface. Nonetheless, compared to Earth, the world ocean is quite small. Its thickness has been compared to the paint on a basketball, an analogy that brings home the point that Earth is primarily a rocky planet, the largest rocky planet in our solar system. Despite the blue marble appearance of its exterior, most of Earth's water may in fact exist in its interior, perhaps as many as ten world ocean volumes.

Though its mass and volume in comparison to Earth's mass and volume is small (0.025% and 0.12%, respectively), the world ocean's proportions in comparison to those of the continents are impressive. The average depth of the ocean is about 3782 m (12,408 feet), while the average height of the continents is only slightly more than 840 m (2756 feet). The deepest known location in the world ocean—the Challenger Deep in the Mariana Trench—was measured at 10,911 m (35,798 feet). Earth's tallest mountain above sea level, Mount Everest, measures 8848 m (29,029 feet). However, the tallest mountain from its base to its crest is found in the world ocean, Mauna Kea, Hawaii, which measures 10,203 m (33,474 feet).

The dimensions of the world ocean make it the largest habitat for life on Earth, although oceanographers would hardly treat it as a single habitat. About 90% of the world ocean's volume exists beneath a depth of 200 m (656 feet), a uniformly dark and cold region known as the abyss. Until the late 1970s, all life on Earth was believed to depend on sunlight. The discovery of hydrothermal vents on which giant worms and clams flourished changed our perspective of life on Earth. In fact, chemosynthetic-based life at the bottom of the ocean has raised speculation about the existence of life elsewhere in the cosmos. Nonetheless, the search for life even on our own planet—and especially in the world ocean—remains a formidable challenge. Until 2004, one of the largest invertebrates in the world ocean, the giant squid, known only from corpses on beaches or in the stomachs of sperm whales, had never been seen alive. That an animal reaching a length of 13 m (37 feet) could elude human observation for centuries reminds us just how big the world ocean really is. Who knows what other unknown creatures lurk beneath the waves?

Some people say we know more about the surface of the Moon than the interior of the world ocean. This may be true, but a concerted global effort to create permanent robotic observatories in the world ocean may soon deliver unprecedented views of the ocean to your home via the Internet. With the combination of Earth-observing satellites, specialized aircraft, oceanographic research vessels, and floating and moored platforms equipped with an array of sophisticated sensors, the twenty-first century promises to be an exciting one for oceanographers. In light of the threats of global warming, declining fish stocks, deteriorating water quality, and a myriad of other problems, knowledge of the world ocean and its inhabitants has never been more critical in human history. Perhaps the twenty-first century will come to be known as the century of the ocean, when humankind finally embraces the ocean planet on which we live. Until then, we can do our part to share what we know and marvel at the discoveries that may lie ahead **(Figure 1a)**.

FIGURE 1a The world ocean remains a place of tranquil beauty and profound mystery.

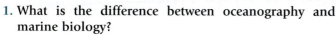
YOU Might Wonder

1. What is the difference between oceanography and marine biology?

The distinctions between oceanography and marine biology are somewhat arbitrary, and the two disciplines naturally overlap. In general, oceanographers are concerned with the dynamics of ocean systems, whereas marine biologists focus on the organisms within those systems. Oceanography emphasizes the study of ocean systems and the interaction of geological, physical, chemical, and biological processes. Marine biology typically gives greater attention to the biology, ecology, and evolutionary adaptations of marine organisms.

2. What kinds of careers are available in oceanography?

Among the fields of study associated with oceanography are (in no particular order) physics, chemistry, biochemistry, geology, geophysics, geochemistry, biogeochemistry, geography, biogeography, meteorology, climatology, biology, marine biology, ecology, marine ecology, benthic ecology, systems ecology, evolutionary biology, zoology, archaeology, marine archaeology, paleontology, cultural anthropology, hydrological optics, bio-optics, hydrodynamics, hydrology, agriculture, aquaculture, mathematics, mechanical engineering, electrical engineering, aeronautical engineering, chemical engineering, architecture, marine architecture, cosmology, law, and lots more. This leaves a whole lot of opportunity for combining a field of study in which you are interested with a career in oceanography. Oceanography offers possibilities in pure and applied research, academics and industry, government and nonprofit organizations, design and engineering, technical and support roles, education and public advocacy. You are limited only by your imagination should you wish to pursue an ocean-related career.

Key Concepts

- The world ocean encompasses all of the waters of the ocean basins.
- Oceanographers view the world ocean as a single, interdependent system.
- Oceanography is a multidisciplinary, international, team endeavor.
- Oceanography requires the knowledge and skills of an enormous number of people with various backgrounds, training, and interests.
- Oceanography serves practical and basic goals driven fundamentally by a desire to understand how the ocean works.
- Humans negatively impact the ocean on global and local scales.
- Educating yourself about the ocean will prepare you to make informed and intelligent decisions regarding the conservation and protection of ocean resources.

Terms to Remember

abrupt climate change, 6
activity model of the scientific method, 5
climate, 5
global warming, 5
greenhouse effect, 5
interdisciplinary science, 3
marine communities, 6
oceanographers, 3
oceanography, 3
scientific inquiry, 3
spatial scales, 5
temporal scales, 5
world ocean, 3

Critical Thinking

1. What is the difference between a world ocean perspective and a view of the ocean as a series of independent oceans?
2. In what ways does oceanography exemplify the activity model of the scientific method? Why is the scientific method more than a simple linear sequence of observation, hypotheses, experiments, results, and conclusions?
3. Describe another field of study that relies on teams of scientists, technicians, and support personnel working together toward a set of common scientific goals.
4. Make a list of ocean resources that you use in your daily life. Discuss the positive and negative impacts of your use of two of those resources.
5. Why is oceanography important to human society, and why is it important to you personally?

Explore Online

 Visit www.mhhe.com/chamberlin1e for access to chapter quizzing, key term flash cards, video clips, interactive activities, and more. Further enhance your knowledge with web links to chapter-related material!

Exploration Activity 1-1

Exploring Our Troubled Ocean Through Critical Thinking

Exploration Activities require you to dive deeper into ocean topics of current scientific and societal interest. Based on the principles of inquiry, they are designed to help you make connections between what you already know and what you are learning. They help you construct new understandings of scientific concepts, models, and ideas, and to apply those understandings to solving complex problems. Inquiry-based activities have many possible answers; what's important is that your statements and solutions are supported by evidence or data. The lack of a clearcut answer often leaves students feeling very uncertain of their work. For that reason, we provide rubrics at the end of this chapter for assessing the depth and quality of your work. As you complete more activities, you will discover the way of organizing, analyzing, and synthesizing ideas that works best for you.

Question for Inquiry

How are humans impacting the world ocean?

Summary

In this chapter, we introduced the idea that human activities are having significant and harmful impacts on the world ocean. Although we explore several of these impacts in greater detail in the chapters that follow, it is instructive (and important) to gain a deeper appreciation for the magnitude of these problems. In this activity, you will become an "expert" on one of the human impacts on the world ocean presented in Table 1.2.

Learning Objectives

By the end of this activity, you should be able to:

- Distinguish between reliable and unreliable sources of scientific information
- Analyze scientific information and summarize its key evidence
- Present scientific evidence that a particular human activity is harming the ocean

Part I: Developing Analytical Tools

Review the following statement from the Pew Ocean Commission's report *America's Living Oceans: Charting a Course for Sea Change* (May 2003).

> Just as the 20th century brought us into knowledgeable contact with outer space, the 21st will almost certainly connect us more intimately to our ocean. In fact, it is imperative because—as much as we love our oceans—our ignorance has been destroying them. We love clean beaches, but what we discharge into the ocean befouls

them. We destroy the very coastal wetlands we need to buffer storms and filter fresh water. A nation of seafood lovers, we are careless about how we treat the ocean's "nurseries" and brood stocks that replenish our fish supply.

> Furthermore, the size of the world's human population and the extent of our technological creativity have created enormously damaging impacts on all of the oceans. We are now capable of altering the ocean's chemistry, stripping it of fish and the many other organisms which comprise its amazingly rich biodiversity, exploding and bleaching away its coral nurseries, and even reprogramming the ocean's delicate background noise.

> We love our freedom to move about the ocean surface where no streets, signs, or fences impede us, yet our sense that no one owns this vast realm has allowed us to tolerate no one caring for it.

> During the 20th century our nation has come to regard the air we breathe, the fresh water we drink, and the open lands as "common goods," part of our public trust. Now we must acknowledge that the oceans, too, are part of our common heritage and our common responsibility.*

In teams or individually, attempt to answer the following questions based on what you already know and have just read:

1. What is meant by the statement "our ignorance" is destroying the ocean?
2. What kinds of things do humans discharge into the ocean?
3. Why are coastal wetlands important?
4. What are the ocean's "nurseries" and why are they important?
5. How have humans altered the ocean's chemistry?
6. What is biodiversity?
7. How have humans reprogrammed the ocean's background noise?
8. Why is it difficult for humans to recognize the world ocean as a "common good"?

Review your answers and identify areas where your knowledge is uncertain or the author's meaning is unclear. What kinds of information and data would you need to understand and verify the author's statements and arguments? Brainstorm a list of missing information and unclear points and write it on a separate piece of paper.

---

* *America's Living Oceans: Charting a Course for Sea Change* (May 2003): © 2003 The Pew Charitable Trusts. Reprinted with persmission.

Part II: Preliminary Research

From Table 1.2, find a particular human impact on the world ocean that interests you. Then do some preliminary research on your topic using the materials in this textbook (the index is a great place to start), references in the library (ask your librarian for help), or on our website (see the Resources section). Be sure to use verifiable sources of scientific information. If you are uncertain whether a particular source is valid, check with your instructor. During this initial phase of your research, make notes, write down any questions, and list terms or concepts that are unclear to you. In essence, apply the same techniques you used to analyze and identify the key topics and missing knowledge in Part I. You may find that the topic you picked is not what you expected, overly difficult, or lacking information. If so, then refine or change your topic. Return to your previous sources and find new ones. Be sure to note the sources of your information. Extract the scientific evidence that the impact you have chosen is harmful. Be careful to distinguish between fact and opinion. Compare and cross-check evidence from different sources. Does it add up? You may have to do additional research if your original sources are less than reliable or your information is incomplete.

Part III: Defend Your Data!

Using your notes, write a position statement (a thesis sentence or paragraph) focused on the issue and problem you chose. You may style it after one of the sentences in the paragraph above. Review your thesis statement with a classmate or your instructor to refine it. Then write a two-page paper that presents three to four key pieces of evidence to support your thesis statement. Your arguments should be logical, self-consistent, and supportive of your thesis. Use the scoring rubric below as a guide to writing and improving your paper.

Rubric for Analyzing and Presenting Scientific Data and Information

	Exemplary	Adequate	Minimal
Quality of sources or data	Cites mostly primary scientific literature, peer-reviewed or validated web sites, other reviewed and substantiated sources or data.	Cites some reviewed and substantiated sources and some unreviewed or unvalidated sources or data.	Cites mostly unreviewed and unvalidated sources or data.
Analysis of sources or data	Offers clear presentation of facts/evidence, distinguishes facts/evidence from opinion/interpretation, provides multiple sources or lines of evidence	Presentation of facts/evidence not always clear, mixes some facts/evidence with opinions or interpretations, relies mostly on single sources of data and information.	Facts/evidence and opinions/interpretations not clearly distinguished, sources not reliable
Summary of key points and concepts	Synthesizes key points and conclusions; makes connections between concepts in multiple sources and data	Less synthesis and fewer and unclear connections between sources and data	Lists facts/evidence with little organization and interpretation
Organization and presentation	Arguments presented in a thoughtful and logical manner; evidence supports arguments; no grammatical or syntactical errors	Some arguments presented well; evidence does not always support arguments; some grammatical and syntactical errors	Poor arguments; lacking good evidence; many grammatical and syntactical errors

2

World
Ocean
Origins

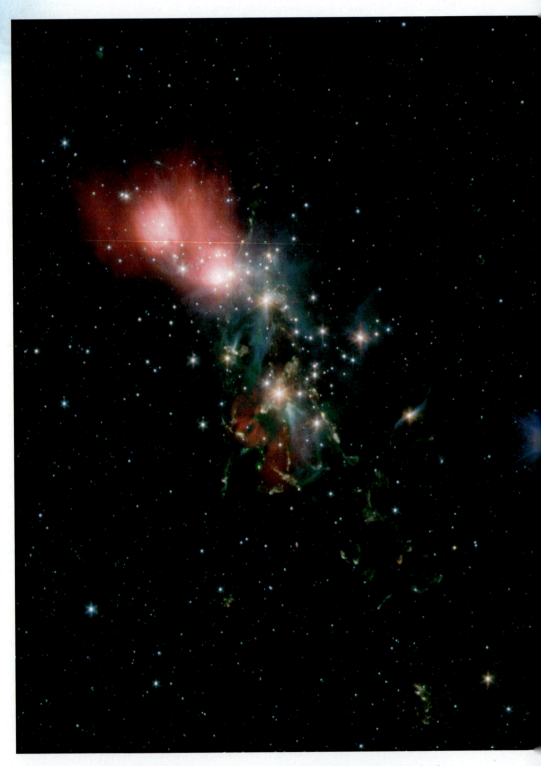

A star is born. The bright objects shrouded in gases represent young stars in the constellation Perseus, about 1000 light-years from Earth. By studying star-forming regions, astronomers seek clues for understanding the origins of our solar system and its planets.

Questions to Consider

1. Why is Earth the only planet in our solar system with an ocean on its surface?

2. How have scientists arrived at various models of Earth's formation?

3. What are the possible extraterrestrial sources of Earth's water, and are these hypothesized sources sufficient to explain Earth's ocean?

4. How do studies of the solar system and universe increase scientific understanding of the impacts of humans on Earth?

The Ocean Planet

While astronomers have discovered enough water in the universe to fill millions of world oceans, the source of Earth's water remains a mystery. Yet important clues exist. The origin of the world ocean must certainly be linked to the birth of our solar system. Scientific studies of composition, structure, and evolution of stars and planets seek evidence of conditions at the time of Earth's formation. They also illustrate the means by which scientists test ideas and hypotheses using observations and models (see Chapter 1). Though not the domain of mainstream oceanographic research, these studies provide a convenient starting point for understanding the nature of Earth and the world ocean.

The Foundations of Modern Geology

The first well-developed ideas about Earth's origins date to the seventeenth and eighteenth centuries, when two opposing schools of thought emerged. One group, the Neptunists, founded by the German naturalist Abraham Gottlob Werner (1750–1817), believed that all the rocks and minerals on Earth originated in a universal ocean, consistent with the biblical teachings of a Great Flood. The other group, the Plutonists, founded by the Scotsman James Hutton (1726–1797), proposed that Earth underwent an endless cycle of formation and destruction with "no vestige of a beginning, no prospect of an end." Though Hutton sought the hand of a Creator in his observations, his anti-Neptunist views were inconsistent with the Bible. Consequentially, Hutton was virtually ignored until after his death. Nevertheless, Hutton planted the suggestion that Earth's features and processes were very old and had operated for very long periods of time—infinitely, in Hutton's view **(Figure 2-1)**. For his contributions, Hutton is now recognized as the "father" of modern geology.

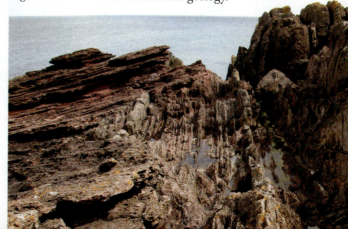

■ **FIGURE 2-1** Siccar Point, Scotland. The upheaved and discontinuous rock layers here brought James Hutton to the realization that Earth processes had been operating for a long time. This realization established the principle of uniformitarianism.

Eras	Hadean	Archean				Proterozoic						
Eons	?	Eoarchean	Paleoarchean	Mesoarchean	Neoarchean	Paleoproterozoic				Mesoproterozoic		
Periods	?	?	?	?	?	Siderian	Rhyacian	Orosirian	Statherian	Calymmian	Ectasian	Stenian

4560?	3800?	3600	3200	2800	2500	2300	2050	1800	1600	1400	1200

Million years ago

■ **FIGURE 2-2** Geologic time scale. Geologists partition the time since Earth's formation into four major divisions called eons. Each eon is further subdivided into eras, periods, and epochs. The geologic timeline provides a guide to major events that have occurred in Earth's past. Modifications of the timeline often accompany new findings and a better understanding of the 4.6 billion year history of Earth.

Charles Lyell (1797–1875), an Oxford geologist, developed Hutton's ideas further and promoted the **principle of uniformitarianism,** the idea that processes acting today are the same processes that acted in the past. In Lyell's view, there was no need to invoke cataclysmic events to explain Earth's features; modern geological processes were sufficient to explain past and present features. Equally as important, **geologic time,** the timeline of the Earth from its origin, gave a framework for the history of Earth **(Figure 2-2).** The acceptance of geologic time by the mid-1800s marked the birth of modern geology.

The fundamental concept of uniformitarianism remains today, but modern geology acknowledges a role for episodic events such as climate change and meteorite impacts. This modern view of uniformitarianism, called **actualism,** encompasses three principles: 1) geologic processes observed today occurred in the past; 2) some past geologic processes no longer occur; and 3) the rate and intensity of geologic processes may vary through geologic time. Using rocks and their features as clues, a geologist attempts to reconstruct the series of events that gave rise to a particular landscape, such as a mountain range or rocky coastline. Through a knowledge and understanding of present-day geological processes, geologists can interpret events that occurred in Earth's past. At the same time, the past and present provide clues (predictions) about the future geologic state of our planet. This approach—using the present to interpret the past and predict the future—proves useful in many fields of science, including oceanography, where scales of time (and space) have special relevance (**see inside front cover**).

The Solar System

To understand the origins of the world ocean, we go back in time to events that led to the formation of our solar system. The sequence of events that led to formation of our solar system has direct bearing on the sources of water found on Earth. At the same time, these astronomical events and their by-products, namely the existence of the Sun, planets, and the Moon, have importance implications for modern Earth processes, especially ocean currents and tides.

Origins of the Solar System

The most widely accepted model of solar system formation, the **solar nebula model,** states that solar systems originate when clouds of interstellar gas and dust are disturbed by the explosion of a nearby star or some other cosmic event **(Figure 2-3).** The cosmic distur-

■ **FIGURE 2-3** Artist's concept of a newly formed solar system, based on observations from the Spitzer Space Telescope.

■ **FIGURE 2-4** A star nursery in the Orion Nebula. More than 150 proplyds have been identified within this mosaic of images. The widespread occurrence of proplyds suggests that solar systems may be abundant throughout the universe. Proplyds were first discovered by the Hubble Space Telescope in 1992.

bance causes the gases and dust to form a spinning disk. Disks such as these, called **proplyds** (short for protoplanetary disk), are found throughout the universe today, lending support to the solar nebula model **(Figure 2-4)**.

Gravitational compression of the mass of materials in the central bulge of the disk cause it to heat up more than the surrounding material. The central bulge at this stage is a **protostar,** the precursor to a star. At a certain temperature, hydrogen atoms fuse into helium atoms. The onset of nuclear fusion, however, generates a tremendous explosion of particles that vaporize any nearby dust and gases, including water vapor. Scientists think that this **solar ignition** creates a "dry zone" in the region surrounding the star.

From the center of the proplyd to its boundary with interstellar space is a thermal gradient of high to low temperatures. The hot gases of the proplyd cool as they radiate heat to surrounding space. Eventually, the cooling gases reach their condensation point and coalesce into solid bits of dust. (Because of the low pressure of space, solids form directly from condensed gases.) Through gravitational attraction, these solid bits of matter form marble- and baseball-sized particles. Subsequent collisions create larger and larger rocky fragments ranging in size from boulders to asteroids. These rocks comprise the embryonic materials of planets, or **planetesimals.** Eventually, planetesimals collide and coalesce to form larger planetary bodies **(Figure 2-5)**.

An important consequence of the condensation of gases and accretion of planetesimals is the order in which various compounds appear. Compounds such as metallic iron condense at a temperature of 1050° C (about 1900° F), while compounds rich in silicates condense at temperatures between 950 and 800° C (about 1750 and 1475° F). At 0° C (32° F), water condenses into ice. Because the proplyd cools more rapidly at its outermost boundaries, gases in this region condense into solids first. Hot gases closest to the protostar condense later. The **condensation sequence** of compounds helps to explain the chemical composition of the planets in our solar system **(Figure 2-6)**. It is no accident that Mercury, the planet closest to Sun, is largely composed of iron with few silicates. Temperatures nearest the Sun prevented condensation of silicates and water when Mercury formed. Earth, on the other hand, accreted from planetesimals containing a mix of iron and silicates. Beyond Earth, temperatures were sufficiently cooler to permit the condensation of water vapor, methane, and ammonia.

According to **Table 2.1**, the first bits of solid planetary materials formed about 4.56 billion years ago (bya). The first distinguishable "proto-planets" accreted about 100,000 years later. Earth reached 64% of its present mass at 4.55 bya, approximately 10 million years after formation of the solar system. The world ocean likely existed by 4.3 billion years ago. Throughout this textbook, we use 4.56 bya for the age of the Earth because the evidence on which this age is based (radiometric dating of meteorites) is subject to the least amount of interpretation.

Disk of gas and dust
spinning around young sun

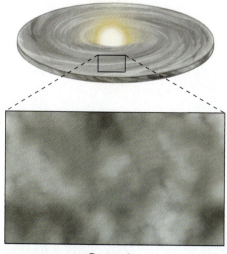

Dust grains

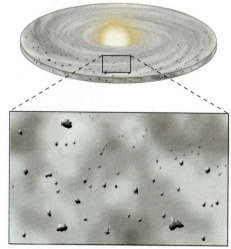

Dust grains clump into planetesimals

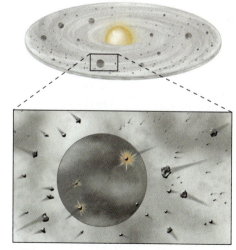

Planetesimals collide and collect into planets

■ **FIGURE 2-5** Planetesimal hypothesis. As dust grains progressively grew in size, they clumped together to form larger fragments called planetesimals. Eventually, through collision and accretion of these planetesimals, planets formed.

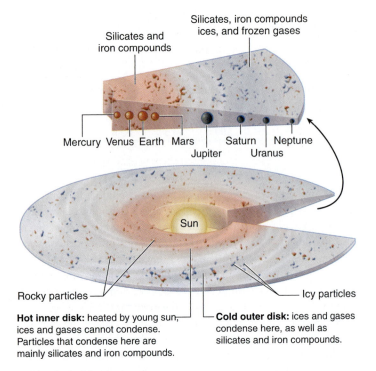

Silicates and iron compounds

Silicates, iron compounds ices, and frozen gases

Mercury Venus Earth Mars Jupiter Saturn Uranus Neptune

Rocky particles — Sun — Icy particles

Hot inner disk: heated by young sun, ices and gases cannot condense. Particles that condense here are mainly silicates and iron compounds.

Cold outer disk: ices and gases condense here, as well as silicates and iron compounds.

■ **FIGURE 2-6** Condensation model for the formation of rocky and ice particles in the solar system. Note that heat from the Sun prevented gases and ices from forming in the inner disk. Condensation of gases and ices proceeded in the outer disk, and eventually, the inner disk cooled sufficiently to allow gases and ices to develop.

The Sun

The modern Sun makes up 99% of the total mass in the solar system. The Sun supplies **solar energy** (light, heat, and other forms of electromagnetic radiation) and a stream of charged particles called the **solar wind.** At the time of formation of the Sun, the output of solar radiation was 25–40% less than it is today. This reduced solar output meant that Earth was either frozen over like a snowball or that heat-trapping greenhouse gases in the atmosphere retained heat released from Earth's interior. The presence of greenhouse gases could explain the existence of liquid water on Earth at 4.3 bya. (See section on Water and the Primordial Earth.)

The Rocky Planets and the Gas Planets

Progressive accretion of planetesimals eventually formed the nine planets in our solar system **(Figure 2-7)**. The four inner planets—Mercury, Venus, Earth, and Mars—are called the **terrestrial or rocky planets** because they are primarily composed of silicate rocks with an iron-nickel core. The outer four planets—Jupiter, Saturn, Neptune, Uranus—make up the **Jovian or gas planets** because they are primarily composed of helium and hydrogen-based gases, much like our Sun **(Figure 2-8)**. In fact, Jupiter is

Time (billions of years ago)	Event
4.5647 + 0.6	The first solid bits of raw planetary materials formed from the solar nebula.
4.5646	The embryonic protoplanets were distinguishable as objects accreting faster than surrounding objects.
4.5547	About 10 million years after formation of the solar system, at least 64% of Earth's mass was accreted (also known as the mean age).
4.54 - 4.44	The Moon solidified following the impact of a Mars-sized or larger meteorite with Earth.
~ 4.3	Liquid water was present on Earth, probably in the form of an ocean.

TABLE 2.1 Timeline of Earth's Formation, Based on Radiometric Dating Using the Hafnium Isotope

Source: Jacobsen, 2003

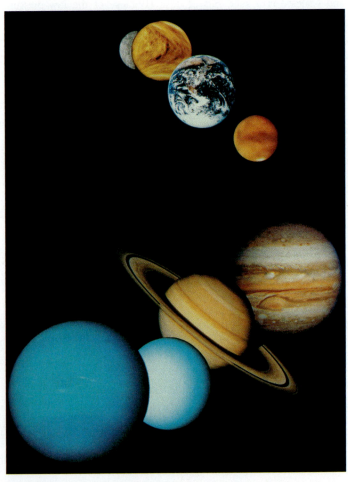

■ **FIGURE 2-7** The rocky planets (Mercury, Venus, Earth, Mars) and the gas planets (Jupiter, Saturn, Neptune, Uranus).

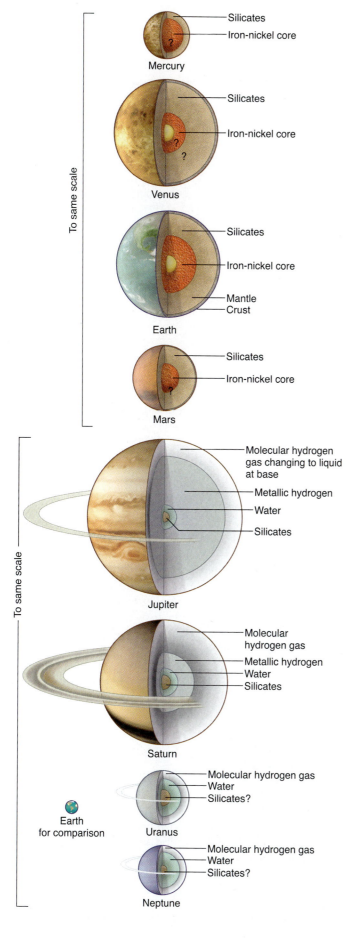

FIGURE 2-8 The composition of the planets. The inner rocky planets have an iron core surrounded by a silicate shell. The outer gas planets feature a rocky core of unknown composition surrounded by a large shell of hydrogen gas. Models of planetary formation seek to explain the different compositions of the inner and outer planets.

thought to be a protostar that failed to ignite. Pluto is so unique in its characteristics that astronomers now classify it as a different planetary object altogether, a dwarf planet.

The Asteroid Belt

Between the rocky and gas planets lies the **asteroid belt,** a rag-tag bunch of oddly shaped rocks whose combined mass is no greater than that of the Moon. Asteroids are thought to represent fragments of planetesimals. In 1996, the Near Earth Asteroid Rendezvous (NEAR)–Shoemaker mission observed more than one-hundred thousand craters on the surface of Eros, a large asteroid **(Figure 2-9)**. The cratered surface of Eros and other asteroids provides evidence of a violent past—a cosmic bowling alley—during which collisions between planetesimals and planets occurred frequently. These observations lend support to the idea that Earth formed from the accretion of planetesimals.

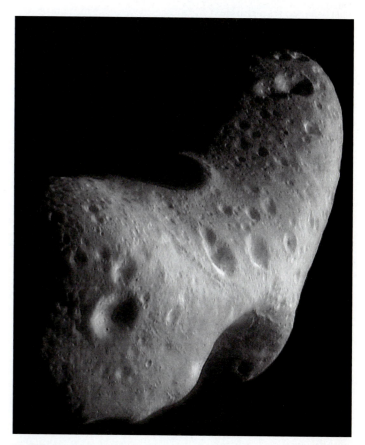

FIGURE 2-9 Composite image of the northern hemisphere of the asteroid Eros. From February 14, 2000, to February 12, 2001, the robotic spacecraft *Shoemaker* orbited, analyzed, and eventually landed on Eros, the first and only time a spacecraft has landed on an asteroid. This mission yielded important and intriguing discoveries on the properties of asteroids.

Seismology and the Birth of Earth Science

S P O T L I G H T

2.1

How did scientists arrive at a picture of Earth's interior? What is the evidence on which our understanding of Earth's structure is based?

For centuries, philosophers and scientists have speculated on the structure of the interior of the Earth. Some thought it solid, some thought it hollow, and some deduced that parts of it must be molten and that it must have an iron core (to explain both its density and active volcanism). The first quantitative effort to describe the interior of Earth came in 1897, when German-born Emil Wiechert developed a theoretical model to describe the structure of Earth that took into account astronomical observations that clearly showed Earth was not homogeneous (in terms of its density). Wiechert proposed that Earth consisted of a metallic core surrounded by a *mantel*, a German word meaning "cover."

In 1898, Wiechert constructed an inverted-pendulum seismometer, an instrument for measuring ground motion associated with earthquakes. While not the first seismometer ever built (various types of ground motion instruments and true seismometers originate from the 1840s; crude instruments for measuring earthquakes date to a Chinese philosopher, Change Heng, in 132 AD), Wiechert's seismometer is allegedly the earliest seismograph that is still used today.

Shortly after, Richard Dixon (R. D.) Oldham, an Irish-born Britisher acting as superintendent for the Geological Survey of India, conducted a careful analysis of seismograms and identified S waves, a type of slow, transverse (up-and-down) seismic body wave that had been predicted since 1824 but never measured. The P wave, a faster, longitudinal (back-and-forth) body wave, had already been detected; discerning the S wave signal was harder with the types of instruments and techniques they were using. (P and S simply stand for primary and secondary, respectively, because P waves propagate more quickly and arrive first at a seismometer, whereas S waves arrive afterward. Both waves travel through solid rock, but S waves cannot pass through a liquid.) Oldham's pioneering work and his 1906 publication of the arrival times of P and S waves provided convincing proof that Earth was not homogeneous.

A prominent geologist by the name of Edward Seuss (no relation to Dr. Seuss, whose real name was Theodor Seuss Geisel) seized upon these data and published in 1909 his hypothesis for the internal structure of Earth. Seuss proposed that Earth consisted of three "zones" to which he gave *Lion King*-sounding names: Nife, the nickel-iron (Fe) core; Sima, the silicate-magnesium middle layer; and Sal, the thin silica-aluminum crust. Although Seuss had their dimensions wrong, the basic idea of a three-part Earth was established.

Subsequent seismic studies revealed more layers. In 1909, Croatian scientist Andrija Mohorovicic (mo-ho-ROV-i-chich)

discovered a discontinuity—a change in the velocity of seismic waves—at the boundary of Earth's crust and mantle. This boundary still bears his name, although it has been shortened (thankfully) to the Moho.

In 1914, a German scientist, Beno Gutenberg, who would later serve as the director of the California Institute of Technology (Cal Tech) in Pasadena, verified through seismic studies the precise depth of the core-mantle boundary at 2900 km (1800 miles). This discontinuity, the Gutenberg discontinuity, though less well-known, still bears his name.

A Danish scientist, one of the few women in science at the time, Inge Lehman, provided the final broad brushstrokes to our view of the internal structure of Earth. In 1936, she published data that demonstrated the existence of an inner core distinct from an outer core. This discontinuity between the inner and outer core was named the Lehman discontinuity.

The result of these early studies formed what may be called the simplified model of Earth's interior, consisting of the inner and outer core, the mantle, and the crust. At the same time, these studies gave birth to the field of seismology, the study of earthquakes. Seismology represents one of the first "whole Earth" disciplines and has contributed significantly to our view of Earth as a single, interdependent system.

Seismology remains an active field of study in the twenty-first century. The USArray-Earthscope program is an ambitious program to map the three-dimensional structure of the interior of Earth beneath the continental United States. A dense grid consisting of hundreds of seismometers is being deployed across the western United States as part of this program. The USArray-Earthscope study will provide the most detailed seismic images of the continental United States ever obtained. These images will help seismologists better characterize seismic hazards, such as earthquakes, help locate new geologic resources, and contribute to scientific and public understanding of plate tectonics and earth processes.

The Moon

In 2001, scientists published a reanalysis of moon rocks brought back by astronauts who landed on the Moon. Using a new, more accurate technique for identifying various forms of elements in rocks, they found that moon rocks were nearly identical to basalts found on oceanic ridges (where seafloor spreading occurs: see Chapter 3). These data support the idea that the Moon was created when a large planetlike object struck the infant Earth and ejected the Moon. (The impact was not unlike what happens when you punch a jelly donut and a glob of jelly squirts out.) The impactor was comparable in size to Mars and probably struck the Earth between 4.54 and 4.44 billion years ago **(Figure 2-10)**. The **Giant Impact model** predicts identical compositions for Earth and the Moon. More than thirty years after humans first landed on the Moon, scientists appear to have finally confirmed that Earth and the Moon were formed from the same materials.

Dating the Solar System

Geologists and other scientists (including oceanographers) often seek information on the ages of rocks for clues about the history of the Earth. One technique, **radiometric dating,** determines the ages of rocks based on the products of radioactive decay within the rock. Radiometric dating relies on knowledge of **isotopes,** the different forms of a single element **(Figure 2-11)**. All atoms of elements have

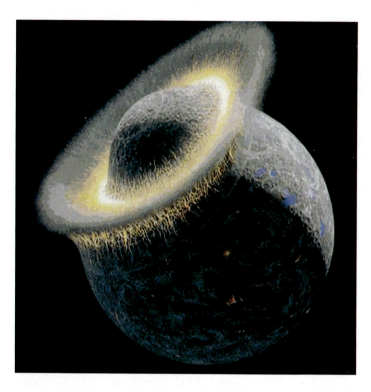

■ **FIGURE 2-10** Formation of the Moon. Although not all scientists agree on its origins, general consensus exists that the Moon was created when a Mars-sized object struck Earth in the early years of its formation.

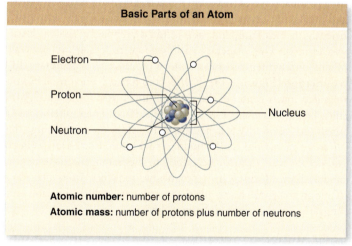

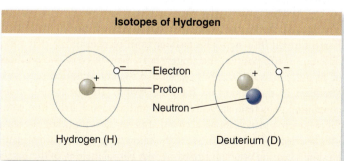

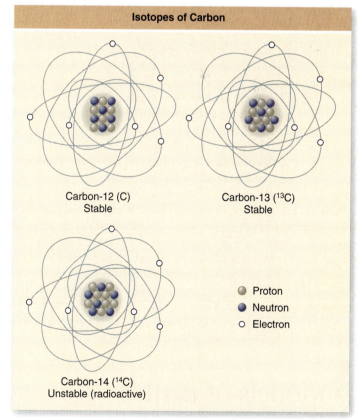

■ **FIGURE 2-11** Understanding isotopes. Isotopes have an identical atomic number but a different atomic mass. Deuterium (D) is the stable isotope of hydrogen (H) and contains an extra neutron in its nucleus. Carbon occurs in multiple isotopic forms, both stable and radioactive. In its most abundant form, carbon has an atomic number of 6, meaning it has 6 protons and 6 neutrons in its nucleus. The radioactive isotope carbon-14, designated as ^{14}C, has 6 protons and 8 neutrons but reverts to the most common form of carbon (C) when it emits radioactivity.

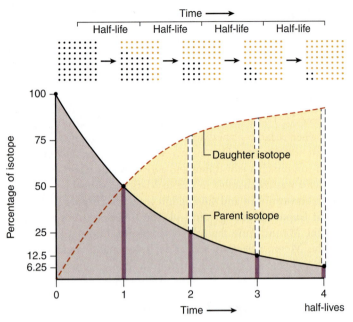

■ **FIGURE 2-12** Parent isotopes decay into daughter isotopes at a rate that follows a curved line. Mathematically, this curved line is described by a decreasing exponential function, that is, e^{-x}. The exponential decay of radioisotopes allows geologists to calculate very precisely the age of rocks. The half-life of the element, indicated by dark blue bars, indicates the amount of the parent isotope remaining after one, two, three, and four half-lives. The red dashed line indicates the amount of daughter isotope formed over time. Measurements of the ratio of daughter isotopes to parent isotopes allow calculation of the rock's age when the half-life of the element is known. See our web site for a more detailed explanation.

a nucleus that contains protons, positively charged particles, and neutrons, neutrally charged particles, surrounded by a "cloud" of **electrons,** negatively charged particles. The number of protons in an atom defines its **atomic number.** The number of protons and neutrons combined defines its **atomic mass.** Elements of a specific kind (such as carbon, iron, or sodium) have the same number of protons and neutrons. The isotopes of an element have the same number of protons but a different number of neutrons. Two types of isotopes have been identified: **radioactive isotopes,** which emit elementary particles through a process called **radioactivity;** and **stable isotopes,** which have a stable or unvarying chemical and atomic composition. Radioactivity alters the chemical and atomic composition of radioactive elements, while stable isotopes retain their composition under natural conditions. Dating rocks (and other materials) relies on knowledge of the precise rates at which radioactive elements spontaneously disintegrate from the **parent isotope** to the **daughter isotope.** Scientists define the **half-life** of a radioactive isotope as the time it takes for an isotope to decay to half its original abundance. From this information, a rock's age can be determined **(Figure 2-12).**

Models of Earth's Formation

The sequence of events that formed our solar system also led to Earth's formation. To "make" an Earth, a "recipe" is required that

accounts for all of the features that we observe on Earth. This recipe, what scientists call a **model,** provides a conceptual or mathematical framework for developing and testing hypotheses about Earth's formation.

One of the features that must be explained by a model of Earth's formation is the presence of distinct layers in Earth's interior: 1) the **crust,** a thin surface layer (5–35 km or 3–22 miles thick); 2) the **mantle,** the thickest layer occupying 80% of Earth's total volume (2885 km or 1793 miles thick); and 3) the **core,** the innermost part of the earth with a size slightly larger than Mars (3486 km or 2166 miles thick) **(Figure 2-8)**. These layers exhibit differences in **density,** the mass of a material within a specified volume. The core consists of the densest materials, while the crust is made up of less dense materials. A model of Earth's formation must explain this layering and the presence of Earth's atmosphere and ocean, as well.

Homogeneous Accretion

Two contrasting models have been proposed for Earth's formation **(Figure 2-13)**. The first model, the **homogeneous accretion model,** assumes that the planetesimals that formed Earth were nearly identical (homogeneous) in composition; all of the elements that we find on Earth today were present in these planetesimals. The accretion of the planetesimals led to a "proto-Earth" (incompletely formed Earth) that was everywhere the same from its surface to its core. Subsequently, the proto-Earth began to heat up from the frictional force of meteorite impacts, gravitational compression, and radioactive decay. Eventually, the interior of Earth heated sufficiently to melt rock, and, as a result, a completely molten Earth was formed. This planetary meltdown permitted heavy minerals to sink and lighter minerals to rise, a process called **planetary differentiation (Figure 2-14)**. Earth's core formed when molten iron migrated to the center of the Earth. Earth's crust formed as lighter silica-bearing elements rose to Earth's surface. The molten state of Earth also allowed water and gases to escape by **outgassing,** the release of gases from Earth's interior. As a result, Earth's atmosphere and oceans were formed. According to the homogeneous accretion model, Earth formed in a progressive sequence from planetary accretion to whole-Earth melting to planetary differentiation and outgassing of the ocean and atmosphere.

Heterogeneous Accretion

The second model, the **heterogeneous accretion model,** states that Earth formed in a step-wise fashion from planetestimals that were different in composition (heterogeneous). The different composition of planetesimals came about because of the condensation sequence discussed in the section on the solar system. In this model, iron-containing planetesimals accreted first and formed Earth's iron core. Silica-bearing planetesimals accreted afterward, surrounding the core in a layer of lighter elements. At the same time or soon after, planetesimals bearing water and gases arrived. Partial melting of the Earth promoted outgassing of the atmosphere and ocean. Water-bearing planetesmials and comets may also have arrived as a **late veneer,** a late-stage bombardment of wet planetesimals. Upon collision with the Earth's surface, their volatile contents escaped through melting.

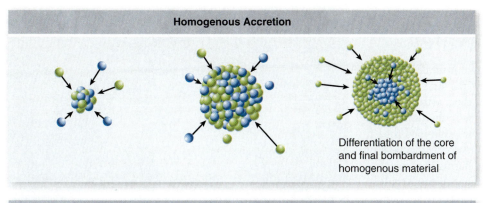

Homogenous Accretion

Differentiation of the core and final bombardment of homogenous material

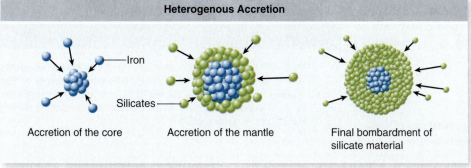

Heterogenous Accretion

Iron

Silicates

Accretion of the core

Accretion of the mantle

Final bombardment of silicate material

■ **FIGURE 2-13** Competing models for the formation of the Earth. In the homogeneous model, Earth formed from planetesimals that were similar in composition. Whole-planet melting caused iron to sink toward the interior of the Earth, forming the core. Silicate metals floated toward the surface, resulting in a layered Earth. In this molten state, Earth outgassed water vapor and gases to create the ocean and atmosphere. In the heterogeneous model, iron-bearing planetesimals accreted into the core first, and silicate-bearing planetesimals accreted afterward to form Earth's mantle and crust. Partial melting of the outer layer of Earth and possibly late-arriving wet planetesimals led to formation of the ocean and atmosphere.

Iron and silicates mixed

Heating melts iron and silicates

Silicates

Iron

Gravity

a.

Undifferentiated

Iron sinks to core

Differentiated

b.

■ **FIGURE 2-14** Planetary differentiation: (a) One model of the formation of Earth's layers involves melting of the Earth and sinking of dense minerals such as iron to form Earth's core. (b) This process is like what would happen if you allowed chocolate chip ice cream to melt.

Sources of Earth's Water

The launch of the European Space Agency's Infrared Space Observatory in 1996 enabled astronomers for the first time to detect water outside Earth's atmosphere. Since then, astronomers have discovered vast amounts of water throughout the universe, almost 99% of it in the form of ice. In the regions examined thus far, water ranks third in abundance among all molecules present. A great deal of this water appears to be formed during the violent explosions accompanying the birth of stars. The Orion Nebula may produce up to sixty volumes of ocean water every day. Water has also been found in comets and meteorites. With all this water in the universe, you might expect that scientists would have little trouble identifying where Earth's water came from. As it turns out, the answer is not so simple.

Water from Cosmic Gases?

The observation of water throughout the universe demonstrates that water was most likely present at the time of formation of our solar system. However, as we learned in the section on the solar system, ignition of the Sun blasted away any vaporous substances within the vicinity of Earth. If this is true, the planetesimals that created the inner planets would have been stripped of their water and gases during formation of the solar system. Observations from NASA's Chandra X-Ray Observatory support the violent beginnings of stars and lessen the likelihood that cosmic gases were a source of water for our planet **(Figure 2-15)**. Thus, if the solar nebula model is correct, cosmic gases probably did not supply Earth's water.

■ **FIGURE 2-15** The violent birth of stars. In March 2002, the Chandra X-Ray Observatory captured this image of X rays "exploding" from the Omega Nebula (M17) in the Orion Nebula.

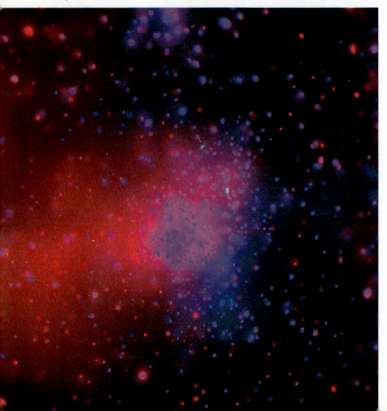

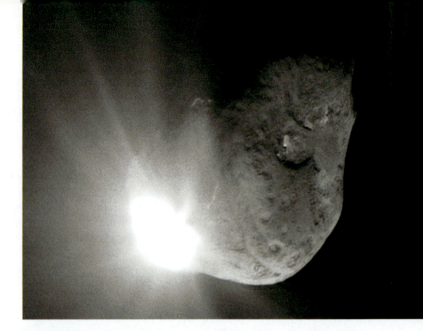

■ **FIGURE 2-16** Deep Impact. On July 4, 2005, NASA scientists slammed a washing machine-sized probe into comet Tempel 1. The impact generated a spectacular plume of dust and debris. Scientists hope that such experiments may reveal clues about conditions during the formation of the solar system.

Water from Comets?

In 1986, Louis Frank and John Sigwarth, physicists at the University of Iowa, proposed the **small comet hypothesis,** the idea that small comets might be a major source of Earth's water. The researchers' challenge was to demonstrate that such comets actually exist, a difficult task given their small size (6–9 m or 20–30 feet in diameter). Despite many critics, the small comet hypothesis got a boost in 1997, when an ozone-detecting satellite observed abundant water in Earth's upper atmosphere. In 2001, Frank published a study using ground-based telescopes suggesting that twenty or so 20- to 40-ton comets (the size of a small house) break apart in Earth's atmosphere every minute. This number of comets would be sufficient to fill the world ocean over billions of years. However, when scientists compared the isotopes of water molecules (heavy [D] and normal [H]) in ocean water with three well-known comets—Halley, Hyakutake, and Hale-Bopp—they found a composition that was quite different. They concluded that comets were not a major source of Earth's water. In addition, observations from the Deep Impact mission in July 2005 demonstrated that comets may contain less water than once believed **(Figure 2-16)**. Of course, if astronomers discover a new set of comets with a D/H ratio that matches Earth's water or a group of comets with more abundant water, then the possibility of comets as a source of Earth's water may be revisited.

Water from Meteorites?

Despite these seeming setbacks, measurements of the D/H ratios of other planetary materials have yielded another possible candidate for Earth's water, a type of meteorite called a **carbonaceous meteorite.** As their name implies, carbonaceous meteorites contain a significant percentage of carbon. In addition, they contain water or minerals that have been altered by water. This class of

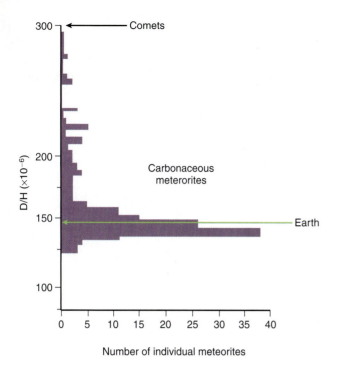

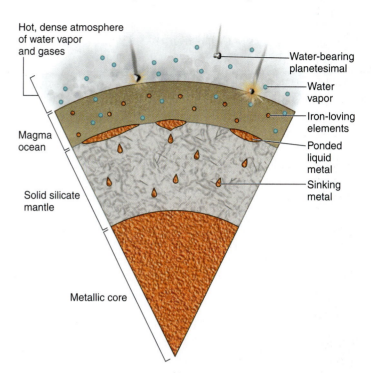

■ FIGURE 2-17 Comparison of hydrogen isotope ratios, D/H (deuterium/hydrogen), in meteorites, comets, and Earth's water. While the D/H ratios of carbonaceous meteorites and Earth's water are similar, ratios of other isotopes (not shown) in these meteorites indicate that their contribution to Earth's water was minor (data from Francois, 2001).

■ FIGURE 2-18 Illustration of wet magma ocean. Scientists envision an Earth that, rather than melting completely, had a surface that was maintained in a partially molten state, perhaps as deep as 700 km (400 mi) or more. The arrival of water-laden meteorites and comets during this period supplied Earth's water. The molten layer held onto gases and water vapor until it cooled sufficiently to outgas an atmosphere and ocean.

meteorites is believed to represent the most primitive materials present during the formation of the solar system. Measurements of the D/H ratio on more than a hundred of these meteorites include the range of values observed on Earth **(Figure 2-17)**. Are carbonaceous meteorites the source of water in the world ocean? Planetary scientists calculate that even under the most generous of circumstances, carbonaceous meteorites could only provide one volume of ocean water. Thus, their contribution appears limited.

The Wet Magma Ocean Hypothesis

Some scientists propose that wet planetesimals arrived after solar ignition but before the solidification of Earth's crust. This hypothesis is called the **wet magma ocean hypothesis** because it assumes that water was temporarily stored in the Earth's molten interior **(Figure 2-18)**. Heat trapped by a thick atmosphere melted or maintained the molten state of the Earth's surface and created an "ocean" of magma in which significant quantities of water remained dissolved. When the atmosphere cooled, the wet magma ocean released its water vapor via outgassing and created a planet submerged in water. The wet magma ocean hypothesis represents a kind of hybrid between the homogeneous and heterogeneous accretion models. According to this model, Earth formed from planetesimals whose composition was fairly similar (as in the homogeneous model), but whole-Earth melting did not occur (as in the heterogeneous model). Instead, partial melting

and formation of an ocean of wet magma in Earth's mantle led to outgassing of the ocean and atmosphere, cooling of Earth's surface, and formation of the crust. Partial melting also permitted formation of Earth's iron core and segregation of other layers in Earth's interior.

Like many endeavors in science, models of Earth's formation are works in progress. Support for one model or the other will require additional observations and experiments. Nonetheless, it is the pursuit of knowledge that drives science. Toward that end, we can be hopeful that some day in the future, scientists will uncover the evidence that answers the question of where Earth's water came from.

Water and the Primordial Earth

Although the source of Earth's water remains uncertain, its presence on Earth has been revealed through studies of ancient minerals called **zircons**. These tiny crystals (like the zircons you find in jewelry stores) reveal that liquid water was present as early as 4.3 bya. This result surprised scientists. How is it, with a solar output that was 25–40% weaker than it is today, that the world ocean was not completely frozen over? Scientists call this phenomenon the **early faint Sun paradox.**

The Mars Rovers and Martian Water

While robotic missions to nearby planets may seem far afield of earthbound oceanography, such studies play an important role in our understanding of Earth. The formation of Earth, the source of Earth's water, the evolution of Earth's geology, the changes in Earth's climate, and the impact of humans on Earth gain insights from studies of other planets. At the same time, the development of technology for exploring space offers many spin-offs to technology for exploring the ocean. The success of a mission to space or the world ocean depends on scientists' ability to design and build technology that delivers the desired set of observations or measurements.

NASA's mission to Mars provides one example. In January 2004, two robotic geologists, the Mars Exploration Rovers *Spirit* and *Opportunity*, successfully landed on Mars and began their drive into the history books. Slated for a ninety-day mission, the twin explorers exceeded all expectations by continuing to operate years beyond their presumed lifetime. Their observations and measurements of the Martian terrain yielded a wealth of scientific data on the geology and climatology of Mars. While *Spirit* traversed a crater dominated by a lava flow, *Opportunity* struck out across a landscape enriched in minerals deposited by interactions with water. The blueberry-sized hematite, a product of the chemical interaction of iron and water, resembled similar deposits found in Utah **(Figure 2b)**. Bedform deposits, like those found along streams, were also observed by *Opportunity*. Nevertheless, the bulk of the data points to a very limited presence of water on Mars in the past. Analyses of atmospheric dust and volcanic soils and rocks reveal minerals untouched by water. Mars appears as a sulfur-rich planet whose geochemistry resembles something in between Earth and Io, a sulfur-spewing volcanic moon that orbits Jupiter. As scientists learn more about our Martian neighbor, oceanographers may better understand the origins and nature of water on Earth as well.

FIGURE 2b "Blueberries" on Mars (right). The hematite observed by *Opportunity* on Mars resembles hematite-rich structures found in Utah (left). Because hematite only forms in water, scientists conclude that liquid water must have once existed at this site on Mars.

One possible answer to the faint Sun paradox involves the greenhouse effect **(Figure 2-19)**. When solar radiation, made up of energy with different wavelengths, falls on the exterior of a greenhouse, only the shortwave radiation passes through it. The glass roof of the greenhouse blocks longwave radiation. Objects inside the greenhouse, including the plants and the soil, absorb the shortwave radiation and re-emit heat, a form of longwave radiation. Because glass blocks longwave radiation, the heat is trapped inside the greenhouse. A number of gases in the atmosphere act like the glass in a greenhouse. Carbon dioxide, water vapor, ammonia, and methane are a few of the greenhouse gases that trap heat, albeit with varying efficiencies. Under natural conditions, there is a balance between incoming solar radiation and outgoing radiation such that a constant atmospheric temperature is maintained. However, changes in the output of solar radiation or the concentration of greenhouse gases can affect this balance.

Thus, one means for maintaining a warmer Earth with a weaker Sun is to increase the concentration of greenhouse gases. Is this what happened on early Earth?

Although carbon dioxide is one of the strongest greenhouse gases, its concentration in Earth's early atmosphere was about twenty times too low to maintain conditions above freezing. For that reason, some scientists have suggested that methane, if produced in sufficient quantities, would be capable of maintaining the warm conditions of early Earth. Some evidence suggests that **methanogens,** a methane-producing microbe found in oxygen-free environments on Earth today, may have been present about 3.5 bya. While not sufficient to explain liquid water at 4.3 bya, methane produced by one of Earth's earliest inhabitants may have provided the greenhouse effect required to overcome a weak Sun. Scientists hope that further research may shed clues on the causes of global warming on early Earth and in modern times.

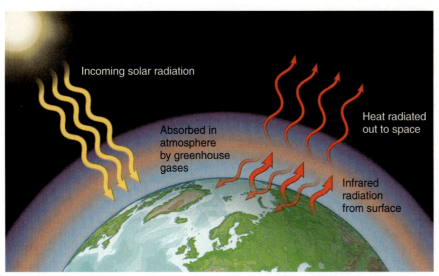

a. Greenhouse effect under modern Sun

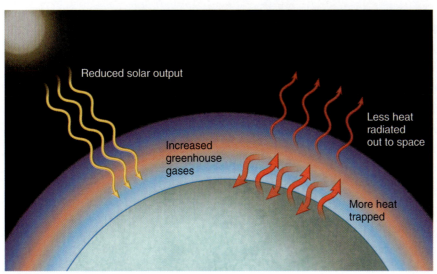

b. Greenhouse effect under primordial Sun about 4 billion years ago

■ FIGURE 2-19 The greenhouse effect and the early faint Sun paradox. (a) Under the modern Sun, greenhouse gases trap heat and maintain a comfortable climate on Earth. (b) How did the Earth stay warm when the Sun was weaker? Scientists speculate that the increased concentrations of greenhouse gases in Earth's primordial atmosphere kept Earth warm despite a weaker solar output. Thus, at a very early stage in Earth's history, interactions between Earth's atmosphere and the Sun were important to Earth's climate and the physical state of water.

YOU Might Wonder

1. **How do we know that Earth is 4.56 billion years old?**
Much of what we know about the age and chemical composition of primitive Earth comes from our study of meteorites. Meteorites may be found in the Antarctic, the African Sahara, or the Mohave Desert in California, where conditions are favorable for the preservation of meteorites. One class of meteorites, the chondrites, is among the oldest rock in our solar system. Chondrites contain minerals within their stony matrix that have been dated at about 4.5647 + 0.6 billion years old using lead isotopes. The simplest explanation for the origin of our planet assumes that these chondrites represent the same stuff from which our planet formed.

2. **How did life originate on our planet?**
Scientists continue to seek an answer to that question. In 1977, new forms of life were discovered off the Galapagos Islands. Six-foot worms, plate-sized clams, and other oddities were found flourishing on hydrothermal vents (see Chapter 4). For the first time, life did not solely depend on light-driven photosynthesis to derive primary organic matter. Instead, chemosynthesis, the production of organic matter using chemical energy, was recognized as an alternative means for organisms to acquire energy and matter. The realization that ecosystems may flourish on chemical energy led to a flurry of interest in hydrothermal vents as a possible site for the origin of life. The research that followed placed geologists, physicists, chemists, biologists, and even space scientists side by side in an unprecedented interdisciplinary effort to understand the chemical processes that support life on hydrothermal vents.

Key Concepts

- Modern geology arises from the idea that the key to Earth's geologic past lies in an understanding of processes occurring in modern times. This scientific philosophy acknowledges a role for episodic events and allows for the possibility that some past processes no longer operate on Earth.
- The solar nebula model explains the origins of the solar system as the condensation of cosmic gases into a flattened, whirling disk whose gravitational forces at the center of the disk created a star that became our Sun.
- The condensation sequence for the transition of gases into planetesimals led to rocky planets in the inner orbits of the solar system and gas planets in the outer orbits.

- Earth may have accreted from planetesimals that were homogeneous in composition or in a sequence from planetesimals that were heterogeneous in composition. Both homogeneous and heterogeneous accretion can account for the layered structure of Earth's interior, but they have different implications for the timeline of Earth's formation and the appearance of water on Earth.
- Several sources of extraterrestrial water have been identified, but none of them completely explains the abundance and isotopic composition of Earth's water.
- The greenhouse effect may have played an important role in maintaining the liquid state of water on early Earth.

Terms to Remember

actualism, 14
asteroid belt, 17
atomic mass, 20
atomic number, 20
carbonaceous meteorite, 22
condensation sequence, 15
core, 20
crust, 20
daughter isotope, 20
density, 20
early faint Sun paradox, 23
electrons, 20
geologic time, 14
Giant Impact model, 19

half-life, 20
heterogeneous accretion model, 20
homogeneous accretion model, 20
isotopes, 19
Jovian or gas planets, 16
late veneer, 20
mantle, 20
methanogens, 24
model, 20
outgassing, 20
parent isotope, 20
planetary differentiation, 20
planetesimals, 15
principle of uniformitarianism, 14

proplyds, 15
protostar, 15
radioactive isotopes, 20
radioactivity, 20
radiometric dating, 19
small comet hypothesis, 22
solar energy, 16
solar ignition, 15
solar nebula model, 14
solar wind, 16
stable isotopes, 20
terrestrial or rocky planets, 16
wet magma ocean hypothesis, 23
zircons, 23

Critical Thinking

1. Two opposing models have been proposed to explain Earth's formation. Summarize each model and include a sketch that illustrates when and where water might have arrived on Earth. Use clay or Playdough to create your own model of Earth's formation and present it to a classmate, family member, or friend.

2. Some scientists suggest that an "as-yet-undiscovered" source of extraterrestrial water might explain Earth's abundant water. What conditions must be met for a particular planetary object to qualify as a source of Earth's water?

Explore Online

 Visit www.mhhe.com/chamberlin1e for access to chapter quizzing, key term flash cards, video clips, interactive activities, and more. Further enhance your knowledge with web links to chapter-related material!

Exploration Activity 2-1

Exploring Earth's Formation Through Physical Models

Question for Inquiry

How do models help scientists test hypotheses about Earth's formation?

Summary

In this chapter, we introduced several models to explain various aspects of the formation of Earth and the solar system. As you will discover in the coming chapters, models are an integral part of scientific research. They enable scientists to synthesize their ideas and test hypotheses. In this activity, you will explore a physical model for understanding and comparing hypotheses on Earth's formation.

Learning Objectives: By the end of this activity, you should be able to:

- Identify differences in the composition of Earth's layers
- Relate similarities and differences in two contrasting models of Earth's formation
- Postulate the implications of each model for the formation of the ocean and atmosphere
- Evaluate the advantages and disadvantages of using models to explore hypotheses

Materials and Preparation

You will need:

1. Play-Doh in three contrasting colors;
2. several sheets of newspaper;
3. a knife sufficient for cutting Play-Doh

Hint: You can easily make your own sweet-smelling ersatz Play-Doh:

- 2½ cups flour
- 1 cup salt
- 3 tablespoons vegetable oil
- 2 cups boiling water
- 2 packages unsweetened Kool-Aid®
  (red, blue and yellow work best)
- 1 1-quart plastic bag

Mix dry ingredients, then add oil and water. Knead with hands for 10 minutes (use gloves if you don't want your hands to be dyed with Kool-Aid). Alternatively, if a bread-making machine is available, you may put the mixture in the machine and mix for 10 minutes on the dough setting. Put in plastic bag and let cool in refrigerator overnight (for best results).

Classroom Adaptation

Divide into groups of four. Each group prepares either the homogeneous or heterogeneous model (see below). Members from each group present their models to the class. Invite classroom debate and discussion on thought questions. Each student writes a five-minute paper on the clearest points and muddiest points to complete the exercise.

Part I: Overview

Planetary scientists debate two models for Earth's formation. In the **homogeneous accretion** model, Earth formed from homogeneous planetesimals and subsequently melted and differentiated, through the iron sink, into the core, mantle and crust. In the **heterogeneous accretion** model, Earth formed from planetesimals with different compositions: iron-containing planetesimals formed the core while silicate-containing planetesimals arrived later and formed the mantle and crust.

In this activity, you will simulate the two accretion models using Play-Doh and compare and contrast the implications of these models for the formation of Earth's layers, the world ocean, and the atmosphere.

Part II: Prepare the Planetesimals

1. Create several 1-inch-diameter balls from each of the three colors of Play-Doh.
2. Do not mix the colors at this time.
3. Assign the following color scheme to the balls that you have created: red=iron; yellow=silica-magnesium (Si-Mg); and blue=silica.
4. If working in groups, use the same color scheme to avoid confusion.

Part III: The Homogeneous Model

1. Take one ball of each color and stick them together.
2. Create a single ball of the three-color mixture but avoid mixing the colors. Each individual color should be recognizable in the single larger ball. This single tri-color ball represents a homogeneous planetesimal.
3. Create five to six of the homogeneous planetesimals.

Write a description of the shape, size and color of your homogeneous planetesimal.

4. You are ready to create your planet. Form a single large ball from the five to six planetesimals you just created. Avoid mixing the colors but create a sufficiently spherical ball that will not fall apart.

Write a description of the shape, size and color of your planet.

5. Use the knife to carefully cut your planet into two equal halves.

Write a description of the interior of your planet.

Part IV: The Heterogeneous Model

1. Each colored Play-Doh ball represents a planetesimal with a different composition.

Write a description of the shape, size and color of each of the different planetesimals.

2. You are ready to create your planet. Take three to four of the red (iron) planetesimals and create a single larger proto-planet. This ball represents the core of the emerging planet.
3. Apply the yellow (ψ-mg) planetesimals to the exterior of the core until it is completely surrounded. You should now have a large yellow protoplanet with no visible signs of its red interior.
4. Apply the blue (silica) planetesimals to the exterior of the yellow protoplanet. You have now completed your planet.

Write a description of the shape, size and color of your planet.

5. Use the knife to carefully cut your planet into two equal halves.

Write a description of the interior of your planet.

Part V: Complete These Thought Questions

1. Compare and contrast the interior of your planet to Earth's interior. How is it similar or different?

2. Why is your planet similar to or different from Earth's interior?
3. What step is missing from the homogeneous Play-Doh model that would be needed to make it more similar to Earth's interior?
4. How would this missing step facilitate the formation of an ocean and atmosphere?
5. What step is missing from the heterogeneous Play-Doh model that would be needed to form an atmosphere and ocean?
6. Postulate the relative timelines of the two models. Explain the sequence of events from the formation of planetesimals to a fully formed planet with an atmosphere and ocean in each model and indicate whether that event would be faster, slower or identical in each model.

Part VI: Five-Minute Paper

1. Take five minutes to write a short paper that summarizes the points that were clearest to you in this exercise and the points about which you are still unclear.
2. What kinds of additional study tools or modifications of this activity might improve your understanding of Earth's formation and the formation of the ocean and atmosphere?

Part VII: Explore and Discuss Further

1. Why do scientists use models to help them explore hypotheses?
2. What are the disadvantages of models?
3. How are observations and models combined to refute a particular hypothesis?
4. How is the homogeneous-heterogeneous accretion debate a good example of science in action?

Use the rubric on next page to self-assess and refine your work.

Rubric for Physical and Conceptual Models

	Exemplary	Adequate	Minimal
Preparation of model	Reproduced model with no major flaws; includes appropriate level of detail.	Model lacking in one or two characteristics and some details.	Model not a suitable representation
Description of model	Offers clear and accurate description of the model, including all of the appropriate details.	Description of the model lacking in a few important details.	Description of the model inadequate for describing its major features
Summary of key points and concepts	Synthesizes key points and conclusions; makes connections between model and concepts.	Less synthesis and fewer and unclear connections between model and concepts.	Few or no connection between model and concepts; conceptual understanding absent.
Organization and presentation	Arguments presented in a thoughtful and logical manner; evidence supports arguments; no grammatical or syntactical errors	Some arguments presented well; evidence does not always support arguments; some grammatical and syntactical errors	Poor arguments; lacking good evidence; many grammatical and syntactical errors

3

Plate Tectonics Theory and Evidence

Lava from Kilauea flows into the Pacific Ocean on Hawaii's Big Island, the tallest mountain on Earth, as measured from its base to its crest.

Questions to Consider

1. How did the Earth acquire its present structure and configuration of ocean basins and continents?

2. What evidence supports the theory of plate tectonics?

3. Why did it take nearly 50 years for scientists to accept the theory of plate tectonics?

4. What are the similarities and differences between convection-driven plate tectonics and plate-driven plate tectonics?

5. Why is the study of plate tectonics an integral part of oceanographic research?

A Theory of Earth

For centuries, writers, philosophers, and scientists have speculated on the structure of the interior of Earth. Some thought it solid, some thought it hollow, and some deduced that parts of it must be molten with an iron core. Summing up the sentiment of his time, Jules Verne's Professor Lidenbrock in the science fiction novel *Journey to the Center of the Earth* (1864) exclaims, "Neither you nor anybody else knows anything about the real state of the earth's interior. All modern experiments tend to explode the older theories." So, too, the theory of plate tectonics supplanted many theories before it, but this theory was vigorously debated for nearly 50 years before it was accepted.

The **theory of plate tectonics** states that Earth's crust is divided into several large and rigid plates that move independently of one another in response to heat flow through the crust. As a result, continents emplaced within those plates move about on the face of our planet, coming together and drifting apart in a phenomenon known as **continental drift.** Continental drift is just one aspect of plate tectonics, but it was the idea that set in motion the long and difficult history that finally led to acceptance of plate tectonics theory.

Plate tectonics theory represents one of the pillars of modern scientific achievement. Scientists and historians consider the theory of plate tectonics on a scale equal to Darwin's theory of evolution for biology, Bohr's theory of the atom for chemistry, and Einstein's theory of relativity for physics. The theory of plate tectonics unifies an immense body of geological observations and phenomena under a common principle: the Earth's crust is composed of plates that move. Many features of our planet—mountains, oceans, deserts, earthquakes, tsunami, volcanoes—have their origins in plate tectonics. As a unifying scientific principle, it explains observations in geology, geophysics, seismology, oceanography, space science, and more. Plate tectonics serves as a model for a wide number of Earth systems processes.

Historical Development of Plate Tectonics Theory

Though numerous scientists and geographers had proposed drifting continents since the sixteenth century, the first comprehensive treatment did not appear until 1915, with the publication of *The Origin of Continents and Oceans,* by German astronomer-turned-meteorologist Alfred Wegener

Seismic Tomography Reveals Our Inner Earth

The theory of plate tectonics revolutionized a number of scientific disciplines by explaining the features and motions of Earth's surface. Yet some scientists would argue that the "beauty" of this theory is only skin deep. Observations of Earth's crust—direct and indirect—have provided abundant evidence of plate motions and tectonic interactions, yet our understanding of processes beneath the crust is another story altogether. Without direct observations of Earth's interior, questions such as about the origins of seafloor features (e.g., swells and rises), the existence or nonexistence of plumes (and hot spots), and the nature of circulation within the mantle, among others, remain unanswered. Fortunately, increasingly sophisticated seismometers, more powerful computing and modeling techniques, and new efforts to deploy high-density grids of land- and ocean-based instruments promise to resolve more details of the inner Earth.

In the early 1970s, Professor Adam M. Dziewonski of Harvard University transformed the way scientists visualize seismic data. By combining digitized seismic data from different locations on Earth and for different earthquake events, Dziewonski produced a type of "CAT scan" of Earth's mantle, a type of analysis called seismic tomography. This technique basically works on the principle that the density of materials within the mantle alters the speed at which seismic waves travel through it. Colder, dense materials speed up seismic waves, while hotter, less dense materials slow them down. By comparing the arrival times of seismic waves at different locations and from different events, a three-dimensional picture of Earth's interior can be obtained.

Seismic tomography has revealed exciting new insights into the nature of Earth's interior, allowing geophysicists to visualize subducting slabs of crust and mantle circulation. It has also provided a tool for testing hypotheses regarding the nature of mid-ocean ridge spreading centers. In the Mantle Electromagnetic and Tomography (MELT) experiment, oceanographers deployed ocean-bottom seismometers along the East Pacific Rise as part of an effort to determine whether mantle flow in this region is active or passive. In the active model, a plume of actively upwelled material (driving its motions) is expected to be narrowly confined to a vertical column beneath the ridge. In the passive model, the "passively" diverging plates drag the mantle material into a broad region of melted material. Tomographic data supported the passive model with a few caveats, namely an asymmetry in the region of melt, with the most pronounced melting occurring off-axis. Future efforts are aimed at deployments of dense grids of ocean-bottom seismometers around Hawaii to resolve questions concerning the nature of the Hawaiian hot spot. Such experiments demonstrate the utility of seismic tomography for understanding plate tectonics and mantle processes.

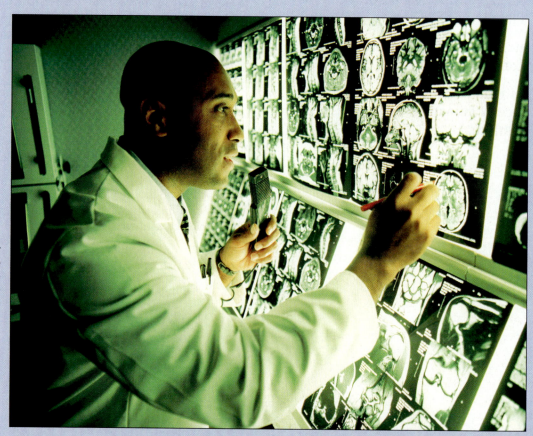

FIGURE 3a Like a medical doctor who uses CAT scans to observe the insides of humans (without cutting them open), seismologists use seismic tomography to reveal the internal structure of the Earth.

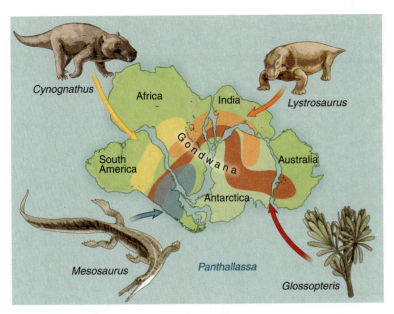

■ FIGURE 3-1 Fossils support the existence of the supercontinent of Gondwana. Alfred Wegener cited the similarity of fossils and the "fit" of the continents as evidence that the continents were once joined together. Unfortunately, alternative hypotheses could explain these observations so Wegener's arguments were not persuasive.

(1880–1930). Wegener sought to explain changes in the geographic distribution of fossil plants and animals through geologic time (**Figure 3-1**). One explanation was that climate had changed over time. But why did climate change occur? Wegener's answer: continental drift. If the continents moved from the tropics to the poles or vice versa, then the plants and animals associated with those continents would change. Wegener's ideas were strongly contested. Most scientists held to the idea that the Earth was rigid and solid and that continental drift was simply impossible. Since Wegener employed a European approach to science, one in which multiple lines of evidence were used to support a single hypothesis, American scientists ridiculed Wegener's approach. The American scientific method, the so-called "method of colorless observation" emphasized testing of multiple, independent, working hypotheses. By the time of his death in 1930, Wegener and his ideas had been pushed into obscurity.

Nonetheless, a small group of scientists working outside Europe and America kept the theory alive. Seismological evidence strongly favored a layered Earth. Even as early as 1914, a few scientists had accepted the idea that Earth's rigid crust was underlain by a weak and plastic interior. This configuration of the crust and upper mantle supported the idea that the continents floated in a molten fluid. With this in mind, Arthur Holmes (1890–1965), a prominent British geologist, proposed in 1928 that **mantle convection** might serve as a possible driving mechanism for continental drift. Materials rising from the mantle were hypothesized to spread laterally beneath the crust and drag the continents horizontally. When this material cooled and sank, fragments of the crust would be dragged with it. Holmes' idea, while widely respected, did not sway mainstream scientific opinion for nearly 30 years.

During this time, newly developed technologies for observing the seafloor provided evidence in support of continental drift. Sonar-generated maps of the Atlantic seafloor revealed a ridge of mountains extending along its length, a mid-ocean or **oceanic ridge.** Heat flow through the ocean's crust was greater along these ridges, an observation at odds with the idea of a static Earth. At the same time, ocean sediments were found to be far thinner than expected, given the Earth's age. A picture of a seafloor much different than imagined was beginning to emerge. In the early 1960s, two scientists, geophysicist-oceanographer Robert Dietz (1914–1995) and geologist Harry Hess (1906–1969), independently proposed the idea of **seafloor spreading.** Building on Holmes' model of mantle convection, Hess proposed that the crust was being driven apart by convection currents. Dietz, who originally coined the term *seafloor spreading*, proposed that oceanic ridges represented eruptions of basalt, a hypothesis later confirmed by direct observations of the seafloor using submersibles.

Although seafloor spreading could explain the observed properties of the seafloor, direct evidence was lacking. Studies of the magnetic properties of rocks indicated they retain a record of Earth's magnetic field at the time of their formation. Pierre Curie (1859–1906), a French physical chemist and Nobel Prize co-winner for his work in radioactivity, had shown in 1895 that magnetic particles in molten rocks align with Earth's magnetic field when the rocks cool to a particular temperature called the **Curie point.** By studying the orientation of magnetic particles in rocks, a field of study called **paleomagnetism,** scientists could determine the position of the magnetic pole at the time the rock formed (**Figure 3-2**). In this way, scientists made two important discoveries: the magnetic pole moves, a phenomenon called **polar wandering;** and the magnetic pole undergoes reversals every few

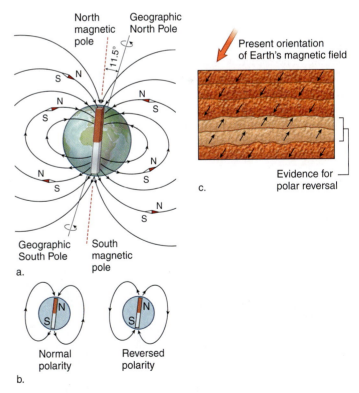

■ FIGURE 3-2 The concept of paleomagnetism. (a) Earth's magnetic field acts like a magnet that orients certain types of metallic particles, like iron, in the direction of the magnetic pole. (b) When molten rocks containing iron solidify, their iron particles become "frozen" in alignment with the magnetic field. (c) In this way, rocks "record" reversals or changes in the location of the magnetic poles.

million years, switching from the North Pole to the South Pole and back again, a process known as **polar reversals.** In 1963, two groups of scientists, one in Australia and the other in the United States, published the first paleomagnetic timescale, a history of polar reversals, in this case, over the past 4 million years. This was exactly the type of information required to conduct a direct test of seafloor spreading.

Between 1952 and 1956, Arthur Raff and Ron Mason at Scripps Institution of Oceanography towed a magnetometer across thousands of miles of seafloor in the northeast Pacific, making measurements of rock magnetism. In August 1961, they published what was to them an unusual pattern of strong and weak magnetism on the seafloor, the zebra stripes, as some called it **(Figure 3-3)**. Lawrence Morley, a young geophysicist with the Canadian Geological Survey, recalls "freaking out" when he saw the zebra stripes. Morley was convinced that seafloor spreading was the best explanation for these magnetic patterns. Two years later, when Dietz's paper was published, Morley set out to convince the scientific world that the zebra map was proof of seafloor spreading. Morley's paper was rejected twice in 1963, with one reviewer suggesting that his ideas were "most appropriate over martinis," certainly not worthy of publication. At the same time, Fred Vine, a graduate student at Cambridge University, noticed similar magnetic patterns. Vine, whose data spanned the Carlsberg Ridge in the Indian Ocean, realized that the patterns resulted from seafloor spreading during polar reversals. Along with his advisor, Drummond Matthews (1931–1997), an established senior scientist, Vine was able to publish his results in September 1963. Fortunately, historians have corrected the hasty dismissal of Morley's work, and the explanation of the magnetic patterns across oceanic ridges has come to be known as the **Vine-Matthews-Morley hypothesis.**

With strong evidence for seafloor spreading, the development and acceptance of continental drift and plate tectonics rapidly followed. The idea that Earth's surface was composed of plates was first proposed by John Tuzo Wilson. Wilson (1908–1993), a Canadian geophysicist with a remarkable talent for visualizing elegant solutions to complex problems, wrote about a new class of **strike-slip faults** on the ocean floor called **transform faults.** These features were "connected into a continuous network of mobile belts about the Earth which divide the surface into several large rigid

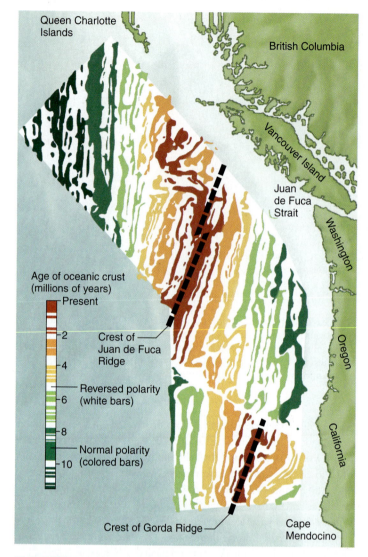

■ FIGURE 3-3 The zebra stripes. As magma erupts on the seafloor and cools, magnetic particles align with Earth's magnetic field. Through time, the spreading sea floor records intervals of normal (colored) and reversed (white) polarity. Using the paleomagnetic timescale of polar reversals, scientists can predict what the magnetic patterns would look like assuming seafloor spreading. When the predicted patterns were compared with the observed patterns, the similarity was astounding. Seafloor spreading is the only explanation which fits the observed patterns of magnetism on the seafloor (data from USGS).

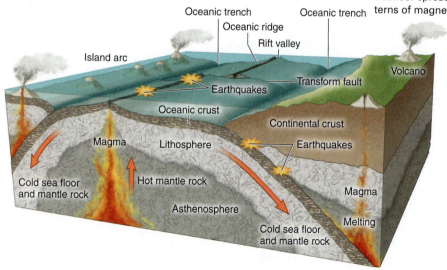

■ FIGURE 3-4 Major features of modern plate tectonics. Seafloor spreading creates new oceanic crust and subduction takes it away. Because of this process, oceanic ridges form, continents move, volcanoes erupt, earthquakes occur, and oceanic trenches line the edges of continents and islands. Plate tectonics theory accounts for many of the features and processes observed on Earth.

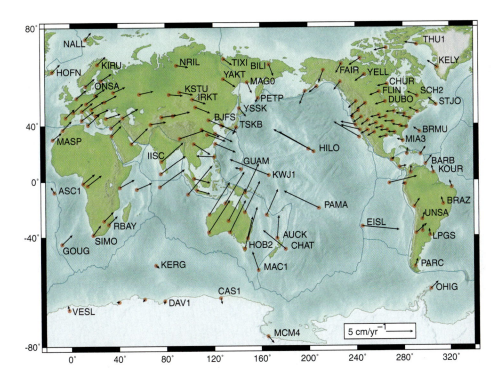

■ FIGURE 3-5 Global positioning systems (GPS) enable precise determinations of the location of objects on Earth's surface. By identifying changes in the location of objects (such as mountaintops) over time, GPS provides a direct measurement of the direction and rate of motion of plates. In this figure, the length of the arrow corresponds to the rate of movement. With a more quantitative understanding of plate movements, geologists hope to better understand the processes that drive plate motions (data from NASA).

plates." His classic paper described the role of transform faults in connecting major structural features of the Earth. It also included the first mention and illustration of Earth's tectonic plates.

Seismology also contributed important observations in support of Wilson's ideas. Lynn Sykes at Lamont Doherty Earth Observatory examined global seismic waveforms and found that the movement of faults along oceanic ridges was just what would be expected if transform faults were present. His work, along with others, provided evidence for **subduction,** the downthrusting of one plate beneath another, thereby providing the means by which oceanic crust is recycled back into the mantle **(Figure 3-4)**. Subduction produces the deepest features on our planet, the **oceanic trenches,** deep and narrow troughs that border the edge of continents and other locations where subduction occurs.

The fully developed theory of plate tectonics was advanced by Jason Morgan of Princeton University at a meeting of the American Geophysical Union in spring 1967. Following shortly after were Dan McKenzie and Robert Parker at Scripps with their publication in *Nature* in late December 1967, and three months later, Morgan in the *Journal of Geophysical Research* in March 1968. These authors introduced a quantitative description of the motions of the plates. Their work essentially showed how it was possible for rigid plates to move about on a spherical Earth. With these publications, a model of plate tectonics was born.

In the late twentieth century, the movements of plates were observed directly using satellite-based global positioning systems (GPS) that pinpoint precise locations on Earth's surface. Using GPS, measurements of the directions and rates of plate movements could be obtained **(Figure 3-5)**. In some cases, changes in elevation could be inferred, enabling geologists to witness the buildup of tectonic stresses between two plates. Direct observation of plate movements provided direct evidence in support of plate tectonics and left little doubt that the continents move.

A Closer Look at Earth's Layers

The three-layer description of Earth's layers introduced in Chapter 2 basically follows from a description of the chemical composition of the layers. However, a better understanding of plate tectonics is achieved by considering the divisions of Earth's interior based on their **mechanical strength,** their susceptibility to movement and deformation. Mapping of the mechanical properties of Earth's interior is based on studies of **seismic waves,** waves of energy produced by earthquakes and other displacements of Earth's crust. Seismic waves propagate through the Earth in different ways. Interpretation of their movements as recorded by seismometers provides information on Earth's interior.

The core consists of two layers: a solid **inner core** and a liquid **outer core (Figure 3-6)**. Scientists believe that the inner core is composed of pure iron with perhaps a trace of nickel. The outer core consists mostly of iron and small percentages of other elements. These slight differences in composition as well as differences in heat and pressure cause the core to separate into liquid and solid parts. Recent studies suggest that the inner core may even have a crystalline structure.

Wrapped around the core, we find the **lower mantle** (sometimes called the mesosphere). Composed of silicate rocks, this layer exhibits great strength and a high density. Though classified as a solid, it flows like a liquid over long timescales, much like glass window panes that thicken on the bottom edge as a result of gravity. At a depth of 2900 km (1802 miles), the lower mantle interacts with the outer core at a boundary called the **core-mantle boundary.** Scientists are keenly interested in the core-mantle boundary for clues it may

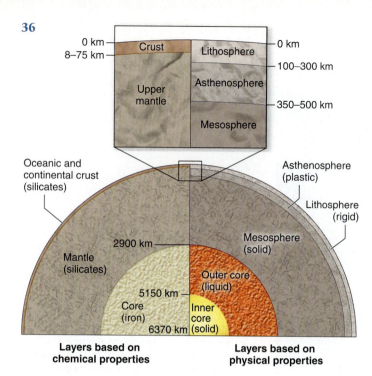

FIGURE 3-6 Structure of Earth's interior based on chemical properties (composition) and physical properties (mechanical strength). The physical properties are important to understanding plate motions. The rigid outer shell, or lithosphere, includes the crust and part of the upper mantle. The plastic asthenosphere includes the rest of the upper mantle. The solid lower mantle is called the mesosphere.

beneath the continents. For scientists studying plate tectonics, the asthenosphere represents the "lubricant" over which the plates glide.

Earth's outermost shell is called the **lithosphere.** This outer layer includes both the crust and a portion of the upper mantle that sticks to the crust. In fact, the tectonic plates, the jigsaw-puzzle-like pieces that make up Earth's outer shell, are often referred to as **lithospheric plates.** Beneath the continents, the lithosphere may extend to a depth of 125 to 250 km (78–155 miles). Beneath the oceans, the lithosphere is thinner, about 70 km (43 miles) deep. While the lithosphere includes all of Earth's crust, important differences exist between the continents and ocean basins. Ocean basins are formed from **oceanic crust,** composed largely of basaltic rock. The continents represent **continental crust** a granite-type rock that varies widely in mineral composition. Some geologists believe that the early Earth was entirely composed of oceanic crust until tectonic processes led to separation of the two types of crust. The density of oceanic crust (~ 3 g cm^{-3}), exceeds the density of continental crust (~ 2.7 g cm^{-3}). This difference has important implications for the motions of the plates and the shape of Earth's surface.

provide for a number of Earth processes, including the nature of Earth's rotation, Earth's magnetic field, and possibly the movements of crustal plates. Surrounding the lower mantle is the upper mantle or **asthenosphere.** Rocks here appear to be close to their melting point, which makes them plastic and weak. The asthenosphere resides at a depth starting somewhere between 100 and 250 km (62–155 miles) deep and extending to 670 km (416 miles) deep. The uncertainty over where the asthenosphere begins stems from disagreements among scientists over the depth of the asthenosphere under the continents. Studies using **seismic tomography,** a technique that transforms seismic data into a three-dimensional image of the Earth, favor a deeper asthenosphere or even no asthenosphere

Plate Kinematics: The Study of Plate Motions

Plate kinematics is the study of the motions of the plates. Using geometric models for spheres, scientists can construct models of the plates using bathymetric, topographic, or other data to define the plate boundaries. Such models enable geologists to test ideas concerning plate motions. They also guide oceanographers toward a better understanding of the impact of long-term temporal and spatial variability in the world ocean (see **inside front cover**).

The Major Plates

One feature common to the major plates is that they include both continental and oceanic crust, albeit in different proportions **(Figure 3-7)**. The Pacific Plate consists mostly of oceanic

FIGURE 3-7 The major tectonic plates. Geophysicists generally recognize 14 major plates: African, Antarctic, Arabian, Australian, Caribbean, Cocos, Eurasian, Indian, Juan de Fuca, Nazca, North American, Pacific, Philippine and South American; and 38 minor ones: Aegean Sea, Altiplano, Amur, Anatolia, Balmoral Reef, Banda Sea, Birds Head, Burma, Caroline, Conway Reef, Easter, Futuna, Galapagos, Juan Fernandez, Kermadec, Manus, Maoke, Mariana, Molucca Sea, New Hebrides, Niuafo'ou, North Andes, North Bismarck, Okhotsk, Okinawa, Panama, Rivera, Sandwich, Scotia, Shetland, Solomon Sea, South Bismarck, Sunda, Timor, Tonga, Woodlark, Yangtze. Plate boundaries are more complex than shown here.

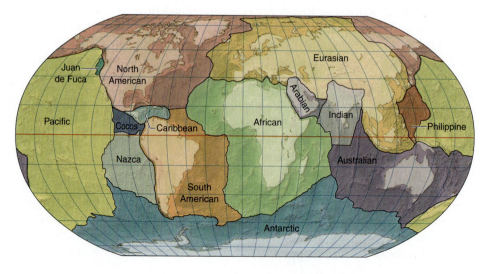

Zone of divergence: plates separate
(example: Mid-Atlantic Ridge)

a.

Zone of convergence: plates collide
(example: Aleutian Trench)

b.

Transform fault: plates slip past each other
(example: San Andreas Fault)

c.

■ **FIGURE 3-8** Plate boundaries illustrated. The three major plate boundaries give rise to different features depending on the type of crust involved in the interaction. (a) Divergent boundaries may produce oceanic ridges or continental rift zones. (b) Convergent boundaries may produce mountain ranges, islands, or submarine trenches. (c) Transform boundaries may create offset ridges and fracture zones along the seafloor.

crust with a small portion of continental crust. On the other hand, the Eurasia Plate is nearly entirely continental. Plates may be large, such as the Pacific Plate, or small, such as the Cocos Plate. Plates and their dimensions are far from static. The Indo-Australian Plate, once joined, is now separating into two plates. **Plate boundaries** mark the narrow regions where plates meet.

Types of Plate Boundaries

Three types of plate boundaries are defined according to the plate tectonics model: 1) **divergent boundaries,** where plates are moving away from each other; 2) **convergent boundaries,** where plates are moving toward each other; and 3) **transform boundaries,** where plates are moving parallel to each other **(Figure 3-8).**

Oceanic ridges occur at divergent plate boundaries where new seafloor is produced. The rate at which new seafloor is created and, hence, the rate at which spreading occurs defines the type of oceanic ridge. Spreading rates affect the morphological, tectonic, and geological characteristics of an oceanic ridge **(Table 3.1).** Slow-spreading ridges tend to have a distinct **axial valley** along their length. These features are reduced or absent on fast- to superfast-spreading ridges.

Convergent boundaries are defined by the type of crust that joins them. The Mariana Trench, an oceanic trench off the coast of the Philippines, provides the best example of an **oceanic-oceanic convergent boundary (Figure 3-9).** The collision of oceanic crust on both sides of the boundary has created the deepest location on Earth. **Oceanic-continental convergent boundaries** also produce oceanic trenches **(Figure 3-10).** The Peru-Chile Trench is a good example where denser oceanic crust subducts beneath less dense continental crust. Melting of the underthrust plate produces buoyant blobs of magma, much like a lava lamp, that rise through the crust to create volcanoes. The Andes along the west coast of South America were formed in this way. **Continental-continental boundaries** give rise to the highest features on our

TABLE 3.1   Types of Oceanic Ridges

Type	Approximate spreading rates	Characteristics	Example
Ultraslow	<15 mm yr^{-1} (~ 0.5 inch)	Staircase faults; lack of transform faults; widely spaced seamounts	Arctic Ridge
Slow	15–50 mm yr^{-1} (~ 0.5–2.0 inches)	Large number of small seamounts	Mid-Atlantic Ridge
Intermediate	50–75 mm yr^{-1} (~ 2.0–3.0 inches)	Linear seamount chains	Juan de Fuca Ridge
Fast	75–150 mm yr^{-1} (~ 3.0–6 inches)	Numerous lava flows, hydrothermal vents, near-axis seamounts; overlapping spreading centers	Upper East Pacific Rise
Superfast	150–200 mm yr^{-1} (6–8 inches)	High rate of crustal formation and hydrothermal heat loss	Lower East Pacific Rise

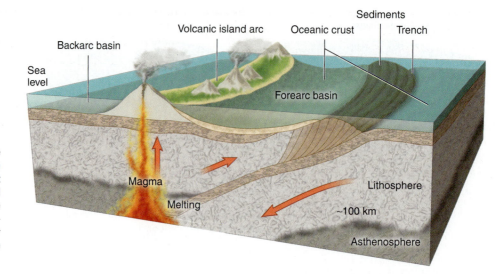

■ **FIGURE 3-9** Oceanic-oceanic convergence of the type that produced the Mariana Trench and the Philippines. Melting of the underthrust plate at depth results in the formation of island arcs. The region on the trench side of the island arc is called the forearc basin; the opposite side is called the backarc basin. Seafloor sediments scraped from the underthrust plate accumulate at the plate boundary.

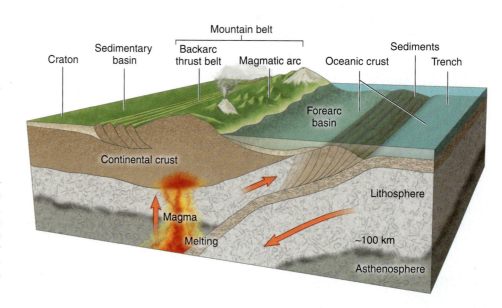

■ **FIGURE 3-10** Oceanic-continental convergence of the type that produced the Chilean Andes and Pacific Northwest Cascades. Melting of the underthrust plate creates a magmatic arc, a string of volcanoes along the plate boundary. Wedges of sediments border a forearc basin on the trench side and a sedimentary basin on the continental side of the magmatic arc.

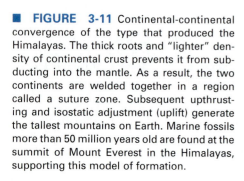

■ **FIGURE   3-11** Continental-continental convergence of the type that produced the Himalayas. The thick roots and "lighter" density of continental crust prevents it from subducting into the mantle. As a result, the two continents are welded together in a region called a suture zone. Subsequent upthrusting and isostatic adjustment (uplift) generate the tallest mountains on Earth. Marine fossils more than 50 million years old are found at the summit of Mount Everest in the Himalayas, supporting this model of formation.

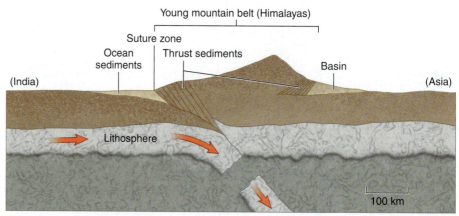

Continental-continental collision

planet **(Figure 3-11)**. The collision between the India Plate and Eurasian Plate created the Himalayas. Here the ocean floor that once existed on the Indian Plate subducted beneath Asia. As the continents of Asia and India collided, some of this ocean floor was trapped between them, much like a smaller person being trapped beneath two larger persons on a subway. As a result, we can find remnants of oceanic crust and fossils of marine organisms among the highest peaks of the Himalayas.

Transform boundaries occur where two plates are moving horizontally past each other **(Figure 3-12)**. As their name implies, transform boundaries modify or transform segments of plate boundaries. For example, the San Andreas Fault Zone, one of the best known transform boundaries, connects the East Pacific Rise and the Gorda and Juan de Fuca Ridges. Transform boundaries can also connect oceanic trenches, or ocean ridges and oceanic trenches.

Hotspots and Mantle Plumes

Before the publication of papers confirming plate tectonics theory, Tuzo Wilson grappled with the origins of volcanic islands, such as the Hawaiian Islands. Nearly a hundred years earlier, James Dwight Dana (1813–1895), a geologist on the U.S. Exploring Expedition (1838–1842), had observed that island chains such as Hawaii progressed age-wise from north to west. Wilson, building on the concept

of seafloor spreading, proposed the idea of **hotspots,** stationary locations in the asthenosphere where magma continuously breaks through the lithosphere. As plates moved over these hotspots, volcanic islands would be created, giving rise to island chains such as the Hawaiian Ridge-Emperor Seamount Chain **(Figure 3-13)**. To explain his idea, Wilson imagined a person lying on their back and blowing bubbles at the bottom of a shallow stream. As the stream (e.g., plate) moved over the person, the bubbles (e.g., the Hawaiian Islands) rose to the surface and were transported away. This explanation satisfied early critics of plate tectonics theory, and the idea of hotspots proved useful.

In 1971, Jason Morgan expanded on the concept of hotspots and hypothesized the existence of **mantle plumes,** a volume of buoyant mantle material that rises to the Earth's surface. Plumes existed in regions where molten material rose from the lower mantle in any of a variety of forms (mushroom-shaped, chimney-shaped, bulbous, etc.). Morgan originally proposed 20 plume-hotspot regions. Since that time, as many as 5200 plume-hotspots have been proposed, but new observations and a narrower definition of hotspots has reduced that number to 50 and perhaps as few as seven. Some scientists now even question the existence of plumes. For example, Yellowstone National Park, with its geothermal pools and geysers, was once considered a hotspot with an active mantle plume, but seismological studies have failed to find a plume. Consequentially, Yellowstone is no longer considered to be a hotspot. Rather, it appears to represent

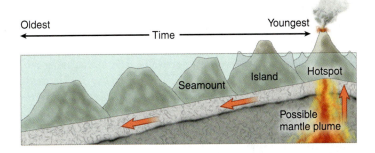

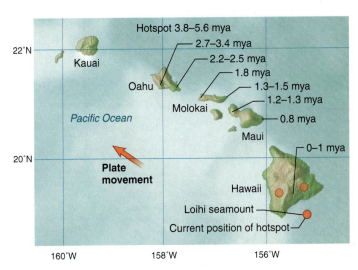

■ **FIGURE 3-12** Transform boundary along the western coast of the United States. The San Andreas fault connects the East Pacific Rise with the Gorda and Juan de Fuca ridges. Fracture zones—extended zones of faulting perpendicular to oceanic ridges—often occur in association with transform faults (data from USGS).

■ **FIGURE 3-13** A hotspot is thought to be responsible for the formation of the Hawaiian Islands and the Emperor Seamount chain from 1–6 million years ago (mya). The progressively increasing age of the Hawaiian Islands from Hawaii to Kauai supports this mechanism of formation.

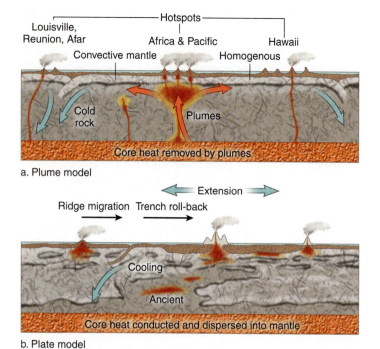

a. Plume model

Hotspots

Louisville, Reunion, Afar

Convective mantle

Africa & Pacific

Hawaii

Homogenous

Cold rock

Plumes

Core heat removed by plumes

Extension

Ridge migration Trench roll-back

Cooling

Ancient

Core heat conducted and dispersed into mantle

b. Plate model

■ **FIGURE 3-14** Plumes or no plumes? (a) Morgan's plume hypothesis has been applied to explain the existence of hot spots, large igneous provinces, and other characteristics of Earth's surface and interior. (b) An alternative school of thought argues that most if not all of the features ascribed to plumes can be explained by other mechanisms. The arguments and evidence for and against plumes are complex but the important thing to remember is that alternative hypotheses continue to be debated.

a region where local melting processes produce its features, although this conclusion remains tentative and controversial **(Figure 3-14)**.

Besides hotspots, other characteristics of the seafloor attributed to plumes are **swells and superswells**. The Darwin Rise and the South Pacific Superswell in the Pacific basin and the Bermuda Rise in the Atlantic basin are prominent examples. These features are thought to result from the "capture" of a plume head by the lithosphere. The plume does not penetrate the lithosphere but

displaces it upward, causing it to swell. Swells in ocean basins may reach diameters of 2000 km (1242 miles) with elevations above the seafloor up to 1000 m (3280 feet). Superswells are larger and may reach diameters more than 10,000 km (6213 miles).

The Slow Dance of the Continents

One of the more striking features of plate tectonics is that the continents move slowly across our planet's surface. Yet the continents do not remain intact during these movements. Geoscientists interested in the history of movements of Earth's crust—especially Earth's ancient crust—must commence on a worldwide hunt for the "pieces." The first step in this puzzle requires the identification and careful description through geologic mapping of **Archean cratons**, pre-2.5-billion-year-old pieces of Earth's crust **(Figure 3-15)**. Geologic mapping reveals that pieces of the same craton can be found in different locations around the globe, demonstrating that some force broke apart the original cratons and moved the pieces in different directions. Fragments of the North Atlantic Craton can be found in the northeastern United States along the Hudson River, in Greenland, and in Britain. Scientists interpret these observations as evidence for the formation of the Atlantic Ocean when a **supercontinent**, an assemblage of primary continents, broke apart.

Scientists generally agree that about 250 million years ago, all of Earth's modern-day continents were grouped together into one supercontinent called **Pangaea**, a name first proposed by Alfred Wegener **(Figure 3-16)**. A reconstruction of Pangaea reveals that North and South China were separated from the other continents and from each other. The continents of Australia and India resembled their modern-day counterparts, but other continents—Africa, North America, South America, and Eurasia—were quite different. **Gondwana**, a supercontinent composed of modern-day Southern Hemisphere continents, was formed around 550 million years ago, near the start of the Cambrian. Somewhat later, the Northern

■ **FIGURE 3-15** Ancient pieces of Earth's crust. By mapping the distribution of Archean cratons (older than 2.5 billion years) and Precambrian crust (before 540 million years ago), geologists can piece together a picture of the configuration of the continents in Earth's past. For example, pieces of the North Atlantic Craton can be found on both sides of the Atlantic. From this observation, geologists infer that these regions were once joined.

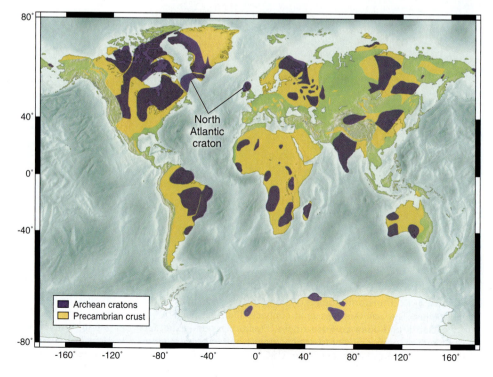

North Atlantic craton

Archean cratons
Precambrian crust

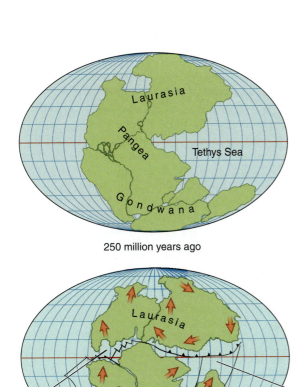

250 million years ago

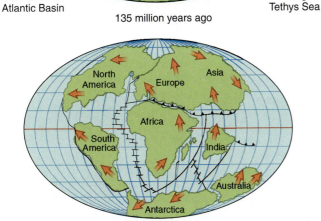

Formation of
Atlantic Basin

Closing of
Tethys Sea

135 million years ago

65 million years ago

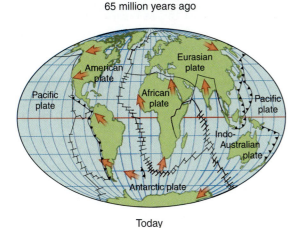

Today

■ **FIGURE 3-16** Continental drift since 250 million years ago. The supercontinent Pangaea represents the last time the modern-day continents were joined.

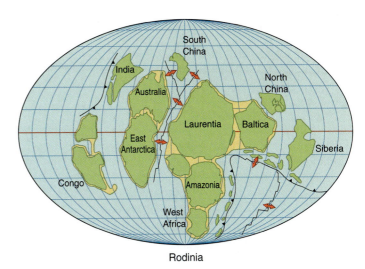

Rodinia

■ **FIGURE 3-17** The breakup of the supercontinent Rodinia. Using paleomagnetic data, geologists have been able to reconstruct the configuration of the continents approximately 750 million years ago. Some 500 million years later, the continents would rejoin as Pangaea.

Hemisphere continents became part of **Laurasia.** The "collision" of Laurasia and Gondwana is what eventually formed the supercontinent Pangaea surrounded by an ocean called **Panthallassa.**

Several of the pieces of Pangaea were previously part of another supercontinent, **Rodinia (Figure 3-17).** Evidence for Rodinia accumulated during the 1990s with the publication of several papers based on paleomagnetic and other geochemical and geologic data. While the exact configuration of Rodinia is still debated, it is generally accepted that Rodinia came together from 1.0–1.1 billion years ago and began to break apart by 850–800 million years ago. As with Pangaea, the consolidation and fragmentation of Rodinia was more of a slow dance than a single occurrence.

In 1996, John Rogers at the University of North Carolina proposed a 3-billion-year history of supercontinent cycles **(Figure 3-18).** Rogers identified and pieced together the oldest fragments of continents, the 2- and 3-billion-year-old cratons that represent Earth's oldest crust. The result was Earth's oldest "supercontinent," which he named **Ur,** a German word for origin or beginning. The diagram adapted from his paper illustrates the history of supercontinent formation from the Archean to modern times. New data may change the details, but this conceptual model for the "slow dance of the continents" provides an important framework for understanding the movements of the continents and interpreting the potential consequences of these movements on the ocean and its life.

The Opening and Closing of Ocean Basins

The evolution of ocean basins from their opening in rift zones to their closing in convergent boundaries occurs in a series of stages known as the **Wilson cycle (Figure 3-19).** In its simplest form, the Wilson cycle involves the rifting or breaking apart of a continent. As the crust is pushed up or pulled apart, a rift valley may form. At some point, seafloor spreading initiates and the rift valley becomes a mature spreading center. The Atlantic Basin, the youngest ocean basin, represents a

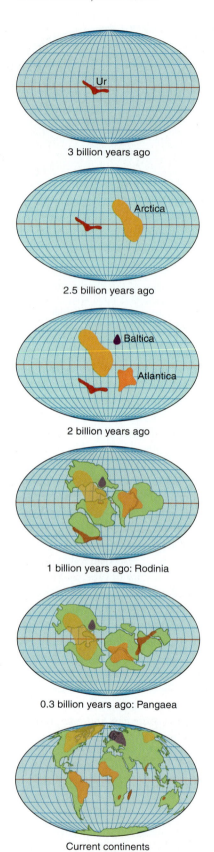

3 billion years ago

2.5 billion years ago

2 billion years ago

1 billion years ago: Rodinia

0.3 billion years ago: Pangaea

Current continents

■ **FIGURE 3-18** The supercontinent cycle from Ur to modern times. Ur is hypothesized to be the oldest continent. Other continents followed and through continental drift became supercontinents. Continental drift assembled and dissembled supercontinents several times in Earth's past.

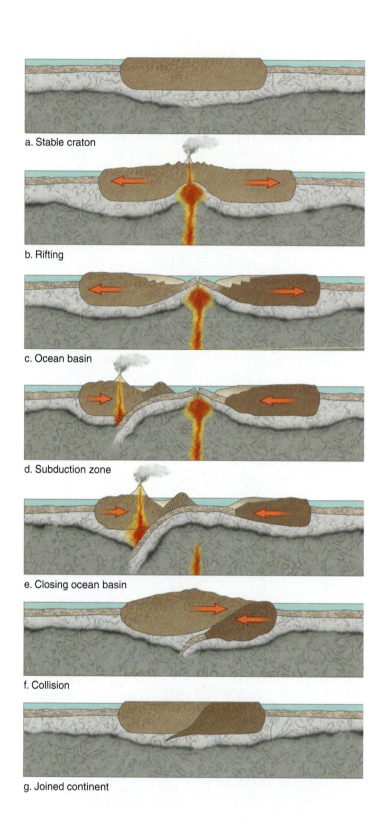

a. Stable craton

b. Rifting

c. Ocean basin

d. Subduction zone

e. Closing ocean basin

f. Collision

g. Joined continent

■ **FIGURE 3-19** The Wilson cycle, the birth and death of an ocean basin. Ocean basins form through rifting of a continent and subsequent seafloor spreading (a–c). Subduction of seafloor eventually closes the ocean basin (d–e). Further collision creates a suture zone and the continents join (f–g). Compare figures 3.9–3.11 to the panels above to see how plate processes contribute to the opening and closing of ocean basins. Try to identify the stage of the Wilson cycle in figures 3.16–3.18.

mature stage basin. Eventually, an ocean basin closes. The Pacific Basin is a good example of a declining ocean basin and is destined to close completely in another 200 million years. Eventually, the entire basin closes, leaving behind faint geological traces of its previous existence. The Himalaya mountain chain represents the "scar" where an early Tethys Sea once existed. In actuality, the formation and closing of ocean basins is rarely so simple. Any of a number of processes may alter or stop the progression of the Wilson cycle and its stages. These stages represent a conceptual model of the life cycle of an ocean basin.

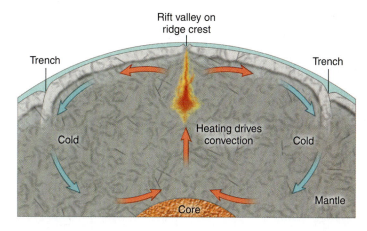

■ **FIGURE 3-20** The convection model of plate dynamics proposes that convection currents within the asthenosphere drive plate motions.

Plate Dynamics: The Study of Plate-Driving Forces

Plate dynamics refers to the forces that drive the motions of the plates. Ever since plates were first proposed, scientists have sought evidence for the mechanisms that cause these motions. Few topics within plate tectonics theory have generated such heated controversy as what causes the plates to move. Lack of a plausible mechanism for moving continents was one of the factors that caused the initial rejection of Wegener's hypothesis. Though plate tectonics theory is now widely accepted, the processes that underlie plate motions continue to be debated. Here we present three models, each of which has varying degrees of support among geologists.

Convection Model

Arthur Holmes' concept of mantle convection, modified by Harry Hess to incorporate seafloor spreading, envisions the plates as slow-moving "icebergs" floating in a "sea" of magma. This model, the **convection model of plate dynamics**, remains a widely depicted model for the driving force of plate movements **(Figure 3-20)**. The convection model works much like pasta sauce heating in a pot on the stovetop. As the sauce heats at the bottom of the pot, it becomes less dense and rises to the surface. At the surface, the sauce cools by releasing heat to the air and sinks again. A convective or circular motion of the fluid is achieved as it alternately rises and sinks because of heating and cooling, respectively. In the interior of the Earth, the core acts as the heat source for mantle materials, which rise to the surface in the region of oceanic ridges. The molten material continues to circulate along the underside of the plates, dragging the plates until the material cools and sinks. In support of this model, Hess observed high rates of heat flow at oceanic ridges

compared to other regions of the seafloor. He also noted the presence of pillow basalts—underwater eruptions of magma—on the seafloor in the vicinity of mid-ocean ridges. On the basis of these observations, mantle convection appears to be supported.

Ridge-Push, Slab-Pull Model

On the other hand, seismic and other types of geophysical and geological observations do not support aspects of whole-mantle convection. An alternative model is the **ridge-push, slab-pull model (Figure 3-21)**. In this model, oceanic ridges, which are elevated above the seafloor, generate a kind of downhill pressure (gravity sliding), causing the plates to move away from ridges. At the same time, cooling of the plates causes them to thicken. Because a plate cools as it moves away from the oceanic ridge, the plate becomes thicker with greater distance from the ridge. Deposition of sediments from the overlying water column onto the seafloor (see Chapter 4) also increases the mass of the plate; the older the seafloor, the greater the accumulation of sediments. As a result, oceanic crust has its greatest mass at submarine trenches. The thick edge of the plate is subsequently "pulled" by gravity into the Earth's mantle. In this model, slab pull is thought to be the dominant mechanism. In a variation of this model, if the sinking slab is cold and dense, it may plunge quite rapidly and at a steep angle, and it may even draw the overriding plate toward it, creating what is called **trench suction**. An important difference between the convection model and the ridge-slab, trench-suction model is the role that the plates play in generating the forces that move them. In the convection model, the plates passively ride atop the mantle, moved along by currents of molten rock. In the ridge-slab, trench-suction model, plate movements are caused by the plates reacting to the force of gravity.

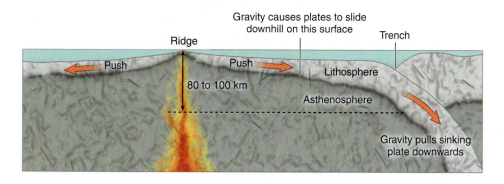

■ **FIGURE 3-21** The ridge-push, slab-pull model of plate dynamics. A variation of this model includes a role for trench suction in driving plate motions.

2004 Indian Ocean Tsunami, Part 1: The Earthquake

The tsunami that struck the Indian Ocean on December 26, 2004, ranks as one of the world's worst natural disasters. More than 283,000 people from dozens of nations were killed within hours. Millions of people lost homes, jobs, and loved ones. For the survivors, the tsunami brought irreparable grief. The tsunami took its toll on nature as well. Where forests once stood, the land was eerily bare. Where fertile soils once provided food, a salty hash of sand and mud covered the ground. Familiar shorelines vanished (**Figure 3b**). The Indian Ocean tsunami, like most tsunami, was caused by a powerful undersea earthquake that caused a vertical displacement of the seafloor. Here we describe the tsunami-causing or **tsunamigenic** event as it provides insights into the nature of tectonic forces that may be unleashed on the seafloor. In Chapter 10, we explore in greater detail the wave-like nature of tsunami and the path and impact of the Indian Ocean tsunami.

The megathrust earthquake that caused the tsunami resulted from subduction of the India Plate beneath the Burma Plate along the Sunda Trench, near the Andaman and Nicobar Islands and Northern Sumatra. The initial rupture set off a series of subsequent earthquakes that shook the Earth for the next 12 hours. Reverberations from the quake—like the slowly subsiding ring of a bell—were detected by seismologists for weeks. The intensity and duration of the quake has made estimation of its magnitude difficult, but scientists generally agree that it measured 9.0 to 9.3 on the moment-magnitude (Mw) scale.

The sequence of events that unfolded during the first moments of the quake set the stage for the devastating tsunami that followed. Analyses of seismic data reveal a complex event, punctuated by three major rupture stages and several minor bursts over a period of minutes. During the first 45 seconds, the rupture propagated about 200 km (125 miles) toward the northwest from the hypocenter at a speed of 4400 m/sec^{-1} (9843 miles/hour^{-1}). Another rupture, this one also toward the northwest, started about 35 seconds later and traveled some 400 km (250 miles). A third rupture broke about 90 seconds later (135 seconds after the initial break) and traveled toward the southwest for about 1000 km (620 miles). These initial stages lasted for at least 4 minutes. The most intense ground motion spanned at least 8 minutes. However, some 25–30% of the total energy of the earthquake was released over periods greater than an hour.

Vertical displacements of the seafloor were estimated between 1 and 5 m (3–16 feet). The total length of the fracture—determined from an analysis of seismograms—extended some 1200–1300 km (750–800 miles), almost the length of California. Observations by remotely operated vehicles and ships equipped with sophisticated seafloor mapping tools revealed cracks, faults, and slumping of the seafloor in the region of the earthquake. It was estimated that nearly 149 trillion tons of rock were set in motion by the earthquake, releasing an amount of energy equivalent to 23,000 atomic bombs. The size, duration, and intensity of this earthquake make it one of the most violent ever recorded by modern science.

FIGURE 3b The tsunami tossed ships ashore and permanently altered the landscape of the regions that felt its impact.

Plate-Driven Plate Tectonics Model

Taking these concepts a step further, some scientists have proposed a dominant role for the plates in organizing the flow of the mantle, a view that has come to be known as **plate-driven plate tectonics** (**Figure 3-22**). The key to plate-driven plate tectonics is to treat the plates and mantle as a coupled system. In this system, the plates and the mantle interact in an active and complementary fashion.

Though not all scientists agree with this model, it offers insights into scientists' ongoing debate of the causes of plate motions. Science works by proposing new ideas and attempting to refute them with observational and experimental evidence. As grand a theory as it is, plate tectonics theory still requires considerable research before the details of how it operates on our planet can be fully understood.

Plate Tectonics and Snowball Earth

One of the more provocative and controversial scientific ideas to reemerge in recent times is the **Snowball Earth hypothesis**, the proposal that Earth's surface was once covered in ice. The idea of a frozen Earth was first proposed by Brian Harland of Cambridge University in 1964. He suggested that globally distributed glacial deposits dating from 580 to 800 million years ago were formed from massive melting following a global deep freeze. Harland's proposal was widely dismissed until California Institute of Technology scientist Joseph Kirschvink

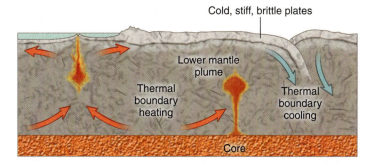

Cold, stiff, brittle plates

Lower mantle plume

Thermal boundary heating

Thermal boundary cooling

Core

■ **FIGURE 3-22** Plate-driven plate tectonics represents a departure from mantle-driven plate motions. In this model, the plates organize the flow of the mantle by creating a warm thermal boundary at the top of the mantle. Heating in the lower mantle causes material to rise while cooling in the upper mantle causes material to sink. In this model, the plates organize the flow of the mantle, in contrast to earlier models in which the mantle drives the motions of the plates. Scientists continue to debate the merits of these models.

presented new data and models in 1992. Kirschvink's data suggested that Earth may have frozen over several times during the Archean-Proterozoic transition about 2,400 million years ago and during the Neoproterozoic between 600 and 800 million years ago. Though widely publicized, Kirschvink's Snowball Earth hypothesis was met with wide scientific skepticism. However, in 2005, an unexpected boost came from studies of deposits of the element iridium, brought to Earth by meteorites. A spike in the amount of iridium found in cores from the Neoproterozoic could be explained if meteorites accumulated on the surface of a frozen Earth and then were suddenly released and deposited when the Earth melted. If the calculations are correct, Earth may have been frozen for as long as 12 million years.

Explanations of a snowball Earth focus on the accelerated growth of continents in the early Proterozoic, about 2,400 million years ago. The emergence of continents caused an increase in rates of sedimentation and increased burial of organic carbon, effectively removing it from the atmosphere. The reduction of atmospheric carbon dioxide, a greenhouse gas, results in cooling because warming via the greenhouse effect is reduced. Continued cooling would cause ice to form even at lower latitudes. When ice forms, sunlight is reflected back into space via an increase in Earth's reflectivity, or **albedo**. The greater reflection of sunlight could accelerate cooling. If the ice sheets reach below 30 degrees latitude, a critical point, then the entire Earth could freeze over.

Kirschvink's data aroused renewed scientific interest, but scientists were still puzzled by how a snowball Earth might end. One possibility is that plate tectonics, through the action of volcanoes and other processes that release greenhouse gases into the atmosphere, might help to melt frozen Earth. The primary means by which carbon is removed from the atmosphere—photosynthesis—would be largely inoperative with a globally glaciated earth. Because volcanoes and other processes would continue to emit carbon dioxide and other greenhouse gases, the atmosphere would gradually warm as greenhouse gases increased.

Many scientists accept that Earth has undergone periods of extreme cooling, but few believe that the Earth ever completely froze over, especially since life has continued to persist uninterrupted for more than 2 billion years. New data, published in 2006, provide unmistakable evidence that cyanobacteria were present 2.4 billion years ago and producing oxygen. Apparently, cyanobacteria and similar organisms survived several episodes of Snowball Earth. For these reasons, some scientists suggest that Earth was more of a "slushball" with regions of open ocean that provided an "oasis" for life. The Snowball Earth hypothesis may continue to be debated for some time. Nonetheless, these provocative ideas illustrate the mechanisms by which Earth systems interact and the ways in which scientists approach the study of climate change.

YOU Might Wonder

1. **What's the deepest anyone has drilled into the Earth?**
   The drilling record for continental crust belongs to the Russians, who reached a depth of 12,242 m (7.6 miles) in 1994. In May 2005, scientists working aboard the *JOIDES Resolution*, an International Ocean Drilling Program ship, came within 1000 m (0.62 miles) of the mantle by drilling 1,415 m (0.88 miles) into the thin oceanic crust in the North Atlantic. As yet, no one has penetrated Earth's crust into the mantle.

2. **When will scientists be able to predict earthquakes?**
   Susan Hough, a USGS geophysicist and author of a delightful book, *Earthshaking Science: What We Know (and Don't Know) About Earthquakes*, calls earthquake prediction the "holy grail of seismology." She quickly points out that our ability to predict earthquakes depends on what we mean by prediction. Ideally, a scientist could provide the time, date, and location of an earthquake, at least with as much accuracy as the arrival of a passenger jet. Realistically, earthquake prediction may be limited to much longer periods of time, over the span of years at best. The near-hopelessness of predicting seismic hazards

begs the question: should we even try? Does it really serve the public interest to sustain the false hope that one day prediction may be possible? While some would argue that we must continue our efforts at all costs, others would argue that the cost of supporting certain-to-fail seismic prediction experiments robs funding from seismic preparation efforts. This line of thinking reasons that we may not be able to predict them, but we sure as heck can be ready when they come!

3. **Is it true that California may one day fall into the sea?**
   The rumors of California's ultimate demise are highly exaggerated. The San Andreas fault forms the boundary between the Pacific and North American plates. Only a small portion of California—southern California—resides on the Pacific Plate. Southern California moves horizontally in a northwesterly direction relative to the rest of California and the United States, which reside on the North America Plate. At a speed of about 5 cm (~ 2 inches) per year, Los Angeles may eventually subduct into the Aleutian Trench off Alaska but not for more than a 100 million years.

Key Concepts

- The outer shell of the Earth is divided into mechanically strong and weak layers that interact to form a dozen or so tectonic plates.
- The theory of plate tectonics provides a unifying explanation for most of Earth's features, such as mountain ranges, submarine trenches, and mid-ocean ridges, and many processes, including earthquakes, volcanoes, and hydrothermal vents.
- Oceanographic evidence for plate tectonics includes paleomagnetic features of the seafloor, age of the seafloor, spreading rates

of the seafloor, thickness of marine sediments, seafloor age-depth relationships, transform faults, and hotspot volcanism.
- Theories to account for plate tectonics include convection currents in the mantle, ridge-push, slab-pull, and plate-driven processes.
- Plate-driven plate tectonics emphasizes the role of the plates in generating the forces that cause them to move.
- Plate motions are a natural result of heat flow through the Earth as it cools.

Terms to Remember

Critical Thinking

1. In some ways, the theory of plate tectonics owes its credibility to the evidence provided by oceanographers. Defend this statement. In what ways has oceanography contributed to the acceptance of plate tectonics as a theory?
2. Thomas Kuhn, a well-known philosopher, argues that scientists often resist change, stubbornly holding to their ideas until overwhelming evidence to the contrary overturns them. Why do you think it is important for scientists to be skeptical and cautious when interpreting new evidence?
3. Compare and contrast convection-driven plate tectonics and plate-driven plate tectonics. Why is it important to understand the mechanisms driving plate motions?
4. Why are oceanographers actively involved in studies of plate tectonics? After all, plate tectonics is a geological discipline. In what ways does knowledge of plate tectonics contribute to an understanding of oceanography?

Explore Online

 Visit www.mhhe.com/chamberlin1e for access to chapter quizzing, key term flash cards, video clips, interactive activities, and more. Further enhance your knowledge with web links to chapter-related material!

Exploration Activity 3-1

Exploring Plate Boundaries Through Scientific Investigation

Question for Inquiry

How do scientists use seismic information to better understand the nature of plate boundaries?

Summary

Many of the ideas presented in this chapter (and previous chapters) were developed through measurements of ground motions associated with earthquakes, namely measurements of seismic waves recorded by seismometers. Analysis of the resultant seismograms and mapping of the locations of earthquakes reveal the location of plate boundaries and their behavior. With more sophisticated analyses, seismograms let scientists visualize Earth's interior and, ideally, identify potentially dangerous seismic hazards. In this activity, you will access near-real-time seismic data and explore a few tools for analyzing seismic waveforms.

Learning Objectives

By the end of this activity, you should be able to

- Locate and navigate Internet sources of near-real-time seismic information
- Identify P- and S- waves in a seismogram
- Calculate the location of an earthquake epicenter by triangulation
- Analyze data on the location and depth of earthquake sources to make inferences about the type of plate boundary
- Articulate the importance of seismology in understanding our planet

Materials and Preparation

For this activity, you will need access to the Internet and a printer. Alternatively, your instructor may provide a printout of figures sufficient for completing the activity. You may want to review Spotlight 2-2 and the section on plate boundaries in Chapter 3 prior to starting the activity. Some of you undoubtedly are expert web surfers with little need for instructions. Others of you may be relatively new to browsing the Internet. We strongly encourage the "experts" to help the "newbies." The first part of the activity asks you to explore a few web pages to become familiar with the ways in which earthquakes are presented and the terms and concepts used to describe them.

Part I: Exploring Earthquakes

Visit http://earthquake.usgs.gov/ and complete the following:

1. What is the USGS Earthquake Hazards Program? (Hint: Click on About Us.)

2. How many earthquakes above magnitude 4.0 (M4.0+) have occurred worldwide in the past seven days? (Hint: Click on the world map. Identify the date and time you are making this estimate.)

3. How many earthquakes above magnitude 1.0 (M1.0+) have occurred in the United States in the past seven days? (Hint: Click on the US map. Identify the date and time you are making this estimate.)

4. Choose one earthquake from the world map and one from the US map and report their characteristics (Hint: Keep clicking on a colored square until you come to the Earthquake Details page.)
   a. magnitude
   b. date-Time
   c. location in degrees latitude and longitude
   d. region
   e. distances
   f. location Uncertainty

5. On the Earthquake Details page, click on the MAPS tab and select Google map. Characterize the terrain you see on the Google map (oceanic, mountainous, hilly, desert, river valley, flatland, swamp, urban, etc.).

6. Click on the Learning and Education link on the menu bar at the top of the page. Find the page titled "Common Myths About Earthquakes" and summarize one common misconception about earthquakes. (Hint: Explore!)

7. From the left menu, click on Earthquake Topics and find the page titled "Magnitudes." Scroll down and find the page titled "Measuring the Size of an Earthquake."

8. Briefly describe the Richter scale (ML).

9. Briefly summarize these four other scales for measuring earthquake properties: Mb, MS, Mw, and Me. (Note that most earthquake intensities are now reported on the moment magnitude scale, Mw, not the Richter scale, which is why this part is important!).

10. Define *amplitude, period, distance,* and *focal depth.* (Use any resource you need to find these definitions.)

Part II: Learning How to Analyze Seismograms

In this part, we will explore seismograms and learn how to interpret them.

1. From the left menu, select Earthquake Topics and find the Learning and Education page on Seismic Waves. Click on the Seismic Waves link from UPSeis, University

of Michigan. Summarize the two types of seismic waves (body waves and surface waves).

2. Find the "Magnitudes" page again (see Earthquake Topics.) Click on The Richter Scale from Scripps Institution of Oceanography. Read the instructions. Then use the electronic nomograph to analyze and "calculate" the following characteristics for each of the four seismograms provided:
   a. time interval between arrival of P and S waves (s)
   b. distance from seismometer (km)
   c. Richter magnitude
   d. nomograph % error
   e. maximum amplitude of seismic wave

3. Which earthquake had the greatest magnitude? Which earthquake was furthest from the seismometer that recorded it?

Part III: Exploring Near-Real-Time Data

Go to http://www.iris.edu and click on the link Seismic Monitor. A world map with earthquake locations and magnitudes should appear.

1. Read the About... link. How is magnitude represented on the map? How is time represented on the map? What do purple dots represent on the map? How often is the seismic monitor updated?

2. From the world map view, click on the Indian Ocean. Find and describe one seismic station. (Hint: Look for blue triangles). Provide its name, location, and a description of the location of its seismometer. If the seismometer location is not described, choose a different station. Return to the world map and do the same thing for stations in the Pacific and Atlantic basins.

3. Click on one of the circles on the world map. Locate an individual earthquake event and click on it. Find an event that has a date or rows of dates that are underlined (hyperlinked). Report the following:
   a. date and time of the event   d. depth
   b. latitude and longitude   e. region
   c. magnitude

4. Click on one of the hyperlinked date links for the earthquake you chose. A list of seismic stations will appear arranged according to their distance from the earthquake. Click on one of them. A series of seismograms will appear. Likely, they will look nothing like the ones you encountered above. You are now witnessing the world of seismology as experienced by seismologists. Study the various squiggles for a few moments and write down a few observations on one of the seismograms (noting the channel and other relevant factors). These are the data on which the various characteristics of an earthquake (magnitude, depth, location) are derived by seismologists.

5. Return to the world map. Print the world map page (or save it if you know how). Trace the "lines" of purple dots using a highlighter (or trace them in an image editing program). Use a pen and make crosshatches (/////) across the "thick" purple lines. What do your traces represent? Make a list of hypotheses to explain "thin" and "thick" lines. Make a list of hypotheses to explain the "individual" purple dots that do not appear close to any "line."

6. Use your newly acquired skills at navigating the Seismic Monitor to complete the following three tables. Choose five different earthquakes in continental locations and in two types of oceanic locations (thin purple, thick purple). Calculate averages.

Table 1

Continental Earthquakes	1	2	3	4	5	Average
Date and Time						N/A
Latitude and longitude						N/A
Magnitude						
Depth						
Region						N/A

Table 2

Oceanic earthquakes on "thin" purple lines	1	2	3	4	5	Average
Date and Time						N/A
Latitude and longitude						N/A
Magnitude						
Depth						
Region						N/A

Table 3

Oceanic earthquakes on "thick" purple lines	1	2	3	4	5	Average
Date and Time						N/A
Latitude and longitude						N/A
Magnitude						
Depth						
Region						N/A

Part IV: Analyzing Seismic Data

Use the three tables above to complete the following:

1. Compare earthquake magnitudes in the three locations. Where did the strongest earthquakes occur? Why? (Use your book and internet sources to research this question.)
2. Compare earthquake depths in the three locations. Which earthquakes were the deepest? Why? (Use your book and internet sources to research this question.)
3. Review your list of hypotheses concerning thin and thick lines. Do the data in your table support or refute any of them? What did your research on the scientific literature reveal? Revise your hypotheses to take into account what you learned from your data.

Part V: Interpreting Seismic Data

Write a short scientific paper that includes the following:

1. Brief introduction on the scientific and practical applications of seismic data for characterizing plate boundaries (based on scientific literature)
2. Summary of your interpretation of the purple dots on the world map, including your hypotheses
3. Summary of your earthquake data, including your tables
4. Discussion of how your data support or refute your hypotheses.
5. Discussion of how your conclusions fit within the context of published ideas about plate boundaries and how they behave.
6. Discussion of the limitations of making interpretations on a data set based on five earthquakes
7. brief summary of additional data analyses you could perform to refine your ideas and hypotheses
8. Use the Rubric for Analyzing and Presenting Scientific Data and Information to self-evaluate and improve your work. (see Exploration Activity 1-1.)

4 Seafloor Features

The geologic wonderland of the seafloor. The manipulator arm of a submersible moves into position to sample the sulfur-rich chimney of a hydrothermal vent.

Questions to Consider

1. How have scientists overcome the technological challenges of mapping the seafloor, and what limitations and challenges remain?

2. What is the relationship between plate tectonics and features we observe on the seafloor?

3. What role do sedimentary processes play in shaping the seafloor?

4. What are the similarities and differences between the five ocean basins?

5. Why is the discovery of hydrothermal vents important to our understanding of life on Earth and perhaps elsewhere in the cosmos?

6. What are some examples of the ways in which studies of the seafloor have contributed to the economic and recreational resources of nations?

The Land Beneath the Sea

Drained of the ocean, our planet would appear quite different than the blue marble it otherwise resembles. Beneath the water lies a geologic wonderland. Jagged mountains rise from the seafloor, vast muddy plains stretch across thousands of kilometers, and deep trenches plummet into the mantle. All are encircled by the sloping sides of the continents, some steep, some gentle, and most cut by impressive canyons. From historical sketches based on soundings to eye-popping representations obtained by satellites, our view of the seafloor has changed dramatically over the centuries, largely through the development of new technologies for exploring the seafloor **(Figure 4-1)**. Each new endeavor brings seafloor features into better focus and reveals new insights into tectonic and geologic processes. Seafloor features play a significant role in ocean circulation, upwelling and mixing of water masses, the path and propagation of tsunami, and the availability of habitat for marine organisms. At the same time, detailed mapping of the seafloor remains essential to exploration for energy resources, including petroleum and methane hydrates, and mineral resources, including diamonds and manganese nodules. Seafloor mapping has many practical applications from the navigation of ships and submarines to the deployment of human-made structures and submarine cables. Although our exploration of the seafloor in this chapter is largely descriptive, you should keep in mind that the shape and composition of the seafloor are expressions of plate tectonics. Your familiarity with seafloor features will guide your understanding of the role of the seafloor in physical, chemical, and biological processes in the world ocean.

From Lead Lines to Satellites

Seafaring literature abounds with descriptions of ingenious methods for measuring the distance to the sea bottom using a sounding or **lead line,** a rope with a heavy weight attached. By slowly lowering the rope until it slackened (not always an easy thing to tell),

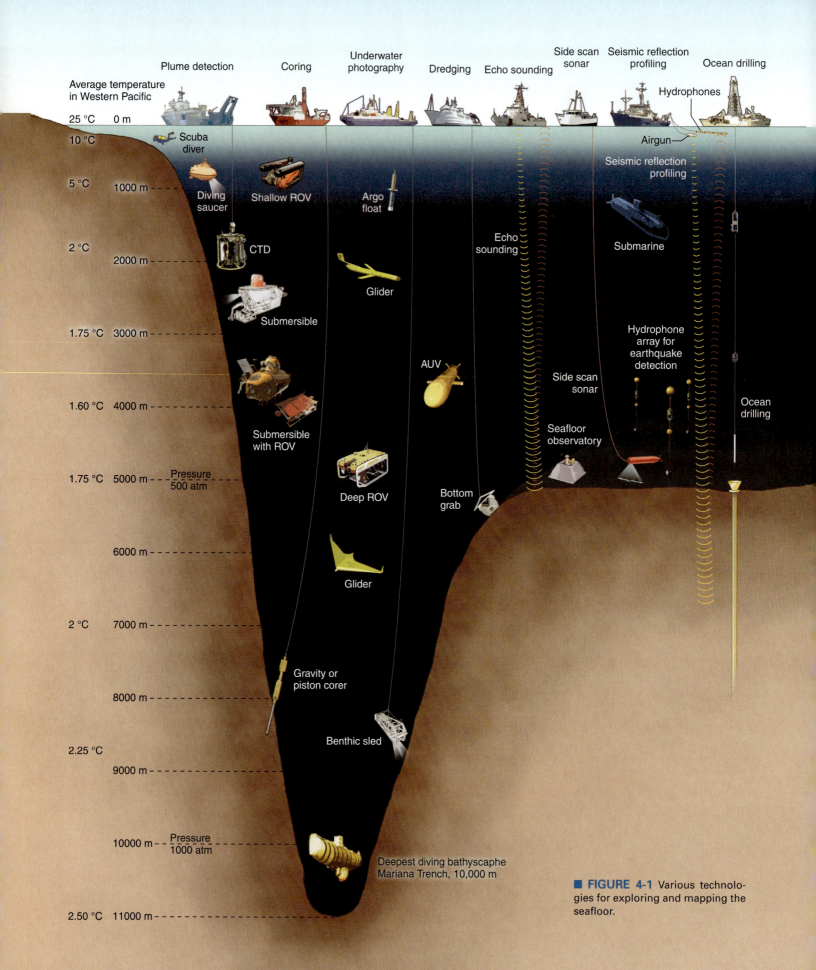

Plume detection   **Coring**   **Underwater photography**   **Dredging**   **Echo sounding**   **Side scan sonar**   **Seismic reflection profiling**   **Ocean drilling**

Average temperature
in Western Pacific

Hydrophones

25 °C   0 m

10 °C   Scuba diver

Airgun

5 °C   1000 m   Diving saucer   Shallow ROV   Argo float   Seismic reflection profiling

2 °C   2000 m   CTD   Glider   Echo sounding   Submarine

1.75 °C   3000 m   Submersible

AUV   Hydrophone array for earthquake detection

1.60 °C   4000 m   Submersible with ROV   Side scan sonar   Ocean drilling

1.75 °C   5000 m   Pressure 500 atm   Deep ROV   Bottom grab   Seafloor observatory

6000 m   Glider

2 °C   7000 m

8000 m   Gravity or piston corer   Benthic sled

2.25 °C   9000 m

10000 m   Pressure 1000 atm   Deepest diving bathyscaphe Mariana Trench, 10,000 m

2.50 °C   11000 m

■ **FIGURE 4-1** Various technologies for exploring and mapping the seafloor.

mariners could determine with some accuracy the depth of the sea bottom. One clever trick was to add a small amount of stickum (usually fat or tallow) to the bottom of the weight. Evidence of mud or sand adhering to the stickum indicated that the weight had touched bottom.

Among the most ambitious undertakings to determine depth by sounding were those of the **US Exploring Expedition** (1838–1842), which surveyed the Atlantic and Pacific basins, and the *Challenger* **Expedition** (1872–1876), which took measurements throughout the world ocean. These early expeditions rank among the first purely scientific voyages dedicated to a study of the ocean and have been recognized as a turning point in modern oceanographic research. The *Challenger* Expedition reportedly carried about 232 km (144 miles) of hemp rope and a 45-kg (100-pound) weight for sounding but managed to obtain only 300 deep-water measurements in its three and a half years at sea. One of those measurements was performed at a location that bears the ship's name to this day, the **Challenger Deep,** a segment of the Mariana Trench measuring 10,911 m (35,798 feet or 6.78 miles) deep, the deepest spot in the world ocean.[1]

Despite limitations, a number of important observations were obtained using lead lines. Matthew Fontaine Maury (1806–1873), the "father" of modern physical oceanography, published a **bathymetric chart** of the Atlantic Ocean in 1854 based on lead line measurements. Maury's chart depicts the first hints of an undersea mountain range, the Mid-Atlantic Ridge. Data from the *Challenger* Expedition also revealed the presence of oceanic ridges in the Atlantic and Indian oceans. However, it would be nearly 100 years before the significance of these early observations would be appreciated as the theory of plate tectonics took form.

In 1922, the US Navy developed the first electronic **echo sounder,** a device that uses sound to determine water depth. Echo sounders rely on the same principle as radar except that sound waves are used **(Figure 4-2)**. One of the earliest applications of the echo sounder came aboard the US Coast and Geodetic Survey ship *Guide* during its 1923 voyage from the eastern coast of the United States through the Panama Canal to the western coast of Mexico. The *Guide* tested the echo sounder against a sounding line over a series of depths between 183 and 8444 m (600–27,702 feet). The favorable comparison subsequently led the Navy to outfit all of its survey ships with echo sounders. A decade later, the *Guide* mapped an underwater mountain, the Davidson seamount, off the coast of Morro Bay, California. This seamount, officially named in 1938, represents the first underwater mountain designated as a seamount on the seafloor. In 1957, oceanographer Bruce Heezen (1924–1977) and cartographer Marie Tharp at the Lamont-Doherty Geological Observatory used echo sounding data to produce a map of

[1]A search of the Internet and print publications will reveal a number of slightly different depths for the Challenger Deep. Some of these differences result from differences in measurement techniques (pressure sensors versus sonar-type instruments), differences in the zero point (the geoid versus sea level), and simply local differences in the shape of the seafloor within the Challenger Deep.

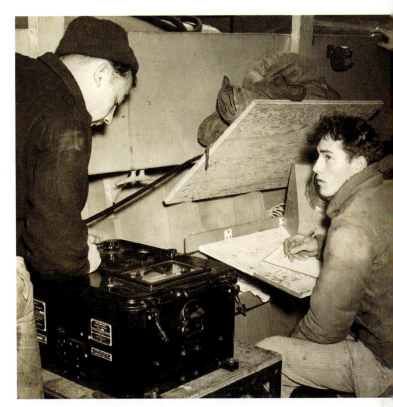

■ **FIGURE 4-2** Determining depth with an echo sounder. A sound transmitter produces a sound that bounces off the seafloor and returns to a listening device. The time interval between the produced sound and the echo detection is determined and transformed into depth. Readings are displayed on the instrument at left. Prior to electronic storage devices, data were recorded in notebooks.

the North Atlantic Basin. Their map was the first to reveal in unmistakable detail the central rift valley that runs along the top of the Mid-Atlantic Ridge. Eventually, they compiled enough data to publish in 1977 a map of the seafloor across the entire world ocean. The Heezen-Tharp world ocean floor map remains in wide use today.

By increasing the intensity of the echo sounding source or altering its frequency, oceanographers can produce "booms" that reflect off features below the seafloor, a technique called **seismic reflection profiling.** This technique reveals the layers of rock and sediments, information that is especially useful in oil exploration. Developed in the 1950s, oceanographers first used half-pound blocks of dynamite to produce a suitable sound source. Since then, less dangerous and more accurate methods have been developed **(Figure 4-3)**. A significant advance has been the deployment of **multichannel seismic systems,** long streamers of air guns and **hydrophones,** essentially underwater microphones, towed behind a ship to obtain more accurate, higher quality records of subbottom features. In 2003, oceanographers discovered that multichannel seismic systems could also be used for measuring the density layers of seawater within the ocean (see Chapter 7).

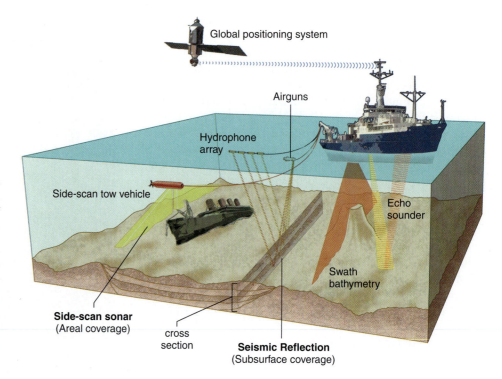

■ FIGURE 4-3 Acoustic technologies for mapping the seafloor. All the technologies shown here depend on the production of sound and detection of the reflected sound signal. They basically differ in the intensity, frequency, and swath of emitted sound signal and the way in which the return echo is detected and processed. Seismic reflection uses an air gun or a chirper to detect the various reflective layers of the seafloor. Swath beam mapping and side-scan sonar employ multiple frequency sound sources to provide high-resolution details of the seafloor and sunken objects.

Advances in seafloor mapping also came with the invention of **multibeam echo sounders.** Instead of multiple receivers and narrow-frequency sound sources, multibeam echo sounders employ multiple sources of sound at different frequencies to resolve the fine structure of the seafloor. Multibeam echo sounders and techniques based on them, such as **swath beam mapping** and **side-scan sonar,** have provided a view of the ocean floor unlike any previously witnessed. Many of the images in this chapter were generated using multibeam bathymetry. While air gun systems can penetrate the seafloor down to 1–2 km (3281–6562 feet), they are not suitable for discerning seafloor features smaller than about 0.5 m (20 inches). To resolve the small-scale features useful for identifying types of sediments, oceanographers employ high-frequency **chirp subbottom profilers,** so-called for the chirping sound they make when active.

In what seems like a most unusual approach to mapping the sea bottom, oceanographers have taken their instruments to outer space. **Satellite altimeters** emit pulses of microwaves and detect their return to obtain very precise measurements of the bumps and depressions in the sea surface **(Figure 4-4)**. The bumps and depressions, on the order of millimeters to a few meters, arise from the variations in gravity caused by mountains and valleys on the seafloor. For the first time, entire regions of the seafloor can be mapped to their full extent and over a much greater portion of the world ocean than can be obtained using ships. Though higher resolution maps of specific locations are available, the Smith-Sandwell map of world seafloor bathymetry is the most comprehensive global map available to modern oceanographers **(Figure 4-5)**. A pullout map of the seafloor and its features is included inside the back cover of this book.

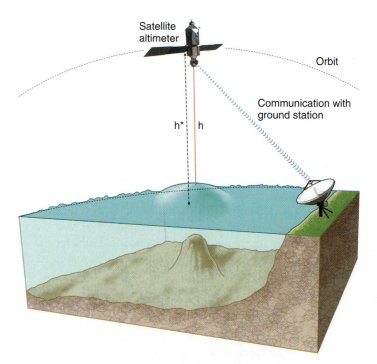

■ FIGURE 4-4 Principles of satellite altimetry for mapping the seafloor. The dotted line represents Earth's ideal surface, the ellipsoid (h*). The gravitational attraction of seafloor features cause the sea surface to deviate from the ellipsoid. Using very precise radar, a satellite can discern these features (h).

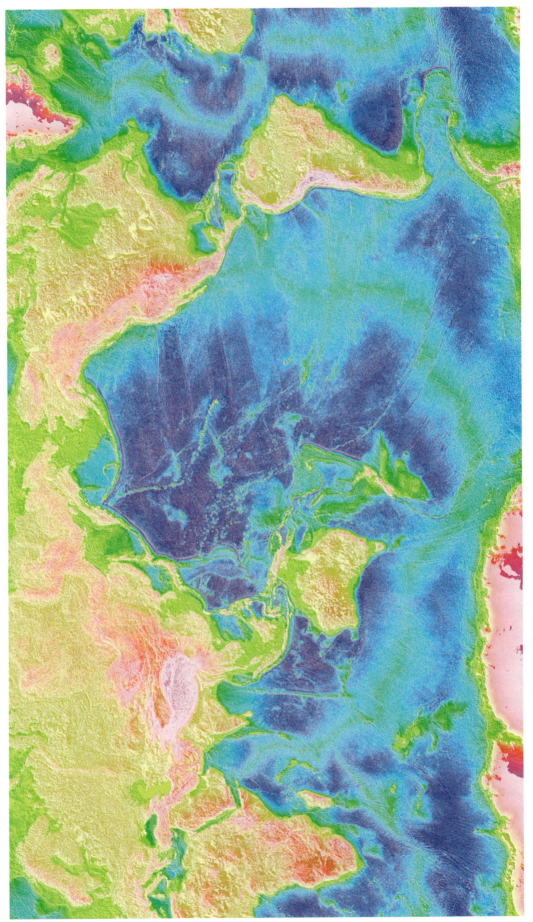

■ **FIGURE 4-5** The most comprehensive map of seafloor bathymetry yet produced for the entire world ocean. Gravity measurements using satellite altimetry were combined with ship depth soundings to estimate water depths.

Sound in the Sea

An understanding of the propagation of sound through seawater helps us better understand the means by which oceanographers use sound to map the seafloor and explore other properties of the world ocean. Seawater is an ideal medium for the transmission of sound **(Figure 4-6)**. The study of sound in the ocean, called **underwater acoustics,** encompasses a wide range of scientific and practical activities. **Passive acoustics** involves the study of ambient underwater sounds using hydrophones. Passive acoustics may involve the study of marine animals that produce sounds, listening for earthquakes, or tracking submarines. **Active acoustics** relies on artificially produced sound to determine various properties of seawater or to detect the seafloor or submerged objects. Active acoustics may be used to study seafloor features, track various marine organisms, or transmit information and data from point to point.

Sound travels through the ocean about five times faster (1500 m s^{-1} or 4921 feet second^{-1}) than through the atmosphere (340 m s^{-1} or 1115 ft second^{-1}). The speed of sound in the ocean varies as temperature, salinity, and pressure change **(Figure 4-7)**. In general, increases in water pressure cause an increase in the speed of sound from the surface to the thermocline (a region of rapid decrease in temperature). Below the thermocline, decreases in temperature

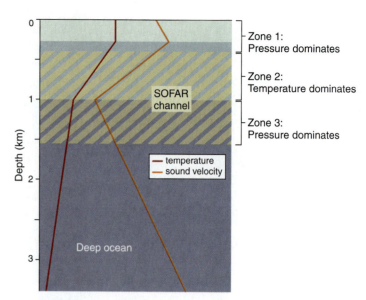

■ **FIGURE 4-7** The relationship between sound speed, temperature, and depth. Pressure effects dominate in zones 1 and 3 but temperature becomes important in zone 2 in the region of the thermocline. The shaded region indicates the SOFAR channel.

cause a decrease in the speed of sound in a special sound-channeling region called the **SOFAR channel** (*sound fixing and ranging*). Here the speed of sound reaches a minimum. The SOFAR channel occurs between water depths of approximately 600 and 1000 m (1969–3281 feet). Within the channel, sound speeds reach a minimum. Below the channel, pressure effects begin to dominate, and the speed of sound increases. These variations in sound speed with depth create interesting behaviors. Sound waves bend (refract) downward above the sound minimum depth and upward below the minimum velocity depth. As a result, sound is "confined" within the SOFAR channel. The refraction of sound in the SOFAR channel has the effect of channeling sound waves over great distances. Since its discovery in the 1950s, the SOFAR channel has been widely used for military applications, ship communications and tracking, and research on currents and undersea earthquakes. Evidence exists that blue and finback whales use low-frequency sounds within the SOFAR channel to communicate over thousands of kilometers of ocean. The properties of sound must be taken into account when making determinations of water depth using sound. Variations in sound with seawater temperature may even provide evidence of global warming as the world ocean heats up, a field of study called **acoustic thermometry.**

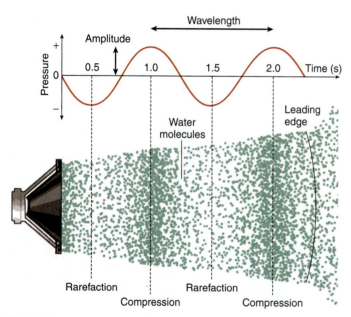

Period of one cycle = 0.5 seconds
Frequency = 2 cycles per second = 2 Hertz (Hz)

■ **FIGURE 4-6** Sound waves act like pressure waves that move through the medium in which they occur. The movement of sound through air or water involves compression and rarefaction (expansion) of the air or water propagating outward from the source. The motion of a Slinky in your hand illustrates the concept. Sound energy may be described in terms of: 1) its wavelength, the distance between successive pressure waves; 2) its amplitude, the maximum distance a vibrating particle is displaced; and 3) its frequency, the number of oscillations or cycles of a vibrating particle in a given time interval, usually measured in cycles per second or Hertz (Hz).

The Bimodal Crust and Isostasy

Studies of the surface of our planet reveal two fundamental features: the **continents,** the elevated regions of the Earth's surface, and the **ocean basins,** the depressions in the Earth's surface. If

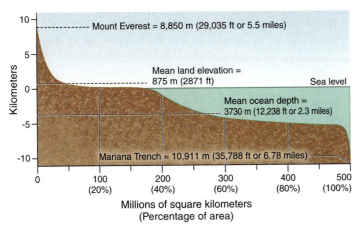

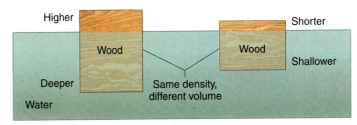

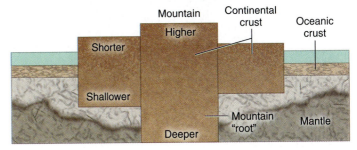

FIGURE 4-8 The cumulative hypsographic curve for Earth. Note that this figure is not a profile of Earth (otherwise, you would see oceanic ridges). Rather, it represents the cumulative area of crust at, above, or below any given elevation on Earth. Two points on Earth lie at the extremes of this curve, Mount Everest, the highest elevation, and the Mariana Trench, the lowest elevation. Submerged continental slopes, oceanic ridges, seamounts and plateaus make up a region at elevations roughly between −0.25 to −2.75 km (−0.15 to −1.7 miles).

FIGURE 4-9 Principles of buoyancy and isostasy. a. Floating objects displace a volume of water equal to their own weight. Both the density of the object and its volume affect its buoyancy. Objects with the same density but different volumes, such as two different-sized wood blocks, will ride at different heights above the sea surface. b. Continental crust and oceanic crust act like "blocks" whose height depends on their density and how far they penetrate into the mantle. Thus, higher mountains have deeper "roots."

we examine the area of land at different elevations above and below sea level, the unique **bimodal distribution** of Earth's surface becomes apparent, namely, the presence of oceanic crust and continental crust, introduced in Chapter 3. About 60% of Earth's surface is occupied by ocean basins, while continents make up approximately 40%. No other planet in our solar system exhibits such a distribution of crustal features on its surface. The cumulative distribution of Earth's crust may be viewed using a graph called a **hypsographic curve (Figure 4-8)**. Hypsographic curves represent the cumulative frequency of the area or percentage of Earth's surface at a given depth or elevation.

The difference in density between oceanic and continental crust affects the buoyancy or **isostasy** of the crust. In essence, isostasy involves the application of **Archimedes principle,** which states that the buoyancy of an object is equal to the weight of the fluid it displaces. Isostasy refers to the balance between the weight of the crust (a downward force) and pressure forces in the fluid mantle (an upward force) **(Figure 4-9)**. Like icebergs in the ocean, the continents and the ocean basins "float" within the malleable upper mantle. This model of isostasy, known as Airy isostasy after British astronomer George Biddell Airy (1801–1892), helped advance the theory of plate tectonics. The deposition of sediments, the advance or retreat of glaciers, or weathering of the crust will cause **isostatic adjustment,** vertical movements of the crust in response to changes in loading. For some regions of the world, the crust responds regionally to loading (or unloading) by flexing, a response called flexural isostasy. The effect is similar to what happens as a group of people walks across a suspension bridge; the bridge bends and flexes as the load on it changes. Isostatic adjustments are important for understanding movements of the crust in response to ice ages. Changes in sea level may occur as a result of isostatic adjustment (see Chapter 11).

Provinces of the Seafloor

A brief survey of the major features of the seafloor illustrates the way in which plate tectonics and other geologic processes shape Earth's crust. The seafloor refers to the portion of the solid Earth that is submerged beneath the ocean. Scientists generally recognize three major parts or provinces of the seafloor: the continental margins, the ocean basins, and the oceanic ridges. Five ocean basins are officially recognized: the Pacific, Atlantic, Indian, Arctic, and Southern Ocean basins. The Pacific Basin **(Figure 4-10)** is approximately twice as large as the Atlantic and Indian basins **(Figure 4-11)**. Features of the three largest basins are listed in **Table 4.1.**

Continental Margins

Continental margins literally are the submerged edges of the continents. They represent a transition zone between continents and ocean basins and, thus, a region of transition between continental and oceanic crust. Continental margins vary greatly, depending on the tectonic and sedimentary processes that shape them. As early as 1885, Eduard Suess (1831–1914) noted the

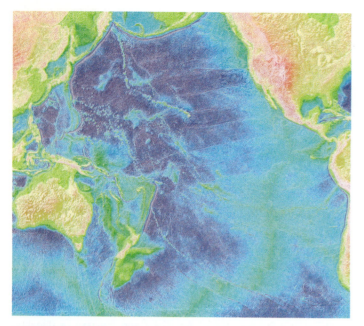

■ **FIGURE 4-10** The Pacific basin, the largest of the ocean basins. The deepest regions are shown in dark blue and purple.

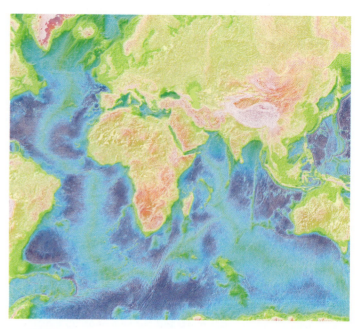

■ **FIGURE 4-11** The Atlantic and Indian basins. Note the presence of oceanic ridges in light blue and green that stretch along the middle of these basins.

differences between the continental margins of the Atlantic and Pacific. Suess classified them as **Atlantic-type margins** and **Pacific-type margins.** With the development of plate tectonics theory, scientists refined and expanded the definitions of continental margins. Atlantic-type margins are more generally called aseismic or **passive margins.** They are commonly found along the margins of the Atlantic Basin but occur in many other locations as well. Passive margins result from rifting and seafloor spreading where an ocean basin forms between two continents.

As a result, the margins ride passively on the moving plate. The aseismic nature of passive margins allows for accumulations of thick layers of sediments that depress the continental edges and oceanic crust through isostatic adjustment. The sedimentary environments of passive margins play a large role in shaping their features. Pacific-type margins are more generally called seismic or **active margins.** While typical of the Pacific Basin, they are by no means confined to the Pacific. Active margins occur where plates converge and may be found at any of the three different

TABLE 4.1 Dimensions of the World Ocean and Its Three Major Basins

	World Ocean	Pacific Basin	Atlantic Basin	Indian Basin
Total Earth area* (Mm³)	510			
Continental area (Mm³)	201			
Oceanic area (Mm³)	309 – 335	155.5	76.7	68.5
Volume (Mm³)	1349	714	337	284
Average depth (m)	3730	3940	3310	3840
Continental shelf and slope	55.2	23.6	20.8	6.7
Continental rise	19.1	4.9	9.1	4.2
Deep-ocean floor	151.2	77.22	40.8	36.5
Oceanic trenches	6.2	5.2	0.7	0.2
Oceanic ridges	118.0	64.6	33.4	22.3
Continental submarine plateaus	14.94			
Oceanic plateaus	29.14			
Oceanic swells	16.55			

* M = mega = 10^6

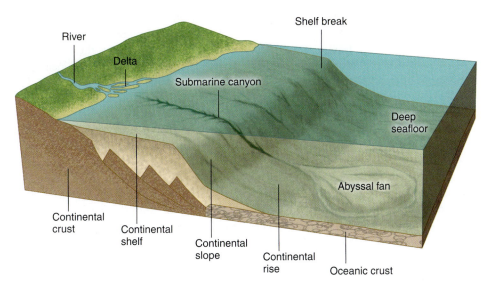

River

Delta

Submarine canyon

Shelf break

Deep seafloor

Abyssal fan

Continental crust

Continental shelf

Continental slope

Continental rise

Oceanic crust

■ **FIGURE 4-12** Features typical of a passive continental margin. Note the association between rivers, submarine canyons, and abyssal fans. Sediments transported by rivers accumulate in submarine canyons where turbidity currents carry them to the seafloor.

types of convergent boundaries. The Peru-Chile Trench and the Mariana Trench are good examples of active margins in subduction zones along a continent and within an ocean basin, respectively. The Himalayas represent an active margin along a continental-continental convergent boundary. Considerable accumulation of sediments occurs in active margins. Sediments scraped off the downthrust plate pile up into wedges that may exert isostatic pressure on the oceanic crust. Sediments may also be subducted where they melt and interact with rising magma or simply become assimilated into the mantle.

Despite differences in their tectonic origin, both passive and active margins exhibit three common features. The submerged flat portion along the edge of a continent is called the **continental shelf (Figure 4-12)**. The shelf has a gentle slope (less than 1:1000) and extends from the shore to the shelf break, where the slope of the seafloor descends rapidly (more than 1:40). Along the coast of central California, the continental shelf is less than a mile wide. Along the western coast of Florida, the shelf extends for more than 400 km (250 miles), well into the Gulf of Mexico. The widest continental shelf, the Siberian Shelf, spans more than 1448 km (900 miles) of the Arctic Ocean!

The steeply sloping edge of the continental block is called the **continental slope.** Technically, the slope ends where the seafloor crosses from continental crust to oceanic crust, but this boundary is not easy to identify. **Submarine canyons,** deep chasms that cut into the continental slope, bring impressive and complex patterns of deposition to the seafloor. The Monterey Canyon off central California is one of the largest submarine canyons in the world. Nearly 483 km (300 miles) long, more than 11 km (7 miles) wide and 1.6 km (1 mile) deep, the Monterey Canyon is equivalent in size to Arizona's Grand Canyon. Along the coast of New York and New Jersey lies the longest submarine canyon in the world, the Hudson Canyon, which extends for more than 643 km (400 miles) offshore. Both sedimentary and tectonic processes sculpt submarine canyons. Sedimentary processes include factors that erode canyons and sediment transport, among others. Tectonic

processes include faulting that leads to offsets of channels in the canyon or earthquakes that cause canyon walls to collapse. Any of these processes may result from or cause a mass wasting event known as a **turbidity current,** a density-driven flow of suspended sediments along the seafloor **(Figure 4-13)**. When sediments accumulate at the head of a submarine canyon, a point is reached at which they become unstable. If disturbed, the sediments collapse and rush down the canyon, entraining water as they flow. The key characteristic of a turbidity flow is the greater density of the turbulent suspension versus the surrounding fluid. The most famous turbidity current occurred on November 19, 1929, when an earthquake along Newfoundland's Grand Banks triggered a

Source area of sediments

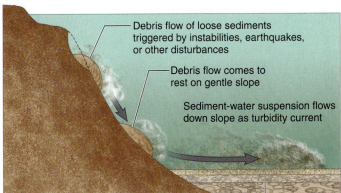

Debris flow of loose sediments triggered by instabilities, earthquakes, or other disturbances

Debris flow comes to rest on gentle slope

Sediment-water suspension flows down slope as turbidity current

Layers of sediment from previous turbidity currents (turbidites)

■ **FIGURE 4-13** Idealized model of debris flow and turbidity current down the continental slope. Whereas debris (sediments and other materials) flows downslope like a mudslide, a turbidity current (sediment-water suspension) continues downslope, often with great velocity. The sediment suspension settles according to grain size (density). Larger sediments are overlain with finer sediments to form a deposit called a turbidite. Successive layers of deposits represent a series of turbidity currents over time.

massive flow, snapping several undersea telegraph cables on the seafloor over a 13-hour period. The event received worldwide attention and brought the importance of turbidity currents, then largely unknown, to the attention of scientists.

Turbidity flows move massive amounts of sediments from the continental shelf to the deep-ocean basin. They may be recognized by the very characteristic type of sedimentary deposits they bring to the deep-seafloor. Called **turbidites (Figure 4-14)**, these deposits typically consist of a mixture of sand, silt, and clays, and occasionally larger sediments, such as gravel and boulders. Because these sediments flow under the influence of gravity, they tend to sort themselves according to size and mass. The layered sediments exhibit upward sorting from coarse to fine particles, resulting in a geologic characteristic known as graded bedding. Some great turbidites can be found exposed along the beach cliffs of California and Oregon.

At the base of submarine canyons, we find **abyssal fans,** large structures very similar to the deltas and alluvial fans formed by rivers or erosion of sediments from mountains. These submarine fans

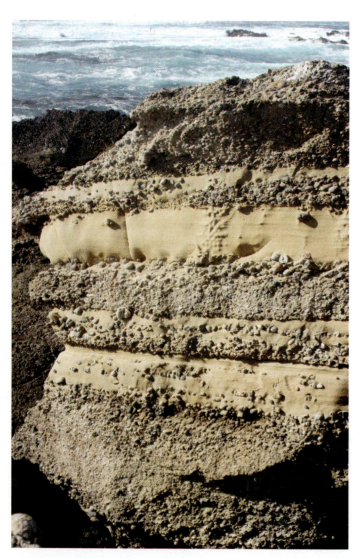

■ **FIGURE 4-14** Turbidite sequences at Point Lobos, California. Notice the alternate layering of coarse and fine sediments.

and other accumulations of sediments at the base of the continental slope define what is known as the **continental rise.** Though not a feature of all continental margins, where present, these accumulations may extend over hundreds of kilometers. The continental rise forms a transition zone between continents and ocean basins. The landward portion of a continental rise may rest on continental crust, while the seaward portion may lie on oceanic crust.

Whereas the "shelf-slope-rise" model describes most continental margins, some margins exhibit greater complexity. Margins may include **carbonate platforms,** relatively flat regions composed of carbonates (e.g., limestone) that lie within or just beyond the continental shelf and slope. Carbonate platforms may form from coral reefs or from the "shells" of marine organisms deposited and lithified on the seafloor (see Chapter 5). A good example of a carbonate platform is the Blake Plateau, located off the southeastern coast of the United States. Continent-attached plateaus, such as the Blake Plateau, are sometimes called marginal plateaus to distinguish them from their deep-sea counterparts, oceanic plateaus.

The California Continental Borderland represents another complex shelf environment. This geologically important region bordering southern California features more than 20 ocean basins (some more than a mile deep) and 32 submarine canyons **(Figure 4-15)**. What makes the borderland especially unique is that these basins and islands all occur within the boundaries of the continental shelf! It is easy to see why a person might mistake this region as a deep-ocean basin, yet the true continental shelf and slope off southern California is found nearly 290 km (180 miles) seaward at the Patton Escarpment.

Deep Ocean Basins

Deep ocean basins encompass all regions of the oceanic crust that are not included in the continental margins or oceanic ridges. The feature most commonly associated with ocean basins is the **abyssal plain,** a flat and relatively featureless region of the seafloor with a slope less than 1:1000 that stretches toward the continents. These sediment-covered regions of the seafloor are best recognized on bathymetric maps as splotches of uniform color, owing to their characteristic uniformity in depth (see Figure 4-5). Abyssal plains receive enormous amounts of sediments from the continents and from biological processes in the overlying water column. Surprisingly, turbidites can be found on some abyssal plains, a testament to the force of strong turbidity currents. The net effect of these processes is horizontal layers of sediments that bury the uneven structure of the rocky seafloor basement. In places where the seafloor basement is especially rough from faulting and volcanic processes, **abyssal hills** may be found. These sediment-covered features rise vertically from the seafloor to heights up to 1000 m (3281 feet). One scientist describes their washboard appearance as similar to the Appalachian Mountains along the US eastern seaboard. Abyssal hills are the most common feature on the face of our planet, covering from 60–70% of the seafloor.

Submarine trenches encircle the Pacific and Indian basins and can even be found in the Atlantic Basin. The Puerto Rico Trench

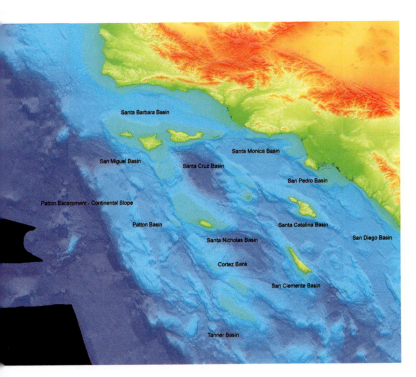

■ FIGURE 4-15 The southern California continental borderland. Note that the continental slope occurs well offshore at the Patton Escarpment.

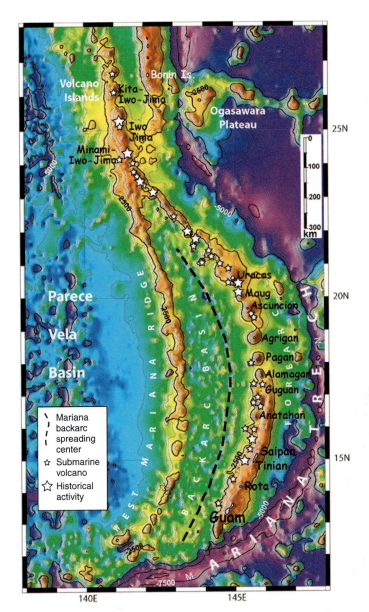

■ FIGURE 4-16 The Mariana submarine trench and associated island arc system. The islands emerge as the downthrust plate melts and buoyant plumes of magma rise through the crust of the overthrust plate. An arc of volcanoes is formed along the perimeter of the subduction zone.

extends for 280 km (175 miles) and reaches a depth of 8.37 km (5.2 miles), the deepest location in the Atlantic. Perhaps the most famous of all submarine trenches is the 2575 km-long (1600 mile) Mariana Trench off the coast of the Philippines **(Figure 4-16)**. The bottom of this trench—the deepest location on Earth—was visited in 1960 by US Navy Lieutenant Don Walsh and Swiss engineer Jacques Piccard, who descended in the bathyscaphe *Trieste* to the Challenger Deep. A remotely operated vehicle (ROV), the Japanese ROV *Kaiko*, explored this region several times, but no humans have visited since Walsh and Piccard. In May 2003, an accident to *Kaiko* put the Challenger Deep once again beyond the reach of humans and their robots.

The largest submarine trench in the Indian basin is the Java Trench. Traversing more than 3000 km (1800 miles) off the coast of Java, this trench features the deepest location in the Indian Ocean at 7.13 km (4.43 miles). In 2006, the Java Trench generated a tsunami up to 7 m (30 feet) high, taking the lives of more than 500 Java residents and destroying thousands of homes. This region experienced a 9 m (30 feet) tsunami in 1994 that killed 223 people. Nearby, the Sunda Trench spawned the world's worst tsunami disaster on December 26, 2004. Subduction zones often lock up, causing tectonic stresses to build over a period of decades or centuries. When these plates release their energy, they generate **megathrust earthquakes,** the most powerful earthquakes on our planet. In some instances, they can generate **megatsunami,** extreme waves produced by vertical displacements of the seafloor (see Spotlight 3.1 and Chapter 10).

Seamounts and guyots dot the landscape of deep ocean basins. More than 800 seamounts have been catalogued in the North Atlantic, although the real number is thought to be much higher.

The Pacific basin encompasses an estimated 50,000 seamounts and flat-topped seamounts called **guyots** (pronounced GEE-oh with a hard "g" as in *get*), discovered by Harry Hess in 1945. Most seamounts and guyots are thought to originate as volcanoes on the flanks of oceanic ridges. As seafloor spreading occurs, these undersea volcanoes move away from the ridge with the seafloor, where they become dormant or inactive. Seamounts whose tops rise above the sea surface from isostatic adjustment or sea level change may become guyots. Their flat tops were thought to result exclusively from erosion by waves, winds, and atmospheric processes. However, evidence from the Ocean Drilling Program (see Chapter 5) suggests that guyots may originate from constructional

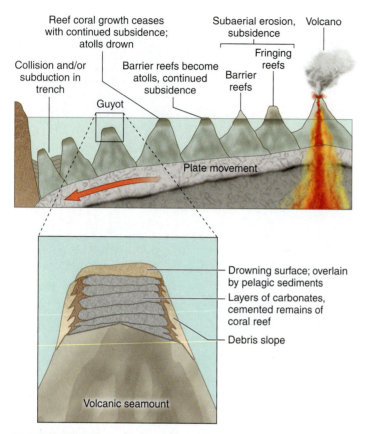

FIGURE 4-17 A new model for guyot formation. The gradual subsidence of seamounts as they move away from oceanic ridges is slow enough to allow for the growth of corals and the accumulation of shallow-water carbonates. The end result is a flat top constructed by corals, rather than a flat top produced by wind and wave erosion. Source: Flood, 1999.

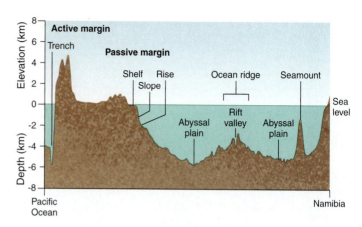

FIGURE 4-18 Cross section of the seafloor from the Andes to Africa. The Mid-Atlantic Ridge is flanked on both sides by abyssal plains. Along the top of the ridge lies a central rift or axial valley. These features are typical of slow-spreading ridges.

processes, such as the building of carbonate platforms by coral and algal reefs **(Figure 4-17)**. Unique in their biota and productive for fisheries, seamounts have become a focal point for international conservation efforts to protect marine ecosystems.

Oceanic Ridges

Oceanic ridges (also known as mid-ocean ridges) form on plate boundaries where seafloor spreading occurs (see Chapter 3). Globally, they form a continuous mountain range with a total length of more than 80,000 km (49,000 miles). A number of important features are associated with oceanic ridges, including transform faults and hydrothermal vents. The classic picture of an oceanic ridge system basin is based on the Mid-Atlantic Ridge, which bisects the Atlantic Ocean Basin from the Southern Ocean to Iceland, where it rises above sea level. Characteristic of slow-spreading ridges, the Mid-Atlantic Ridge features a well-developed axial valley **(Figure 4-18)**. Transform faults and fracture zones also occur along the entire length of the Mid-Atlantic Ridge.

The most prominent oceanic ridge within the Pacific Basin is the East Pacific Rise, a fast-to-superfast-spreading ridge, which extends from the Southern Ocean to Baja California. Fast-spread-

ing ridges do not exhibit well-developed, if any, axial valleys. At its northern terminus in the Gulf of California, the East Pacific Rise becomes a series of offset transform faults. The faults are responsible for disconnecting the Baja peninsula from mainland Mexico about 5 million years ago. These faults also connect the East Pacific Rise to the San Andreas fault system, which runs along the boundary of the Pacific and North American plates from the Salton Sea to Cape Mendocino. North of Cape Mendocino lies one of the most seismically dangerous features in the continental United States, the **Cascadia subduction zone (Figure 4-19)**. Here, the Juan de Fuca Plate is being thrust beneath the North America Plate. The Cascadia subduction zone exhibits features similar to the Sunda Trench. Studies of **paleodeposits,** ancient deposits of sediments, reveal that this region has historically produced megathrust earthquakes and megatsunami and will likely do so in the future.

The Indian Ocean Ridge system consists of three oceanic ridges that come together at the Rodriguez Triple Junction in the center of the Indian Basin. These ridges divide the Indian Basin into three smaller basins. The Southwest Indian Ridge exhibits a rift valley and fracture zones much like the Mid-Atlantic Ridge. The Indian Basin also features **aseismic ridges,** undersea mountain ranges on which no seismic activity occurs (also found in the Atlantic and Pacific basins). The Ninetyeast Ridge and the Chagos-Laccadive Plateau bracket the path of India as it moved rapidly northward between 80 and 50 million years ago. The Ninetyeast Ridge at 4506 km (2800 miles) long is the longest and straightest aseismic ridge in the world ocean.

Hydrothermal Vents

In 1977, scientists deployed a **towed camera sled system** to photograph and take temperature measurements on the seafloor near the Galapagos Islands. The sled system encountered elevated water temperatures—an observation relayed immediately to the ship—but scientists had to wait until the sled returned from its 2.4-km (1.5-mile) deep journey to view the photographs. When the film was developed, the oceanographers could hardly believe their eyes. Giant clams and

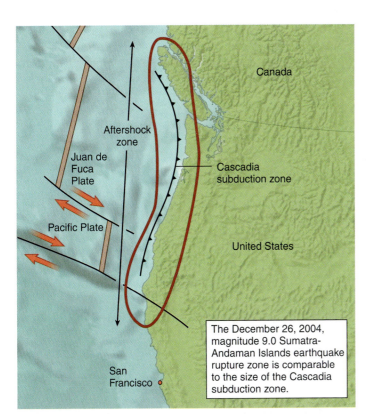

FIGURE 4-19 The Cascadia Subduction Zone is a locked subduction zone that is capable of producing megathrust earthquakes with magnitudes greater than 8.0. The area of the Sumatra Earthquake rupture zone is shown for comparison. The Cascadia Subduction Zone remains a serious threat to the western United States for its potential to create powerful earthquakes and tsunami.

The December 26, 2004, magnitude 9.0 Sumatra-Andaman Islands earthquake rupture zone is comparable to the size of the Cascadia subduction zone.

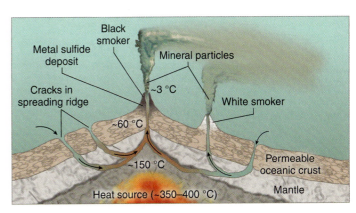

FIGURE 4-20 Seawater circulation in a hydrothermal vent system. Cracks in the spreading ridge permit seawater to enter into the crust. As the seawater circulates and heats under pressure, it loses oxygen and some minerals, and gains other minerals. As the hot, mineral-laden vent water is forced upward and encounters frigid oceanic water, the minerals precipitate, forming large chimneys of rock where the vent fluids carry mostly dark minerals. These chimneys are called black smokers. Vents emitting lighter-colored mineral particles are called white smokers. Plumes of hot water may be detected at some distance from the vent, depending on the rate of flow.

6-foot-long tubeworms lived on the seafloor! A descent by the submersible *Alvin* two days later confirmed the presence of **hydrothermal vents,** a kind of hot-water springs on the seafloor. While oceanographers had predicted that seawater circulated through cracks in the seafloor at mid-ocean ridge, they were not prepared for the discovery of new life-forms. The specimens they collected on this historic expedition were preserved in vodka, the only preservative available.

From 1977 through 1979, oceanographers continued to investigate the hydrothermal vents off the Galapagos Islands. These studies confirmed that hydrothermal vents were associated with seafloor spreading at oceanic ridges **(Figure 4-20).** Seawater enters on the periphery of the vent through subsurface fractures that reach depths of 2 km (about 1 mile) or more. Here, seawater encounters a high-temperature reaction zone where it is heated by a subsurface lens of magma (in fast-spreading ridges). Under high pressure, the seawater becomes superheated and exchanges minerals with the surrounding rocks. Eventually, the superheated fluid rises to the surface, where it exits as a plume of hot, mineral-laden water at temperatures ranging from 10° C (50° F) to more than 350° C (662° F), depending on the nature of the vent. As the exiting vent water comes into contact with the cold ocean water (2° C; 35° F), the minerals precipitate and form deposits on the surrounding seafloor. Some of these deposits accumulate as great mounds of material and may attain great heights. During an expedition along the tip of Baja California, oceanographers discovered

a 20-m (65-feet) high, chimney-shaped structure spitting a plume of dark particles resembling smoke, and the term **black smokers** was born. Black smokers often emit particles that are rich in sulfides, lead, cobalt, zinc, copper, and silver, elements that may make them attractive for mining one day. Another type of vent, a **white smoker,** emits light-colored streams of gypsum and zinc.

The types of vents and their bizarre shapes generate interesting names from the oceanographers who discover them. The first fields discovered at the Galapagos site were named Garden of Eden, East of Eden, and Rose Garden, after the giant red tubeworms found there. The Snake Pit was named for the profusion of white, eel-like fish that live there. One of the black smokers at Snake Pit was named Saracen's Head, after one oceanographer's favorite British pub. A new type of hydrothermal vent system, a carbonate system, was found on the flank of the Mid-Atlantic Ridge at a location called the Lost City **(Figure 4-21).** Carbonate vent systems produce structures composed of carbonates and silica. The Lost City features columnar chimneys that rise 55 m (180 feet) above the seafloor, the largest such features ever observed.

FIGURE 4-21 A view of The Lost City, a carbonate vent system.

Other Important Ocean Basins and Seas

The Pacific, Atlantic, and Indian basins include most of the seafloor of the world ocean. However, a few smaller ocean basins and seas deserve mention for their importance to deep ocean circulation and other processes that we will explore in later chapters.

The Arctic Basin

Surrounded by land on nearly all sides, the Arctic Basin is the smallest of the major ocean basins **(Figure 4-22)**. Studies of the Arctic Basin ridge system have revealed surprising discoveries that may alter our understanding of seafloor spreading. In 2003, scientists announced a new type of oceanic ridge, the **ultraslow ridge,** which spreads at a rate of less than 15 mm (0.5 inches) per year. Although parts of these ridges are non-volcanic, they still produce

new seafloor, albeit in a "cold" fashion. Several sections of the Arctic ridge system and parts of the Southwest Indian Ridge have now been identified as ultraslow ridges. The Arctic Ocean plays a key role in the formation and transport of deep water masses in the world ocean. Because of its sensitivity to climate change, the Arctic Ocean has been called "the canary in the coal mine" for global climate change (see also Chapters 8, 11, and 14).

The Southern Ocean Basin

The Southern Ocean is a relative newcomer in terms of its "official" status as an ocean. In 2000, the International Hydrographic Organization designated the Southern Ocean as a distinct ocean. The Southern Ocean Basin can be subdivided into three major basins that encircle Antarctica **(Figure 4-23)**. The Weddell Sea, confined to the western part of the Atlantic-Indian sector, produces the densest water mass found in the world ocean (see Chapter 9). The Weddell Sea also borders the Ronne Ice Shelf, the second largest ice shelf in the world. In recent decades, this ice shelf has spawned some of the world's largest icebergs. To

■ **FIGURE 4-22**
Seafloor provinces of the Arctic basin.

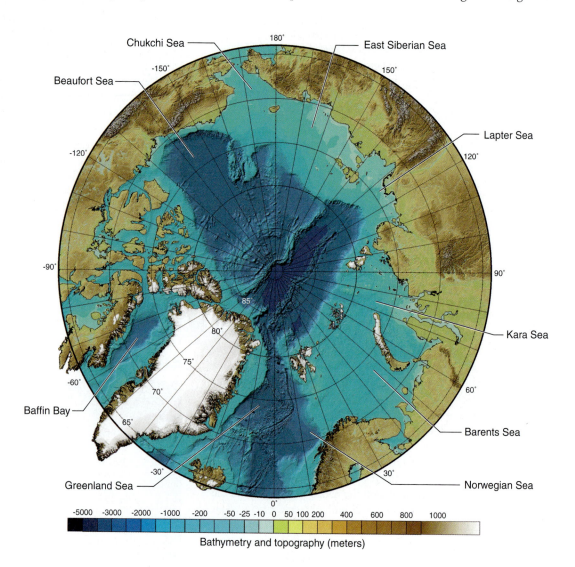

Bathymetry and topography (meters)

productive region stimulated by the mixing of Antarctic and Indian Ocean water masses. Tucked into the Antarctic continent at the western edge of this sector are the Ross Sea and the Ross Ice Shelf, the largest ice shelf in the world. The importance of Antarctica for scientific and other purposes has led to the establishment of three permanent US Antarctic research stations. McMurdo Station resides on the shore of the Ross Sea. Palmer Station was constructed on Anvers Island midway down the Antarctic Peninsula on the Pacific side. The third station, the Amundsen-Scott South Pole Station, is located right at the South Pole (90° S).

The Mediterranean Sea

The Mediterranean Sea represents the last remnants of the Tethys Sea, the sea that once divided the supercontinents of Gondwana and Laurasia (see Chapter 3). The Mediterranean Sea formed as the Africa and Eurasian plates collided, a convergence that continues to this day, as evidenced by the numerous volcanoes in this region (e.g., Mount Etna, Mount Vesuvius). The term **mediterranean sea** refers to a type of circulation pattern in a basin or basins with limited exchange of waters with the world ocean. In these Mediterranean-type basins, density differences largely control the movement of water. For that reason, oceanographers sometimes refer to the original Mediterranean Sea as the Eurafrican Mediterranean Sea to distinguish it from other Mediterranean-type seas. The Eurafrican Mediterranean Sea Basin has limited exchange through the Straits of Gibraltar, where a submarine sill limits the outflow of deep water **(Figure 4-24)**. It also has limited exchange with the Red Sea via the Suez Canal, a human-made opening. Changes in the tectonic interactions in this region some 5–6 million years ago are believed to be responsible for closing the Mediterranean

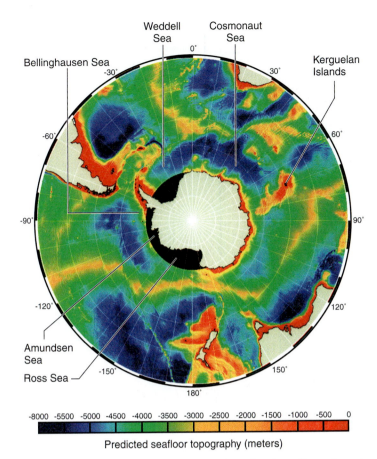

■ FIGURE 4-23 Predicted bathymetry of the Southern Ocean based on satellite altimetry.

the east lies the Antarctic-Australian Basin, which includes the Kerguelen Plateau, a large igneous province formed during the Cretaceous. This geologically complex and tectonically diverse plateau produced the Kerguelan Islands, a harsh and biologically

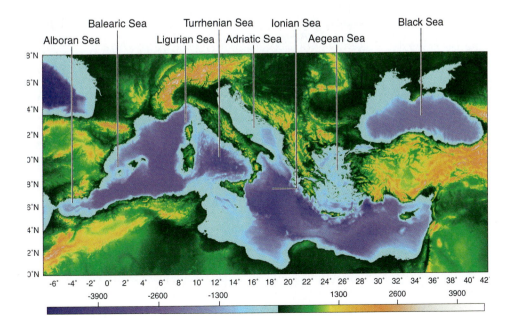

■ FIGURE 4-24 Bathymetry of the Mediterranean Sea.

Seafloor Observatories: Telescopes to Inner Space

While space-based sensors provide an unprecedented view of the ocean surface, they tell us little about processes within or at the bottom of the world ocean. Expeditions aboard research ships can fill in the gaps, but even these are limited to a few locations over short periods of time. Permanent buoys and a host of other advanced sampling platforms exist for studying processes *within* the ocean over long temporal scales but provide limited spatial coverage. As a result, there remains an urgent need to obtain long-term measurements over great expanses of the seafloor, that is, at the *bottom* of the ocean. Plans are under way to create permanent seafloor observatories at tectonic plate-wide scales. These seafloor observatories will include networks of sophisticated instrumentation and sampling tools for monitoring, measuring, and collecting data on a number of geophysical and oceanic processes over a wide range of temporal and spatial scales (see **inside front cover**).

Professor John Delaney of the University of Washington has championed the vision of a plate-scale cabled observatory in a project known as "NEPTUNE: Telescope to Inner Space" **(Figure 4a)**. This project calls for the installation of a network of 25 experimental instrument packages or nodes connected by 3000 km (1800 miles) of undersea cables that span the Juan de Fuca Plate off the northwest coast of the United States. Each node will collect geological, physical, chemical, and biological data and send those data to shore in real time for viewing and downloading by users connected to the Internet. On-site autonomous underwater vehicles (AUVs) will stand ready to collect seismic data, images, and water samples during undersea eruptions of **megaplumes,** hot geysers of water ejected from the seafloor that spread in large plumes within the water column.

An example of the type of event where NEPTUNE will significantly enhance observational capability occurred in February through March 2005 at the Endeavor segment of the Juan de Fuca Ridge. Over a period of 5.5 days, a swarm of 3742 earthquakes was picked up by the US Navy's Sound and Surveillance System (SOSUS), an array of hydrophones installed by the Navy in the 1950s and now used for earthquake detection. A rapid-response team of geoscientists was sent to investigate the region for possible eruptions of magma and megaplumes. The scientists lowered instruments and water collection devices into the ocean but detected no plume activity. A deep-sea camera aboard a remotely operated vehicle failed to find any eruptive event. Nonetheless, the detection of this event by SOSUS and rapid shipboard response demonstrated both the success and the challenges of monitoring plate processes in a timely manner.

In addition to their scientific value, seafloor observatories bear many societal benefits. Seafloor observatories may provide a real-time warning system for earthquakes and tsunami. In addition, students may one day "view" the seafloor from their classrooms or homes via the Internet and high-definition televisions. Students may get a chance to "fly" robotic rovers over the seafloor to get a better look at bottom-dwelling organisms or even sample hydrothermal vent fluids. Who knows, perhaps one day a lucky student will make a discovery that contributes to our knowledge of the seafloor and the world ocean!

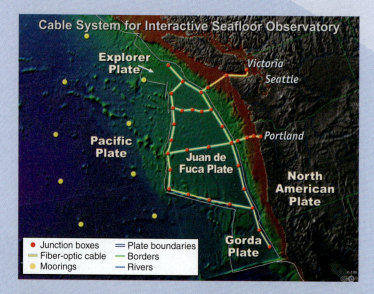

FIGURE 4a Project Neptune involves deployment of instrumentation, moorings, and autonomous vehicles linked to an 1800-mile network of electro-fiber-optic communications cables.

Sea at the Straits of Gibraltar. Cut off from the world ocean, the Mediterranean Sea dried up, an event known as the **Messinian Salinity Crisis.** Extensive subterranean deposits of **evaporites,** the salts remaining after seawater evaporates, have been known since the 1970s. A study of the age and composition of volcanic rocks in the Mediterranean published in 2003 suggests a tectonic origin for the crisis.

The Indonesian Seas

The Indonesian Seas (also called the Australasian Mediterranean Sea) encompass a tectonically complex region with deep basins enclosed by submarine ridges, island chains, and continents. The unique topography and climate of this region influence the exchange of waters between the Pacific and Indian basins and may influence climate cycles, such as El Niño. Recognizing its importance, oceanographers now refer to this region as the **Indonesian Throughflow,** the series of currents that flow through this region. We will explore the Indonesian Throughflow and its significance for world ocean circulation in greater detail in Chapter 9.

The Gulf of Mexico

The Gulf of Mexico is a partially enclosed, roughly oval-shaped basin with a broad continental shelf that extends from Florida to the Yucatán Peninsula. The tectonic origins of the Gulf of Mexico remain uncertain, but it likely formed some time during the opening of the Atlantic Ocean when North America sepa-

rated from Europe. When movements of land masses formed the Central American "land bridge," the Atlantic Ocean, along with the Caribbean Sea and Gulf of Mexico, was permanently cut off from the Pacific. This event played a large role in the circulation of the world ocean, climate change, and the evolution of marine organisms. Along with the Caribbean Sea, this region is sometimes referred to as the American Mediterranean. The Gulf of Mexico features some fascinating undersea landscapes. Limestone shelves rim the continental shelf along Florida and the Yucatán. Subterranean salt deposits underlie much of the US Gulf Coast and the southern portion of the Yucatán. The presence of these salt deposits suggests that evaporites were deposited in the Gulf of Mexico during the Jurassic Period, when the gulf was isolated from other oceans. Subsequent burial of these deposits by sediments caused them to mobilize and migrate, similiar to toothpaste being squeezed from a tube. This salt migration, or **salt tectonism,** has led to extreme deformation of the gulf's seafloor, giving it a hummocky (dune-like) appearance **(Figure 4-25)**. Salt tectonism, combined with deposition of vast amounts of organic sediments by the Mississippi River, has produced rich petroleum resources. In some places, mounds of **gas hydrates,** basically frozen blobs of methane, support a rich variety of chemosynthetic life on the seafloor. In 1997, a new species of worm, the **ice worm,** was found living on these frozen hydrocarbons **(Figure 4-26)**.

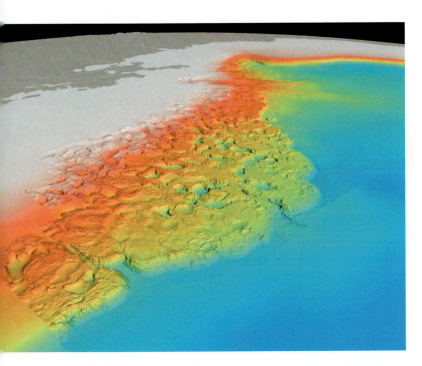

■ **FIGURE 4-25** Salt diapirs, domes of upwardly migrating salt, dot the seafloor along the continental margin of the Gulf of Mexico. Where the source of salt collapses, craters appear.

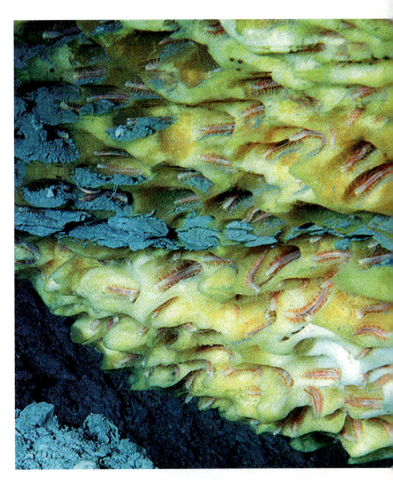

■ **FIGURE 4-26** Gulf of Mexico ice worms, living on frozen methane popsicles.

YOU Might Wonder

1. What is the largest earthquake ever recorded?

The magnitude 9.5 earthquake off the coast of Chile in 1960 still ranks at the top of the list of the world's largest earthquakes. The magnitude 9.2 off the coast of Alaska in 1964 ranks second. By most geophysical measures, the 1964 Alaska earthquake was larger than the 2004 Sumatra earthquake, but instrumentation available at the time did not permit the detailed and precise estimates of magnitude that were available for the Sumatra earthquake. For that reason, the USGS has taken a conservative estimate of the 2004 Sumatra earthquake and recorded its intensity at 9.0 on the moment magnitude scale.

2. Will commercial "flights" to the bottom of the sea ever be available?

If Graham Hawke and others like him have their way, you might just be able to take your own "jetlike" submersible to the Marianas Trench! Hawke hopes to build a version of his Deep Flight submersible capable of reaching more than 11,278 m (37,000 feet). Until then, you might want to check out his underwater flight school at http://www. deepflight.com.

3. Why do space scientists think that hydrothermal vents may exist on other planets or their moons? Aren't they too far from the Sun?

Planets with sufficient mass and an internal core generate internal heat much as Earth does. Smaller planetary bodies may be heated as the gravitational pull of the planet they orbit creates "tides" that regularly displace their mass to generate internal heat, not unlike the effect of rapidly bending and heating a wire coat hanger back and forth (don't touch the bend if you try this!). If a planet or its moon has sufficient water, then the internal heat could be enough to maintain liquid water. Some scientists think that Jupiter's moon, Europa, may conceal an ocean beneath its icy crust. Until we go there, we won't know for sure.

Key Concepts

- Sophisticated sonar and satellite technologies now provide highly detailed maps of the seafloor that continue to build upon and modify our knowledge and understanding of seafloor processes.

- The speed and propagation of sound energy through the ocean depends on temperature, salinity, and pressure, which create a sound-minimum zone called the SOFAR channel.

- The seafloor is composed of both continental crust, which is less dense and contains higher concentrations of silica, and oceanic crust, which is denser and contains higher concentrations of iron.

- The lithosphere "floats" on the plastic asthenosphere and responds with vertical adjustments in elevation to changes in isostasy.

- Passive, "Atlantic-type" continental margins occur on the trailing edge of continental plates and tend to be broad and tectonically quiet, whereas active, "Pacific-type" continental margins occur on the leading edge of continental plates and tend to be narrow and tectonically active.

- Turbidity currents caused by sediment slumping, earthquakes, or other tectonic processes carve spectacular submarine canyons in the continental slope.

- Seamounts are formed along the edge of oceanic ridges and subsequently moved off-ridge, where they subside and become extinct.

- Guyots, flat-topped seamounts, are constructed through the activities of marine organisms.

- Hydrothermal vents represent regions where seawater circulates through the ocean crust, usually on oceanic ridges; recent studies have shown them to be far more diverse than once believed.

- Though smaller in volume, the Southern and Arctic oceans are important sites of deep water mass formation and play an important role in deep ocean circulation.

Terms to Remember

abyssal fans, 60

abyssal hills, 60

abyssal plain, 60

acoustic thermometry, 56

active acoustics, 56

active margins, 58

Archimedes principle, 57

aseismic ridges, 62

Atlantic-type margins, 58

bathymetric chart, 53

bimodal distribution, 57

black smokers, 63

carbonate platforms, 60

Cascadia subduction zone, 62

Challenger Deep, 53

Challenger Expedition, 53

chirp subbottom profilers, 54

continental margins, 57

continental rise, 60

continental shelf, 59

continental slope, 59

continents, 56

deep-ocean basins, 60

echo sounder, 53

evaporites, 67

gas hydrates, 67

guyots, 61

hydrophones, 53

hydrothermal vents, 63

hypsographic curve, 57

ice worm, 67

Indonesian Throughflow, 67

isostasy, 57

isostatic adjustment, 57

lead line, 51

mediterranean sea, 65

megaplumes, 66

megathrust earthquakes, 61

megatsunami, 61

Messinian Salinity Crisis, 67

multibeam echo sounders, 54

multichannel seismic systems, 53

ocean basins, 56

Pacific-type margins, 58

paleodeposits, 62

passive acoustics, 56

passive margins, 58

salt tectonism, 67

satellite altimeters, 54

seismic reflection profiling, 53

side-scan sonar, 54

SOFAR channel, 56

submarine canyons, 59

swath beam mapping, 54

towed-camera sled system, 62

turbidites, 60

turbidity current, 59

US Exploring Expedition, 53

ultraslow ridge, 64

underwater acoustics, 56

white smoker, 63

Critical Thinking

1. What are the physical principles that underlie measurements of the seafloor using sonar or multibeam techniques? What properties of the seafloor and the overlying water column can interfere with precise determinations using these methods?

2. Describe the similarities and differences in the three types of oceanic ridges and the processes that create them.

3. Compare and contrast active continental margins with passive continental margins by describing in detail an example of each. Discuss the limitations in our definitions of these margins and the terminology used to describe them.

4. What processes make the Gulf of Mexico somewhat unique among seafloors in the world ocean? How have these processes contributed to the rich petroleum reserves in this region?

5. How do studies of hydrothermal vents provide one of the best examples of interdisciplinary oceanography? Describe the contributions of geologists, physicists, chemists, and biologists to research on hydrothermal vents.

Explore Online

 Visit www.mhhe.com/chamberlin1e for access to chapter quizzing, key term flash cards, video clips, interactive activities, and more. Further enhance your knowledge with web links to chapter-related material!

Exploration Activity 4-1

Exploring Seafloor Observatories Through Concept Maps

Question for Inquiry

How are oceanographers using seafloor observatories to better understand the geology, physics, chemistry, and biology of the world ocean?

Summary

As you probably have discovered, technology plays a large role in oceanographic research. New and improved methods yield better and often unexpected results. In many ways, progress in oceanography depends on the development of improved techniques for observing the world ocean. New seafloor observatories and similar oceanographic platforms are enabling oceanographers to observe and study the world ocean as never before. In this activity, you will develop a conceptual understanding of the role that seafloor observatories play in advancing oceanographic research.

Learning Objectives

By the end of this activity, you should be able to:

- Identify the major components of a seafloor observatory
- Explain the types of data streams and information obtained from seafloor observatory instrumentation
- List the applications for data and information obtained from seafloor observatories
- Summarize the ways in which particular data streams and information contribute to scientific understanding of the world ocean
- Summarize the ways in which seafloor observatories contribute to public health and safety and public understanding and awareness of the world ocean

Materials and Preparation

For this activity, you will need access to the Internet. Alternatively, your instructor may provide a printout of figures sufficient for completing the activity. You may want to review Spotlight 4-1 prior to starting the activity. Some of you undoubtedly are expert web surfers with little need for instructions. Others of you may be relatively new to browsing the Internet. We strongly encourage the "experts" to help the "newbies."

Part I: Introduction to Concept Mapping

Concept maps were introduced in the 1970s by Joseph Novak at Cornell University to improve student understanding of scientific concepts. Broadly, they help you to assimilate new ideas into existing knowledge and ideas. By identifying the various peoples, places, or things involved in a concept, you can better visualize, interpret, and appreciate the relationships between them. Concept maps provide a tool for constructing a framework of understanding and for exploration and discovery of concepts. Concept maps take many forms. A brief summary of their forms and uses can be found at http://en.wikipedia.org/wiki/Concept_map. Excellent examples and free concept mapping software can be found at http://cmap.ihmc.us/. The first part of this activity helps you gain a basic understanding of concept maps and how to create them.

Concept maps consist of concepts (usually nouns, such as people, places, and things) connected by actions (usually verbs or verb phrases, such as actions, effects or descriptions). For example:

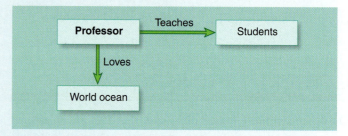

Concepts may be arranged hierarchically (most important concept on top with less important concepts underneath), in a radial pattern (like the spokes on a bicycle wheel with the main concept in the center), or in just about any other pattern you can imagine. *Concept maps are a tool for understanding and connecting ideas.* Thus, they will change as your conceptual understanding and knowledge connections grow. Let's try a few.

On a blank sheet of paper, make a simple concept map for each of the following questions:

- What do you do with the things in your room? (Draw a concept map of your relationships with various objects in your room and how you use them.)
- What courses are you taking and what are their demands on you? (Draw a concept map of your relationships with your courses and their requirements.)
- How do you interact with the natural world? (This is a harder one. Draw a concept map of all the things provided to you by the natural world, including air, water, food, housing, transportation, recreation, and anything else you can think of.)

Part II: Explore a Seafloor Observatory

As a new tool in oceanographic research, seafloor observatories are in various stages of development from concept to full implementation. We have listed one of the more complete ones here but you may wish to explore other examples as more observatories come online.

Visit VENUS: Victoria Experimental Network Under the Sea, http://venus.uvic.ca/index.html and complete the following:

1. List the major components of the observatory. For example, the observatory features imaging systems, sound systems, and instrument platforms. Include other important components that make observations, acquire data, and transmit that information to storage devices or the Internet.
2. Make a list and short description of the various types of instruments that are part of the observatory. For example, the still image system includes an Olympus C8080 wide zoom camera, a lighting system and other parts. List all of them separately for each component. Again, you will have to do some clicking around to find everything.
3. List and describe the various types of "data" provided by the observatory, the so-called "products" or "deliverables." For example, still images of life on the seafloor would be one of the deliverables for the still imaging system. Complete your list for each component of the observatory.
4. Use your lists to construct a concept map that illustrates the various components that support those goals, the instrumentation included in each component, and the type of data and information products provided by those instruments. Don't forget to draw arrows between your boxes with labels describing the relationships between the parts. Your map may be quite large so feel free to use a large sheet of paper, a poster board, or butcher paper. Just be sure you draw it on something that is easy to cut apart for your work in Part III. At the top, add the label Venus Observatory Components and Instrumentation Concept Map.

Part III: Linking Technology with Applications

Go back to the website and look for information on the major scientific and applied goals of the observatory. You may have to explore several different pages, such as About Us, News, and Publications. Make a list of your findings. In addition, use your imagination and creativity to brainstorm applications for the observatory that might be of interest to you. For example,

maybe you see it as a place to find live images of marine animals for your art or biology class. You may also list some ideas of things you wish the observatory could provide.

1. Draw a new concept map with a box at the top of your paper labeled VENUS OBSERVATORY. Beneath this box, draw two boxes on the same line. Label one of them Scientific Goals and the other Practical Goals. Connect them with labeled arrows. For example:

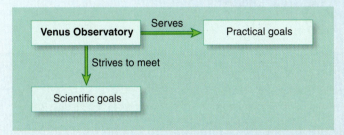

2. Create another set of boxes beneath these boxes each of which corresponds to the list of goals you compiled above. Don't forget the arrows and labels.
3. Now, combine your two concept maps. You may redraw your Components and Instrumentation concept map underneath the new one or you may cut and paste (quite literally).
4. Draw arrows from the goals to the components and instruments. In some cases, one instrument may serve multiple goals. If so, draw several arrows. The idea here is to illustrate the many connections between technology and its applications.
5. Use the Rubric for Physical and Conceptual models to self-evaluate and improve your work. (See Exploration Activity 2-1.)

Part IV: Defend Your Concept Map!

Focus on one scientific and one practical goal that interest you. Create an oral or written summary that answers the following question: How do seafloor observatories contribute to scientific and public understanding of the world ocean? Use your concept map to defend your statements. For example, you may state: "Observatories make the public aware of otherwise hidden marine life." You might support this statement by describing the various visual and auditory instruments and their products that enable the public to observe and experience marine life. Complete your presentation with a summary of your opinion on the importance of seafloor observatories.

5 Ocean Sediments

An Atlantic stingray emerges from its hiding place in the calcareous sediments between the coral reefs of the Florida Keys.

Questions to Consider

1. Why are seafloor sediments called the "memory" of the ocean, and how is that memory being "read" to obtain longer and more accurate records of the world ocean's past?

2. What processes cause sediments to form, and once formed, how do sediments arrive at the seafloor?

3. How does climate affect the origins, transport, and deposition of sediments in the world ocean?

4. What role do biological processes play in the origins, transport, and deposition of sediments?

5. How do oceanographic studies of sinking particles and seafloor sediments help scientists to better understand the possible consequences of global warming?

The Ocean's Memory

Any solid fragment of inorganic or organic material may be termed **sediment.** Familiar ocean sediments include those you find along the coast: rocks and cobbles at the beach, fragments of seashells, or sand and mud at the bottom of a bay. What you may not appreciate about those sediments is the story they tell to a trained observer. Sediments reveal the processes that form beaches, shape coastlines, carve submarine canyons, and control the abundance of marine organisms, to name a few. Sediments also tell a story of Earth's past, what it was like on land and in the sea thousands and millions of years ago. In fact, sediments reveal a history of the ocean over a wide range of temporal and spatial scales (see inside front cover).

Over geologic time, the sinking and settling of sediments—both constant and episodic—can build thick deposits. If undisturbed, the deepest layers will represent the oldest sediments. Those on the surface will be the youngest. Recognition of this process provides geologists with an important tool for interpreting the history of the Earth. It is formalized in the **principle of superposition,** which states that in a deposit of undisturbed sedimentary rock layers, the oldest rocks are at the base of the deposit and the youngest rocks are at the top (**Figure 5-1**). Knowing this, geological oceanographers can determine the relative ages of sediment layers even if their true age is unknown. They can also infer the relative ages for deposits far away from each other as long as the deposits have at least one sediment layer in common.

The study of Earth's history in the ocean's sedimentary record is called **paleoceanography.** As a chronicle of ocean and climatic conditions, marine sediments serve as the ocean's memory. In that sense, they offer one of the rarest of insights: a chance to see what conditions were like on Earth before humans. To study ancient oceans, paleoceanographers typically obtain **sediment cores,** long, narrow-diameter cylinders of sediments, some of which may be hundreds of feet long. Sediment cores dating back to 145 million years ago have been extracted from the ocean floor. The information they provide about Earth's past enables oceanographers and climate scientists to infer patterns of change in Earth's climate. Although the story remains incomplete, at least some evidence exists that human activities are altering our planet as never before.

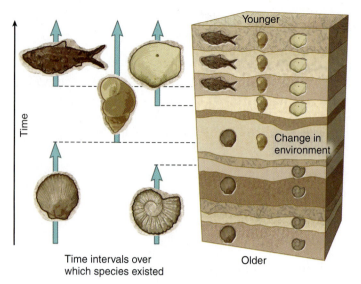

Time intervals over
which species existed

■ **FIGURE 5-1** The principle of superposition and the types of fossils present in sediment layers are tools that paleoceanographers use to interpret the ocean's history. Changes in fossil species over time can reflect changes in their environment. Chemical characteristics of their skeletons also hold clues to ocean conditions at the time they formed.

■ **FIGURE 5-2** Single-celled protists, like the multi-chambered foraminifera, live planktonically or on the seafloor within a specific range of ocean temperatures. Because their shells are naturally preserved in sediments, their occurrence and abundance provide a record of ocean temperatures in the past, as first realized by Wolfgang Schott. Chemical isotopes, like oxygen-18, also provide a record of the temperature conditions when the shell was formed.

The Foundations of Paleoceanography

The first oceanographic expedition to study the global distribution of sediments in the world ocean was the *Challenger* Expedition (1872–1876). Attached to the sounding line was a metal tube surrounded by a donut of weights (called a Baille rod) that would drive the tube into the seafloor and capture a small sample of sediment. From these modest beginnings, a catalog of seafloor sediment types and their descriptions was begun. Expeditions following the *Challenger* Expedition obtained cores of 1–2 m (3–6 feet) lengths, which revealed variations in the abundance of the skeletal remains of **foraminifera,** microscopic marine organisms found throughout the world ocean (**Figure 5-2**). Analyzing sediment samples from the German *Meteor* Expedition (1925–1927), Wolfgang Schott (1905–1989) in 1935 reasoned that the absence of foraminifera skeletons in the lower layers of the core was caused by cold seawater temperatures during the Ice Age. In effect, the presence or abundance of foraminifera skeletons could be used to trace changes in seawater temperature and, by inference, Earth's climate. Schott is credited with establishing the first links between the abundance of marine organisms and climate.

The major advance in paleoceanography came when Cesare Emiliani (1922–1995) demonstrated that measurements of ^{18}O, a stable isotope of oxygen in the shells of foraminifera, could be used as a **proxy** of seawater temperature. The abundance of

^{18}O relative to ^{16}O, the normal form of oxygen (see Chapter 2), varies with temperature, providing a convenient indicator of the seawater temperature in which the foraminifera lived. Plotted as **delta ^{18}O,** or $\delta\ ^{18}O$ (the difference in the ratio of $^{18}O/^{16}O$ relative to a standard), Emiliani found that colder waters tend to have higher values (+3.5‰), that is, they are ^{18}O-enriched, while warmer waters have lower values (0 to −2‰), that is, they are ^{18}O-depleted. Measuring the ^{18}O of foraminifera shells from different sediment layers revealed variations in seawater temperature. Emiliani's work was the first to demonstrate that ocean conditions were not constant through geologic time. The trend in his three-point graph agrees with modern plots based on hundreds of data points and reveals that the world ocean has cooled by about 12° C (about 21° F) over the past 100 million years. Because of his work, Emiliani is considered by many to be the "father" of paleoceanography.

The biggest boost to an understanding of the history of the oceans came with the invention of a new type of coring device, the **Kullenberg piston corer.** This instrument enabled oceanographers on the Swedish Deep-Sea Expedition (1947–1948) to obtain 10–20 m (33–66 feet) sediment cores representing 1–2 million years of deposition. Application of traditional piston core technology culminated during the Climate: Long-Range Investigation, Mapping, and Prediction (CLIMAP) project (1971–1982), organized by the National Science Foundation to study and understand changes in Earth's climate during the past 700,000 years. CLIMAP demonstrated that ice ages result from millennial-scale deviations in Earth's orbit around the Sun, a pattern known as the **Milankovitch**

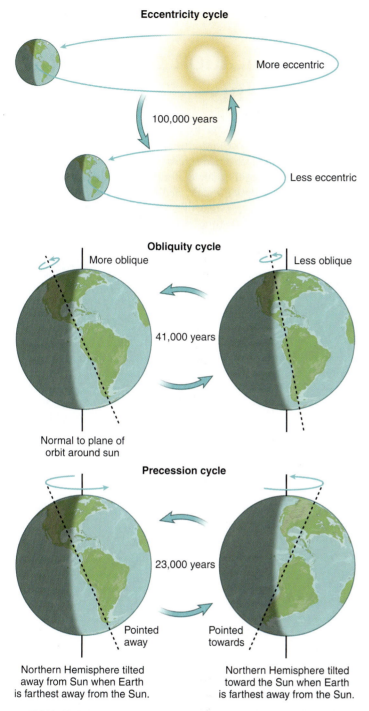

Eccentricity cycle

More eccentric

100,000 years

Less eccentric

Obliquity cycle

More oblique

Less oblique

41,000 years

Normal to plane of orbit around sun

Precession cycle

23,000 years

Pointed away

Pointed towards

Northern Hemisphere tilted away from Sun when Earth is farthest away from the Sun.

Northern Hemisphere tilted toward the Sun when Earth is farthest away from the Sun.

■ **FIGURE 5-3** The 100,000-, 41,000-, and 23,000-year patterns of the Milankovitch cycle. These variations slightly alter the amount of solar radiation that reaches Earth. The result is up to 10-degree C changes in Earth's temperature, sufficient to trigger Ice Ages when other climate factors favor glaciation. Discovery of the influence of the Milankovitch cycle on Earth's temperatures came from studies of microscopic shells in marine sediments and other paleotemperature indicators. See Fig. 5-4.

cycle (**Figures 5-3 and 5-4**). Recognition of the Milankovitch cycle and its effect on climate represents an important milestone in scientific understanding of temporal scales of variability in Earth processes. Another important result of sedi-

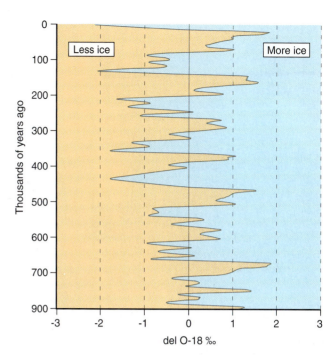

■ **FIGURE 5-4** Studies of oxygen isotopes in foraminifera shells found in sediment cores reveal a sawtooth pattern that matches the patterns of the Milankovitch cycle. Peak values represent periods of glaciation.

ment studies was the discovery of **Heinrich events,** abrupt climate change in the form of rapid cooling within decades to centuries. Six Heinrich events have occurred within the past 75,000 years. Scientists are concerned that human activities may initiate another Heinrich event or other types of abrupt climate change, unleashing a scenario like that depicted in the popular movie *Day After Tomorrow,* albeit at a much reduced pace and scale.

By the 1960s, it became apparent that drilling of the seafloor was the only means to obtain a sufficient length of core for studying past ocean conditions beyond 2 million years. In July 1968, the newly built *Glomar Challenger* set sail on an era of deep-sea drilling that continues to this day. The first drilling program, the Deep-Sea Drilling Project (DPSP), operated from 1968 to 1983. However, in 1983, the project obtained a new and larger vessel, the *JOIDES Resolution* (**Figure 5-5**) and became the Ocean Drilling Program (ODP). The accomplishments of the DPSP and ODP contributed significantly to our understanding of plate tectonics, climate change, seafloor processes, and hydrothermal vents, among other research topics. Results compiled from these programs fill more than 100 volumes! In October 2003, the ODP transferred its scientific objectives to the International Ocean Drilling Program (IODP), a cooperative effort between a nonprofit consortium and academic institutions. Initial phases of the project will be conducted aboard the *JOIDES Resolution,* until a new Japanese vessel, the *Chikyu,* commences operations in 2007. The *Chikyu* is part of a multivessel fleet intended to expand drilling capabilities into shallow-water and ice-covered environments.

■ **FIGURE 5-5** Drilling operations aboard the *Joides Resolution*.

Classification of Marine Sediments

Marine sediments originate from a variety of sources, including continental and oceanic crust, volcanoes, microbes, plants and animals, chemical processes, and outer space (**Figure 5-6**). However, identifying the source of a particular deposit of marine sediments often proves difficult. Sediments may be altered from their original condition by any of a number of physical, chemical, and biological transformations that take place after the sediment is formed. Sedimentologists employ a number of different classification and analysis techniques to characterize sediments. Visual analysis of the texture and composition of a sediment sample, or **descriptive classification,** is often the first step in differentiating sediments. **Size classification,** based on visual, mechanical, or laser-based sizing of sediments, aids in understanding physical and chemical changes in sediments that occur during transport and deposition. **Genetic classification** includes a more complete description of the physical, chemical, and biological properties of sediments. Like forensic scientists working a crime scene, oceanographers search for any clues to piece together an understanding of the history of sediments from the time they were formed to the time they were deposited.

Descriptive Classifications

Descriptive classifications prove highly useful for work at sea and for describing the patterns of sediment distribution on the seafloor. Oceanographers employ a visual classification system

■ **FIGURE 5-6** A few of the sources of sediments in continental margins and in ocean basins. A number of processes may produce sediments from a number of different sources, including rock, organisms, and cosmis debris.

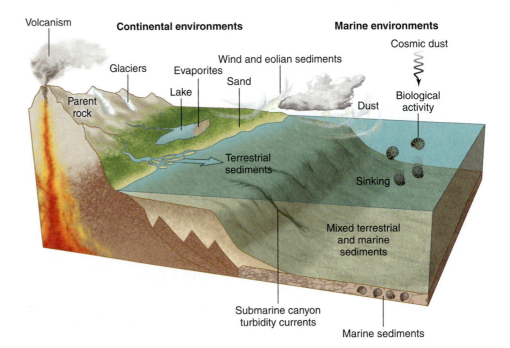

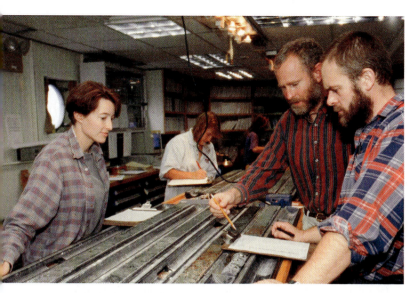

■ **FIGURE 5-7** Scientists visually examine a sediment core during a leg of the Ocean Drilling Program.

called **logging** to create a written description of sediments obtained while at sea (**Figure 5-7**). Occasionally, microscopic observations and other analyses will accompany logging, but in general, the idea is to quickly develop a written description that can be used to guide more detailed analyses on shore at a later time.

In the logging system adopted by the Ocean Drilling Program, sediments are separated into two broad categories: **granular,** resulting from the fragmentation of inorganic or organic parent materials; and **chemical,** forming directly from dissolved compounds in seawater (**Figure 5-8**). Granular sediments include familiar examples such as mud, silt, and sand, and unfamiliar ones, such as the microscopic shells of marine organisms. Granular sediments may be formed by mechanical processes (weathering and erosion), biological processes (secretion of shell or mineralized cell wall), or volcanic processes (above- or below-water ejection of ash and particles). Granular sediments may also be categorized as **biogenous,** originating from biological processes, or **lithogenous,** originating from rocks. Chemical sediments have a variety of pseudonyms and may be referred to collectively as **hydrogenous** (coming from water) or **authigenic** (forming in place). Chemical sediments include the fragments of limestone and limestone-like rocks (dolomite, chert, chalk, etc.), which come from the compacted, buried, and mineralized remains of marine organisms whose cell walls are composed of calcium carbonate and other chemical substances. They also include evaporites, such as those found beneath the seafloor in the Gulf of Mexico (see Chapter 4). As well, metalliferous compounds produced by hydrothermal vents and manganese nodules are included among chemical sediments.

Biogenous sediments originate from organisms that secrete mineral skeletons. In general, they may be subdivided into calcare-

a.

b.

c.

■ **FIGURE 5-8** Examples of major types of granular and chemical sediments. Top: Southern California beach sand (lithogenous granular sediments). (b) Marine shells on Sanibel Island, Florida (biogenous granular sediments). (c) Salt crystals or halites (chemical sediments).

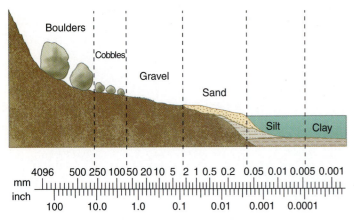

■ FIGURE 5-9 The Udden-Wentworth size classification of sediments.

ous or siliceous and, if known, named after the type of organism that formed the sediment. For example, calcareous sediments originating from foraminifera may be called foraminifera oozes. **Diatoms,** single-celled, photosynthetic microbes whose cell walls are constructed of silica, produce diatom oozes. The particular modifier chosen is usually at the discretion of the person classifying the sediment and often arises from a desire to emphasize the type of study performed. Thus, sediments rich in silica may be called siliceous sediments by a geologist studying chemical processes, while a biologist may call these sediments diatom oozes to emphasize their origins. These biogeneous granular sediments make up a significant fraction of seafloor sediments. Foraminifera oozes constitute about 50% of the deep-sea floor, while diatom oozes comprise about 15%.

Two types of rare sediments are occasionally encountered during logging of sediments using microscopic analysis. These sediments have origins in extraterrestrial processes. **Tektites** were once thought to be the fragments of meteorites that disintegrated in Earth's atmosphere. We now know, however, that tektites originate from the ejection of melted fragments of terrestrial rocks during meteorite impacts. These glassy fragments appear in a dazzling array of shapes and colors, and may settle directly into the ocean or be transported by wind or water. Tektites are considered to be a special class of granular lithogenous sediments that have undergone what is known as shock metamorphism. **Interplanetary dust particles** (also known as cosmic dust) represent a type of **cosmogenous sediment.** These small particles, captured by Earth's gravitational field, originate from asteroids, active comets, and other planetary bodies in the solar system. The tiny fragments, less than 3–35 microns (mm) (0.000118–0.001378 inches) in diameter, avoid vaporization on entering Earth's atmosphere. Interplanetary dust particles are enriched in ^{3}He, an isotope of helium, making them easy to distinguish from terrestrial sediments. Interplanetary dust has received increasing attention in recent decades because it provides information on the "dustiness" of space with possible implications for millennial-scale climate changes.

Size Classifications

Another way to classify sediments is by the average size of sediment particles, called **grains.** The most frequently used size classification system is the **Udden-Wentworth scale.** Using this system, sediments are classified according to their diameter (**Figure 5-9**), often determined using a series of sieves with different-sized pores. The logarithmic scale of grain sizes ranges over six orders of magnitude. Clay and silt represent the smallest diameter sediments, while cobble and boulder make up the largest grain size classification.

Sedimentary Processes

Collectively, processes that involve the production, transport, and deposition of sediments are called sedimentary processes, or simply **sedimentation** (**Figure 5-10**). Sediments are produced through the erosion of rocks, mostly continental, and through the activities of organisms, especially microorganisms that form skeletons or shells. The transport of sediments occurs by winds, rivers, ocean currents, ice flows, glaciers, and a number of other geologic processes. Deposition involves a number of physical, chemical, and biological processes that trap sediments and allow them to accumulate. Ultimately, sediments may be transformed through metamorphosis or melted during subduction, although some evidence exists of a net gain in sediment accumulations

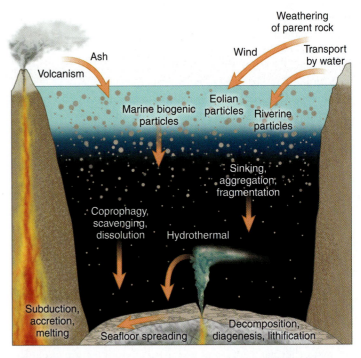

■ FIGURE 5-10 Idealized diagram representing major sedimentary processes. Note that physical, chemical, and biological processes affect the production of sediments and their ultimate fate on the seafloor.

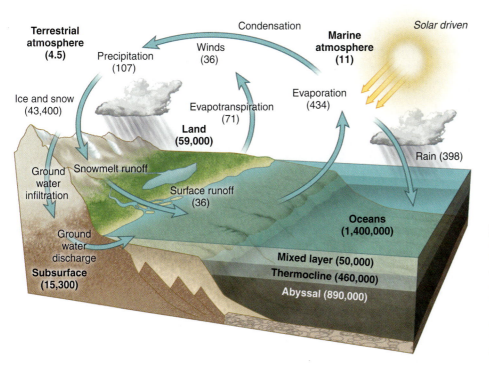

Terrestrial atmosphere (4.5)

Condensation

Solar driven

Marine atmosphere (11)

Precipitation (107)

Winds (36)

Ice and snow (43,400)

Evapotranspiration (71)

Evaporation (434)

Land (59,000)

Rain (398)

Ground water infiltration

Snowmelt runoff

Surface runoff (36)

Oceans (1,400,000)

Ground water discharge

Subsurface (15,300)

Mixed layer (50,000)

Thermocline (460,000)

Abyssal (890,000)

■ **FIGURE 5-11** The hydrologic cycle, including major reservoirs and rates of exchange. Movements of water across and within Earth's crust dominate the origination, transport, alteration, and deposition of marine sediments. Bold numbers in parentheses indicate size of reservoirs in 10^{15} kg. Other numbers in parentheses indicate rates of exchange in $10^{15}yr^{-1}$. Source: Based on data from Leeder (1999) after Chahine (1992).

through geologic time. In a global sense, the origination, transport, deposition, and destruction of sediments describes what is known as the **sediment cycle.** Let's explore various aspects of the sediment cycle with the goal of understanding how and where sediments move about on our planet.

The Hydrologic Cycle

A natural starting point for the sediment cycle is the **hydrologic cycle,** the solar- and gravity-driven exchange of water between various reservoirs (**Figure 5-11**). The hydrologic cycle plays a central role in the formation, transport, and deposition of sediments. Precipitation of water over landmasses leads to dissolution and fragmentation of rocks through weathering. Surface runoff, rivers, and groundwater flow transport sediments and dissolved substances into the ocean. Glaciers, ice, and snow form sediments through mechanical weathering and transport sediments and dissolved substances into the ocean through their movements and when they melt. Ultimately, sediments are transported to continental basins, continental margins, or the deep-ocean floor (i.e., oceanic crust), where they are deposited.

Continental Weathering

The hydrologic cycle plays a major role in **continental weathering,** the dissolution, fracturing, or chemical alteration of rocks through physical, chemical, or biological processes (**Figure 5-12**). **Physical/mechanical weathering** refers to the breaking apart or

fragmentation of rocks due to rock slides, debris flows, and other processes, such as earthquakes. **Chemical weathering** arises due to oxidation of minerals or elements within minerals, dissolution by natural acids, and dissolution of rocks in water. **Biological weathering** stems from the activities of organisms that fracture, dissolve, or chemically alter rocks. Weathering varies according to

■ **FIGURE 5-12** Weathering breaks down solid rock into fragments that become sediments.

■ FIGURE 5-13 The movement of sediment grains under the influence of flowing water in a river or on the seafloor. Sediment grains may roll, slide, jump (saltation), or become suspended. An understanding of the movements of sediment grains under a given set of conditions is important for interpreting sediment deposits and their history.

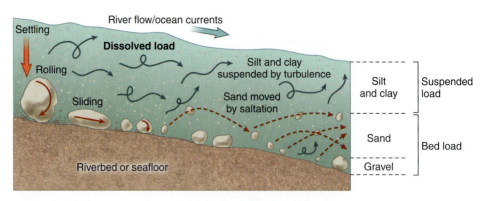

the chemical makeup of the rock (some rocks dissolve more easily than others), climate (temperature, precipitation, humidity), presence of soil (inhibiting rock erosion or promoting plant growth), and length of exposure of the rock at the surface. Weathering plays an important role in the cycling of elements and the salinity of seawater, subjects we explore in Chapter 6.

Sediment Transport

Sediments move when they interact with moving fluids, including air and water. Whether a given sediment grain moves as a result of fluid flow depends on a complex number of variables (**Figure 5-13**). For our purposes, grain size serves as a useful indicator. Grain size determines the **threshold velocity** of a particle, defined as the speed of fluid flow (air or water) that causes the sediment to move. In general, three types of movement occur: 1) **rolling,** the tumbling of a grain along a surface; 2) **saltation,** the hopping of a grain along a surface; and 3) **suspension,** the floating of a grain within the fluid. Small-sized grains, such as dust, are easier to move and suspend than large-sized grains, such as cobble. Similarly, more energetic flows (e.g., fast-moving winds and currents) will move larger grains, whereas less energetic flows (e.g., slow-moving winds and currents) will move only small-grained sediments. By examining the proportions of different sizes—the **grain size distribution**—oceanographers can learn something about how the sediments were transported and the conditions under which they were deposited. Samples that exhibit a narrow range of sizes (i.e., when the grains are nearly all the same size) are said to be **well-sorted,** whereas samples with a wide range of sizes are **poorly sorted.** Sorting occurs because grains of different sizes and density behave differently under a given flow; some particles are carried away, while others are left behind.

Sediment Sinking

Grain size is also an important factor in the sinking or sedimentation of particles. Large grains may settle quickly, but silts and clays may not be deposited for days or months. This means that fine particles may remain suspended in the water and carried much farther from their site of origin than large particles. The rate of sinking of fine particles (less than 0.125 mm or 0.005 inches) is described mathematically by **Stoke's law.**

Oceanographers often estimate the sinking rates of small particles and marine microorganisms using Stoke's law. Alternatively, direct measurements of particle sinking rates may be obtained (with certain restrictions) using an instrument called a **sediment trap** (**Figure 5-14;** see also **Spotlight 5.2**). Determination of

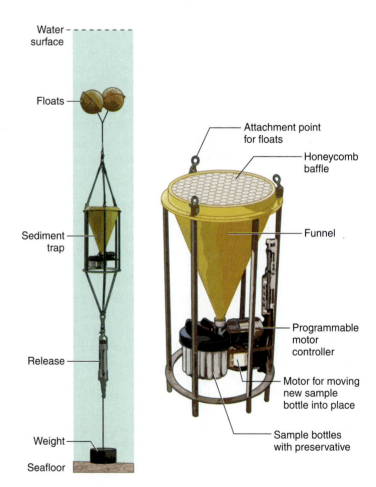

■ FIGURE 5-14 Sediment traps deployed at fixed depths in the ocean act like giant funnels to collect sinking particles in sample bottles at regular time intervals. By moving a new sample bottle into place after a period of time (days to weeks) the rates of particle sinking can be determined and compared. At the end of an experiment (weeks to months), an acoustic signal from the ship releases the trap from its weight and it floats to the surface where it can be retrieved. Sediment traps have proven highly useful for understanding processes that affect the flux of carbon and other elements to the deep sea.

Dawn of the Anthropocene?

At the start of the twenty-first century, it has become clear that human activities affect many facets of the earth system. So pervasive is the human influence on geological, physical, chemical, and biological processes that many scientists believe that our planet has entered a new geologic age, the **Anthropocene**. The Anthropocene marks an era when human activities rival natural processes in their scale and scope of influence on Earth's environment. The beginning of the Anthropocene may be traced to the first signs of increased carbon dioxide in Antarctic ice cores dating back to the eighteenth century. Some scientists place its origins as far back as 100,000 years ago, when human civilizations began to alter the landscape and change the course of rivers. Whether the Anthropocene as a true geologic age will resonate with the earth science community remains to be determined. Yet already several major studies provide clear evidence for the dominance of human activities in many earth systems.

One study, published in 2005, used models and observations to compare prehuman and modern discharges of sediments from 4462 river basins around the world. Globally, rivers discharge some 40,000 km³ (9,596 cubic miles) of water to the world ocean annually. However, water flow and sediment flux to these rivers have been altered by land-use practices and the construction of dams. More than 45,000 dams (above 15 m or about 49 feet high) now control river flows.

Reservoirs associated with these dams have a capacity to store at least 15% of the global annual discharge. At the same time, human activities have accelerated land erosion and increased the rate of sediment transport by rivers. While a greater amount of sediment runoff from land has been estimated for modern times (an increase of nearly 2 billion tons annually over prehuman times), trapping of sediments behind reservoirs has reduced the global discharge of sediments to the world ocean by about 1.5 billion tons annually. Significant regional differences in sediment discharge are also evident. Africa and Asia discharge a reduced volume of sediments, while much of India transports larger volumes of sediment to the world ocean.

The overall effects of reduced sediment discharge to the world ocean are difficult to ascertain. On a local level, reduced sediment transport by the damming of rivers has caused severe beach erosion along many parts of the US coastline. Perhaps the most significant outcome of human alteration of river basins has been increases in the amount of dissolved and toxic substances discharged to the world ocean. Further research aimed at better quantifying the regional and seasonal transports of sediments and identification of issues related to increased or reduced loads will allow us to determine with greater certainty the consequences of human activities on sediments in the world ocean.

FIGURE 5a Though heralded as a triumph of technological achievement, the construction of dams, such as the Hoover Dam shown here, has significant and unforeseen consequences for the earth systems on which human survival depends.

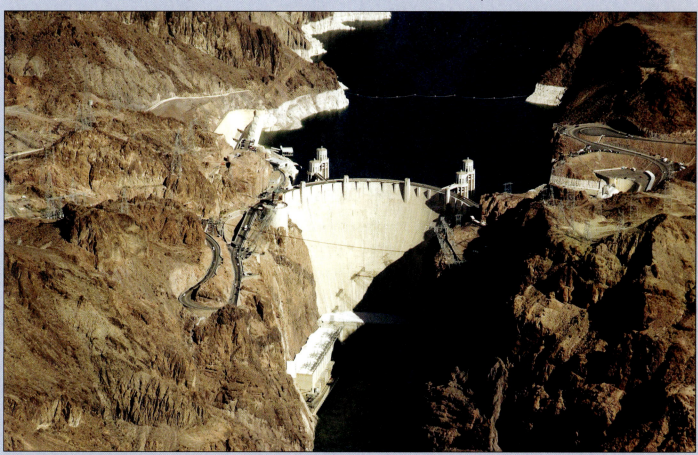

increase the rate at which carbon (as biogenous sediments) sinks into the ocean, a process called the **biological pump** (**Figure 5-15**). In general, the biological pump is affected by three principal factors: 1) those factors that affect the productivity of sediment-producing organisms; 2) those factors that alter the rate of sinking of biogenous (and other) sediments; and 3) those factors that affect the solubility of biogenous (and other) sediments in seawater. The productivity of biogenous sediments is intimately connected with the physical, chemical, and biological processes in the water column that promote and inhibit the growth of planktonic organisms. Regions of high productivity produce the greatest abundance of biogeneous sediments, whereas regions of low productivity produce fewer biogenous deposits. Factors that affect the sinking rate involve processes that alter the size distribution of particles and, thus, the rate at which they sink. Activities that aggregate particles will cause them to sink more quickly, while activities that fragment particles will cause them to sink more slowly. Solubility factors determine which types of biogeneous sediments may dissolve before their deposition on the seafloor. We will explore factors affecting biological productivity and sedimentation in greater detail in Chapters 12, 13, and 14.

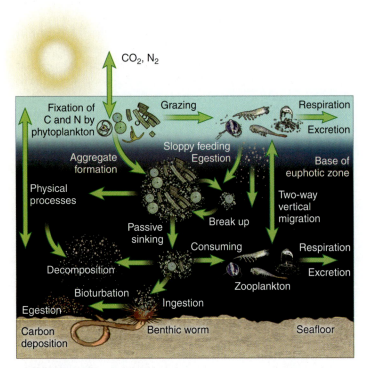

■ **FIGURE 5-15** Conceptual model of the biological "pump." The transport of carbon from the upper ocean to the seafloor is not simple! Nonetheless, a small but significant percentage of carbon is delivered to the seafloor via the biological pump. Many of these processes are discussed in detail in Chapters 12, 13, and 14.

particle sinking rates is especially useful for estimating carbon flux to the seafloor, an important component of the global carbon cycle and climate change.

Biological Sedimentation

A number of *in situ* (in place) biological processes affect the transport and deposition of sediments. An important aspect of biological sedimentation are those activities of organisms that

Calcium Carbonate Compensation Depth

Below a certain depth in the ocean, sediments composed of calcium carbonate (mainly, the shells and skeletons of marine organisms) begin to dissolve rapidly. This depth, called the **lysocline,** is the point at which seawater becomes undersaturated in dissolved calcium carbonate (**Figure 5-16**). That means that solids composed of calcium carbonate begin to go into solution. The factors that determine the solubility of calcium carbonate in seawater are complex and not fully understood, but in general, lower temperatures and higher pressures increase its solubility. Thus, warm surface waters are more easily saturated, while cold deeper waters remain undersaturated. As a result, carbonate sediments that sink beneath the lysocline begin to dissolve.

■ **FIGURE 5-16** The solubility of calcium carbonate in seawater affects the distribution of calcium-containing (calcite) sediment on the seafloor. The surface waters of the world ocean are generally saturated in calcium carbonate and calcite sediments. The surface waters of the world ocean are generally saturated in calcium cabonate and and calcite sediments may accumulate. Below the lysocline, seawater is undersaturated in calcium carbonate, and the shells of calcium-carbonate-secreting organisms dissolve.

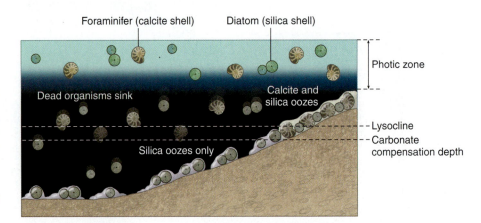

Whether a carbonate sediment makes it to the seafloor and survives intact once deposited depends both on the rate of production of the sediment and the depth of the lysocline. High rates of production are generally accompanied by high rates of sinking that allow carbonate shells to accumulate on the seafloor before they are dissolved. Alternatively, a shallow lysocline in deep water increases the likelihood that carbonate sediments will dissolve before they reach the bottom. Oceanographers define the depth at which the rates of supply and dissolution of carbonate sediments are equal as the **calcium carbonate compensation depth** (CCD), that is, the depth at which there is no net accumulation of carbonate sediments (**Figure 5-17**). If a given region of the seafloor is above the CCD, carbonate sediments may accumulate; below the CCD, carbonate sediments do not accumulate. Note the important distinction between the lysocline and the CCD: in regions of low productivity, their depths may be equal, but in regions of high productivity, the CCD will be at a greater depth than the lysocline. Regional differences in the depth of the lysocline may also be important. The depth of the lysocline in the Pacific is shallower (about 3.5 km or 2.2 miles) than the Atlantic (~5 km or 3.5 miles).   Note that regions of high productivity promote a deeper calcium carbonate compensation depth.

Deposition and Fate of Sediments

When sediments reach the seafloor, a number of processes may occur before their burial as sedimentary layers or rocks. Immediately upon settling, sediments may be consumed or processed by organisms living on the seafloor. A host of benthic or bottom-dwelling animals extract meager supplies of energy and nutrients from biogenous sediments. Sea cucumbers and sea urchins plow the surface of seafloor sediments in search of nutrition. Deposit-feeding worms burrow through seafloor sediments much like earthworms in soil. The reworking and processing of sediments by organisms is called **bioturbation** (**Figure 5-18**). Bioturbation disturbs the layering and chemistry of sediments because of the activities of organisms. The degree of bioturba-

tion and the depth to which it occurs depend on the kinds of species present and their abundance.

Another important process on the seafloor is **diagenesis,** the physical, chemical, or biological transformation of sediments or sedimentary rocks that occurs once they are deposited (in any environment). Microbial degradation, in particular, alters the chemical composition of sediments soon after they are deposited. The chemical environment of interstitial waters—waters in the spaces between sediment grains—determines the elements that may be released or retained by sediments. Digenesis is a complex process that varies considerably over the broad expanse of the seafloor from the continental shelves to the deep basins.

Eventually, sediments become buried and preserved. The pressure of overlying sediments may cause compaction and cementation of sediment grains. Eventually, sediment grains crystallize into sedimentary rocks, a process called **lithification.** Most sedimentary rocks result from the consolidation of sediment grains that were once part of solid rocks themselves. Sandstones and shales are good examples of sedimentary rocks formed in this manner.

Once deposited, sediments and sedimentary rocks form successive layers whose depth and properties reveal the history of their formation, transport, and deposition. In some cases, tectonic processes may uplift and expose these sediments to the atmosphere. Alternatively, sediments and sedimentary rocks may be carried by plate motions to subduction zones, where they may be scraped off the oceanic crust into accretionary wedges or subducted and melted. In many cases, the composition of subducted sediments may contribute to the chemical composition of magmas generated by subduction. The silica-rich magmas along the western United States result, in part, from subduction of silica-rich sediments on the seafloor of the Pacific.

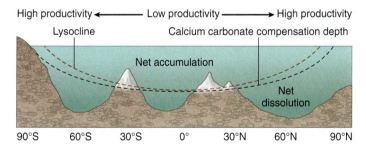

High productivity ◄———— Low productivity ————► High productivity

FIGURE 5-17 A generalized schematic of the calcium carbonate compensation depth. Note that in high productivity regions the CCD may be below the lysocline.

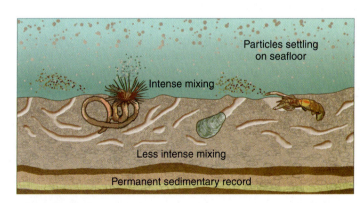

FIGURE 5-18 Bioturbation affects rates of exchange of materials and gases between the seafloor and the overlying water column. They also affect the rate at which carbon and other elements are permanently buried. These processes may be important to biogeochemical cycles and climate change.

Carbon Flux to the Deep Sea

To improve our understanding of global climate and especially the role of the world ocean in climate change, oceanographers want to know how much carbon is delivered to the seafloor and under what circumstances. Ideally, such studies will help to establish a carbon budget for the world ocean that will allow scientists to assess its role as a source or sink of carbon dioxide, an important greenhouse gas. To answer such questions, oceanographers employ an ocean-going instrument called a sediment trap (**Figure 5b**). In use since the late 1970s, sediment traps have been deployed to catch the rain of ocean sediments, or sediment flux (the amount or mass of material that falls through a horizontal area per unit time, i.e., mg m^{-2} day^{-1}). The sediments found in these traps consist of a variety of sinking sediments, but scientists are most interested in sediments containing carbon, also known as particulate organic carbon (POC). Modern sediment traps are designed with multiple containers attached to a wheel that sequentially rotates beneath the trap funnel on a pre-timed schedule. Particles that fall into the funnel sink into these containers which hold preservatives to prevent their decay. The preserved samples are later weighed and carefully analyzed to identify the chemical and isotopic composition of the particles and the species of organism that produced them (if possible).

Sediment traps have been deployed from moorings in practically all ocean basins. In some cases, instruments attached to the sediment trap or nearby moorings collect meteorological, physical, biological, and chemical data that allow oceanographers to establish causal effects for the changes in carbon delivery to sediments. The Ocean Flux Program (OFP), established in 1978, represents one of the longest sediment trap time-series studies ever conducted. The program was initiated in the Sargasso Sea off the coast of Bermuda by Werner Deuser of the Woods Hole Oceanographic Institution. Since 2000, the OFP has been led by Maureen Conte of the Bermuda Biological Station for Research. The decades-long record of the OFP illustrates the importance of long-term and continuous oceanographic observations. During this time, oceanographers have observed unexpected variability related to seasonal cycles, interannual climate phenomena, passing ocean eddies, and extreme storm or hurricane events.

One example of the importance of these studies occurred when the OFP sediment traps captured an extreme sediment flux event during Hurricane Fabian, which passed over the OFP mooring in 2003. The amount of sediment collected by a sediment trap deployed at 1500 m (4921 feet) was greater for a single two-week period than the total amount of sediments collected at that depth since 1978. What caused this record-setting pulse of sediment into the traps? One hypothesis was that the hurricane mixed the ocean and stimulated the growth and sinking of phytoplankton to depth. Another hypothesis was that the sediment arrived from a source some distance away from the site. As it turned out, the chemical composition of the posthurricane sediments was largely calcium carbonate instead of phytoplankton carbon. In addition, a color satellite image showed a large plume of material originating from the Bermuda archipelago and drifting very near the OFP mooring. Thus, a plume of sediment appears the most likely explanation. These results demonstrate the importance of keeping a close and constant eye on the ocean. Episodic events like these have already altered the way oceanographers think about carbon delivery to the ocean. The steady rain of particles from above has now become a storm!

FIGURE 5b Deployment of a sediment trap from the stern of the Weatherbird II off Bermuda.

TABLE 5.1 Proportion of Sediment Types Covering the Deep-Sea Floor

Sediments	Atlantic Basin (%)	Pacific Basin (%)	Indian Basin (%)	World Ocean (%)
Calcareous oozes	65.1	36.2	54.3	47.1
Pteropod oozes	2.4	0.1	—	0.6
Diatom oozes	6.7	10.1	19.9	11.6
Radiolarian oozes	—	4.6	0.5	2.6
Red clays	25.8	49.1	25.3	38.1
Relative size of ocean	23.0	53.4	23.6	100.0

Source: Schulz and Zabel, 2000; after Berger 1976.

Global Distribution of Seafloor Sediments

The distribution of sediments on the seafloor represents the integration of a number of geological, physical, chemical, and biological processes. By examining the types of sediments on the seafloor in the major basins, we can draw a few generalizations about the processes that brought them there. According to **Table 5.1**, carbonate sediments occupy a greater percentage of the seafloor in the Atlantic basin (67.5%) versus the Pacific basin (36%). It has been hypothesized that lower overall rates of microorganism productivity (on a per area basis) may be responsible. A contributing factor is the shallower lysocline in the Pacific.

Although maps of sediment distributions are generally depicted as broad swaths of single types (defined on the basis of the sediment type whose concentration exceeds 30%), in fact, a mixture of types are present at most locations. The global patterns of surficial sediments may be explained by the processes that dominate in a given location (**Figure 5-19**). For example, the "ring" of siliceous sediments surrounding Antarctica comes from diatoms associated with ice-edge processes. Siliceous sediments along the equatorial Pacific have been attributed to **radiolarians,** a protist with a silica skeleton that appears to thrive in the upwelling of cold waters in this region. Terrigenous deposits dominate continental margins, especially at the mouths of major rivers.

The thickness of sediments typically corresponds to the age of the seafloor (**Figure 5-20**). Because all seafloor originates at mid-ocean ridges, it only gradually acquires a blanket of sediments. Newly formed seafloor will be practically devoid of sediments, while the seafloor farthest from the spreading center will have the thickest sediments. Sediments accumulate slowly in most of the ocean basins, except near large rivers and where biological processes are active. Plate tectonics plays a role in the distribution of sediments by altering the speed at which the seafloor spreads. In fast-spreading ridges, such as the East Pacific Rise, sediment layers may be shallow over a broad stretch of the basin. In the North Atlantic, where seafloor spreading is slow, the sediment layers increase in thickness at shorter distances from the ridge. Tectonic processes also trap sediments in submarine trenches. Some of the thickest sediment layers in the ocean can be found in the submarine trenches along Alaska and Chile.

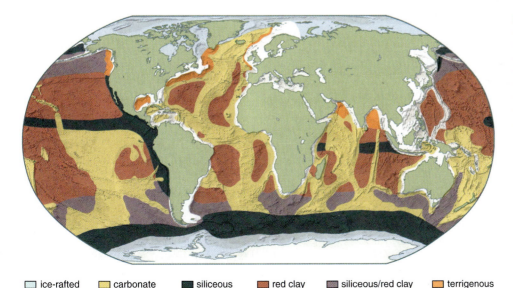

■ **FIGURE 5-19** Global distribution of surficial sediments in the world ocean. Equatorial upwelling in the Pacific and ice-edge processes in the Antarctic contribute to the productivity of diatoms and, hence, the accumulation of siliceous sediments. Carbonate sediments are generally confined to shallower regions of the world ocean. Terrigenous sediments dominate near the mouths of major rivers.

☐ ice-rafted   ☐ carbonate   ■ siliceous   ■ red clay   ■ siliceous/red clay   ■ terrigenous

■ FIGURE 5-20 Thickness of sediments on continental margins and ocean basins. The thickest sediment layers occur near the mouths of major rivers or in submarine trenches where sediments are scraped off the subducting plate. The thinnest sediment layers occur on oceanic ridges. Moderately thick layers occur in the Southern Ocean and the equatorial Pacific.

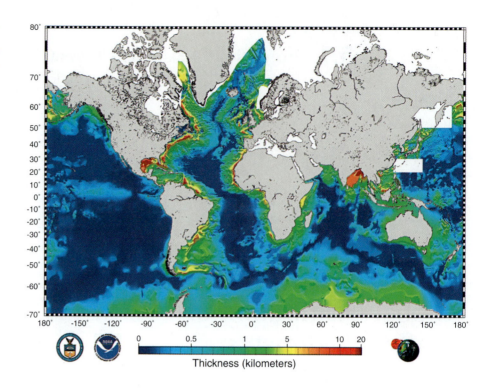

Thickness (kilometers)

YOU Might Wonder

1. **What are the oldest marine sediments ever recovered?**
   In December 1989, ODP scientists drilled Hole 801C and recovered Jurassic-age rocks and sediments, about 170–165 million years old, from the Pigafetta Basin in the western Pacific, near the Mariana Islands. Sediments of nearly identical age had been found previously from the Deep Sea Drilling Project Hole 534A, located on the Blake-Bahama basin in the central Atlantic. These sediments and the rocks on which they rest are among the oldest oceanic crust found in the world ocean.

2. **What are the oldest sedimentary rocks ever found?**
   Before about 1998, scientists thought that the Isua supracrustal belt in Greenland represented the oldest sedimentary rocks, about 3.8 billion years old. Reanalysis and reinterpretation of these rocks indicate that they are metamorphic in origin and not sedimentary. This finding was significant because claims were made for chemical signs of life in these rocks. Their metamorphic origin excludes the possibility that life ever existed in these rocks. The next candidates are sedimentary rocks in Western Australia and South Africa, dating to about 3.5 billion years ago. Some scientists question their origins as well, so the question of the oldest sedimentary rocks on Earth remains open to interpretation.

3. **If the Earth is getting cooler, why are we worried about global warming?**
   While it is true that our natural climate has been transitioning from "hothouse" to "icehouse" conditions over the past 5 million years, considerable evidence exists that human activities have reversed that trend and created unprecedented warming in the past 100 years. Some scientists argue that if our climate had not been cooling, then global warming would be even more severe than it is now. Still other scientists suggest that human activities might counteract the effects of cooling. A great deal of uncertainty remains concerning the long-term effects of anthropogenic releases of carbon dioxide, which is why studies of marine sediments and other climate research studies are so important.

Key Concepts

- Sediments can provide insights into past ocean and atmospheric climates, a field of study known as paleoceanography.

- Sediments may be classified according to their origins, their properties, or their size.

- Sedimentary processes include the origin, transport, deposition, and lithification of sediments.

- Biological processes play an important role in the formation, transport, and deposition of sediments.

- The global distribution of sediments on the seafloor depends on the factors and processes that form, transport, and deposit sediments at a particular location.

Terms to Remember

Anthropocene, 81

authigenic, 77

biogenous, 77

biological pump, 82

biological weathering, 79

bioturbation, 83

calcium carbonate compensation depth, 83

chemical, 77

chemical weathering, 79

continental weathering, 79

cosmogenous sediment, 78

delta ^{18}O, 74

descriptive classification, 76

diagenesis, 83

diatoms, 78

foraminifera, 74

genetic classification, 76

grain size distribution, 80

grains, 78

granular, 77

Heinrich events, 75

hydrogenous, 77

hydrologic cycle, 79

interplanetary dust particles, 78

Kullenberg piston corer, 74

lithification, 83

lithogenous, 77

logging, 77

lysocline, 82

Milankovitch cycle, 74

paleoceanography, 73

physical/mechanical weathering, 79

poorly sorted, 80

principle of superposition, 73

proxy, 74

radiolarians, 85

rolling, 80

saltation, 80

sediment, 73

sediment cores, 73

sediment cycle, 79

sediment trap, 80

sedimentation, 78

size classification, 76

Stoke's law, 80

suspension, 80

tektites, 78

threshold velocity, 80

Udden-Wentworth scale, 78

well-sorted, 80

Critical Thinking

1. Visit the Ocean Drilling Program website, www.oceandrilling.org/, and explore one of the preliminary reports under Publications. Prepare a brief summary of the drilling expedition, including drill sites and major findings. Discuss the importance of the study in terms of improving our understanding of climate change, plate tectonics, or the nature of marine sediments.

2. Find a sample of sand, mud, clay, or some other type of sediment. Examine it using a magnifying glass and describe its characteristics. Classify your sediment as terrigenous, biogenous, or chemical. Postulate on its origins, using your observations as evidence. Write a story or scientific essay describing a possible path for your sediment from the location where you found it to the bottom of the sea.

3. Compare and contrast the weathering of rock and the transport of sediments under dry and wet conditions. Perform an identical comparison with opposite conditions during weathering and transport (i.e., dry weathering-wet transport and wet weathering-dry transport). Discuss the clues a paleoceanographer might find millions of years later that would suggest what kind of climate produced the sediment and what kind of climate transported it to the site where it was found.

4. Based on what you have learned about the role of biological processes on sedimentation in the ocean, discuss the ways in which biological processes might affect sediment processes on land. Provide a few examples based on your own outdoor experiences. (Hint: what do tree roots do to sidewalks?)

5. How is the past or present a key to the future where climate is concerned? Write an essay that discusses the role of paleoceanography or modern-day sediment-related research in helping us to better understand the causes and consequences of increased atmospheric greenhouse gases.

Explore Online

 Visit www.mhhe.com/chamberlin1e for access to chapter quizzing, key term flash cards, video clips, interactive activities, and more. Further enhance your knowledge with web links to chapter-related material!

Exploration Activity 5-1

Exploring the Anthropocene Through Critical Thinking

Question for Inquiry

How do scientists use records of the past prior to the existence of humans to understand the impacts of humans on natural processes in the present?

Summary

In Spotlight 5.2, we introduced the Anthropocene, a newly proposed geologic time period that recognizes human influence on Earth systems. Like our Exploration Activity for Chapter 1, we examine here in greater detail the impacts of humans on natural systems. As before, our goal is to illuminate the scientific evidence for these impacts. However, we also wish to provoke your thinking about the role and limitations of studies of past climates in the interpretation of present-day climate change.

Learning Objectives

By the end of this activity, you should be able to:

- Compile evidence of human activities and impacts since the mid-1800s
- Explain the limitations of evidence for supporting claims of human impacts
- Propose a set of observations or experiments that would strengthen the case for or against human impacts on natural systems
- Evaluate scientific attempts to determine impacts on natural systems

Part I: Sharpen Your Analytical Skills

The quotation below comes from the Preface to *Man and Nature* (re-published as *The Earth As Modified by Human Action*) written by George Marsh in 1864. Carefully review these paragraphs and make notes on the key points. Identify sections that pertain to geological, physical, chemical, and biological processes. Record your thoughts and opinions, too.

. . . In the rudest stages of life, man depends upon spontaneous animal and vegetable growth for food and clothing, and his consumption of such products consequently diminishes the numerical abundance of the species which serve his uses. At more advanced periods, he protects and propagates certain esculent vegetables and certain fowls and quadrupeds, and, at the same time, wars upon rival organisms which prey upon these objects of his care or obstruct the increase of their numbers. Hence the action of man upon the organic world tends to derange its original balances, and while it reduces the numbers of some species, or even extirpates them altogether, it multiplies other forms of animal and vegetable life.

The extension of agricultural and pastoral industry involves an enlargement of the sphere of man's domain, by encroachment upon the forests which once covered the greater part of the earth's surface otherwise adapted to his occupation. The felling of the woods has been attended with momentous consequences to the drainage of the soil, to the external configuration of its surface, and probably, also, to local climate; and the importance of human life as a transforming power is, perhaps, more clearly demonstrable in the influence man has thus exerted upon superficial geography than in any other result of his material effort.

Lands won from the woods must be both drained and irrigated; river-banks and maritime coasts must be secured by means of artificial bulwarks against inundation by inland and by ocean floods; and the needs of commerce require the improvement of natural and the construction of artificial channels of navigation. Thus man is compelled to extend over the unstable waters the empire he had already founded upon the solid land.

The upheaval of the bed of seas and the movements of water and of wind expose vast deposits of sand, which occupy space required for the convenience of man, and often, by the drifting of their particles, overwhelm the fields of human industry with invasions as disastrous as the incursions of the ocean. On the other hand, on many coasts, sand-hills both protect the shores from erosion by the waves and currents, and shelter valuable grounds from blasting sea-winds. Man, therefore, must sometimes resist, sometimes promote, the formation and growth of dunes, and subject the barren and flying sands to the same obedience to his will to which he has reduced other forms of terrestrial surface.

Besides these old and comparatively familiar methods of material improvement, modern ambition aspires to yet grander achievements in the conquest of physical nature, and projects are meditated which quite eclipse the boldest enterprises hitherto undertaken for the modification of geographical surface. . . .

In teams or individually, complete the following. Refer to sections in your textbook and other sources when necessary.

1. For each paragraph above, list and briefly describe in a table (see the example) of the types of mid-1800 human activities on the next page. For example, agricultural activity in the mid-1800s involved plows that were pulled by animals, and so on. Consult historical sources to expand and refine your list and descriptions.

2. For each item you entered in the table, list and briefly describe its modern analogue. For example, farming in the 21st century involves combustion-driven farm machinery.

3. Complete the table by listing the impacts of 19th and 21st century practices on geological, physical, chemical, and biological processes.

Examples of 19th and 21st Century Human Activities and Impacts

19th century activity	Analogous 21st century activity	19th century impact	21st century impact
Farming with plows	Farming with tractors	Conversion of natural lands to agricultural ones.	Release of carbon dioxide and expansion of agricultural land use
Fertilization with manure	Fertilization with chemicals.	Possible local impact on water quality	Eutrophication and other water quality impacts.

Add your own . . .

Part II. Gathering Evidence of Human Impacts

Spotlight 5.2 offers one example of the kinds of scientific studies that are being conducted to quantify the impacts of humans on natural systems. Yet determination of human impacts can be understood at least qualitatively by a number of other different means. Complete the following:

1. Brainstorm and list the various ways that humans recorded their activities in the 18th century and in modern times. (Hint: how do you keep "memories" of important events for future enjoyment and reference.)
2. Use any available resources to document at least three 21st century modifications to natural systems that have been or could be attributed to human activities. (A google search on glacier retreat will bring up an obvious example of the effects of global warming on glaciers. We're confident you can think of many others.)
3. Write a short comparison of your 18th century and 21st century evidence. Describe as objectively as possible the differences between the two centuries (as revealed by your evidence). Can you conclusively state that humans are the cause or do other possible causes exist? What information is missing and how might you gather additional or alternative evidence to better support human-caused change?

Part III. If You Could Turn Back Time . . .

Late one night while surfing the Internet, you accidentally stumble upon a blueprint for a time machine, which you quickly build. Unfortunately, just as you are about to get away from it all, your oceanography professor pulls the plug. "Bring me scientific evidence of natural processes before the time of humans and I'll give you an A in the class," your professor offers. You agree and are soon on your way. To pass the time as the centuries roll backwards, you begin to think about where you will go and what you will do when you get there. You prepare a written proposal that includes the following:

- The geologic time periods you would visit and the rationale for visiting them
- The evidence you would obtain and your rationale for obtaining it
- The ways in which your evidence would complement modern-day evidence

Use the Rubric for Analzying and Presenting Scientific Information to self-evaluate your proposal. (See Exploration Activity 1-1.) Discuss it with your instructor and classmates and revise it as needed.

6 Ocean Chemistry

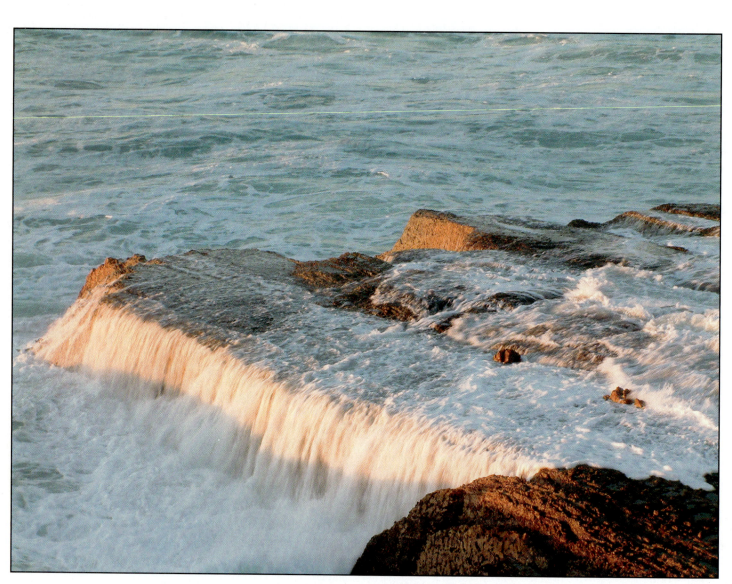

The land enriches the sea, the sea nourishes the land. The chemistry of the world ocean encompasses billions of years of interactions between land and sea.

Questions to Consider

1. How does the hydrologic cycle control short- and long-term changes in the salinity of the world ocean?

2. What are "shifting baselines," and how do they make it difficult to assess the effects of human activities on the world ocean?

3. Why is calibration of oceanographic instrumentation important?

4. What processes control the residence times of conservative versus nonconservative elements?

5. Why is knowledge of the carbonate buffering system important to an understanding of the effects of anthropogenic carbon dioxide on marine organisms?

The Chemistry of the World Ocean

Chemistry is defined as the science of the composition, structure, properties, and behavior of matter. The study of the chemistry of the world ocean, **chemical oceanography,** requires not only a basic understanding of chemical processes but also of geological, physical, and biological processes. As we learned in Chapter 5, the chemistry of the ocean is an important factor in the distribution of sediments on the seafloor. Ocean circulation and the mixing of water masses also involve chemical processes. Many chemical processes affect ocean life, and at the same time, ocean life affects ocean chemistry. In fact, it is difficult to discuss ocean chemistry without discussing ocean life, so we explore a few aspects of ocean life in this chapter, too.

A number of practical uses of the sea require an understanding of ocean chemistry. The engineering of ships, bridges, and other coastal structures requires special attention to the corrosive power of salts. Mining of seafloor minerals, extraction of energy resources, and production of freshwater from saltwater are additional examples of the importance of chemical oceanography. If you have ever kept a saltwater aquarium or visited a marine-related attraction, you have witnessed the careful application of chemistry. Organisms enclosed in seawater tanks have little tolerance for variations in the makeup of their seawater. Almost every aspect of oceanography, in some way, involves chemistry.

In this chapter, we explore the chemical properties of water as they pertain to chemical oceanography. **Chemical properties** refer to characteristics of a substance on an atomic (microscopic) level that involve changes in the composition of a substance. For instance, when a candle burns, the compounds in the candle undergo a chemical change that releases heat, carbon dioxide, and water. The addition or removal of dissolved elements and the exchange of heat in seawater by hydrothermal vents are examples of chemical and physical processes, respectively, in the ocean. In Chapter 7, we explore the physical properties of water, those traits that occur on an environmental (macroscopic) level without a change in composition, such as heating and cooling.

The Foundations of Chemical Oceanography

The poet and philosopher Empedocles (495–435 BCE) called seawater "the sweat of the Earth," an apt analogy if we allow that water was outgassed from Earth's interior. The earliest threads of scientific research on the chemistry of the sea can be traced to Aristotle (384–322 BCE), who postulated that the salts in seawater came from "some earthly mixture." Aristotle correctly ascertained that rain resulted from the evaporation of the ocean, an early reference to the hydrologic cycle (see Chapter 5). Robert Boyle (1627–1691), best known for his gas laws, developed analytical methods for testing the presence of salts. Boyle discovered that river and lake waters contain dissolved substances. The hydrologic cycle caught the interest of comet-discoverer Edmund Halley (1656–1742), who provided some of the first estimates of rates of precipitation and evaporation.

One of the earliest shipboard chemical oceanographers was Count Luigi Ferdinando Marsigli (1658–1730). In the early 1700s, Marsigli measured the "saltness" of seawater samples and determined that the ocean was slightly alkaline. Unfortunately, his yen for sea life brought about his capture by Barbary pirates who sold him into slavery. In the latter half of the eighteenth century, Antoine Laurent Lavosier (1743–1794) produced the first semiquantitative identification of a few of the **major constituents** of seawater, those compounds present in the highest concentrations. Lavosier recognized the presence of sodium chloride (NaCl), the most abundant compound in seawater, and "Epsom salt" ($MgSO_4$, $MgCl_2$), which gives seawater its bitter taste (**Figure 6-1**).

Another early contributor to chemical oceanography was chemist Joseph Louis Gay-Lussac (1778–1850). He was the first to determine that water was composed of one part oxygen and two parts hydrogen. Gay-Lussac coined the word **salinity** (*salure*, in French), the concentration of salts in seawater. He also introduced the term **water column,** a widely used expression for denoting an undefined volume of water from the surface to depth, that is, a column of water (**Figure 6-2**).

One of the founding principles of chemical oceanography is the relative constancy in the ratios of major salts, regardless of the total abundance of salts. This observation was noted in 1819 by Swiss physician-chemist Alexander Marcet (1770–1822) and in 1862 by Danish mineralogist-chemist Johann Georg Forchhammer (1794–1865). However, it wasn't until 1884 that the ratios of salts in seawater were firmly established. Seawater samples collected during the *Challenger* Expedition were painstakingly analyzed by German-born chemist William Dittmar (1833–1894). Dittmar's thorough and now legendary analyses were largely unmatched until the 1960s. Many people consider Dittmar's work the beginning of modern chemical oceanography.

Some scientists hypothesized that organisms control the concentrations of elements. In the mid-1800s, Justus von Liebig, a German chemist (and a "useless" student, according to his early schoolmaster), formulated one of the most important agricultural, oceanographic, and ecological principles in use today: Liebig's law of the minimum. Liebig recognized that plants are limited by the resource that is in least supply relative to their needs (**Figure 6-3**). In modern times, this law applies to our understanding of **limiting factors,** factors that limit the growth of organisms, including temperature, light, and dissolved substances. While Liebig's law did not explain the constant ratios for major ions in seawater, it did provide important insights into the interactions between dissolved elements and

■ **FIGURE 6-1** Sea salt evaporated from a lagoon in Maio Island, Cape Verde, Africa. Sea salt contains sodium chloride and trace concentrations of other elements. Identification of the elements present in sea salts and their abundances represented a major advance in chemical oceanography.

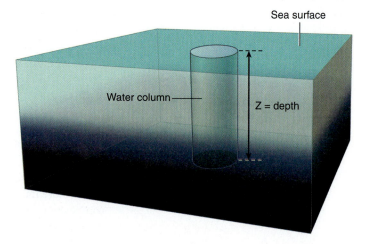

■ **FIGURE 6-2** The concept of a water column. For non-oceanographers, the idea of a column of water might seem a bit abstract. Yet, for oceanographers, the term denotes the depth-dependency of many of the ocean's properties.

Shortest plank

■ FIGURE 6-3 The concept of a limiting factor. Liebig's Law of the Minimum states that the growth of plants is controlled by a single element, one that is required by the plant but which is available at the lowest rate of supply. If we use the analogy of a bucket as illustrative of the growth of a plant, the amount of water the bucket will hold (i.e., the amount of growth of the plant) is limited by its shortest plank. If we lengthen the shortest plank, another plank becomes limiting. So, too, the availability of elements will limit plant growth.

organisms. Both agriculture and oceanography have benefited by understanding the role played by dissolved substances in the growth and distribution of plants and phytoplankton (see the section in this chapter on biologically important nutrients).

In 1908, James Johnstone (1870–1932) described the exchanges of chemical elements in plants and animals, and the cycling of matter between the biotic and abiotic world. A global conceptual model of the cycling of matter emerged in the recognition of **biogeochemical cycles,** the exchanges and cycling of matter through biological, geological, and chemical processes. The birth of the biogeochemical model for understanding Earth as a system is attributed to a series of lectures given in 1922 and 1923 by Ukrainian geoscientist

Vladimir Ivanovich Vernadsky (1863–1945), often referred to as the "father" of biogeochemistry. Vernadsky reinforced the importance of living processes in the cycling of the elements by integrating the **biosphere,** the living part of the world; the **hydrosphere,** the watery part of the world; the **geosphere,** the rocky part of the world; and the **atmosphere,** the gaseous part of the world. He brought earth systems science to the forefront and stimulated a number of important studies in the ensuing decades.

In 1934 and the ensuing decade, Alfred Redfield (1890–1983), an oceanographer from Harvard University and Woods Hole Oceanographic Institution, published data indicating a fixed ratio of carbon (C), nitrogen (N), and phosphorus (P) in particulate and dissolved matter in the ocean. The **Redfield ratio**—the near-constant proportions of C:N:P in seawater—remains an important concept in modern chemical and biological oceanography today. Redfield's pioneering work also inspired the study of **ecological stoichiometry,** a branch of science that examines and interprets the ratios of elements in ecosystems. Though beyond our scope here, ecological stoichiometry holds great potential for linking the activities of organisms to global biogeochemical cycles.

Forms of Matter

Substances are matter that has a definite or constant composition and that exhibits distinctive properties (**Figure 6-4**). Substances can be either elements or compounds. **Elements** are substances comprised of atoms of a single type that cannot be divided into other substances. **Compounds** are substances made up of the atoms of two or more elements that are chemically united in constant proportions. Compounds can be further divided into other substances. Water, a compound, can be separated into two elements, hydrogen and oxygen. While all compounds are **molecules,** substances composed of two or more atoms, not all molecules are compounds. Atmospheric oxygen is a molecule because it normally exists in a **diatomic state,** meaning that two oxygen atoms are bound to each other in the gaseous state. Oxygen by itself, however, is an element.

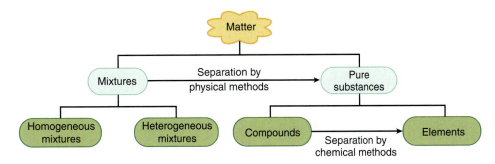

■ FIGURE 6-4 Classification of matter illustrating relationships between substances, elements, and compounds.

The Periodic Table and Valence Electrons

The **periodic table of the elements** is a listing of the known elements that highlights their common chemical and physical properties (**Figure 6-5**). The rows, called periods, and the columns, called groups, represent intervals where elements exhibit similar properties. At least 116 elements are known to chemists, although the existence of some of these elements remains disputed. Nearly all of these elements have been detected in the world ocean. A complete description of the periodic table is beyond the scope of our discussion, but a few points merit our attention. For example, groups of elements share a common trait in the way they form chemical bonds. In the language of a chemist, they have the same number of electrons available for bonding, or the same **valence.** As we learned in Chapter 2, elements are composed of atoms with a certain number of protons, neutrons, and electrons. The protons and neutrons form the nucleus of an atom, which is surrounded

by a "cloud" of electrons. However, the electrons are not randomly arranged around the nucleus. They occupy very specific positions, called **electron orbitals.** Each electron orbital permits a fixed number of electrons, but multiple orbitals can exist. It is the outer electron orbital that participates in chemical bonding and determines the valence of an element. Thus, hydrogen (H), a Group 1A element, has one electron in its outer shell that is available for bonding. In fact, all of the elements in Group 1A (lithium [Li], sodium [Na], potassium [K], etc.) have one valence electron.

Water and Its Unique Properties

An understanding of the chemistry of the world ocean begins with water. More than 96% of the mass of seawater is water so it follows that the water and seawater share common properties. Nonetheless,

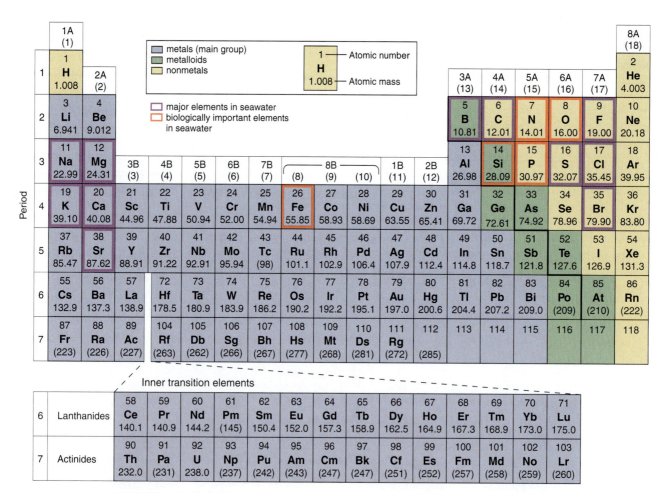

■ FIGURE 6-5 The Periodic Table of the Elements. Major elements in seawater and biologically important elements are highlighted.

the dissolved elements, representing about 4% of the mass of seawater, exert their own influence over the properties of seawater. As you study this section, keep in mind that the global behavior of the world ocean emerges from the properties of its molecules.

The Structure of Water

The chemical bonding of atoms into molecules results in a three-dimensional structure that may be simple, like oxygen gas (O_2), or highly complex, like chlorophyll. Because atoms and molecules are microscopic, chemists resort to models (physical or illustrative) to represent their structure. Viewed in this way, a water molecule consists of two "balls" of hydrogen and one "ball" of oxygen connected to each other by "sticks" that represent their chemical bonds. The chemical bond between the oxygen atom and the hydrogen atoms is called a **covalent bond,** a type of bond that results from the sharing of electrons between hydrogen and oxygen. Each hydrogen atom shares its electron with the oxygen atom and the oxygen atom shares an electron with each of the hydrogen atoms (**Figure 6-6**). Four unshared electrons remain around the oxygen atom. Because of differences in the electrochemical properties of hydrogen and oxygen, the electrons in the covalent bond are not shared equally. Oxygen attracts electrons to a greater degree than hydrogen; in chemical terms, it is highly **electronegative.** The resultant electron clouds cluster around the oxygen atom. This leaves the hydrogen atoms slightly exposed and slightly positive in charge, or **electropositive.** Because of the separation of negative and positive charges, water is classified as a **polar molecule.** The structure of a water molecule resembles a four-sided pyramid called a tetrahedron. The oxygen atom occupies the center of the tetrahedron and the hydrogen atoms reside at two corners. The two pairs of unshared electrons from the oxygen atom make up the other two corners of the tetrahedron.

The structure and polarity of the water molecule enables it to form a special kind of bond called a **hydrogen bond** (**Figure 6-7**).

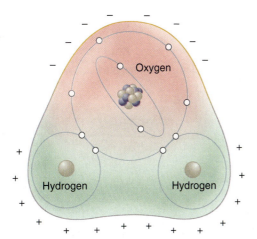

FIGURE 6-6 The water molecule, showing the distribution of its electrical charges. The polarity of the molecule results from the covalent bonds between the oxygen atom and the two hydrogen atoms.

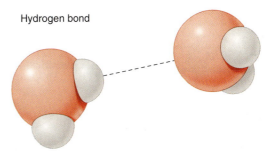

FIGURE 6-7 Hydrogen bonding. Note that the electronegative part of the water molecule (oxygen) attracts the electropositive part (hydrogen) of other water molecules. Hydrogen bonds are about 1/10th as strong as covalent bonds.

Because the electrons of a water molecule reside closest to the oxygen atom, the positively charged proton of each hydrogen atom is available to bond with the negatively charged parts of other substances. Thus, hydrogen bonds result from the attraction of the electropositive hydrogen atoms in a water molecule to negative charges on other molecules, including other water molecules. In fact, a single water molecule may form hydrogen bonds with up to four other water molecules. Hydrogen bonds give an arrangement to water molecules that can be quite complex. Nonetheless, hydrogen bonding contributes greatly to water's unusual physical properties, as we shall learn in Chapter 7.

The Dissolving Power of Water

The process by which solids pass into solution, called **dissolving,** involves chemical reactions that bind water to substances and maintain those substances in a dissolved state. The sugar you stir into your coffee is a good example. In this case, coffee acts as a **solvent,** a substance (usually in the greater amount) into which another substance dissolves, and sugar acts as a **solute,** the substance or substances (usually in the lesser amount) being dissolved. On the other hand, the process of breaking the chemical bonds between water and dissolved substances so that the solid comes out of solution is called **precipitating.** A great example can be found in the making of rock candy. As the heated sugar-water mixture cools, crystals of sugar precipitate on whatever surface you provide, such as a submerged string. The dissolving and precipitating of substances are important to many processes in the world ocean.

Because more substances dissolve in water than in any other liquid, it is often called the universal solvent. However, some compounds dissolve more readily than others. Compounds composed of oppositely charged molecules, called **ionic compounds,** dissolve best in water. When these compounds dissolve in water, they separate into ions and are surrounded by water molecules (**Figure 6-8**). Chemists characterize ions according to their electrostatic charge. Ions with a positive charge are called **cations,** while ions with a negative charge are called **anions.** The electrostatic attraction between water molecules and ions exceeds the attraction of the ions for each other and causes the ions to separate. Ions surrounded

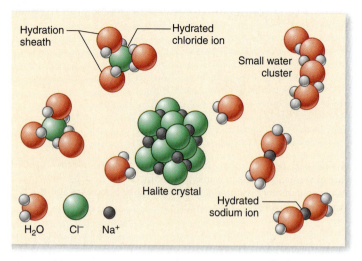

FIGURE 6-8 Dissolution of sodium chloride into cations of sodium and anions of chloride and hydration by water.

by water molecules are said to be **hydrated.** The dissolution of common table salt in water illustrates the process. When dissolved, the sodium and chloride atoms separate into ions. The positively charged sodium cation attracts the electronegative portion of the

water molecule. The electropositive part of the water molecule surrounds the negatively charged chloride anion (**Figure 6-9**).

Not all substances dissolve well in water. Nonionic compounds, such as hydrocarbons and other nonpolar molecules, dissolve very weakly and are typically thought to be insoluble. Oil on water is a good example. However, trace amounts of hydrocarbons can dissolve in water and, in some instances, can be harmful to humans and organisms.

When water dissolves a compound, the solute occupies the space between the water molecules. There is a limit: water can only dissolve a certain quantity of a substance. All substances differ in their **solubility** in water. When the maximum quantity of a substance that can be dissolved in water has been reached, the solution is said to have reached its **saturation concentration,** the total amount that can be dissolved in water (**Figure 6-10**). If the concentration of a dissolved substance is below the saturation concentration, then a solid composed of the substance will dissolve. Under these conditions, the solution is said to be **undersaturated.** Above the saturation concentration, a dissolved substance will come out of solution and precipitate as a solid. Under certain conditions, a substance may remain in solution above its saturation concentration, and the solution is said to be **supersaturated** with regard to that substance.

LOVE AT FIRST ELECTRON: THE STORY OF SALT

FIGURE 6-9 A little chemistry humor: Love at First Electron, the story of bonding in salt.

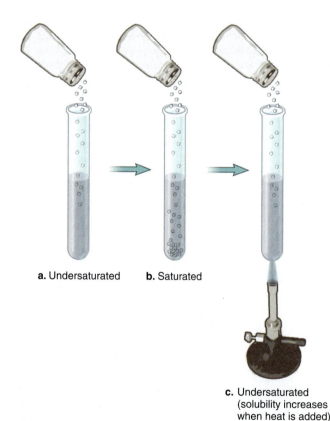

a. Undersaturated   b. Saturated

c. Undersaturated (solubility increases when heat is added)

FIGURE 6-10 Salt, a solute, dissolves in water, a solvent, until a point is reached where no additional salt can go into solution, the saturation point. The saturation point is determined by the properties of the solute and solvent as well as conditions of temperature and pressure. By heating a solution, a greater amount of solute can be dissolved.

When the Dust Settles: Airborne Particles and the World Ocean

Traditionally, sampling and measurement of chemical elements and compounds in the ocean have occurred aboard ships. Despite their usefulness, ships are limited for long-term, continuous sampling of ocean properties. As oceanographers recognize the need for regular and frequent sampling to answer fundamental questions about climate and the world ocean system, they are developing autonomous devices attached to unmanned platforms for sampling and measuring chemical properties. These platforms have allowed oceanographers to document long-term changes in airborne dust that settles on the world ocean, bringing with it chemical contaminants such as lead, and biologically important nutrients, such as iron.

Autonomous samplers face the challenge of preserving samples in a state that does not allow them to degrade or change before they are analyzed in a laboratory. Many important chemical elements and compounds exist in very low concentrations and thus are difficult to detect. Samples are also subject to external contamination by the collecting apparatus. Despite these challenges, chemical oceanographers are now collecting valuable chemical time-series data using autonomous chemical samplers and sensors deployed from buoys and moorings. Autonomous devices can now observe chemical changes associated with many important physical and biological phenomena, including upwelling, ocean mixing, mesoscale eddies, storms, hurricanes, typhoons, dust events, phytoplankton blooms, zooplankton excretion, and nutrient regeneration.

One particularly innovative chemical sampler, the Moored In situ Trace Element Sampler System (MITESS), samples trace elements and metals, including lead and iron. Each MITESS unit includes 12 independent sampling modules that are fabricated of an ultrahigh-molecular weight polyethylene material that ensures "clean" samples. Half-liter Teflon sampling bottles collect samples as the unit opens and closes automatically at predetermined time intervals, typically every week. The samples are preserved with a dilute acid that diffuses from a chamber within each bottle. Two MITESS samplers have been successfully deployed for time-series studies from moorings in Bermuda and Hawaii. MITESS can also be deployed from a ship to collect profile data.

Time-series data obtained over a 16-year period from both ship sampling and the MITESS have documented the history of lead contamination in ocean water in the Sargasso Sea. Lead may enter the atmosphere from the incomplete combustion of leaded gasoline that produces fine particles. Winds carry these particles over the ocean, and currents transport the particles into the deep ocean. Other lead sources include smelting, coal combustion, and cement production. Legislation by the US government in the early 1970s, however, led to the phasing out of leaded gasoline and the introduction of unleaded gasoline. As a result, lead concentrations decreased in the Sargasso Sea until recently. Mean lead concentrations have leveled off since the ban, and the values now recorded may be attributable to emissions from industrial activities in the United States. Large fluctuations in lead concentrations due to atmospheric dust events, passing mesoscale eddies, and other episodic events stimulated the development of MITESS because long-term trends were difficult to perceive without high temporal resolution data.

Iron-containing dust that originated from the African Sahara has been sampled using an above-ocean system developed by Ed Sholkovitz and the "Dust Busters" at the Woods Hole Oceanographic Institution and deployed in the Sargasso Sea. Iron, an essential trace element for certain species of phytoplankton, may limit productivity in as much as 30% of the world ocean. The correspondence between the input of iron from dust, phytoplankton blooms, and climate is of special interest for oceanographers and climatologists. Dry and windy climate conditions promote dust production, which may, in turn, fuel phytoplankton growth and carbon dioxide drawdown. Increases in phytoplankton growth may also increase production of dimethyl sulfides, compounds that participate in cloud formation. Wet climates may limit dust production and reduce ocean productivity.

The feedbacks and interactions between atmospheric and ocean processes remain at the heart of long-term chemical oceanographic studies. As new data from autonomous chemical samplers becomes available, we may better understand the nature of these processes and the impacts of human activities on them.

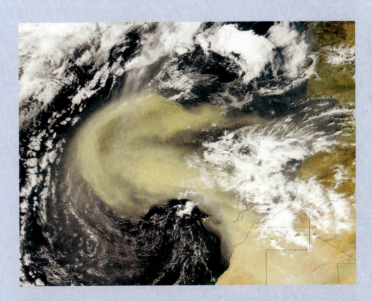

FIGURE 6a A dust storm, several hundred kilometers across, swirled off the western coast of Africa and over the Atlantic on September 4, 2005. The moderate resolution imaging spectroradiometer (MODIS) flying onboard the Terra satellite captured this image the same day. The dust storm covers almost the entire image, its color ranging from tan to pale beige. The dust appears to concentrate in the center of the storm. To the west, clouds fringe the edges of the dust storm. To the east are the coastlines of Mauritania and Western Sahara, and the Canary Islands.

The Dissolved Elements of Seawater

The chemical makeup of seawater has its origins in interactions between water and the solid Earth, from roughly 4.3 billion years ago to the present (see Chapter 2). Knowledge of the dissolved elements in seawater provides clues to the processes that supply and remove them.

Major Constituents

The major constituents, by definition, make up the majority of elements in seawater and exert the greatest effect on its density (**Table 6.1**). These 11 elements have a concentration greater than 1 part per million (ppm) and account for 99.9% (by weight) of the dissolved substances in seawater.

An interesting property of the major constituents is the near-constancy of their ratios. This property was confirmed by William Dittmar, who called it the **principle of constant proportions** (also known as Dittmar's principle, Forchammer's principle, and Marcet's principle). Stated formally, the principle of constant proportions holds that regardless of variations in salinity, the ratio between the amounts of major ions in seawater is constant or nearly constant. In samples of any salinity, chlorine accounts for 55% of the ions (by weight), sodium comprises 30.6%, sulfate makes up 7.7%, magnesium, 4%, and so on (**Figure 6-11**). Because of this constancy, the major ions are called **conservative elements.**

Historically, determinations of salinity were based on an easy-to-measure property, such as chlorinity, that could be extrapolated, using the principle of constant proportions, to estimate salinity. Nowadays, scientists rely on the **electrical conductivity** of seawater, the ability of seawater to transmit an electrical charge. Because seawater contains ions, an electrical charge is transmitted through seawater in proportion to the concentration of those ions (**Figure 6-12**). In 1978, an international organization of scientists defined the **practical salinity scale**, which relies on the conductivity of a known standard as a basis for estimating salinity. Standard seawater, manufactured under very strict specifications, is taken from a specific location in the Atlantic Ocean, filtered, and adjusted with distilled water to a precise salinity. Because its salinity is known, scientists can adjust their salinity-measuring instrument accordingly. Measured salinities are often reported in practical salinity units (psu) or parts per thousand (ppt), though they are not strictly equivalent. Nonetheless, the average salinity of the world ocean is reported at about 35 grams of salts per kilogram of seawater, or S 35.00 psu or ppt, sometimes shortened to S 35.00 without indicating psu or ppt. You may encounter any or all of these units in various scientific publications.

The absolute concentrations of major constituents in the world ocean vary with location (e.g., Atlantic versus Pacific) and with depth (e.g., from the surface to the bottom). An excellent illustration of the geographic and vertical distributions of the major constituents and

TABLE 6.1   Major Constituents of Seawater and Their Origins (S=35.000)

Element	Symbol	Name as found in seawater	Ionic state	Concentration in seawater mmol/kg	% by weight g/kg	Sources
Chlorine	Cl	Chloride	Cl^-	545.88	19.353	Volcanic gases
Sodium	Na	Sodium	Na	468.96	10.781	Crust
Magnesium	Mg	Magnesium	Mg^2	52.83	1.284	Crust
Sulfur	S	Sulphate	SO_4^{2-}, $NaSO_4^-$	28.23	2.712	Volcanic gases
Calcium	Ca	Calcium	Ca^2	10.28	0.4119	Crust
Potassium	K	Potassium	K^+	10.21	0.399	Crust
Carbon	C	Bicarbonate	HCO_3^-	2.06	0.126	Volcanic gases, crust, sedimentary rocks
Bromine	Br	Bromide	Br^-	0.844	0.0673	Volcanic gases
Boron	B	Borate	$H_2BO_3^-$	0.416	0.0257	Volcanic gases
Strontium	Sr	Strontium	Sr	0.0906	0.00794	Crust
Fluorine	F	Fluoride	F^-	0.068	0.00130	Crust

Sources: Pilson, 1998; Open University, 1995; Holland, 1984.

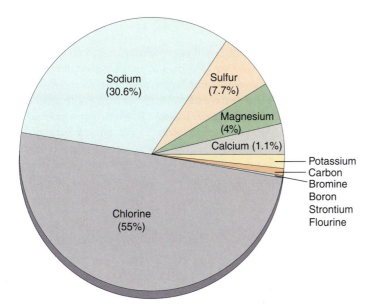

FIGURE 6-11 Relative proportions of the major constituents of seawater depicted as a pie chart. These proportions hold regardless of the salinity. Percentages shown are by weight.

other elements in the world ocean can be found on the interactive website, Periodic Table of Elements in the Ocean, linked on our website.

Minor Constituents

Elements whose concentrations are below 1 ppm are called **minor constituents** or **trace elements.** Despite their low concentrations, many trace elements are essential to marine organisms and useful to humans.

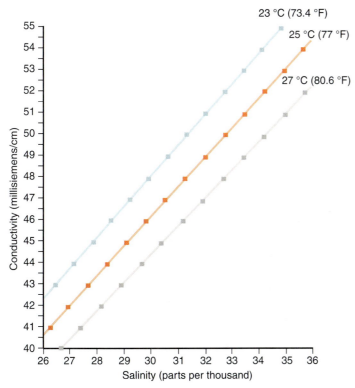

FIGURE 6-12 Salinity versus conductivity at different temperatures. The linear relationship between salinity and conductivity allows oceanographers to use measurements of conductivity to instantaneously determine salinity very precisely at sea, a great savings in time, expense, and effort.

Some marine organisms, such as the cone shell (*Conus sp.*), concentrate trace elements to manufacture toxins for paralyzing their prey. Modified and administered to humans, *Conus* toxins show promise for relieving chronic pain. Other natural marine products may be useful in fighting cancer. Natural compounds have stimulated great interest in commercial development of medicines and other products from the ocean. Most biologically active elements are **nonconservative elements,** meaning they do not obey the principle of constant proportions. Their concentration varies independently of other elements.

Biologically Important Nutrients

A special category of dissolved substances, **biologically important nutrients** or simply nutrients, is defined for photosynthetic marine organisms (**Figure 6-13**). Marine plants, benthic algae (seaweeds and benthic microalgae), and **phytoplankton,** (drifting photosynthetic microorganisms) require trace amounts of nitrogen, phosphorus, potassium, iron, silica, and many other compounds. If they are required in substantial quantities, such as nitrate and phosphate, they are referred to as **macronutrients.** Nutrients required in lesser quantities, such as silicate and iron, are called **micronutrients.** The availability and supply of both macronutrients and micronutrients determine the distribution and abundance of photosynthetic life in the world ocean (see Chapter 13).

Dissolved Organic Matter

Dissolved organic matter (DOM) represents another important category of dissolved substances in the world ocean. DOM consists of dissolved carbon and other organic substances that originate from the bacterial decomposition of organic matter, "leakage" from marine

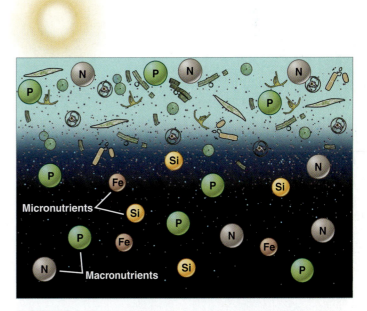

FIGURE 6-13 The biologically important nutrients. Phytoplankton growth depends on an available supply of macro- and micronutrients. Nitrogen compounds (N), like nitrate, may limit growth seasonally but phosphorus (P) may be limiting over geologic time. Iron (Fe) and silica (Si) are micronutrients that may limit certain types of phytoplankton in certain environments. Elements are shown here diagrammatically.

organisms, and terrestrial sources, among others (**Figure 6-14**). Discovery of the microbial food web in the 1970s and its reliance on DOM brought intense research efforts that continue to this day. Because DOM is part of ocean food webs and the carbon cycle, oceanographers are keenly interested in its distribution and fate in the world ocean. Dissolved organic matter often adds color to the ocean (e.g., colored dissolved organic matter, or CDOM), which enables it to be detected by satellites that measure ocean color. We explore the role of dissolved organic matter in ocean food webs in Chapter 14.

Dissolved Gases

Gases, such as oxygen and carbon dioxide, are dissolved in seawater through exchanges with the atmosphere. Gases increase in solubility as the temperature decreases. Consequently, the highest concentrations of gases, such as oxygen, can often be found in polar waters. Biological processes also control the concentration of gases. For example, the concentration of dissolved oxygen represents a balance between oxygen-producing reactions (i.e., photosynthesis) and oxygen-consuming reactions (i.e., respiration). If the rate of photosynthesis exceeds respiration, the dissolved oxygen concentration increases (unless saturated). If the rate at which oxygen is consumed by respiration exceeds the rate at which it is supplied from photosynthesis or from mixing of surface waters in contact with the atmosphere, then dissolved oxygen decreases. The combination of biological processes and the temperature dependency of solubility is critical for bays and estuaries that warm during the summer. The resultant low oxygen concentrations in these waters, a condition called **hypoxia** (dissolved oxygen <5 mgL^{-1}), may exceed the tolerances of marine organisms. Prolonged hypoxia can be fatal. Certain conditions in the water column may promote the complete depletion of oxygen, or **anoxia.** When such conditions are widespread, massive invertebrate and fish kills may occur. Although

human activities have contributed to hypoxia and anoxia in coastal and estuarine waters (see Chapter 15), zones of low oxygen occur naturally in some parts of the world ocean. **Oxygen minimum zones** are found at intermediate depths in the eastern Pacific, eastern South Atlantic, Arabian Sea, and Bay of Bengal. In some locations, seamounts penetrate the oxygen minimum zone, giving rise to unusual patterns in marine organisms (**Figure 6-15**). The largest deep-basin anoxic environment can be found in the Black Sea, where waters below 100 m (328 feet) are permanently anoxic.

Sources and Sinks

The salinity of the world ocean reflects a balance between **sources** of salts, processes that add salts to seawater, and **sinks** of salts, processes that remove salts from seawater. Add more sources or reduce sinks and the salinity goes up; add more sinks or reduce sources and the salinity goes down. The balance of source and sinks ultimately determines the concentration of dissolved elements in seawater.

Continental Weathering: A Source of Elements

In Chapter 5, we explored the role of the hydrologic cycle in the weathering of continental crust and the transport of sediments into ocean basins. The hydrologic cycle also delivers dissolved substances into the ocean as water dissolves the soluble components of the crust. Rivers represent the principal source of dissolved constituents to the world ocean. Comparison of the types of dissolved constituents of continental crust and river water (**Table 6.2**)

■ **FIGURE 6-14** Leached from vegetation upstream, colored dissolved organic matter is visible as the brownish tint in this stream flowing into the Pacific Ocean.

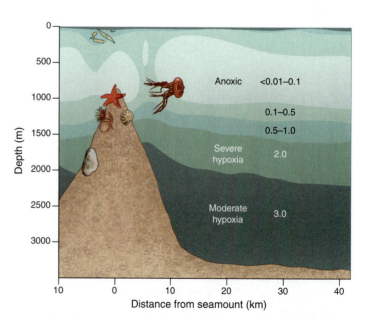

■ **FIGURE 6-15** Oxygen minimum layers occur naturally in many parts of the world ocean. They allow low-oxygen-tolerant organisms to exploit habitats that might otherwise be unavailable to them (data from Levin, 2002).

with seawater (Table 6.1) reveals that they share many substances in common. Yet the concentrations of these substances differ significantly. The world ocean is highly enriched in sodium and chloride relative to continental crust and river water. Can weathering explain their concentration in seawater? According to Table 6.1, there are about 10 gm (0.353 ounces) of sodium in 1 l (0.2642 gallons) of seawater. If the volume of the world ocean is approximately 1.37×10^9 km^3 (3.62×10^{20} gallons), then the total mass of sodium in the world ocean is about 1.33×10^{19} kg (1.47×10^{16} tons). The current continental mass is about 1.93×10^{22} (21.3×10^{18} tons), so the total sodium held in continental rocks is approximately 3.48×10^{19} kg (3.83×10^{17} tons). Taking into account the 4-billion-year history of the continents, we can reasonably assume that weathering accounts for the amount of sodium found in the world ocean. Though admittedly crude, oceanographers use these "back-of-the-envelope" calculations to determine whether a given source of elements is reasonable. The same logic can be applied to other constituents in seawater. Some elements, such as chloride, cannot be explained by continental weathering: the concentration of chloride in continental rocks is too small. It is believed that volcanoes and perhaps hydrothermal vents were the original source of chloride ions. In the present world ocean, chloride ions are "recycled" when they go airborne as salt particles (aerosols produced by breaking waves) and return to the ocean in the form of rainwater. Thus, their rate of disappearance from seawater is low. Chloride and other conservative elements remain in seawater much longer than nonconservative elements.

Hydrothermal Vents: A Source and a Sink

Hydrothermal vents are a recent entry in the ledger of processes that control the concentration of dissolved elements in the ocean. The low concentration of magnesium in seawater relative to its supply in river water was a puzzle until the discovery of hydrothermal vents. Oceanographers now know that hydrothermal vents act as both a source and a sink of dissolved elements (see Figure 4-20). Hydrothermal systems have been shown to serve as a source of some elements (e.g., manganese, iron, lithium, rubidium, and cesium) on a scale equivalent to terrestrial sources. At the same time, hydrothermal vents have been shown to remove other elements (e.g., sulfate and magnesium). While chemical oceanographers generally regard hydrothermal vents as a minor contributor to the overall chemistry of the world ocean, continued studies of these fascinating structures may reveal new surprises in the future.

Residence Times of Elements

When sources and sinks are in balance, the ocean is said to be at **steady-state.** An equilibrium exists between rates of addition and removal of elements. At steady-state, we can define the amount of time it takes for the total mass of an element in the ocean to be replaced. One way to express this replacement time is **residence time,** the average amount of time an element or compound spends in a steady-state ocean (i.e., how long it resides in the ocean). Formally, scientists define residence time as follows:

Residence time = mass of element present/rate of input or removal

Consider the residence time of sodium in seawater. Its total input from rivers is about 1.81×10^{11} kg per year (2.05×10^8 tons of sodium per year). The total mass present in the ocean is about 1.33×10^{19} kg (1.47×10^{16} tons). Therefore, its residence time is:

1.47×10^{16} tons/2.05×10^8 tons yr^{-1}, or 70 million years!

In other words, it takes 70 million years to replenish the supply of sodium present in the ocean. Because the concentration of sodium in the ocean is constant, we have to assume the total amount of sodium in the world ocean is removed every 70 million years. Sodium and chloride have some of the longest residence times in the ocean, which accounts, in part, for their high concentrations in seawater. In general, the shorter the residence time, the higher the reactivity of an element in chemical or biological processes.

TABLE 6.2   Average Quantities of Common Elements in Continental Crust and River Water (In Order of Their Abundance in Seawater)

Name of element (Symbol)	Continental crust form of element, concentration (mol ton^{-1})	River water form of element, concentration (mmol m^{-3})
Chloride (Cl)	Cl, 13.3	Cl$^-$, 233
Sodium (Na)	Na, 1027	Na$^+$, 313
Magnesium (Mg)	Mg, 905	Mg^{++}, 150
Sulfur (S)	S, 21.7	SO$_4^-$, 120
Calcium (Ca)	Ca, 961	Ca^{++}, 367
Potassium (K)	K, 547	K$^+$, 36
Carbon (C)	Many forms, 166	HCO$_3^-$, 869
Silica (Si)	Si, 10,254	SiO$_2$, 173

Source: Pilson, 1998.

Carbon Dioxide and the World Ocean

The world ocean stands center stage in the scientific debate over whether human-released carbon dioxide, or **anthropogenic CO₂**, is changing Earth's climate. As discussed in Chapters 1 and 2, carbon dioxide traps heat in Earth's atmosphere through the greenhouse effect. Because the world ocean exchanges carbon dioxide with the atmosphere, an understanding of the chemistry of carbon in the world ocean is central to determining the effects of human activities on global climate (see Spotlight 6.2).

The Marine Carbonate System and pH

Very little carbon dioxide exists as a gas in seawater. Nearly all of carbon dioxide in seawater is bound in chemical form as part of the **marine carbonate system** (**Figure 6-16**). In an aqueous solution such as the ocean, carbon dioxide reacts immediately with water to form **carbonic acid,** essentially, CO_2 in solution or H_2CO_3. Carbonic acid quickly dissociates into **bicarbonate** (HCO_3^-) and **carbonate** anions (CO_3^{2-}). Most carbon dioxide in the ocean exists as bicarbonate. These reactions are summarized by the following equations:

$$H_2O + CO_2 \leftrightarrow H_2CO_3 \text{ (carbonic acid)}$$

$$H_2CO_3 \leftrightarrow H^+ + HCO_3^- \text{ (bicarbonate)}$$

$$HCO_3^- \leftrightarrow 2H^+ + CO_3^{2-} \text{ (carbonate)}$$

The double-ended arrows indicate that these reactions can proceed in either direction. Under some conditions, the formation of carbonic acid is favored. Under other conditions, carbonate is favored. The condition that most affects the equilibrium state for the marine carbonate system is **pH,** the degree of acidity or alkalinity of a substance. Chemists define an **acid** as any substance that donates a proton when dissolved in water and a **base** as any substance that accepts a proton when dissolved in water. Acids generally taste citrusy or sour, like oranges or lemons, whereas bases taste bitter, like caffeine tablets. Common acids include carbonated soft drinks, vinegar, and lemon juice. Common bases include seawater, ammonia, and lye. The **pH scale** indicates the strength of an acid or base according to its propensity to give up or accept a proton. (**Figure 6-17**). The marine carbonate system acts as a **buffer** to reduce variations in pH in the ocean. The narrow range of pH found in the ocean—from 7.9 to 8.4—can be attributed to this buffering system.

Anthropogenic Carbon Dioxide

The burning of fossil fuels and other carbon-rich compounds combined with widespread deforestation over the past two centuries has added significant amounts of carbon dioxide to the atmosphere. Atmospheric carbon dioxide concentrations have risen from 270 parts per billion in 1800 to 380 parts per billion in 2004, an increase of 30%. Continuous measurements of atmospheric CO_2 concentra-

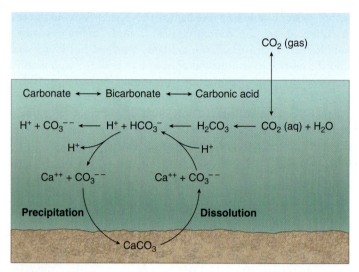

■ **FIGURE 6-16** The carbonate system in seawater. The concentrations of any one of the "species" in the marine carbonate system are pH dependent. At high pH (above 9), the equilibrium shifts towards the right, favoring carbonate ions. At low pH (below 6), the equilibrium favors the left-hand products, namely CO2. At the intermediate pH of the world ocean (~8.2), most of the dissolved carbon dioxide exists as bicarbonate.

tions were begun in 1958 by David Keeling of Scripps Institution of Oceanography at the Mauna Loa Observatory on the Big Island of Hawaii. The **Keeling curve** (**Figure 6-18**) represents the longest record of atmospheric carbon dioxide and the most widely cited graph of atmospheric carbon dioxide in existence. The annual cycle of change in CO_2 shown by this graph reflects the Northern Hemisphere seasonal cycle in which plants consume CO_2 via photosynthesis in warm seasons and plants and animals release CO_2

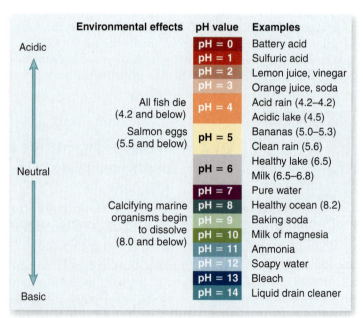

■ **FIGURE 6-17** The pH scale with examples and environmental effects for aquatic organisms. The term pH stands for the "power" (in terms of exponents) of the negative logarithm (log) of the hydrogen concentration of a substance, or formally: pH = −log [H+]. The pH is given as the negative log of the concentration of the hydrogen ion because it simplifies the scale to values between zero (0) and fourteen (14) (data from EPA).

Increasing CO₂ in the World Ocean

Efforts to determine the sink for anthropogenic CO_2 naturally focused on the world ocean. However, early efforts were hampered by a lack of sufficient methods for quantifying dissolved inorganic CO_2. Technological advances, scientific coooperation, and major funding enabled two international research programs, the Joint Global Ocean Flux Study (1988–1998) and the World Ocean Circulation Experiment (1990–2002), to undertake efforts to estimate the oceanic storage of CO_2. Their studies, published in 2004, revealed that the world ocean could account for approximately 48% of the carbon dioxide released through human activities (**Figure 6a**). The greatest storage occurred in deep waters in the North Atlantic and intermediate waters in the Atlantic and Southern Oceans.

The increase of CO_2 concentrations in the atmosphere has resulted in an increase in CO_2 concentrations in the world ocean. At the same time, oceanographers have observed an increase in the acidity of surface waters of about 0.4 pH units. This shift in pH results from the carbonate buffering system, which favors an equilibrium toward the right-side of the equation (i.e., formation of carbonic acid and bicarbonate) and the release of protons (H^+). This lowering of pH in oceanic waters has important implications for biological processes, including the ability of marine organisms to form carbonate skeletons. Ocean acidification levels predicted for the year 2100 have been shown experimentally to dissolve the shells of calcium-secreting organisms. Increased oceanic pH also alters the depth of the calcium carbonate compensation depth through a shallowing of the lysocline. The resultant increase in dissolution rates of carbonate sediments would lead to additional increases in dissolved CO_2 in the ocean and a further lowering of pH. While these consequences are complex and not completely understood, increasing CO_2 concentrations in the world ocean may have far-reaching effects on ocean chemistry and biology. Future studies will enable scientists to better determine the kinds of actions that may be needed to mitigate the negative consequences of anthropogenic CO_2 on the world ocean.

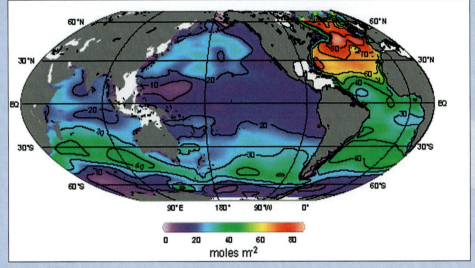

moles m²

FIGURE 6a Water-column integrated anthropogenic CO2 concentrations in the world ocean. Source: NOAA.

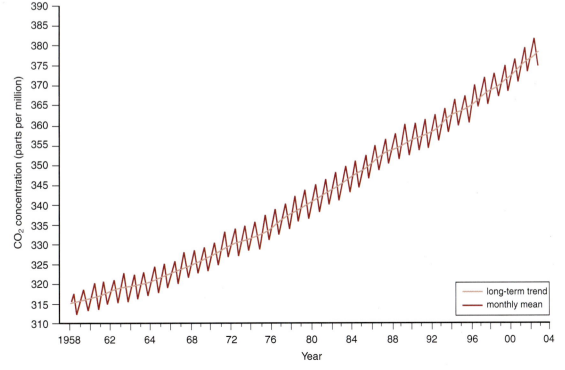

■ **FIGURE 6-18** Monthly average atmospheric carbon dioxide concentration at Mauna Loa Observatory, Hawaii, from the period 1958–2004. The "sawtooth" appearance of the monthly trends can be attributed to the seasonal cycle of photosynthesis and respiration in the northern hemisphere, what some scientists call "the breathing of the Earth". The long-term trend represents the increase in atmospheric carbon dioxide due to the burning of fossil fuels and other human activities (data from Keeling and Wharf, 2005).

through respiration in cold seasons. The most dramatic feature of the time series is the rapid increase in the rate at which CO_2 is being added to the atmosphere in recent years.

The fate of anthropogenic CO_2 once it enters the atmosphere is uncertain. Estimates of the sources and sinks of CO_2 in the atmosphere reveal a discrepancy of nearly 2 gigatons (Gt) of carbon. In other words, there is less carbon in the atmosphere than what should be present, given what we know about the sources and sinks of atmospheric carbon. The **missing carbon problem** represents an additional puzzle in the search to understand the impact of human activities on climate.

The Global Carbon Cycle

Estimation of the sources and sinks of carbon to the atmosphere requires an understanding of the **carbon cycle,** the exchange of carbon between various carbon reservoirs through the action of biogeochemical processes (**Figure 6-19**). Most of the carbon on our planet is stored in sedimentary rocks, such as limestone and dolomite (**Table 6.3**). The second largest carbon reservoir exists as organic carbon buried in sediments. Carbon dioxide dissolved in the world ocean represents the third largest reservoir of carbon on Earth, more than fossil fuels, humic substances, the atmosphere, and living organisms combined. But the reservoirs only tell part of the story. The residence time of carbon in a particular reservoir is equally important (**Figure 6-20**). Geologically speaking, the residence time of organic carbon stored as fossil fuels is on the order of thousands of years.

Small changes in a reservoir may have profound effects and large reservoirs may exert a disproportionate influence over small reservoirs. Thus, scientists are keenly interested in processes that move carbon from one reservoir to the other. However, the temporal scales over which these processes operate range from months to hundreds of thousands of years (see inside front cover). For that

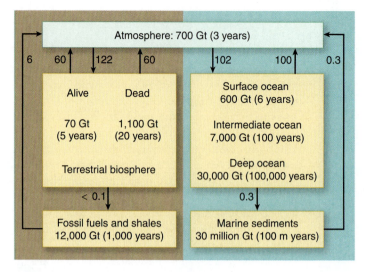

■ **FIGURE 6-20** Reservoirs and residence times for carbon storage on Earth. Because the atmosphere is a relatively small reservoir, it may be easily influenced by large reservoirs. Source: Royal Society, 2005.

reason, scientists recognize three different carbon cycles based on the time scales over which they operate. The **short-term organic carbon cycle** includes seasonal interactions between the atmosphere and the biosphere. This is the cycle observed as seasonal changes in atmospheric CO_2, the sawtooth pattern in the Keeling curve. The **long-term organic carbon cycle** refers to processes that lead to the formation of fossil fuels, that is, the burial of organic carbon in sediments. The extraction of fossil fuels by humans and the release of this stored carbon through combustion is represented by the solid line in the Keeling curve, depicting the increase in atmospheric CO_2 over short- and long-term organic carbon cycles (decades to thousands of years). The **long-term inorganic carbon cycle** (**Figure 6-21**) dictates the storage of carbon in rocks, a process that involves

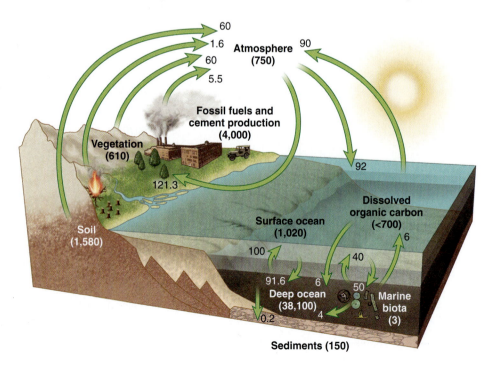

■ **FIGURE 6-19** Scientists recognize different time scales over which the organic carbon cycle operates. Photosynthesis and repiration by organisms drive the short-term carbon cycle. The formation and burning of fossil fuels over tens to thousands of years represents the long-term organic carbon cycle.

TABLE 6.3   Total Carbon in Various Earth Reservoirs

	Total mass of carbon in units of	
	Gt (10^9 tons)	Pmol (10^{15} mol)
Limestone and dolomite (carbonates)	82,370,000	6,860,000
Organic carbon in sediments	15,500,000	1,120,000
Ocean-dissolved CO_2	36,600	3,050
Recoverable fossil fuel	6,500	542
Humic substances in soils and seawater	2,400	197
Atmospheric CO_2	781	65
Living organisms	560	47

Source: Pilson, 1998.

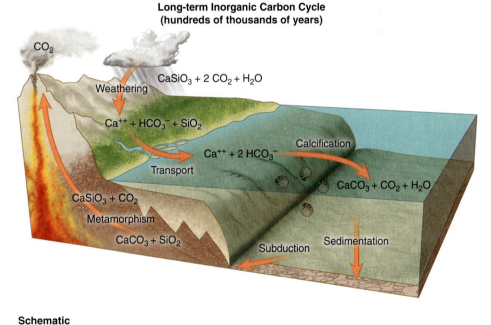

**Long-term Inorganic Carbon Cycle
(hundreds of thousands of years)**

■ **FIGURE 6-21** The long-term inorganic carbon cycle is closely tied to the silica cycle. The carbonate-silicate cycle begins with the dissolution of atmospheric carbon dioxide in rainwater. The resultant carbonic acid causes the weathering of silicate rocks, which combine to form calcium, bicarbonate, and dissolved silica. These silicate weathering products are subsequently transported into rivers and the world ocean. Calcification of carbonate and silica occurs through the activities of marine organisms, like foraminifera. Eventually, some fraction of the carbonates and silicates of the skeletons of these organisms are deposited on the seafloor. For the most part, those deposited in shallow waters above the depth of the lysocline harden into deposits of limestone and other carbonate sediments. Through the action of seafloor spreading, carbonate deposits on the seafloor subduct into oceanic trenches. High temperatures and pressures in subduction zones generate conditions that cause recombination of carbonate and silicate minerals (into quartz) and a release of carbon dioxide. Volcanoes and other conduits for mantle gases release carbon dioxide back into the atmosphere, completing the cycle. The carbonate-silicate cycle controls Earth's climate over time scales of hundreds of thousands of years. Platetectonics plays a major role in this cycle.

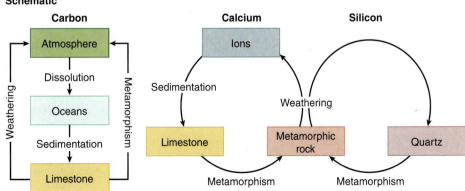

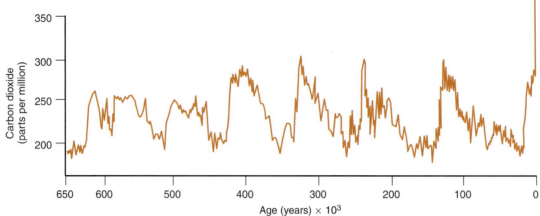

■ **FIGURE 6-22** Concentration of atmospheric carbon dioxide over the past 650,000 years as determined from measurements of gases in ice cores. These fluctuations represent the long-term inorganic carbon cycle.

biological processes such as the formation of biogeneous sediments (see Chapter 5). The long-term inorganic cycle is best represented by measurements of CO_2 trapped in ice cores recording hundreds of thousands of years (**Figure 6-22**). Collectively, all three parts of the global carbon cycle interact and influence atmospheric concentrations of CO_2, albeit over different time scales.

As we encounter this topic in later chapters, the following summary may be helpful:

- Although most of Earth's carbon resides in rocks, the world ocean holds a greater supply of carbon dioxide than the atmosphere; thus, small changes in oceanic concentrations may have large effects on atmospheric concentrations.

- Carbon dioxide in the atmosphere influences Earth's climate via the greenhouse effect.

- Carbon dioxide and oxygen undergo rapid changes as a result of biological activity, namely, photosynthesis, which absorbs carbon dioxide and releases oxygen, and respiration, which consumes oxygen and releases carbon dioxide.

- The storage of carbon in the ocean can only be understood by quantifying the production, sedimentation, and burial of plant material.

- Humans have been adding carbon dioxide to the atmosphere for more than 100 years.

- The world ocean has absorbed 48% of the anthropogenic carbon dioxide, increasing ocean acidity (lowering the pH).

YOU Might Wonder

1. **Just how much gold is dissolved in the ocean?**
   The average concentration of gold is 4×10^{-6} mgL^{-1} of seawater, roughly 1 mg of gold for every ton of seawater. You would need to process about 57 million pounds of seawater to get a single ounce of gold. Needless to say, the cost of extraction far outweighs the worth of the gold. Still, the 25 billion ounces of gold that exist in the entire world ocean keep this question alive in the minds of gold seekers.

2. **Why does the ocean smell like, well, the ocean?**
   That fresh (or pungent, depending on your point of view) smell that you associate with the ocean comes from a group of chemicals called aerosols. Sea spray from breaking waves and bursting bubbles produces small particles (about 1 μm) that contain salts. As it turns out, these aerosols are very important for cloud formation and for Earth's heat budget. Aerosols and their properties is an active area of scientific research.

3. **How can I make sure that I am not polluting the ocean with chemicals I use while fertilizing and spraying my plants and lawn?**
   The single most important thing you can do is never to introduce fertilizers directly into a lake, waterway, or sewer. Careful and repeated applications of small amounts of fertilizers are better than indiscriminate applications of large amounts of fertilizers. Time-release fertilizers and natural organic fertilizers that slowly release nitrogen and phosphorus compounds also benefit your plants and maintain healthy waterways. It's also important to eliminate or reduce the use of detergents that contain phosphorus. Washing your car or boat with "nutrient-rich" cleaners and letting the water drain into a sewer or the ocean negatively impacts our waterways. The US Environmental Protection Agency distributes free documents on the responsible use of fertilizers and pesticides in lawn care and landscaping. Visit their website at www.epa.gov.

Key Concepts

- Major constituents remain in relatively constant proportions, regardless of the salinity.
- Minor constituents, though present at low concentrations, may be limiting to marine organisms or play an important role in chemical processes.
- The biologically required element whose supply is least available and that limits the growth of organisms is called a limiting nutrient.
- Nitrogen is typically a limiting nutrient for phytoplankton in the ocean, although phosphorus and iron can be limiting under some circumstances at certain locations.
- Hydrothermal vents likely played a major role in controlling the concentration of elements in Earth's early ocean. They continue to control the abundance of certain elements in the ocean today.
- The sources and sinks of most salts appear to be in balance globally; thus, the salinity of the ocean remains fairly constant. The marine carbonate system buffers changes in pH in the ocean.
- A major sink for carbon in the ocean is the formation of carbonaceous sediments from the shells of marine organisms supplied to the seafloor.
- The ocean holds a greater supply of carbon dioxide than the atmosphere.

Terms to Remember

Critical Thinking

1. Contrast the short- *and* long-term effects of increased precipitation versus prolonged dry climatic conditions on the salinity of a particular geographic region of the ocean. Under what conditions might wet periods increase the local salinity? Discuss the importance of sources versus sinks in terms of the residence times of elements under varying climate conditions.

2. Design a study in your neighborhood that examines the impacts of a human activity (pick one) on vegetation in a park, lake, forest, or coastal ocean. Discuss your strategy for establishing a baseline from which the effects of your chosen human activity can be measured.

3. You have just invented an instrument that measures the sweetness of candy by the sound it makes when placed in your machine. Make a checklist of all of the tasks that must be completed to ensure that your machine gives accurate readings of the glucose concentration of candy when you take it on its first trial run to the local candy store. What factors might need to be considered to ensure that your sweet machine works properly in different candy stores and at different times of day?

4. Pick a conservative element and sketch on a piece of paper its path from land to sea and back again. Do the same thing for a nonconservative element. Compare and contrast the factors that govern conservative and nonconservative elements and their residence times in the ocean.

5. Write a paper that proposes one possible remedy for slowing down or reversing the release of anthropogenic CO_2. Include in your paper a discussion of the feasibility and benefits of the remedy as well as the challenges and potential drawbacks.

Explore Online

 Visit www.mhhe.com/chamberlin1e for access to chapter quizzing, key term flash cards, video clips, interactive activities, and more. Further enhance your knowledge with web links to chapter-related material!

Exploration Activity 6-1

Exploring the Carbon Cycle Through Simple Models

Summary

An understanding of the carbon cycle is fundamental to understanding human-caused climate change, the structure and functioning of biogeochemical cycles, and the productivity of the world ocean. In Chapter 6, we introduced the global carbon cycle and differentiated between time scales (short- or long-term) and species of carbon (organic versus inorganic). Here we explore conceptual, physical, and mathematical models that will help you grasp the distinctions between the two organic carbon cycles.

Learning Objectives

By the end of this activity, you should be able to:

- Categorize the various reservoirs of carbon on Earth
- Identify the pathways of exchange between the various carbon reservoirs
- Rank the size of reservoirs and their rates of exchange pathways
- Generate simple conceptual, physical, and mathematical models that illustrate different parts of the carbon cycle
- Summarize the distinctions between the short- and long-term organic carbon cycle

Materials and Preparation

This activity may be completed without a computer, however, a spreadsheet program will greatly facilitate the formulation and solution of simple mathematical models.

Part I: Building Simple Conceptual Models

Conceptual models are visual constructs of concepts and ideas. Fundamentally, they provide a means for organizing your thoughts, brainstorming ideas, and framing hypotheses for testing. A simple "to-do" list is a conceptual model of your activities throughout the day. Doodles of a design for an aircraft, a work of art, or a fashion accessory are conceptual models. Typically, they are the first step in the construction of more elaborate physical or mathematical models.

Figures 6-19 to 6-21 represent different types of conceptual models of the carbon cycle. Take a few moments to examine them closely. All of them include reservoirs, storage compartments for carbon, and exchanges, pathways by which carbon moves from one reservoir to another. Figure 6-19 illustrates the organic carbon cycle and lists the size of the reservoirs (in parentheses) and the annual rates of exchange between them (the numbers associated with the arrows). Figure 6-20 is a box model (a type of conceptual model useful for building mathematical models) that depicts the size of the reservoirs and the residence time (see section on residence time in Chapter 6) for carbon in each of them. Figure 6-21 does not include numbers but rather focuses on defining the processes involved in the long-term inorganic carbon cycle. Understandably, these models simplify the world considerably! But they help scientists (and us!) to identify the key components and processes that make up the carbon cycle.

As a first step towards creating your own conceptual models, create a simple box model based on this quote from Jacob Bronowski (1908–1974):

> "You will die but the carbon will not; its career does not end with you. It will return to the soil, and there a plant may take it up again in time, sending it once more on a cycle of plant and animal life." ("Biography of an Atom—And the Universe" *NY Times* 13 October 1968)

1. Label the boxes below with the proper carbon reservoirs based on the quote and your knowledge of the carbon cycle. Label the connecting arrows with the appropriate processes.

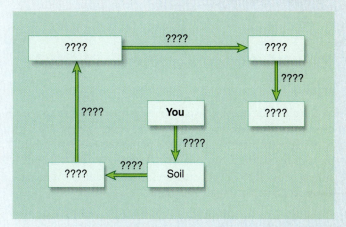

2. Answer the following:
   a. Which "reservoir" of the carbon cycle do you belong to?
   b. Did you identify bacteria and decomposition in your model? (If not, do so.)
   c. How would you make this diagram a "closed" cycle?
   d. Why are some of the elements of the above diagram "missing" from Figure 6-19?

Part II: Conceptualizing Your Local Carbon Cycle

In this part, you will build a more detailed conceptual model of the carbon cycle and your interaction with it. Be sure to use all the resources of this textbook, your instructor, and the Internet to guide you. There is abundant material on the carbon cycle available, but you may have to sift through it carefully.

1. Brainstorm and research the various reservoirs and pathways of exchange for carbon in your local neighborhood. Represent your local carbon cycle in the form of a simple conceptual box model that links to the model in Figure 6-19. Include as many reservoirs and pathways as you can think of. You may also include as few or as many elements of Figure 6-19 as you like. Alternatively, you may wish to simply represent Figure 6-19 as a single box that connects to your neighborhood.

2. Rank your reservoirs by size. Label the biggest reservoir with a "1" and so on. Also rank your rates of exchange from fastest to slowest.

3. Write a detailed description of the path of a carbon molecule from the atmosphere to your home and back to the atmosphere.

4. Identify and describe which reservoirs and pathways in your local carbon cycle belong to the short-term organic carbon cycle and which ones belong to the long-term organic carbon cycle.

Part III: Building a Simple Physical Model

When all of the reservoirs and pathways are taken into account, even a local carbon cycle can become quite complex! That's why scientists often simplify their models or focus on specific parts of the model at any one time. In this part, we will build a simple physical model to better understand the flows of materials in and out of a single reservoir.

1. Let's start by considering a bathtub as a physical model of a reservoir. In this case, the bathtub may represent a lake, an ocean basin, or a storage compartment for carbon (or any other element or compound we choose). The drain hole represents a sink or outflow for our reservoir and the faucet represents a source or inflow. Answer the following questions:

a. If the volume of the bathtub is 40 gallons, how long would it take to fill the bathtub if your faucet delivers water at a rate of 5 gallons per minute and the drain hole is completely plugged?

b. If you fill the bathtub half full and pull the plug on the drain while the faucet is still running, how fast (in gallons per minute), would the bathtub need to drain to keep the water in the bathtub at the same level (i.e., half full)?

c. Describe how you would determine the rate at which your bathtub fills and drains?

2. Construct a two-reservoir system (with inflows and outflows) by connecting two plastic soda bottles (Use any other materials you might need, such as tape, a knife, straws, funnels, a ring stand. The design and construction of the model is part of the learning process).

a. Use measuring cups or graduated cylinders to determine the outflow rate (in liters per minute) for each of the reservoirs. Depending on how you constructed the drain holes for each bottle, these rates could be different.

b. What is the inflow rate for the second bottle?

c. Predict the quantity of water you must add to the system (in liters per minute) to achieve steady state (where inflow = outflow). Try it experimentally and describe what happens. Is it possible to achieve steady state with your system? Why or why not?

d. What happens to the water levels in your reservoirs (the soda bottles) when the rate at which you add water (inflow) exceeds the rate at which water leaves the system (outflow)? Predict how fast the reservoirs will fill if you double the inflow rate. For example, if your first reservoir is half full (say, 1 liter) and you double the steady-state rate (determined experimentally above), how fast will the water rise (in liters per minute) in both bottles? Check it out experimentally.

Part IV: Building a Simple Mathematical Model

The models you just built make an ideal starting point for developing a set of mathematical equations to quantify the rates of movements of carbon between various reservoirs. Such models can be used to test hypotheses concerning the effects of changes in those rates on the sizes of the reservoirs. For example, increases in the rate of burning of fossil fuels have led to increases in the amount of carbon in the atmosphere. Carbon and climate models are an active area of oceanographic research and a difficult one, as you might imagine. Fortunately, we can gain an appreciation for these models by constructing a simple mathematical model based on your physical model of two reservoirs. You may wish to use a spreadsheet program to simplify your calculations and create a model whose output you can graph.

We can write a simple equation that describes the rate of inflow and rate of outflow for the entire two-reservoir system at steady state as

Total Inflow (TI) = Total Outflow (TO), or TI = TO

If we want to know the total quantity of water in the system at any given time, we can write

Total Volume (TV) = Start Volume (SV) + Total
Inflow (TI) – Total Outflow (TO)

For our two-reservoir system, we need a set of equations that describes each reservoir individually. Or

Total Volume Reservoir 1 (TVR1) = Start Volume
Reservoir 1 (SVR1) + Inflow Reservoir 1 (IR1) – Outflow
Reservoir 1 (OR1)

Total Volume Reservoir 2 (TVR2) = Start Volume
Reservoir 2 (SVR2) + Inflow Reservoir 2 (IR2) – Outflow
Reservoir 2 (OR2)

Complete the following:

1. How can we simplify the equation for Inflow Reservoir 2 (IR2)? (Hint: Where does the water come from?)
2. Write an equation for the Total Volume when both reservoirs are half full and there is no water entering or leaving the system.
3. Write an equation that describes the rate of flow for each reservoir individually.
4. Write an equation that describes the total rate of flow through the system when the two reservoirs are identical in volume and their outflow rates are the same.
5. Use some of the numbers you found experimentally to calculate volumes in the half-full reservoirs when you cut the inflow in half. Calculate how long it will take for each reservoir to drain completely under these conditions. Repeat the exercise for a rate that is three times the steady state rate and calculate how long it will take the reservoirs to fill.

Part V: Thinking About Models

Oceanographers may spend years developing and refining their models. Nonetheless, your simple models should already have provided some insights into the carbon cycle. Use the following questions to help you synthesize what you have just learned.

1. Compare and contrast your physical model with your conceptual model. How might you use physical models to test hypotheses about reservoir sizes and rates of exchange in models of the carbon cycle? Propose a set of observations, measurements, and experiments (using a more realistic physical model of the carbon cycle than the one you constructed here) to determine the size and rates of exchange of carbon in your "local" carbon cycle.
2. Compare and contrast your physical model with your mathematical model. What kinds of assumptions did we make in formulating our mathematical model? (Hint: did you pour water into the system at a steady rate or intermittently?) How might these assumptions affect predictions of system performance based on your mathematical model? On the other hand, what are the advantages of a mathematical model as opposed to a physical model (or conceptual model)? How might you improve your mathematical model or expand it to include more reservoirs?
3. Redraw Figure 6-19 as a box model that includes your local carbon cycle model. Then redraw Figure 6-21 as a box model that includes Figure 6-19 and your local model. Write a short paper that highlights the similarities and differences between your model and Figures 6-19 and 6-21. Why are each of these models important (including yours) and how do they help us understand different aspects of the carbon cycle?
4. Use the Rubric for Physical and Conceptual Models to self-assess and refine your work. (See Exploration Activity 2-1.)

7 Ocean Physics

Knowledge of ocean physics is essential to crossing the ocean safely in a ship. It's also fundamental to understanding geological, chemical, and biological oceanography. At its core, oceanography is a physical science.

Questions to Consider

1. In what ways has scientific progress in physical oceanography depended on the development of technology for measuring seawater temperatures?

2. How do the physical properties of seawater influence the heat budget of the world ocean?

3. What kinds of natural phenomena might raise the temperature of Earth and what kinds might lower it?

4. How does the physical structure of the upper ocean change as a result of seasonal variations in heating at midlatitudes, tropical latitudes, and high latitudes?

5. What evidence supports the contention that human activities are warming the world ocean?

An Ocean of Energy

Oceanography as a discipline has its roots in physics. **Physical oceanography,** the study of the physics of the world ocean, provides a framework for interpreting and understanding an enormous variety of geological, physical, chemical, and biological phenomena. Physical oceanography involves the study of the exchanges of energy and matter in the world ocean. Observations and measurements of the physics of the world ocean help oceanographers to understand a myriad of phenomena: how the ocean heats and cools, how energy and momentum move between the atmosphere and the ocean, how the ocean moves and circulates, how light and sound propagate, and how marine organisms behave, reproduce, and survive in the oceanic environment. Physical oceanography provides fundamental information for atmospheric studies, including weather and global climate change. It also contributes to a broad range of engineering, economic, military, and recreational interests.

In Chapter 6, we explored the properties, transformations, and exchanges of *matter* in the world ocean. In this chapter, we focus on the properties, transformations, and exchanges of *energy,* especially heat. Energy drives the movements and exchanges of matter. **Table 7.1** lists the kinds of energy important to oceanographers. Your knowledge of ocean physics will prove very useful in our chapters on ocean-atmosphere interactions, circulation, waves, and productivity. It is also essential to appreciating temporal and spatial variability in the world ocean (see inside front cover). Though the concepts may seem difficult at first, you may find that many of the physical processes described in this chapter occur in your daily life. Use your experiences of the natural world to guide you through this chapter.

TABLE 7.1   Forms of Energy Important to Oceanographers

The **law of conservation of energy** states that energy can be neither created nor destroyed, but it can change forms. Thus, one form of energy may, in principle, be converted to another. For example, when the Sun heats Earth, solar energy is converted to thermal energy. When a rock slide creates waves in a lake or the ocean, potential energy is converted to kinetic energy.

Form of energy	Definition	Example
Radiant or **solar energy**	Energy in electromagnetic radiation that comes from the Sun, important in heating of Earth, the ocean, and the atmosphere, and essential to photosynthesis	Sunlight
Kinetic energy	Energy of motion of moving objects ($= \frac{1}{2} \times$ mass $\times$ velocity^2)	Movement of wind across the ocean surface or collisions of water molecules in a glass
Thermal energy or **heat**	Energy of random motion of molecules, which results in the exchange of heat between objects	Ocean warming the overlying air or vice versa
Chemical energy	Energy stored within the chemical bonds of substances	Chemical reactions that occur within the cells of marine organisms
Potential energy	Energy related to the position of objects or the position of atoms in molecules	A rock sitting on top of a ledge or the spacing between water molecules

A Brief History of Thermometry

Seemingly, no instrument for measuring physical properties could be simpler than a thermometer. Yet construction of a thermometer rugged enough for oceangoing work would prove difficult for early physical oceanographers. In 1714, an ambitious meteorological instrument maker, Daniel Fahrenheit (1686–1736), built the **Fahrenheit thermometer.** Fahrenheit constructed a glass tube filled with mercury and used a mixture of ice, water, and sea salts to set the lower end point of his thermometer scale, 0 degrees Fahrenheit (°F), and the temperature of a human armpit (or mouth) for the upper end point, 96°F. French naturalist Antoine Ferchault de Réaumur (1683–1757) published his own experiments on thermometry in 1730. The Réaumer scale was widely adopted and remained in use until the late nineteenth century. Swedish astronomer Anders Celsius (1701–1744) invented in 1740 the centigrade scale for the mercury-based thermometer. Oddly enough, he set 100 degrees Celsius (°C) as the freezing point of water and

0°C as the boiling point. After his death, scientists inverted the centigrade scale (setting 0°C as freezing) and named it the **Celsius scale** in his honor. The Celsius scale supplanted the Réaumer scale. Another important temperature scale is the **Kelvin scale,** invented by Lord Kelvin in the 1800s and used by physicists working at temperatures beyond the natural range. This scale sets zero as the temperature at which all molecular motions reach a minimum (also called absolute zero). Fahrenheit, Celsius, and Kelvin scales remain in use in modern times (**Figure 7-1**).

Armed with these early thermometers, seagoing captains and oceanographers began to take measurements of sea surface temperatures using various types of buckets to sample surface waters. In 1751, Captain Henry Ellis (1721–1806) of the British slave trader *Earl of Halifax* took the very first temperature measurements of deep waters in the subtropical North Atlantic. The idea had been suggested to him by a British clergyman, the Reverend Stephen Hales (1677–1761), who provided Ellis with a sampling apparatus, essentially a modified bucket. The bucket was outfitted with flaps that opened on descent and closed on ascent, thereby trapping a sample of seawater at the depth to which it was lowered. Using this bucket and a Fahrenheit thermometer,

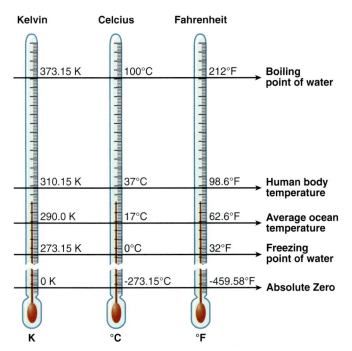

Kelvin   Celcius   Fahrenheit

373.15 K	100°C	212°F	→ Boiling point of water
310.15 K	37°C	98.6°F	→ Human body temperature
290.0 K	17°C	62.6°F	→ Average ocean temperature
273.15 K	0°C	32°F	→ Freezing point of water
0 K	-273.15°C	-459.58°F	→ Absolute Zero

K   °C   °F

■ **FIGURE 7-1** A comparison of the Fahrenheit (F), Celsius (C), and Kelvin (K) scales for measuring temperature.

Ellis performed a series of temperature measurements from depths between 109–1629 m (360–5346 feet). To his surprise, the water from 1188 m (3900 feet) and deeper was uniformly cold, 53°F, by his measure. In a letter to Hales, Ellis noted that the experiments seemed "food for curiosity" at first. He soon realized, however, that the cold water could be used to cool wine and drinking water onboard and to provide a cool bath while in the tropics, for which he was most grateful. The implications of his discovery for oceanography—that the ocean was layered according to temperature—completely escaped him, as it did most scientists of the time.

In 1874, the **reversing thermometer** was invented for measuring the temperature of seawater at depth. Attached to a line, the reversing thermometer turned upside down on a reversing frame when triggered by a brass weight called a **messenger.** The action of reversing cut off the supply of mercury and preserved the exact length of mercury before reversing. The length of mercury isolated in the tube corresponds to the temperature at depth. The reversing thermometer quickly became a mainstay of oceanographic research and can still be found in use aboard oceanographic vessels.

While improvements to thermometers (including the reversing thermometer) offered the needed precision, they only provided measurements at individual water depths. Discrete-depth temperature measurements often missed key features of the water column, a problem of **undersampling.** In 1937, Athelstan Spilhaus (1911–1998) invented the **mechanical bathythermograph,** a torpedo-like device about the size of a water bottle that could be

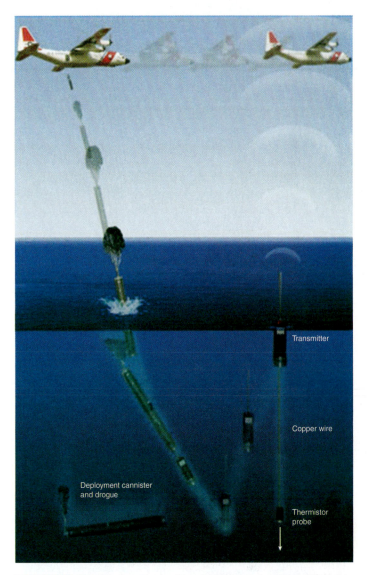

■ **FIGURE 7-2** Deployment of an AXBT or aircraft expendable bathythermograph. The AXBT allow for rapid assessment of the vertical distribution of seawater properties using aircraft, like this one used for the U.S. Coast Guard's International Ice Patrol.

attached to a line and thrown over the side to obtain a continuous record of temperature versus depth. The development of **thermistors,** electronic devices for measuring temperature, enabled the invention of the **expendable bathythermograph (XBT).** XBTs remain widely used aboard research and naval vessels. An airborne XBT, or AXBT, is deployed by hurricane-tracking aircraft or for oceanographic applications in remote regions (**Figure 7-2**).

The availability of thermistors and electronic devices for measuring conductivity and pressure led to the invention of the **salinity-temperature-depth instrument (STD)** by Australian oceanographers Neil Brown and Bruce Hamon in 1955. The STD design was later modified and called a **conductivity-temperature-depth**

a.

b.

■ **FIGURE 7-3** (a) The CTD, the "work horse" of modern ocean-ography. The ring of sampling bottles is called a rosette. They are shown here in their open position prior to lowering. (b) Once on board, water samples for various chemical and biological analy-ses and experiments are quickly taken.

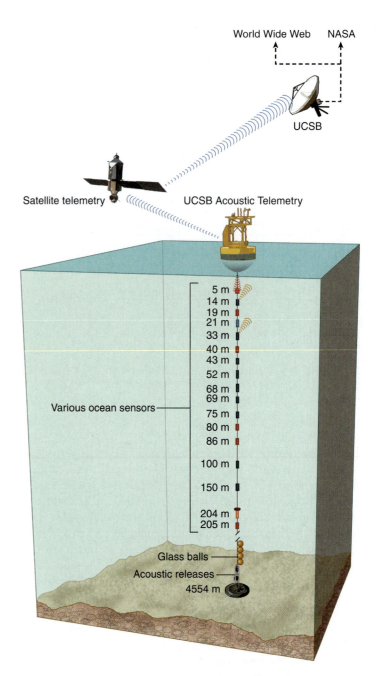

■ **FIGURE 7-4** Anatomy of the Bermuda Testbed Mooring. The cable of the mooring is attached to the seafloor by a large anchor (often the steel wheel of a railroad car) and held in a vertical position by a surface buoy. Attached to the cable at different depths are various sensors, such as thermistors, that can record and store data at preselected time intervals (seconds to hours). In this way, oceanographers obtain a near continuous record of ocean properties at several depths over periods of months.

instrument (CTD) (**Figure 7-3**). The CTD is one of the most essential instruments for shipboard oceanographic use today. In a typical configuration, it can be outfitted with additional sensors and electronically triggered sampling bottles, configured as a **rosette,** that allow an oceanographer to view a virtually continuous profile of water column properties before taking water samples.

Thermistors have also been placed in ships' intake pipes for continuous underway measurements of sea surface tem-peratures (SST) and on moored buoys, such as the Bermuda testbed mooring for recording seawater temperatures for long periods of time (**Figure 7-4**). With the advent of satellite-buoy telecommunications, moored and even drifting buoys have been

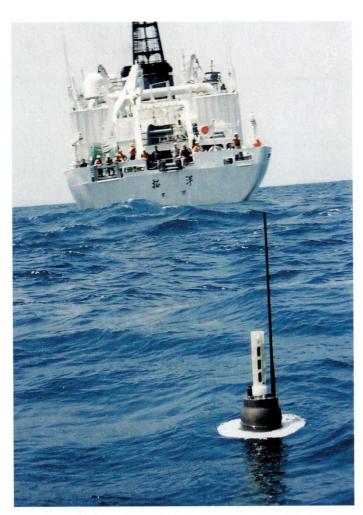

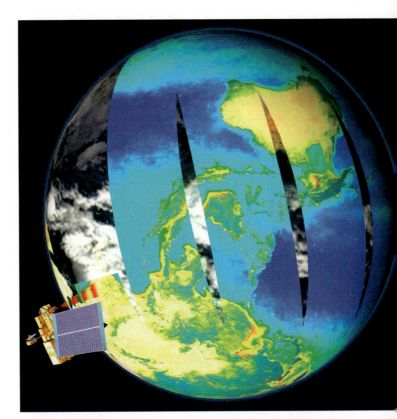

■ **FIGURE 7-6** Satellite determinations of sea surface temperature and other properties rely upon very precise, high resolution measurements of visible, near-infrared, and thermal radiation passively emitted from the surface of the ocean, land, and clouds. They provide information on weather, ocean conditions, fire-detection, land-use, and other Earth properties.

■ **FIGURE 7-5** The Japanese research vessel, Takuyo, deploys an Argo float. This vessel and others like it are part of an international effort to measure salinity and temperature of the upper 2000 meters of the world ocean. A major goal is to determine the effects of global warming on the upper ocean.

outfitted with electronic sensors to record and transmit near-continuous data on physical properties, including temperature. The **National Data Buoy Center** and the **TAO-Triton Buoy Array** obtain data on SST and other oceanographic and meteorological measurements and make the data available through the Internet. One of the most ambitious efforts to characterize ocean temperature and other properties is the **Argo system,** an array of 3000 robotic floats distributed throughout the world ocean. Equipped with a number of sensors on free-floating floats that move autonomously up and down in the water column, the Argo system provides near-continuous monitoring of the world ocean. While the float is at the surface, data are transmitted via satellite and made available in near-real time over the Internet (**Figure 7-5; see also figure 7-15**). The temporal and spatial resolution of vertical scales using this system is unprecedented in the history of oceanography.

The mid-1970s brought some of the first satellite measurements of sea surface temperature using infrared sensors aboard weather satellites (**Figures 7-6 and 7-7**). The vast improvement in spatial coverage brought quite a different picture of the world ocean compared to shipboard measurements. Satellite measurements of SST provided the first global picture of the world ocean and literally changed the way oceanographers view and think about ocean processes. Because satellites only measure the "skin" of the ocean, the upper 10 microns, temperature measurements based on satellite data must be calibrated against *in situ* or in-the-water measurements, such as those from buoys and ships. Satellites are a key sentinel in the measurement of ocean temperature and the monitoring of global climate change.

One of the latest tools for measuring the temperature structure of the ocean is **multichannel seismic profiling,** a technique normally used for mapping the seafloor (see Chapter 4). By carefully analyzing the seismic reflection profiles within the water column, oceanographers can detect layers of water with different physical and chemical properties. Multichannel seismic profiling represents an exciting opportunity for rapid, three-dimensional mapping of water column properties.

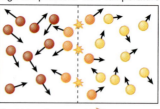
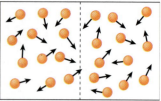

High temperature Low temperature   Thermal equilibrium

Heat transfer   Net heat transfer has ceased

■ **FIGURE 7-8** Heat transfer occurs when two systems have different temperatures. When the two systems reach thermal equilibrium, heat transfer stops and heat no longer exists.

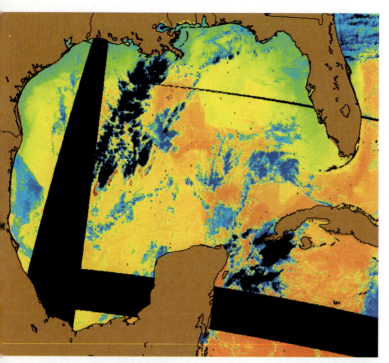

■ **FIGURE 7-7** Sea surface temperature on February 28, 2006, in the Gulf of Mexico from the MODIS instrument aboard Terra. Note that the image is a daily composite, a combination of a few to several measurements taken by the satellite. Images are often combined to remove the effects of clouds, which block measurements of SST. Even so, clouds remain "visible" in this image as black patches of missing data. The black rectangles represent a swath of missing data, possibly indicating an area not covered by the satellite on this day. This image is a good reminder of the many steps involved in the acquisition, processing, and verification of satellite measurements of SST.

When we think of heat, we imagine something warm, such as a stove or a fire. But ice cubes can also "emit" heat if their surroundings are colder than the freezing point of water. It is the temperature difference between two objects or an object and its surroundings that determines the quantity of heat that will flow between them. In fact, temperature differences *cause* heat. Defined in physical terms, ***heat*** *is the spontaneous transfer of energy from a high temperature system to a lower temperature system*. When the temperatures of the two objects or systems are the same, the transfer of heat stops and the two objects or systems can be said to have reached **thermal equilibrium** (**Figure 7-8**). Physicists emphasize that objects and surroundings do not possess heat: it is improper to say that an object contains heat. Heat only refers to the flow of energy, which itself is caused by temperature differences.

Physical Properties of Seawater

Physical properties involve traits of a substance that occur on an environmental (macroscopic) level without changing the identity or composition of the substance. The heat flow out of a pot of boiling water is a physical property measured by temperature. The **change of state** or phase change from a liquid to a solid or gas to a liquid alters the physical properties of water but not its chemical composition. Some physical properties of seawater are listed in **Table 7.2**. Many of the phenomena we discuss throughout this textbook relate to these properties. Note that nearly all of these physical properties relate to heat.

Specific Heat

The quantity of heat absorbed or lost by a unit mass of an object that results in a 1° C change in temperature is called the **specific**

A Few Physical Definitions

To avoid misconceptions over terminology, we define here two familiar terms, *temperature* and *heat*, that have precise scientific definitions that differ from their popular usage. Temperature is often thought of as "the degree of hotness or coldness" of an object or the environment. Indeed, the measurement of temperature by a thermometer relies on a scale of units, called degrees. For physicists, however, temperature refers to a specific measure of a particular type of energy found in objects or the environment. In physical terms, **temperature** is defined as a measure of the average kinetic energy of molecules (see Table 7.1). At higher temperatures, the average kinetic energy of molecules is higher and their motions are faster; at lower temperatures, the opposite is true.

TABLE 7.2   Physical Properties of Seawater

Property	Explanation	Effect on world ocean
High boiling point	Pure water boils at 100° C. If it behaved like other molecules of similar size, it should boil at $-60°C$.	Water exists primarily in a liquid state on our planet.
High freezing point	Pure water freezes at 0°C. This is a higher than predicted temperature based on its molecular properties.	Water forms ice, glaciers, and other forms of solid water within the range of temperatures that exist on our planet. Ice not only acts as a reservoir of stored water but also reflects heat back into space by increasing Earth's albedo.
High specific heat	The amount of heat required to raise the temperature of a given mass of water by 1°C. Except for liquid ammonia (NH_3), water has the highest specific heat of all liquids and solids.	Water tends to resist temperature changes. Where large water bodies are present, fewer temperature extremes exist.
High heat conductivity	Water conducts heat faster than any other liquid.	Heat is distributed rapidly within the cells of organisms.
High latent heat of vaporization	The amount of heat input required to transform water from a liquid to a gas. For liquids, water has the highest latent heat of vaporization.	Evaporation transports tremendous quantities of heat from the ocean to the atmosphere. This property also enables mammals to cool via sweating.
Highest latent heat of fusion	The amount of heat removal required to transform water from a liquid state to a solid state. Except for liquid ammonia (NH_3), water has the highest latent heat of fusion.	Ice forms and melts slowly on the ocean, thereby moderating temperatures in polar regions.
Exists in all three physical states	The physical states of matter are gas, liquid, and solid.	While most of the world ocean is liquid, a large percentage of sea ice forms and melts annually. Water vapor at the air-sea boundary and in the form of clouds plays a large role in many ocean processes.
Maximum density at 4°C	Pure water reaches its maximum density above its freezing point. Dissolved salts remove this effect.	Icebergs float and sea ice forms a layer at the surface.
High surface tension	Surficial water molecules exhibit the strongest cohesive forces of any liquid.	Some marine organisms live within this boundary layer.
Hot water may freeze more quickly than. cold water	Known as the Mpemba effect, its cause is unknown but may relate to supercooling. This, property which has been observed in the laboratory, serves as a reminder that water, beguilingly simple, exhibits many complex and yet unknown behaviors.	What effects this may have or if ever observed in nature is unknown.

heat (**Figure 7-9**). Specific heat is commonly measured in **calories.** A calorie is the heat required to raise 1 gram of water by 1° C. This unit is often confused with the unit used to measure food intake, the Calorie (with a capital C), which equals 1000 calories (with a lowercase c). To avoid this confusion, oceanographers use the energy unit Joule (J). Water has one of the highest specific heats of all substances on Earth, equal to 4.18 J per gram per ° C or 4180 J per kilogram per ° C.

■ **FIGURE 7-9** Specific heat versus heat capacity. Heat capacity refers to the heat need to raise the entire mass or volume of an object or system. Specific heat refers a specific mass or volume. Though often interchanged, these two terms do not mean the same thing. For example, consider a sea lion warming itself on a rock. The heat need to raise the temperature of the entire sea lion by one degree is the heat capacity of this particular sea lion. The heat needed to raise specific parts of the sea lion, like a gram of flipper tissue, is the specific heat. The specific heat of various parts of the sea lion is different allowing it to better regulate its body temperature.

Sensible Heat

Sensible heat can be sensed by your finger or a thermometer. Sensible heat occurs when two systems exchange heat, such as your finger and a hot pan, or the ocean and the atmosphere. In fact, oceanographers and atmospheric scientists more narrowly define sensible heat by restricting it to exchanges of heat by two means only: **conduction,** heating though direct contact; and **convection,** heating through the circular motion of air or water over a surface being heated by conduction or radiation (**Figure 7-10**).

The transfer of heat by direct contact is conduction. You may recognize this process through the experience of carelessly grabbing the hot handle of a metal skillet on a stove. The heat was

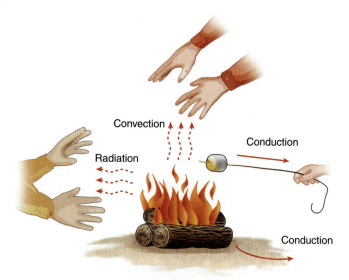

■ **FIGURE 7-10** Heating by conduction, convection, and radiation. Heat may be transferred through conduction between two objects if they are not in thermal equilibrium and if they are in physical contact. Convection transfers heat through the movements of molecules. Radiation, not considered a part of sensible heat by oceanographers, transfers heat through radiation of electromagnetic energy.

conducted from the stove through the skillet to its handle and, unfortunately, to your hand. Conduction tends to be inefficient and is confined to narrow boundaries between objects or systems. If you have ever been to the beach on a hot day, you may have noticed that the surface of the sand is very hot, but several centimeters deeper, the sand is cool. The thermal conductivity of heat tends to be low for sand, water, and air compared to substances with high thermal conductivity, such as copper and iron skillets.

Convection results from the flow of a fluid, such as air or water, over a surface or boundary that is not in thermal equilibrium. Convection causes the transport or mixing of the fluid's properties from a localized region to the entire fluid. A boiling pot of water moves heat by convection from the bottom of the pan to the top of the pan and back again. The baseboard heater in your house relies on the convective motion of air to transfer heat from the floor to the ceiling and back again. Convection plays a large role in heat exchange within the atmosphere and the ocean. Convection transfers heat through the movements of water within the ocean. Convection in the atmosphere occurs as the motion of air. In fact, deep ocean currents and winds are the result of convective movements of the ocean and atmosphere, respectively.

Latent Heat

While the sensible heat of water refers to the energy exchanged by conduction and convection at temperatures between 0° C and 100° C, another type of heat plays an important role in ocean and atmospheric circulation. Water absorbs or loses energy when it undergoes a change of physical state. This

Mesoscale Eddies: Oases in an Ocean Desert

Oceanographers have long recognized the contribution of small-scale turbulent eddies to ocean mixing, hydrodynamic drag, and other physical processes in the ocean. In an analogous fashion, photographs from spacecraft of sea glint illuminating the intricate whorls and swirls of the sea surface provided stunning views of intermediate-scale or mesoscale eddies. Subsequently, images of satellite-derived sea surface temperatures revealed Gulf Stream rings, slow moving ocean "weather" systems with a motion and lifetime independent of the surrounding waters. But it wasn't until the 1960s that oceanographers recognized the scale and complexity of mesoscale eddies. Space-based observations profoundly altered the way oceanographers think about circulation in the ocean and the way they sample the ocean from ships. Today, shipboard sampling often occurs with the assistance of a drifter or satellite image to identify the "surroundings" in which the ship is located. These methods, along with satellite altimetry, aerial photos, coastal radar, ocean color measurements, acoustic current meters, and similar devices, may be used to track and study mesoscale eddies.

Modern-day oceanographic research on mesoscale eddies has focused on their role in the distribution and exchange of heat and other properties in the large-scale ocean circulation. Efforts to model eddies have begun to yield promising results. An understanding of the role of the ocean in climate change depends on our ability to account for mesoscale variability at all scales. At the same time, oceanographers have also begun to focus on the role of eddies as "oases" of biological productivity in the otherwise oligotrophic central gyres of the major oceans. Satellite measurements of ocean color often reveal high concentrations of chlorophyll in association with cold-core eddies. The cyclonic (counterclockwise) circulation of cold-core eddies draws nutrient-rich water toward the surface. Phytoplankton thrive on these pulses of nutrients in what would otherwise be a nutrient-poor environment. As much as 50% of the nutrients supplied to phytoplankton in oligotrophic environments may come from "eddy pumping" and other physical features of eddies that enhance nutrient fluxes. Phytoplankton blooms associated with cold-core eddies fuel microbial food webs and attract a number of migratory predators, such as tuna and swordfish. Thus, mesoscale eddies have considerable importance for fisheries.

Given their importance, oceanographers have begun major efforts to better understand the origins and life cycles of meso-scale eddies. Unfortunately, satellites and computer models lack the spatial resolution needed to fully address these questions. Shipboard studies suffer their own challenges, given that eddies are constantly on the move. To overcome these limitations, a group of oceanographers is conducting a series of experiments in a "natural laboratory," the Hawaiian Islands, where "semipermanent" mesoscale eddies occur regularly. This so-called E-Flux experiment takes advantage of atmospheric conditions that are favorable to the generation of eddies. Northeasterly tradewinds often blow strongly through the Maui Channel, separating Haleakala on Maui and Mauna Loa and Mauna Kea on the Big Island of Hawaii. The presence of the islands creates differences in wind forcing at the ocean surface and generates localized areas of upwelling and downwelling. Not all "eddy seeds" result in mature eddies, but well-developed, if not beautiful, eddies occur with regularity when conditions are right. The goal of the E-Flux experiment is to better define the conditions that generate eddies and to describe the physical, chemical, and biological processes that accompany eddy formation, intensity, and duration, especially as they relate to carbon export and global climate change.

FIGURE 7a Mesoscale eddies are revealed in the reflection of the sun from the sea surface.

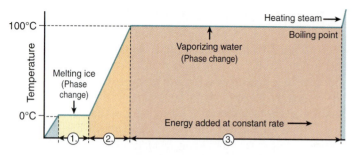

1. Latent heat of fusion = 80 cal/gm or 33 J/g
2. Latent heat of vaporization = 540 cal/gm or 2250 J/g
3. Energy required to heat/cool 1 gm liquid water between 0 and 100°C = 100 cal/gm or 418 J/g

■ **FIGURE 7-11** Changes of state and latent heat in water. Note that state changes or phase transitions require an input of heat that is not expressed as a change in temperature. The latent heat is the heat required to change the physical state of a substance. Latent heat changes the potential energy of a system even though its kinetic energy may remain the same.

additional energy is required to break hydrogen bonds and "rearrange" water molecules into their new form. Because this additional energy does not change the temperature of the water, it is "hidden" or **latent heat** (**Figure 7-11**). The **latent heat of vaporization** refers to the heat required to change water from a liquid to a gas or vice versa. The **latent heat of fusion** includes both freezing and melting and refers to the heat removed or added during the change of state from a liquid to a solid (freezing) or a solid to a liquid (melting).

Water has one of the highest latent heats of vaporization and fusion. To convert 1 gram of water from a 100° C liquid to a 100° C gas requires 540 calories (2250 J), the latent heat of vaporization for water. This amount is considerably more than the 100 calories of energy (418 J) required to heat water from 0° C to 100° C. At the other end of the temperature scale, converting 1 gram of liquid water to 1 gram of ice requires the removal of 80 calories of energy (333 J), the latent heat of fusion for water.

Pressure and Buoyancy

As a fluid, the ocean (and the atmosphere) exerts pressure on objects immersed within it. This pressure, known as **hydrostatic pressure,** arises from the weight of water (or air) pressing down on the object. This pressure acts in all directions on an object. Formally defined, the force of static pressure (P, measured in newtons m^{-2}) is a function of the density of the fluid (ρ, pronounced *rho*, in kg m^{-3}), its height or depth (h or z, in meters) and the acceleration due to gravity (g, in m sec^{-2}), or

$$P = \rho\, g\, h \qquad \text{Equation 7-1}$$

The factor g is constant and equal to 9.8 m sec^{-2}. For a fluid of constant density, pressure is almost solely a function of the

height of the fluid because gravity is constant. Equation 7-1 is known as the **hydrostatic equation.** We can use this equation to calculate the pressure that results from the weight of a fluid acting on an object. The pressure at sea level is defined as one atmosphere. As an object descends into the ocean, it encounters greater pressure because of the weight of seawater above it. In fact, for about every 10 m (33 feet) of depth, we add another atmosphere of pressure, assuming that seawater density remains constant (**Figure 7-12**).

Another useful term when considering pressure in the ocean describes "surfaces" or layers of equal pressure, called **isobars** (**Figure 7-13**). Isobaric surfaces in the ocean are surfaces where the pressure is constant, just as in the atmosphere. The expression of ocean layers in terms of their pressure allows oceanographers to calculate motions in the ocean. In an ocean with no horizontal gradients in density, isobaric surfaces will be parallel to the sea

■ **FIGURE 7-12** For every 10 m (33 feet) a diver descends, an additional atmosphere of pressure is added. If strict procedures are not followed to compensate for the buildup of blood and tissue gases at depth, the effects of hydrostatic pressure on humans can be fatal. Fortunately, well-tested Navy dive tables and dive computers make Scuba a safe and enjoyable way to explore the world ocean.

Sea surface

a. Barotropic conditions

isobars
isopycnals

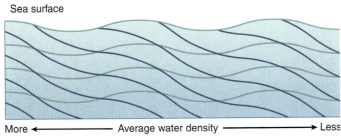

Sea surface

More ← Average water density → Less

b. Baroclinic conditions

FIGURE 7-13 Barotropic versus baroclinic surfaces. (a) Under barotropic conditions, isopycnals and isobars are parallel. Horizontal pressure gradients result from the slope of the sea surface independent of depth. (b) Under baroclinic conditions, isopycnals and isobars are not parallel. Seawater density can cause pressure changes with depth and create horizontal pressure gradients that generate currents (i.e., baroclinic flow). Horizontal pressure gradients are responsible for the winds and currents we observe in the atmosphere and ocean.

surface and parallel to isopycnals, a condition termed **barotropic.** Where horizontal gradients in density occur, the isopycnals and isobars may not be parallel, a condition termed **baroclinic.**

Hydrostatic pressure also comes into play in a property known as **buoyancy,** a concept first demonstrated by Greek mathematician and inventor Archimedes (287–212 BCE). The Archimedes principle states that the buoyant force on a submerged object is equal to the weight of the fluid that is displaced by the object.

Buoyancy refers to the difference between forces acting upward on an object (e.g., pressure) and forces acting downward (gravity) (**Figure 7-14**). An object may exhibit positive buoyancy, meaning it rises in the water column, or it may experience negative buoyancy, meaning it sinks. Objects that neither rise nor sink but remain in place in the water column are said to be neutrally buoyant.

Marine archaeologists, ocean engineers, and oceanographers have long used the principles of buoyancy to their advantage. You may have seen movies of divers who send treasure to the ocean surface by attaching a "balloon" and inflating it with air. The treasure-laden balloon becomes lighter than the surrounding water (i.e., positively buoyant) and rises. A submarine is a good example of a vessel that moves up and down using the principles of buoyancy. The intake or expulsion of water in ballast tanks effectively makes the submarine lighter or heavier than the surrounding seawater, causing the vessel to rise or sink, respectively. Oceanographers have developed a number of sampling platforms that can move up and down in the water column through changes in buoyancy. Floats deployed in the Argo system alter their buoyancy by moving oil from an internal

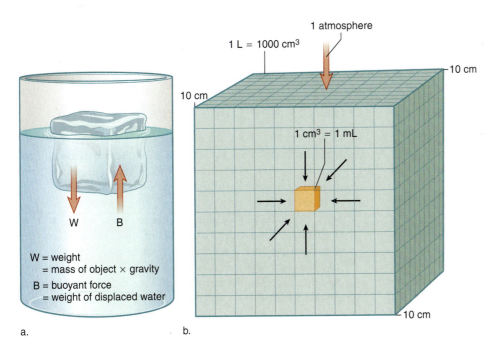

W = weight
  = mass of object × gravity
B = buoyant force
  = weight of displaced water

a.

b.

1 L = 1000 cm³
1 atmosphere
10 cm
10 cm
1 cm³ = 1 mL
10 cm

FIGURE 7-14 (a) A cube of fluid (or an object in the fluid) experiences forces acting upon it from all directions. (b) Vertical forces on an object determine its depth in the fluid layer. The upward vertical force, or buoyant force, is equal in magnitude to the weight of the volume of water displaced by the object. The downward vertical force is equal to the mass of the object times gravity. When an object is neutrally buoyant, the upward and downward vertical forces are equal. Changes in the mass, volume or density of an object may cause it to sink or rise.

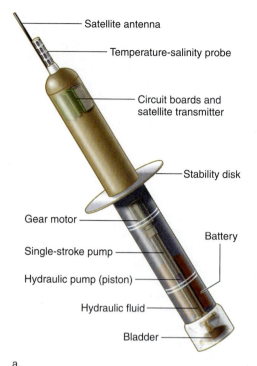

a.

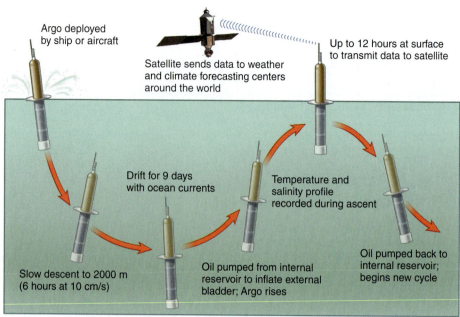

b.

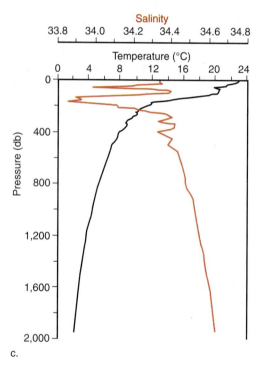

c.

■ FIGURE 7-15 Designed to follow ocean currents at specific depths, the Argo floats obtain near-continuous measurements of temperature, salinity and pressure. (a) Buoyancy is changed by moving oil via a hydraulic pump to/from an external/internal reservoir causing the float to ascend or descend, respectively. (b) A single profile typically consists of measurements from the surface to 2000 meters depth and back again every 10 days. Data are stored in memory chips or on-board hard drives until the platform can be retrieved by a ship and/or until the data can be transmitted via radio or satellite to a ground station. Some oceanographers are adding optical, chemical, and biological sensors to the Argo floats to obtain other types of important oceanographic data. (c) Temperature and salinity profile obtained from an Argo float in the North Pacific in May 2004 (data from Argo).

than the layers immediately below it. An unstable water column will rearrange its layers to become a stable water column, unless other processes continue to make its layers unstable.

Heating and Cooling with Pressure

The temperature of a fluid may change as a result of pressure changes. **Adiabatic changes** are defined as changes in the internal energy of a substance that are not due to exchanges of heat with the surroundings (**Figure 7-16**). In the atmosphere, adiabatic changes may be more important than other mechanisms of temperature change. Adiabatic heating due to compression of air is the force that generates California's Santa Ana winds. Adiabatic cooling may occur as pressure is decreased and air or water expands.

Adiabatic heating and cooling pose special problems for oceanographers. A sample of water may exhibit a different temperature solely as a result of its depth (**Figure 7-17**). Without taking these pressure effects into account, the comparison of water masses at different depths becomes nearly impossible. For that reason, oceanographers define **potential temperature,** the temperature a sample of water would exhibit if it were brought to the surface (i.e., brought to 1 atmosphere or 1000 millibars) adiabatically,

reservoir to an external tank and back again to rise and descend, respectively (**Figure 7-15**).

Oceanographers also apply principles of buoyancy to characterize the layers of seawater in the ocean. Negatively buoyant layers (i.e., more dense) will sink beneath more positively buoyant layers (less dense) and vice versa. Any process that changes the buoyancy of a layer of water will cause that layer to move accordingly. A **stable water column** is one in which the density of the layers increases progressively from the surface to the bottom. An **unstable water column** is one in which the top layer is more dense

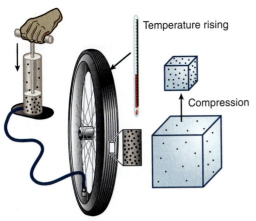

a. **Adiabatic heating**

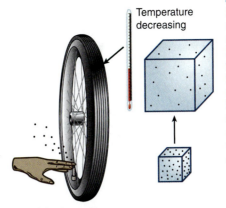

b. **Adiabatic cooling**

■ **FIGURE 7-16** Adiabatic heating and cooling. (a) Compression of air into a tire packs molecules of air into a smaller volume resulting in increased internal heat and the temperature of the air inside the tire rises. (b) Expanding air exhibits the opposite effect. The internal energy of the air is now spread over a greater volume and the temperature of the air decreases. Air exiting a tire cools as it expands.

meaning without heat exchange with the surrounding waters. Potential temperature differs from *in situ* **temperature,** the temperature of a sample of water at the depth where it is measured, by taking into account the adiabatic effects. Potential temperature provides a tool for comparing water masses in an equal fashion by removing the effects of adiabatic heating.

Seawater Density

The effects of density changes in the atmosphere and world ocean are profound, causing changes in atmospheric and ocean circulation, ocean productivity, and a number of other important phenomena. The density of seawater (or any substance) refers to the mass present within a specific volume, typically reported in units of kilograms per

cubic meter ($kg\ m^{-3}$). Temperature and salinity *independently* affect the seawater density, albeit in different ways (**Figure 7-18**). If we add heat to a sample of seawater, its density decreases, that is, there is less mass per unit volume. That's because the addition of heat causes the intermolecular distance between water molecules to expand. Naturally, if we increase the space between molecules, the volume must also increase. Though undetectable in a glass of water, heating of seawater causes thermal expansion of the ocean and an increase in its volume. The opposite happens when we remove heat from a sample of seawater. Cooling reduces the internal energy and shrinks the space between molecules. Now, a greater amount of mass exists within the same volume and the density increases. The relationship between temperature and density is an inverse one: as one quantity

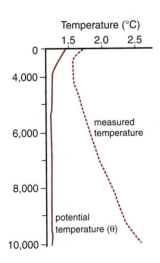

■ **FIGURE 7-17** Vertical profile of measured temperature and calculated potential temperature. Although the measured temperature rises with depth, the potential temperature remains the same. Therefore, we can conclude that the measured temperature rose because of adiabatic heating. The water at depth is merely surface water under greater pressure and not a different water mass, as might be concluded without considering potential temperature. Oceanographers signify potential temperature with the Greek small letter, theta, θ.

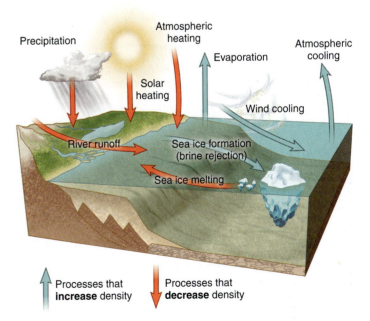

■ **FIGURE 7-18** Effects of temperature and salinity on seawater density. Heating of seawater at the ocean surface causes it to expand slightly and decrease in density. Cooling has the opposite effect. Evaporation of seawater at the ocean surface leaves salts behind and increases its density. Ice formation has a similar effect as salts are rejected from the freezing ice. Precipitation and freshwater runoff lower the density of surface waters.

goes up, the other goes down and vice versa. For the same volume of water, warmer water is less dense than colder water. Salinity has a direct relationship with density. By adding salts to seawater, we increase the mass contained within a specific volume. Thus, an increase in salinity results in an increase in density. Over short temporal scales (days to months), changes in salinity result from evaporation and precipitation of water, which has the same effect as adding or removing salts, respectively. Changes in salinity (and density) due to the addition or removal of salts over longer temporal scales (decades to millennia) usually occur in response to changes in the sources and sinks of dissolved elements, as we learned in Chapter 6.

Solar Radiation

In the mid-1800s, Scottish physicist James Clerk Maxwell (1831–1879) demonstrated that solar energy could be described as a traveling wave that consisted of both electric and magnetic fields. This traveling wave does not require a material medium to exist. In fact, it can travel in a vacuum. Maxwell's discovery led to recognition of

electromagnetic radiation. There are many forms of electromagnetic radiation. Physicists represent the range of types of electromagnetic radiation as the **electromagnetic spectrum (Figure 7-19)**. This spectrum spans short wavelengths, such as gamma rays, to long wavelengths, such as radio waves. It also includes the very narrow band of energy called **visible light,** the type of electromagnetic radiation detected by humans. Although most physicists express the electromagnetic spectrum in terms of wavelengths (the distance between peaks in a wave), they subsequently found that light simultaneously exhibits particlelike properties. The indivisible particle of light, the photon, is also used as a unit for expressing electromagnetic energy.

The total flux of solar energy to the top of the atmosphere is called the **solar constant** (S_o). The solar constant represents the sum of electromagnetic radiation at every wavelength across the entire **solar spectrum.** Each day, the Sun supplies more energy to the atmosphere and the ocean than has been used by all of humanity in the past 100 years! That equates to more than 4000 trillion kilowatt-hours daily. But how does solar energy turn into heat?

The distribution of energy in the solar spectrum behaves according to the theory of **blackbody radiation.** The theory is based on a theoretical object that absorbs and emits radiant energy with 100% efficiency. Because it absorbs perfectly and reflects no light, the theoretical object appears black. Blackbody radiators emit

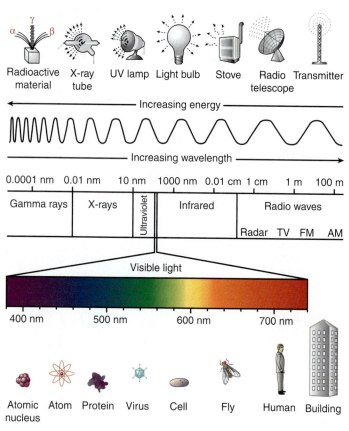

■ **FIGURE 7-19** The electromagnetic spectrum. Short wavelength radiation, like ultraviolet radiation, has important implications for the health of humans and organisms. Long wavelength radiation, like infrared radiation, is involved in heat. Microwaves and radio waves provide important tools for satellite observations of the world ocean. Visible light occupies only a narrow section in the middle part of the spectrum.

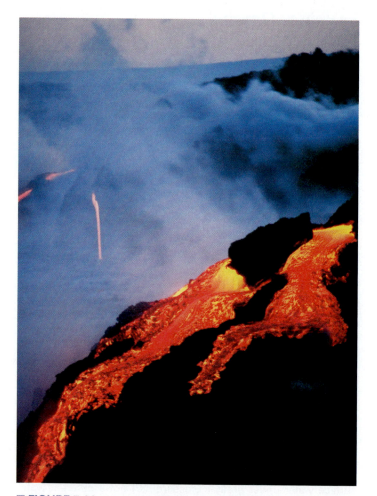

■ **FIGURE 7-20** Hot lava exhibits a range of colors, in accordance with Wien's Law. The hottest parts of the lava glow yellow, while the cooler parts take on shades of red. Cooled lava is black.

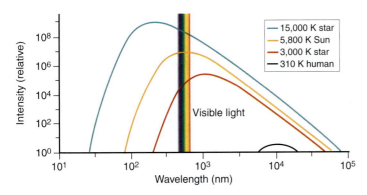

■ **FIGURE 7-21** Wien's Law, expressed graphically. The higher the temperature of an object, the shorter its wavelength of maximum emission.

energy at all wavelengths, regardless of the temperature. However, the wavelength of maximum emission is inversely proportional to the temperature. As the temperature increases, the wavelength of maximum emission decreases, that is, it shifts toward shorter wavelengths. Thus, hot objects emit more energy at short wavelengths, while cool objects emit more energy at long wavelengths (**Figure 7-20**). The wavelength-dependency of radiant energy emission is known as **Wien's law** (Equation 7-2), named after German physicist Wilhelm Wein (1864–1928), who won a Nobel Prize for his research. Formally, Wein's law can be expressed as:

$$\lambda_{max} = 2890/T \qquad \text{Equation 7-2}$$

where λ_{max} (pronounced *lambda max*) is the wavelength of maximum emission and T is the temperature in Kelvin (K = °C + 273). When graphed, you can see that the higher the temperature, the smaller the wavelength of maximum emission (**Figure 7-21**). Wein's law will be helpful to us when we consider Earth's energy budget and the greenhouse effect.

Another useful expression, the **Stefan-Boltzmann formula** (Equation 7-3), converts the temperature of a blackbody radiator to the total amount of energy it emits, such as the solar constant. The Stefan-Boltzmann formula is expressed as:

$$E = \sigma \times T^4 \qquad \text{Equation 7-3}$$

where E is the energy emitted from a given area of the object's surface per unit time, σ (sigma) is the Stefan-Boltzman constant (5.67×10^{-8} W m^{-2} K^{-4}), and T^4 is the temperature raised to the fourth power.

In nonmathematical terms, we can see that the absorption of solar energy (by air, water, or land) converts it into thermal energy. The total amount of energy remains the same: we have simply changed its form from electromagnetic energy to thermal energy.

Earth's Seasons

The intensity, duration, and angle of incidence of sunlight on Earth vary according to the seasons. Fall colors, winter snowfall, spring flowers, and warm summer days characterize the seasons on land. Though less appreciated, the world ocean responds in a similar fashion. Despite its popularity, the seasonal cycle is fraught with misconceptions. One popular misconception is that the seasons are caused by changes in the distance between Earth and the Sun as Earth orbits the Sun. While the distance between Earth and the Sun does change during Earth's orbit around the Sun, it is not in the way that you might expect. In the current Milankovitch cycle (see Chapter 5), *Earth is closer to the Sun in the winter than in the summer.* So, clearly the Earth-Sun distance is not responsible for the seasons.

The seasonal cycle arises as a result of the tilt of Earth on its axis relative to its orbit around the Sun (**Figure 7-22**). As Earth orbits the Sun, the place where sunlight falls directly changes. Where it falls directly, sunlight is intense. Where it falls at an angle, the radiant

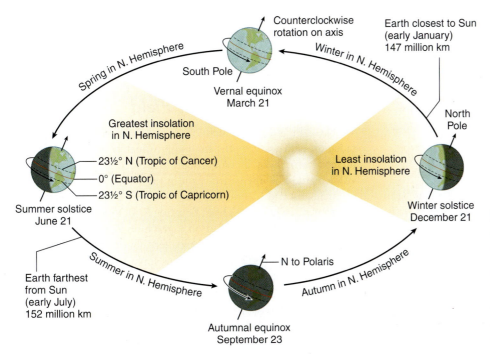

■ **FIGURE 7-22** The annual orbit of the Earth around the Sun. The 23.5° tilt of the Earth on its axis relative to its plane of orbit around the Sun is responsible for the seasons. Changes in the overhead position of the Sun cause variations in the amount of solar radiation incident on a unit area of the Earth's surface. Thus, on a seasonal basis, the area of Earth's surface heated most directly changes (day to day, in fact).

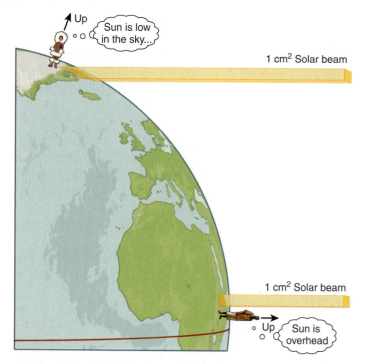

FIGURE 7-23 Solar radiation at an angle. Sunlight striking the Earth at high latitudes has its energy spread over a greater surface area, thereby reducing the intensity of solar energy per unit area. On the other hand, sunlight striking directly on the surface of the Earth is focused in a smaller area, so that the intensity of solar energy is higher per unit area. This differential heating of the Earth causes differences in the climate experienced by different regions of the Earth and plays a key role in the seasonal cycle of heating.

energy is spread over a greater surface area (**Figure 7-23**). Twice a year, the Sun appears overhead at noon at the equator: on March 21, the **spring equinox,** and on September 23, the **autumnal equinox.** At this time, sunlight falls directly on the equator. Between these dates, the apparent location of the Sun directly overhead at noon changes, moving into the Northern Hemisphere from March 21 to September 23 and into the Southern Hemisphere from September 23 to March 21. On December 21, the **winter solstice,** the Sun appears directly overhead at 23.5° S, the latitude of the Tropic of Capricorn. On June 21, the **summer solstice,** the Sun appears overhead at 23.5° N, the Tropic of Cancer. Of course, the Sun is not moving. It is the angle of Earth relative to its plane of orbit around the Sun that changes seasonally. It is important to note that day length and the total daily quantity of sunlight experienced at a given latitude also change seasonally. The seasonal cycle of solar radiation and heating has a profound effect on many earth, ocean, and atmospheric processes.

The Seasonal Thermocline

If you've ever jumped into a lake during summer, you might have been surprised by the cold water just below the warm surface. This boundary between the warm surface waters and the cold deeper waters marks the depth of the **seasonal thermocline,** a region of rapid temperature change that develops seasonally in lakes and the ocean (**Figure 7-24**). The seasonal thermocline forms in response to seasonal variations in incident solar radiation that warm the surface of the ocean and cause **stratification,** or layering, of the upper ocean. At first, stratification may be relatively weak and the thermocline may be deep. However, warming of the upper ocean as a result of increasing

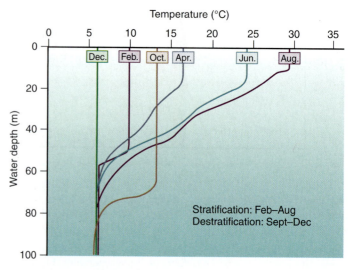

FIGURE 7-24 Idealized view of stratification and destratification of a mid-latitude water column over the seasonal cycle.

day length and higher Sun angles produces greater stratification and a shallowing of the thermocline. Under calm summer conditions, continued warming may produce multiple thermoclines. In the fall, cooling and sinking of surface waters cause weakening and deepening of the seasonal thermocline, a process called **destratification.** Cold winter temperatures combined with an increase in winds and waves bring near complete mixing of the upper ocean. The depth of the seasonal thermocline defines the **mixed layer depth.** The mixed layer by definition is homogeneous in its properties from the surface to the depth where temperature (or some other property) begins to change. It is considered to represent a layer that is well-mixed, although the rate of mixing (overturning) of the water column will vary with winds, buoyancy changes, and other factors that impart energy and momentum to the upper ocean. The seasonal patterns of the thermocline and mixed layer depth also depend on latitude in the world ocean (**Figure 7-25**). The mixed layer and the seasonal variations in its depth control a great number of physical, chemical, and biological processes in the ocean, subjects we explore in greater detail in our chapters on ocean productivity and food webs.

The Submarine Light Field

Visible light is formally defined as the range of electromagnetic radiation with wavelengths from approximately 400 to 700 nanometers (nm). Like humans, many marine organisms can sense visible light. The intensity of light in the water column diminishes or attenuates with depth according to a well-known mathematical formulation called **Beer's law** (Equation 7-4). When light intensity at different depths is plotted on a graph and the points are connected, the resultant shape is an exponentially decreasing curve. The advantage of this result is that it lets us apply a simple mathematical formula to the determination of light at depth. All that is needed is knowledge of the light intensity at the surface and an optical property of the water known as **attenuation.** Attenuation is a factor that expresses the capacity of water and everything in it to reduce light. It represents the combined effects of the major light-reducing

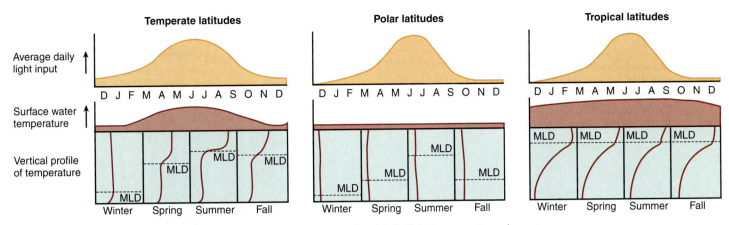

FIGURE 7-25 Idealized seasonal cycle of daily light input, water column temperature structure, and mixed layer depth for winter, spring, summer, and fall in mid-latitudes, high latitudes, and tropical latitudes. MLD refers to the mixed layer depth.

components: 1) water; 2) suspended living particles, including phytoplankton and marine bacteria; 3) suspended nonliving particles, including particulate organic matter and inorganic particles; and 4) colored dissolved substances.

Two optical processes contribute to the attenuation of light in the water column (**Figure 7-26**). **Absorption** is the process by which light is absorbed and removed by any of the light-reducing components. **Scattering** is the process by which the direction of light is changed by the light-reducing components. Scattering does not remove light, but it can prevent the vertical penetration of light. It can also lengthen the path of light and thereby increase the probability that it will be absorbed. The sum of absorption and scattering in the downward direction in the water column is expressed in a quantity related to attenuation known as the **extinction coefficient,** symbolized by K_d. A simpler, albeit crude, means for determining K_d is through use of an instrument called a **Secchi disk** (pronounced *SEK-ee*) (**Figure 7-27**). This simple circular disk provides a rough estimate of extinction by relating it to the depth at which the disk just disappears, a depth known as the Secchi depth, Z_{SD}, when it is lowered in the water column. More accurate measurements

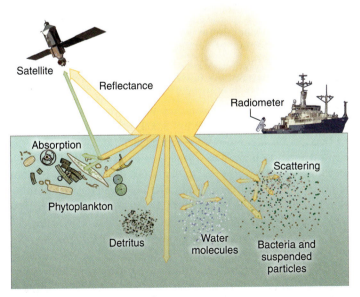

FIGURE 7-26 Absorption, scattering, and reflection of the submarine light field.

FIGURE 7-27 A Secchi disk measurement through the ice. This simple device was invented around 1860 by the Italian Astronomer Pietro Angelo Secchi, a Jesuit priest. Secchi worked for the Vatican aboard the papal ship, S.S. Immaculate Conception, studying the transparency of the Mediterranean sea with the original secchi disk, a 16-inch all-white china plate. A few decades later, George Whipple, a professor of Sanitary Engineering at Harvard University, made the plate smaller and added alternating black-and-white quadrants. Interestingly, both types of "Secchi" disks are in use today. Oceanographers use the papal one while limnologists use Whipple's. Many classrooms use Secchi or Whipple disks as a tool for understanding water transparency and underwater optics.

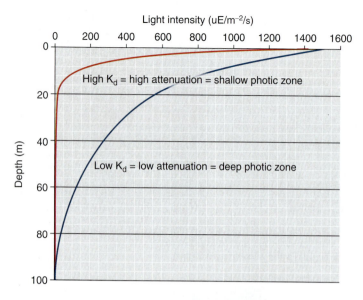

FIGURE 7-28 Predicted profiles of visible light intensity versus depth for two oceanic environments. Each of the profiles represents water columns with differing amounts of attenuation.

of K_d can be obtained by lowering a light sensor, called a **radiometer,** to compare the intensity of light at different depths. Spectroradiometers, instruments that measure the intensity of specific wavebands of downwelling and upwelling light, are routinely deployed in studies of the optical characteristics of the water column.

Estimates of the attenuation coefficient allow us to predict underwater light intensity using Beer's law. According to Beer's law, the light intensity at a given depth (I_z) is equal to the light intensity at the surface (I_o) multiplied by the exponential attenuation factor ($exp^{-K_d \times Z}$). Expressed mathematically:

$$I_z = I_o \times exp^- \ K_d \times Z \qquad \text{Equation 7-4}$$

where Z is depth and exp is the exponential function. Beer's law allows us to compare the properties of the light field at different locations and depths in the world ocean (**Figure 7-28**). The maximum depth of light penetration defines the **photic zone.**

The Color of the Ocean

The absorption properties of water provide an answer for one of the most commonly asked questions about the ocean: why is the ocean blue? Measurements of the absorption coefficients of pure water at different wavelengths provide the answer. Inspection of **Table 7.3** reveals that pure water (close enough to seawater for our purposes) has the lowest values of absorption for blue and violet light. Water strongly absorbs red light (and infrared light). Because blue light is absorbed the least, blue light is scattered from the ocean and the water appears blue. In any location or depth, the color of the ocean is an expression of the wavelength-varying absorption of visible light by seawater and its contents (**Figure 7-29**). The difference in ocean color between places such as Hawaii (blue) and California (green) can be attributed to suspended particles, phytoplankton, and dissolved substances, each of which have their own absorption properties Where the ocean is blue, water dominates the absorption of light, and the

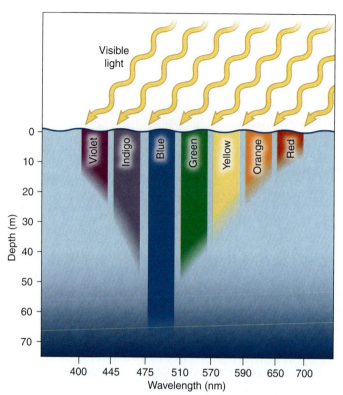

FIGURE 7-29 The penetration of different colors of light through the water column of typical oceanic waters. Note that red light is strongly absorbed compared to other wavelengths. Blue light penetrates the deepest. Living and non-living particles and dissolved colored substances will alter the spectral distribution of the submarine light field. Knowing this, oceanographers can infer a great deal about ocean processes by studying ocean color.

concentration of dissolved and suspended particles is minimal. By contrast, where the ocean is green, we can expect a significant contribution to absorption from other components, such as phytoplankton, which contain green photosynthetic pigments (see Chapter 13).

The color of the ocean varies in response to geological, physical, chemical, and biological processes that alter the abundance of light-absorbing particles and dissolved substances. In fact, this is the basis for a new subdiscipline of oceanography called **bio-optical oceanography.** One of the first satellite missions to study ocean color, the Coastal Zone Color Scanner (CZCS), exceeded the wildest expectations of oceanographers. Intended as a "proof-of-concept" mission, the CZCS operated over a period of 91 months (October 1978 to June 1986), capturing more than 68,000 images of ocean color. It provided for the first time a detailed picture of ocean color across the world ocean over periods of several days to several years. Although more than 13 years would pass during which no satellite observations of ocean color would be available, the launch of the Sea-Viewing Wide Field-of-View Sensor (SeaWiFS) instrument in August 1997 again enabled an unprecedented view of the ocean. Whereas the CZCS offered five intervals or bands of color measurements from 433–800 nm, the SeaWiFS measures color in eight bands from 402–885 nm. SeaWiFS also benefit from advances in technology leading to greater sensor stability, more careful attention to calibration, and greater resolution. SeaWiFS can cover nearly the entire globe in a day, providing snapshots of ocean color at a greater temporal resolution

TABLE 7.3 Absorption Coefficients for Light at Wavelengths from 380–800 nm in Pure Water

Wavelength (nm)	Absorption coefficient (m^{-1})	Wavelength (nm)	Absorption coefficient (m^{-1})
380	0.023	600	0.245
390	0.020	610	0.290
400	0.018	620	0.31
410	0.017	630	0.32
420	0.016	640	0.33
430	0.015	650	0.35
440	0.015	660	0.41
450	0.015	670	0.43
460	0.016	680	0.45
470	0.016	690	0.50
480	0.018	700	0.65
490	0.020	710	0.839
500	0.026	720	1.169
510	0.036	730	1.799
520	0.048	740	2.38
530	0.051	750	2.47
540	0.056	760	2.55
550	0.064	770	2.51
560	0.071	780	2.36
570	0.080	790	2.16
580	0.108	800	2.07
590	0.157		

Source: Kirk, 1994; based on Smith and Baker, 1981.

than previously available. A new generation of satellite-based instruments, the MODIS instruments aboard the Terra (launched in 1999) and Aqua (launched in 2002) satellites, acquire ocean color information in 36 spectral bands from visible to infrared wavelengths. The goals of Aqua and Terra are to provide multidisciplinary support for studies of climate change in the Earth system, including the atmosphere, hydrosphere, and geosphere. These spacecraft will enable oceanographers to monitor air-sea interactions, radiative transfer, ocean productivity, and marine ecosystem dynamics (**Figure 7-30**).

Earth's Energy Budget

To account for processes that add or remove energy from Earth, scientists construct an **energy budget.** If our energy budget is balanced, we can write it in terms of a simple formula:

$$\text{energy input } (E_i) = \text{energy output } (E_o)$$

or

$$E_i - E_o = 0 \qquad \text{Equation 7-5}$$

FIGURE 7-30 A spectacular bloom of coccolithophorids in the Barents Sea, as captured by the MODIS instrument aboard Terra. Coccolithophorids are a type of phytoplankton with a calcium carbonate skeleton. They are responsible for producing a major percentage of carbonates stored in rocks.

From Equation 7-5, you should be able to see that if E_o is greater than E_i, then Earth will lose energy. Alternatively, if E_o is less than E_i, then Earth will gain energy. Simple in concept, this equation proves useful for understanding climate change.

The energy gained and lost by Earth takes the form of the physical properties discussed earlier. Take a few moments to study the inputs and outputs for Earth's energy budget illustrated in **Figure 7-31.** Light traveling through the atmosphere is subject to absorption and scattering by suspended particles (dust), water vapor (clouds), and other gases. Thus, not all sunlight hitting the outer atmosphere will make it to Earth's surface. According to Figure 7-31, these sources absorb 67/342 W m^{-2}, or about 19.5%, of the incoming solar radiation. Backscattering of sunlight (labeled as

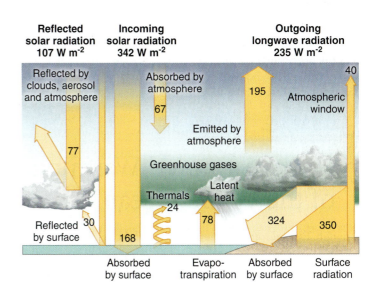

FIGURE 7-31 Earth's energy budget. See text for details.
Source: Kiehl and Trenberth, 1997.

reflection in the figure) also contributes to losses through the atmosphere. Clouds, air, dust, haze, and even the surface of Earth cause backscatter. About 107/342 W m^{-2}, or 31%, of the incoming solar radiation is reduced by backscattering. Note that clouds alone may contribute 22% of the backscattering. (Clouds are a very important and highly variable regulator of solar radiation. Blackbody radiation from clouds may produce both upward and downward fluxes of longwave radiation. The downward flux of heat from clouds is responsible for warm, cloudy nights compared to the cold nights that occur when the sky is clear.) Once sunlight hits the ocean surface, some is reflected (about 8%), and the rest is transmitted through the water, where it is completely absorbed. On the whole, about half (49%) of the sunlight that hits the outer atmosphere, or 168 W m^{-2}, makes it to Earth's surface. Most solar radiation is absorbed by the world ocean because its surface area is greater than the land's.

If we assume that gains and losses are equal (no net gain or loss of heat), then the governing equation for Earth's heat budget can be written as a simple summation:

$$\text{Received shortwave radiation } (Q_{SWNet}) =$$
$$\text{outgoing longwave radiation } (Q_{lw}) +$$
$$\text{sensible heat } (Q_{sens}) + \text{latent heat } (Q_{lat})$$

where Q_{SWNet} is the total amount of shortwave radiation absorbed at Earth's surface, Q_{lw} is the net longwave radiation loss, Q_{sens} is the sensible heat loss, and Q_{lat} is the latent heat loss. Does our budget add up according to Figure 7-31? Let's explore this.

According to Figure 7-40, the value for Q_{SWnet} is 168 W m^{-2}. The value for Q_{lat} is 78 W m^{-2}, and Q_{sens} is 24 W m^{-2}. A value for Q_{lw} is trickier because we need to know the net longwave radiation that leaves Earth's surface. Looking at the difference between surface radiation and trapped radiation, we obtain a value of 66 W m^{-2}. This provides the quantity we need to balance our budget. Entering these values, we get:

$$168 \text{ W m}^{-2} (Q_{SWNet}) = 24 \text{ W m}^{-2} (Q_{lw})$$
$$+ 66 \text{ W m}^{-2} (Q_{sen}) + 78 \text{ W m}^{-2} (Q_{lat}).$$

Heat budgets are not simply math exercises for climate scientists, oceanographers, and their students. They help scientists identify uncertainties in the measurements of where heat is stored and how it moves about on our planet. They help us better understand the impacts of human activities on global climate.

Human-Caused Warming of the World Ocean

Though little recognized outside scientific circles, one of the major accomplishments of physical oceanographers in the twenty-first century is the amassing of evidence that the world ocean is warming. The first suggestion that the world ocean might be an important sink for excess heat due to global warming came as early as 1959 from Carl-Gustav Rossby (1898–1957) of the Massachusetts Institute of Technology. Roger Revelle (1909–1991) at Scripps Institution of Oceanography and others published data in 1965 linking increased ocean temperatures with increasing greenhouse gases. By 2000, Sydney Levitus and coworkers confirmed Rossby's hypothesis. Their data demonstrated unequivocally that the upper ocean (0–3000 m or 0–1.8 miles) was warming (**Figure 7-32**). From 1955–1998, the world ocean heat content increased by approximately 14.5 × 10^{22} J, or about 10 W m^{-2}, per year. Though the mean temperature increase over this 43-year-period appears modest, 0.037°C, significant amounts of heat in the world ocean remain "invisible" as latent heat. As emphasized by Levitus and coworkers, a 0.1°C increase in ocean temperature represents a quantity of heat that if released would instantaneously raise the temperature of the global atmosphere by 100°C! Evidence linking the warming of the world ocean with anthropogenic causes remains circumstantial. However, careful calculations of the amount of ocean heating expected from human-released greenhouse gases closely match the observations of ocean temperature. Models support the conclusion that approximately 84% of the global heat increase over the past 50 years has been absorbed by the world ocean.

In 1959, Rossby wrote that "mankind now is performing a unique experiment of impressive planetary dimensions by now consuming during a few hundred years all the fossil fuel deposited during millions of years." It appears that some of the results from that experiment are now being discovered by oceanographers and climate scientists.

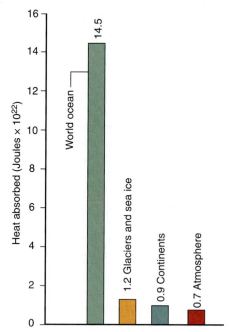

■ FIGURE 7-32 Estimates of the heat input to various components of Earth's heat budget as a result of global warming (1955–1998). The 14.5 * 10^{22} J absorbed by the world ocean represents 84% of the total increase in global heat content.

Are the Polar Ice Caps Melting?

The polar ice caps represent a tremendous reservoir of fresh water to our planet. More than 70% of the world's fresh water supply exists as polar ice. Given this amount of water, it's only natural for people to ask what might happen to sea level if the polar ice caps melted. Scientists estimate that if the entire Antarctic ice sheet melted (highly unlikely, given that most parts of the Antarctic have temperatures well below freezing), sea level would rise by about 90 m (200 feet). Melting of the Arctic ice cap would have little effect because floating ice already displaces its volume of seawater. The biggest concerns for sea level rise are attributed to melting of land-fast ice, namely the Greenland and Antarctic ice sheets. While some fraction of recent sea level rise can be attributed to the melting ice caps, the larger problem has to do with the *effects* of polar ice cap melting. Ice reflects a significant portion of sunlight back to outer space, effectively keeping our planet cool. At the same time, the high heat capacity of ice moderates heating due to increased greenhouse gases. The presence of ice provides a buffer against rapid changes in Earth's temperature by absorbing solar radiation and storing it as latent heat. Without the polar ice caps, Earth will absorb greater quantities of solar radiation, and global warming will accelerate.

Using passive microwave sensors aboard polar-orbiting satellites, oceanographers have documented a decrease in the area and extent of sea ice in the Arctic Ocean since 1979 (**Figure 7a**). In 2002, the September sea ice minimum set a new record at 15% below average. New minimum records were set in September 2003, September 2004, and June 2005. On average, sea ice extent has been decreasing by about 3% per year since the late 1970s. Satellite altimeter and gravity observations in the Antarctic reveal that ice is thinning rapidly in the West Antarctic. Dramatic satellite photos of retreating glaciers and Rhode Island–sized icebergs splitting from the Ross Ice Shelf seem to strengthen the evidence for human-caused global warming. Yet studies of some glaciers in the West Antarctic show that the rate of retreat has been constant for the past 10,000 years, well before the invention of the internal combustion engine. Increased snowfall in the East Antarctic may have caused a

growth in ice by some 45 billion metric tons since 1992 although recent measurements indicate melting here, too. This ice growth may have slowed sea level rise by about 1.2 mm (0.05 inches) over the last decade.

A major question confronting climate scientists and oceanographers is why the ice caps are melting. Starting in the 1920s, the Arctic experienced more than two decades of warming, with an average increase in temperature of 1.7°C over previous decades. Cooling in the decades that followed never reached prewarming levels, but the occurrence of warming events at reduced greenhouse gas concentrations (about 20% of present values) suggests that natural climate and oceanographic variability plays a significant role. Scientists debate the degree to which these warming events and atmospheric circulation patterns can be attributed to natural variability or global warming. Until scientists gain a better understanding of the interactions of atmospheric, oceanographic, and ice processes, we cannot say for certain what processes are causing the ice caps to melt. New satellites and even submarine-based studies may help scientists to determine the fate of the polar ice caps in the near and long-term future.

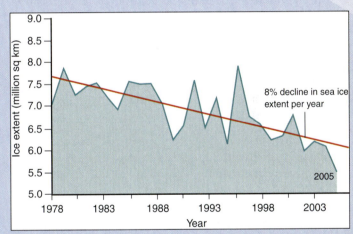

FIGURE 7a The decline in minimum sea ice extent (September) in the Arctic since 1979. The blue line represents an 8% decrease per year. Similiar trends have been observed for the west Antarctic Ice Sheet.

YOU Might Wonder

1. **If water doesn't turn into a gas until it reaches a temperature of 100°C, how can it evaporate from a glass without boiling?**

   The devil is in the details here. On a molecular scale, heat is rarely distributed evenly in a given volume of water. At any given moment, a group of water molecules may contain sufficient energy to transform from a liquid to a gas. When the internal energy reaches the appropriate level, the liquid-phase water molecules become vapor-phase water molecules. In fact, water molecules at the surface of a glass of water are constantly leaving the glass as vapor and simultaneously entering the glass as liquid. Factors such as temperature and humidity alter the equilibrium of the phase change. By the way, this phenomenon also explains how ice cubes "evaporate" (the process is called sublimation) in the freezer.

2. **Can't we just skip this stuff about physics? It's terribly difficult to understand and not very relevant.**

   If you were in a burning building, would you know what to do? Would you think to crawl along the floor? Knowing that hot air rises and cool air sinks can save your life. While this is a dramatic example of an application of physics, it does illustrate one of the fundamental ways that physics affects your life. Riding a bike or snowboard or performing tricks on a skateboard take time to learn as well, but the rewards last a lifetime. If you will give your mind over to it, physics can be a lot like that, too. At the very least, you will begin to appreciate the natural world as never before.

3. **The amount of ocean warming seems ridiculously small. Why all the worry?**

   Perhaps the most difficult thing to convey about the physics of the world ocean is how tiny changes in a property can result in enormous consequences for the ocean as a whole. To some degree, it's a matter of scale: the ocean represents a vast reservoir, so that it takes a lot of anything to alter its mean properties even slightly. But once altered, the sheer size and inertia of the ocean make it difficult to stop. Small changes in density result in changes to the layering of the ocean, which has consequences for ocean circulation and heat exchange. Perhaps a good example is how a 1°C change in sea surface temperatures can mean the difference between a busy hurricane season and a quiet one.

Key Concepts

- Water has one of the highest specific heats of all known substances.

- The latent heats of fusion and vaporization are the energy required to change the physical state of liquid water to or from ice and to or from a gas, respectively. Water exhibits the highest latent heats of all substances.

- Sensible heat is the heat that can be sensed directly, that is, by direct contact between two objects or systems, and results in a temperature change in one or both of those objects or systems.

- Heat is the exchange of energy between systems or objects when they are not in thermal equilibrium, that is, when their temperatures are different.

- Convection is the cyclic movement of fluids as a result of differences in density, generally brought about by differences in temperature.

- Adiabatic heating occurs when seawater is placed under pressure as it descends deeper. Oceanographers take this heating into account by use of the potential temperature when comparing water masses.

- Both temperature and salinity affect the density of seawater. Increases in salinity increase density, while temperature has an inverse relationship with density, that is, increases in temperature result in a decrease in density.

- The ocean layers itself according to density, which can be expressed in terms of buoyancy; positively buoyant water masses rise while negatively buoyant water masses sink. Neutrally buoyant water masses remain in place unless disturbed.

- Different locations on Earth experience seasonal changes in sun angle, daylength, and incident solar radiation because of Earth's tilt relative to its plane of orbit around the Sun. The distance between Earth and the Sun does not affect the seasons!

- Seasonal changes in incident solar radiation drive the formation and disappearance of the seasonal thermocline and affect the physical structure and stability of the water column.

- Light decreases in intensity with increasing depth according to the absorption and scattering properties of water and dissolved and particulate substances within it.

- Water and materials suspended within it selectively remove certain wavelengths of light, resulting in a change in the color of the water.

- Earth's heat budget represents the balance between factors that absorb solar radiation on Earth and factors that reflect, scatter, or allow the transfer of solar radiation back to outer space.

Terms to Remember

absorption, 129

adiabatic changes, 124

Argo system, 117

attenuation, 128

autumnal equinox, 128

baroclinic, 123

barotropic, 123

Beer's law, 128

blackbody radiation, 126

bio-optical oceanography, 130

buoyancy, 123

calories, 119

Celsius scale, 114

change of state, 118

conduction, 120

conductivity-temperature-depth instrument (CTD), 115

convection, 120

destratification, 128

electromagnetic radiation, 126

electromagnetic spectrum, 126

energy budget, 131

expendable bathythermograph (XBT), 115

extinction coefficient, 129

Fahrenheit thermometer, 114

heat, 118

hydrostatic equation, 122

hydrostatic pressure, 122

in situ, 117

in situ temperature, 125

isobars, 122

Kelvin scale, 114

latent heat, 122

latent heat of fusion, 122

latent heat of vaporization, 122

mechanical bathythermograph, 115

messenger, 115

mixed layer depth, 128

multichannel seismic profiling, 117

National Data Buoy Center, 117

photic zone, 130

physical oceanography, 113

physical properties, 118

potential temperature, 124

radiometer, 130

reversing thermometer, 115

rosette, 116

salinity-temperature-depth instrument (STD), 115

scattering, 129

seasonal thermocline, 128

Secchi disk, 129

sensible heat, 120

solar constant, 126

solar spectrum, 126

specific heat, 118

spring equinox, 128

stable water column, 124

Stefan-Boltzmann formula, 127

stratification, 128

summer solstice, 128

TAO-Triton Buoy Array, 117

temperature, 118

thermal equilibrium, 118

thermistors, 115

undersampling, 115

unstable water column, 124

visible light, 126

Wien's law, 127

winter solstice, 128

Critical Thinking

1. Draw a diagram of the ocean's heat budget that depicts the major pathways of heat exchange between the atmosphere and the ocean. Using colored arrows, indicate the changes that you expect would occur if internal heating of the ocean by hydrothermal vents rose by 25%. Using a different color, indicate the changes that might occur as the result of a major volcanic eruption that ejects dust into the upper atmosphere.

2. Illustrate and trace the path of a gram of water molecules at 70 °C as they evaporate from surface of the tropical ocean through the atmosphere to the place where they precipitate as rainfall at 50 °C back into the ocean at midlatitudes. Label the places where latent heat exchange occurs between the ocean and atmo-sphere, and indicate the quantities of energy involved. Do the same thing for the same gram of water molecules evaporating at their midlatitude location and falling as snow at polar latitudes.

3. Jam a pencil through the center of a piece of round fruit (or a Styrofoam ball) and, using a lamp as the Sun, explain to someone (or your pet) the seasonal cycle of day length and light intensity for your location as Earth orbits the Sun.

4. Write a short essay describing the ideal location for a desk-sized solar energy panel on the roof of your home or school. Discuss the advantages and disadvantages of the location you chose. Draw a graph of predicted solar energy produced by the panel versus time of year at your location.

Explore Online

Visit www.mhhe.com/chamberlin1e for access to chapter quizzing, key term flash cards, video clips, interactive activities, and more. Further enhance your knowledge with web links to chapter-related material!

Exploration Activity 7-1

Exploring the Seasonal Cycle Through Graphical Analysis

Question for Inquiry

How does solar radiation vary at different latitudes over seasonal and multidecadal time scales?

Summary

In this chapter, we explored the nature of solar radiation and its variations over the seasonal cycle. We also saw how changes in solar radiation over the seasonal cycle affect the distribution of heat and buoyancy in the water column. These changes have a marked effect on ocean circulation and on the distribution and productivity of ocean life, as we shall learn in subsequent chapters. Here we examine in greater detail the changes in the apparent solar path that cause variations in the amount of solar radiation that reaches Earth's surface.

Learning Objectives

By the end of this activity, you should be able to:

* Generate tables of data on daylength for different locations at different times of the year
* Generate tables of data on solar altitude (angle above the horizon) for different locations at different times of the year
* Prepare and present graphs that compare solar altitude and daylength at different locations over the seasonal cycle
* Discuss the consequences of these changes in terms of Earth's heat budget
* Explain the reasons for the seasonal changes in each location

Materials and Preparation

For this activity, you will need access to the Internet. While you may graph your data by hand, use of a spreadsheet program is highly recommended. Alternatively, your instructor may provide a printout of data sufficient for completing the activity.

Part I: Exploring Daylength Data at the US Naval Observatory Data Service

The US Naval Observatory provides an excellent set of resources for acquiring data on a number of astronomical phenomena. In this activity, we will focus on two data types: solar altitude and daylength. Go to http://aa.usno.navy.mil/data/ and complete the following:

1. Locate and click on the link Duration of Daylight/Darkness Table for One Year. Follow the directions to obtain a table of daylength data for the current year at cities located in a tropical (23.5S to 23.5 N), temperate (23.5 S/N to 66.5 S/N), and polar location (66.5 S/N to 90 S/N). Use a map and the GeoNet's Name Server (linked on the page) to find the latitude and longitude of appropriate cities. Copy and paste the data for each location into a spreadsheet program (or word processor if you are plotting by hand). You may find that the data you pasted into your spreadsheet all fall in a single column. You can correct this by selecting the column and choosing Text to Columns from the top menu. This will put your data into the proper number of columns. You may also save your data as a text file and open it into your spreadsheet program. You will be prompted on how to treat each column of data, insuring that each cell contains only one data point. Label each table with the city name and year. Save your work.
2. Repeat step one for the 50 years previous to and the 50 years following the current year for each city. For example, if the current year is 2008, then obtain data for the years 1958 and 2058.
3. Compute the average daylength for each month at each of the three locations and for each of the three years in step 2.
4. Plot daylength versus month for each city in the current year (for a total of three graphs). On each graph, also plot daylength versus month for the other two years. You

should now have three graphs with three lines each. Get help from a classmate, your instructor, the help files, or the internet if you are not certain how to create graphs by hand or in a spreadsheet program. A partial example is provided here. (See more examples on our website.)

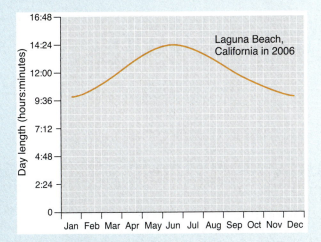

5. For each location, describe the range in the annual day length for each year as shown on each of your graphs. When is day length shortest? Longest? By how many minutes does day length change over the year? How does day length change over multidecadal scales? Speculate why.

6. Compare and contrast day length between locations. Why are they so different? Support your reasoning with your data (i.e., the graphs).

Part II: Exploring Solar Altitude

The height of the sun above the horizon is solar altitude. Together with azimuth (the direction of the sun measured clockwise around the horizon from North), they describe the path of the sun during daylight from the viewpoint of an observer on Earth. Of course, it's not the sun that's changing its location, right?

1. Before you start this part of the activity, predict the variations in solar altitude for your tropical, temperate, and polar locations. Based on what you know, how will solar altitude change over the course of a year at each location? Make a sketch if it helps.

2. Return to the US Naval Observatory Data Services web page. Click on the link Altitude or Azimuth of the Sun or Moon During One Day. Create tables of data for the solar altitude at your tropical, temperate, and polar locations for the dates of the two solstices and two equinoxes in the current year. You will have four tables of data for each location. The dates of the solstices and equinoxes can be found on the website.

3. Plot solar altitude versus hour of day for the four dates at each location. You will have three graphs with four lines each.

4. Describe the range in solar altitude for each day at each location. When is it lowest? Highest? By how much does it change? How much does solar altitude change over the year? When is it lowest? Highest? How much?

5. Compare the variations in solar altitude between locations? When and where is solar altitude highest? Lowest? Which location exhibits the greatest annual variability in solar altitude? Explain why.

6. Assess your predictions of solar altitude. Were your predictions right or wrong? Explain.

Part III. Explaining Day Length and Solar Altitude by the Seasons

Write a paper or prepare a presentation (perhaps even using a model of the Earth and Sun) that describes the annual variations in solar radiation at different locations on Earth. Explain the effects of these variations on Earth's heat budget over the annual cycle at different locations. Discuss the implications for heating of the upper ocean. Use the Rubric for Analyzing and Presenting Scientific Data and Information to self-evaluate and improve your work. (See Exploration Activity 1-1.)

8

The Ocean and the Atmosphere

Hurricane Katrina in the Gulf of Mexico on August 28, 2005 at 12:37 UTC. A hurricane is the ultimate expression of the interaction between the ocean and atmosphere. Warm sea surface temperatures fuel the updraft of air and generate circulation patterns that transform atmospheric disturbances into hurricanes.

Questions to Consider

1. How do the ocean and atmosphere differ, and how are they similar as Earth's geophysical fluids?

2. What are the pathways of exchange between the ocean and the atmosphere, and what are their relative rates and importance?

3. In what ways does the wind affect the upper ocean?

4. What is the role of ice in heat exchange, gas exchange, and light penetration in the ocean?

5. How has our understanding of global atmospheric circulation changed since the 1700s?

6. How do hurricanes form and why are some hurricane seasons more active than others?

7. What are El Niño and La Niña, and what patterns of ocean and atmospheric circulation characterize them?

Weather and Climate

"I wish that I was the weather, you'd bring me up in conversation forever," a popular song goes. Nearly everyone talks about the weather because it affects so many things that we do in our daily lives. We often think of the weather in terms of its observed effects in the atmosphere, such as a hot or cold day, rainy or dry, windy or calm, sunny or cloudy, and so on. But did you realize that the ocean has weather, too? As with the atmosphere, we can observe **ocean weather,** the day-to-day fluctuations in ocean conditions, such as warm or cold water, choppy or calm, fast-moving or slow-moving, murky or clear, and similar sea states. Scientists define **weather** as the conditions of the atmosphere or ocean (and even outer space) at a particular place and time. If these quantities are averaged at a given site or region over a specific period of time, then we can obtain the "average atmospheric weather" or the "average ocean weather." The averaged weather conditions are a quantity known as **climate.** Climate refers to the accumulated weather statistics over a defined period of time, often for a specific location (**Figure 8-1**). The study of atmospheric weather and related phenomena

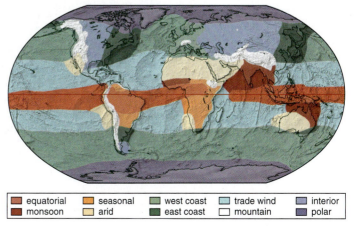

| equatorial | seasonal | west coast | trade wind | interior |
| monsoon | arid | east coast | mountain | polar |

■ **FIGURE 8-1** World climate regions. Each region is represented by the predominant long-term climate processes that contribute to local weather.

is called **meteorology.** Weather studies play a large role in **weather forecasting,** the prediction of weather at a given time and location. **Climatology,** on the other hand, is the study of long-term weather or atmospheric patterns. It seeks to understand the past and future states of the atmosphere. Climatologists look for patterns in climate variability over many thousands and hundreds of thousands of years (see inside front cover). Much like oceanography, meteorology began as an observational science and later applied physical laws and theory to develop predictive models. In modern times, oceanographers and meteorologists have begun to work together to understand the structure and dynamics of the **ocean-atmosphere system,** the mutually interacting, interdependent system of atmospheric and oceanic processes.

A Brief History of Meteorology

Over the centuries, the human desire for weather prediction led to a number of half-myths and pronouncements, such as "red sky at night, sailor's delight; red sky in morning, sailor take warning." **Folk forecasting** of weather has a long and famous history that continues to this day. Nevertheless, human transportation, economic, and military interests require forecasts with much greater dependability.

Aristotle (384–322 BCE) coined the term *meteorology* with the publication of his book *Meteorologica* in 340 BCE. Despite these early beginnings, weather recording did not begin in earnest until the 1400s, when a number of weather-measuring devices were invented (**Figure 8-2**). A very precise rain gauge was invented in 1441 by Korean leader King Seong the Great (1397–1450), who supplied his instrument to every village. Italian architect Leon Battista Alberti (1404–1472) built in 1450 the first **anemometer** to measure wind speed. In that same year, German Cardinal Nicolas of Cusa (1401–1461) measured **humidity** by comparing differences in the weight of wool as water vapor in the atmosphere grew or diminished. Centuries would pass before a systematic and scientific study of the atmosphere would occur. The invention of the barometer in 1643 by Italian physicist Evangelista Torricelli (1608–1647) is often cited as the true beginning of meteorological science.

In 1696, astronomer Edmund Halley (1656–1742) sought to explain the trade winds and the westerlies, patterns of atmospheric circulation long recognized by sailors (**Figure 8-3**). Halley proposed that air warmed by the intense tropical heat would rise and flow northward to the poles, where it would cool and return southward to the equator. Halley envisioned a kind of "giant sea breeze" with two major **atmospheric circulation cells** in each hemisphere. However, as all sailors knew, the trade winds blew in a more easterly direction than predicted by Halley. In 1735, George Hadley (1685–1768), an English lawyer versed in mathematics, modified Halley's model to include the effects of Earth's rotation (what later came to be known as the Coriolis effect). Though Hadley's model could not explain all aspects of the observed wind patterns, it did provide a foundation for the development of more sophisticated approaches. In the 1850s, American meteorologist William Ferrell (1817–1891) published the **three-cell model of atmospheric circulation.** Though oversimplified in many of its aspects, this conceptual model shaped scientific thinking about atmospheric circulation for nearly a hundred years.

a.

b.

■ **FIGURE 8-2** Various instruments for measuring weather. (a) modern rain gauge (b) anemometer and weather station in the Antarctic.

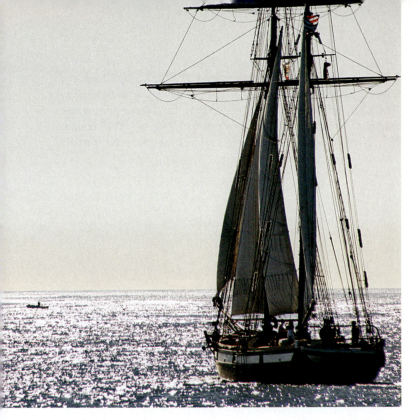

■ **FIGURE 8-3** Scientific efforts of meteorologists in the 18th and 19th centuries were aimed at explaining and understanding global wind patterns, especially the trade winds and westerlies. Sailing vessels depended on these winds for exploration of new lands, international commerce, and warfare.

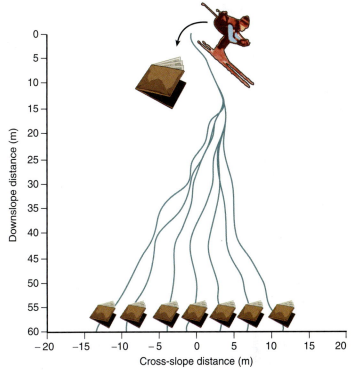

■ **FIGURE 8-4** Chaos theory on a ski slope. Slight changes in the position where a skier loses his wallet dramatically alters the final position of the wallet at the bottom of the slope. In a similar way, small perturbations in one process may lead to significantly different end results in weather or ocean conditions. Chaos prevents long-term prediction of weather and ocean conditions with any degree of certainty.

In 1941, Carl-Gustav Rossby (1898–1957) refined Ferrel's model to include the effects of friction and other physical processes. However, in a good example of how real data can destroy a good theory, Rossby was forced to abandon most aspects of his model when newly invented weather balloons revealed that the upper atmosphere did not behave according to his model. Rossby subsequently proposed the existence of a new type of atmospheric phenomenon: midlatitude atmospheric "waves" that meander back and forth and separate cold polar air from warm tropical air. These long waves, or **Rossby waves,** are thought to be the major means by which heat is transferred between the equator and the poles.

In 1904, Norwegian physicist Vilhelm Bjerknes (1862–1951) began to develop analytical and graphical tools for weather forecasting. Bjerknes proposed the use of mathematical models to predict the future state of the atmosphere using observations of the current state, called the initial state. Bjerknes' work marks the beginning of **numerical weather prediction,** the means of predicting weather by solving equations that describe the behavior of the atmosphere. Unfortunately, a lack of computing power made the large number of computations required for these models impossible. American meteorologist Lewis Richardson (1881–1953) further contributed to weather prediction by developing new mathematical tools to solve equations. He, too, was decades ahead of his time. In his book, *Weather Prediction by Numerical Processes* (1922), Richardson calculated that it would take 60,000 people working continuously to perform the hand calculations required to predict the next day's weather. Despite these limitations, Bjerkenes, Richardson, and many others continued to promote the application of mathematics to weather forecasting, an approach that proved highly successful with the invention of computers in the 1950s.

The complexity of earth systems makes it exceptionally difficult to predict weather more than a few days in advance. **Chaos theory,** advanced by the Massachusetts Institute of Technology's Edward Lorenz in the 1950s, states that dynamical systems are highly sensitive to their initial conditions. In other words, a slight change in one factor can give widely different results. Lorenz provides a simple example of a skier losing his wallet on a hilly ski slope. By changing the location where the wallet is dropped by just 1 mm (0.04 inches) and letting the wallet slide downhill, the final location of the wallet at the bottom of the hill differs by nearly 20 m (66 feet). Slight changes in initial conditions can create large changes in the future state (**Figure 8-4**). According to Lorenz, prediction of future states in chaotic systems may be impossible. Nevertheless, even a few days' warning can be useful, and scientists may yet discover a way to account for chaos in predicting the future state of the atmosphere or ocean.

Earth's Geophysical Fluids

The complex behavior of the atmosphere and the ocean emerges from their properties as **fluids,** substances that flow. Both the atmosphere and the ocean are fluids, which is why they are often referred to as

Earth's geophysical fluids (Table 8.1). The branch of physics dealing with fluid flow on a global basis, **geophysical fluid dynamics,** aims to understand their motions and thermodynamics on molecular to global scales. Compare the properties of the atmosphere and ocean listed in Table 8-1 and think about your own experience with these fluids. Many of the behaviors we attribute to the atmosphere or ocean emerge from the difference in their densities. The average density of seawater ranges from 1030 kg m^{-3} (64 pounds ft^{-3}) at the surface to 1060 kg m^{-3} (66 pounds ft^{-3}) at the seafloor. By contrast, the average density of air at sea level is approximately 1.3 kg m^{-3} (0.08 pounds ft^{-3}). Where they meet at the surface of the ocean, seawater is nearly 800 times denser than air.

TABLE 8.1   Earth's Geophysical Fluids

Characteristic	Atmosphere	Ocean	Comments
Fluid type	Gas	Liquid	Both vital to life
Composition	78.1% N, 20.9% O, 0.9% Argon, 0.0035% CO_2	96.5% H_2O, 2.5% salts	Small amounts of elements and compounds exert large effects, e.g., atmospheric CO_2 as greenhouse gas, NaCl as density-driver in ocean circulation
Relative density	Low	High	Water is about 800 times more dense.
Compressibility	High	Low	Water is nearly incompressible but can support sound waves.
Factors affecting density	Temperature, water vapor, pressure	Temperature, salinity, pressure	Major difference: water vapor versus salinity as complicating factor
Responds to gravity?	Yes	Yes	Causes stratification, surface and internal gravity waves, convection
Internal gravity waves?	Yes	Yes	Seen as rows of clouds and "slicks" on ocean surface
Source of heat	Mostly from below	Mostly from above	Sun energy causes heating of Earth and ocean surface
Examples of convection	Plumes of smoke and clouds	Formation of deep ocean waters	Convection causes upward motion in atmosphere; downward motion in ocean
Vertical boundaries	Land, sea, and ice	Surface winds, seafloor	Creates boundary layer effects
Lateral boundaries	No	Continents, islands	Ocean circulation constrained
Affected by Earth's rotation?	Yes	Yes	Critical to understanding large-scale flows
Tides?	Yes	Yes	Most visible near land-sea boundary for ocean
Turbulent?	Mostly	Mostly	Causes efficient mixing and chaotic motion
Molecular viscosity effects?	Very near boundaries	Very near boundaries; quiescent waters	Important for flows around small organisms
Particles?	Dust and aerosols	Organisms and sediments	Causes very interesting light effects!
Light propagation?	Excellent unless clouds and dust present	Poor in open ocean and worse in coastal waters	Important for heating and photosynthesis
Sound propagation?	Poor	Excellent	Used by marine organisms and oceanographers

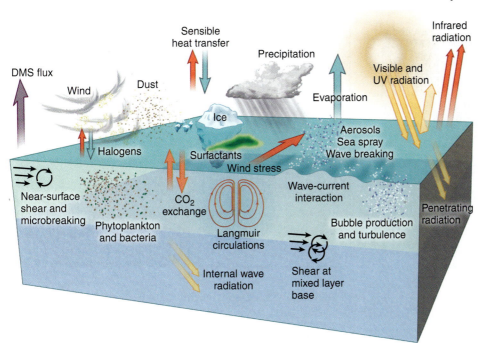

■ FIGURE 8-5 Exchanges of energy and matter at the air-sea interface. A number of different atmospheric and oceanic processes move energy and matter from the atmosphere to the ocean and vice-versa. Scientific studies of the atmosphere and the ocean aim to quantify the rates of exchange and model them under a given set of conditions. Ultimately, an understanding of Earth's climate depends on the success of these efforts.

The Air-Sea Interface

The common boundary between the atmosphere and the ocean is called the **air-sea interface,** the place where the atmosphere and ocean meet and interact (**Figure 8-5**). Through the collision of their molecules, the atmosphere and ocean exchange **momentum,** a quantity variously defined as the quantity of motion or mass in motion. (Formally, momentum = mass × velocity). The wind blowing over the ocean imparts momentum to the ocean surface and causes the ocean to flow or form waves. The ocean and atmosphere also exchange various forms of matter. Gases, salts, water, and a number of volatile organic compounds move across the air-sea interface. Perhaps most important, the atmosphere and ocean exchange energy in the form of latent and sensible heat (see Chapter 7). An understanding of the mutual exchange of matter, energy, and momentum is central to our understanding of the dynamics of the ocean-atmosphere system.

Evaporation and Precipitation

Evaporation and precipitation play key roles in the hydrologic cycle (Chapter 4) and drive variations in ocean chemistry (Chapter 6) and ocean circulation (Chapter 9). An understanding of evaporation and precipitation lies in the concept of **vapor pressure,** the pressure caused by gas molecules (e.g., water vapor). Above any liquid—a glass of water, a cup of coffee, a fish tank—a certain number of molecules escape the surface as a vapor, the familiar process of **evaporation.** Simultaneously, some of the vapor returns to the liquid state as **condensation.** At some point, the rate of evaporation equals the rate of condensation. If we cool the liquid, water evaporates and condenses more slowly. Conversely, rates of evaporation and condensation are faster when we heat a liquid (**Figure 8-6**). The point at which rates of evaporation and condensation are equal is called the **equilibrium vapor pressure.** A graph of equilibrium vapor pressure versus temperature reveals a curvilinear or **exponential relationship** (**Figure 8-7**). The curve becomes steeper at higher temperatures. If the vapor

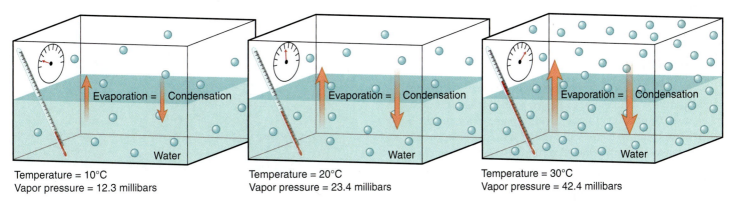

Temperature = 10°C
Vapor pressure = 12.3 millibars

Temperature = 20°C
Vapor pressure = 23.4 millibars

Temperature = 30°C
Vapor pressure = 42.4 millibars

■ FIGURE 8-6 Relationship between temperature, evaporation, condensation, and vapor pressure. Cooler temperatures reduce both rates of evaporation and condensation until the vapor pressure above the liquid reaches equilibrium. Increasing the temperature increases both rates of evaporation and condensation until a new equilibrium is reached. Note, too, that vapor pressure increases as temperature increases.

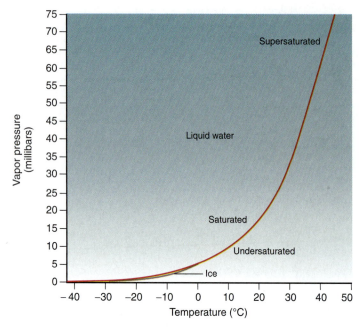

FIGURE 8-7 The relationship between equilibrium vapor pressure and temperature is curvilinear. The curved line represents the equilibrium vapor pressure or the point at which rates of evaporation and condensation are equal. Below this curve, conditions are undersaturated and evaporation may exceed condensation (until equilibrium conditions are reached). Above this curve, conditions are supersaturated and condensation will exceed evaporation. The relationship between equilibrium vapor pressure and temperature plays a major role in heat exchange, cloud formation and changes in salinity in the world ocean.

pressure is lower than the equilibrium point, then conditions are **undersaturated.** If the vapor pressure is higher than the equilibrium point, then conditions are **oversaturated** or even supersaturated. If the air is undersaturated, the rate of evaporation will increase until the equilibrium vapor pressure is reached. If the air is oversaturated, the rate of condensation will increase. The difference between ambient vapor pressure and equilibrium vapor pressure determines whether water evaporates quickly from puddles or condenses as clouds or precipitation. Note that evaporation and condensation of a liquid may occur in a vacuum. Air does not need to be present for the liquid to evaporate or condense. Thus, the common expressions that air "holds" water or that dew forms because the "holding capacity" of air is exceeded are incorrect, scientifically speaking.

Formation of Clouds

Clouds (and other forms of precipitation) form under oversaturated or supersaturated conditions. However, condensation of water vapor requires the presence of a surface. Water vapor at the surface of the ocean or land can condense directly, but water vapor in the free air requires microscopic particles called **condensation nuclei.** These tiny particles, ranging in size from 0.01 to 10 micrometers (μm), come from dust and soil picked up by winds and carried aloft. Volcanoes and fires also contribute particles. Breaking waves and whitecaps supply salt particles to the atmosphere. **Aerosols,** liquid or solid particles less than 1 μm in size, can also act as cloud condensation nuclei. Biological processes also contribute aerosols, especially **dimethyl sul-**

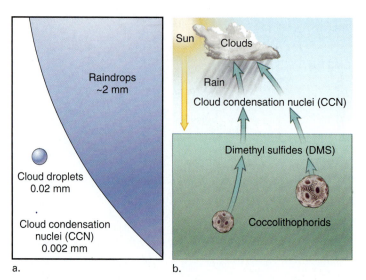

FIGURE 8-8 (a) Cloud condensation nuclei are required for the formation of clouds. (b) Substances like dimethyl sulfides, released by organisms, may contribute to the formation of clouds. Some scientists see this process as a kind of negative feedback. As clouds form, less light is available for photosynthesis, reducing the growth rate of the DMS-producing organisms. With less DMS, fewer clouds form, more light penetrates the atmosphere, growth rates increase, and the cycle repeats.

fides (DMS). Some scientists have proposed that DMS produced by phytoplankton and coral reefs may play a significant role in climate by regulating the formation of clouds (**Figure 8-8**).

Wind Stress and Turbulence

When wind is present, it exerts a force on the sea surface in a near horizontal or tangential fashion, called the **surface wind stress** (**Figure 8-9**). As air molecules impact the surface of the ocean, their momentum is transferred across the air-sea interface to the

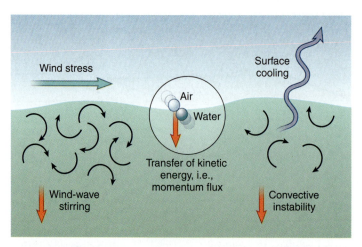

FIGURE 8-9 Wind stress at the air-sea interface transfers kinetic energy from the atmosphere to the ocean through the collision of air and water molecules. Wind stress produces turbulence in the upper ocean. Winds may also increase the rate of evaporative cooling and lead to buoyancy changes at the surface.

■ FIGURE 8-10 Sea ice plays a number of important roles in the ocean-atmosphere system.

Part of Earth's surface	Albedo range (%)
Fresh snow	60–90
Old snow	40–70
Clouds (average)	50–55
Clouds <150 m thick)	25–63
Clouds (150–300 m thick)	45–75
Clouds (300–600 m thick)	59–84
Beach sand	30–50
Forest	5–20
Ocean	5–10

TABLE 8.2   Average Albedo Values for Various Parts of Earth's Surface

Sources: Ruddiman, 2001, and www.geografforlaget.dk/course/ENGLISH/basic/vejrsat.htm.

water column. These collisions transfer kinetic energy from the air molecules to the water molecules. This energy generates surface waves and creates currents and small-scale motions. Some is converted to heat. The amount of energy transferred by wind stress depends on numerous factors, including the roughness of the sea surface, the presence of waves, and wind speed. In general, high wind speeds over rougher sea surfaces transfer the most energy across the air-sea interface.

Unstable, often chaotic, fluid flow, or **turbulence,** determines the depth to which wind energy is distributed in the upper ocean. The wind blowing over the surface of the ocean can cause parcels of water to swirl as **turbulent eddies** of varying sizes whose movements are quite complex. If you have ever swished your hand through a bathtub of water, the swirls of water produced were turbulent eddies. Wind-generated turbulence is the dominant means by which energy and momentum are transferred from the atmosphere to the ocean surface. The surface **turbulent boundary layer** is the region of the upper ocean affected by wind-generated turbulence. The depth of this layer depends on the amount of energy imparted to the upper ocean and will vary according to existing atmospheric and oceanographic conditions.

The Role of Sea Ice

Sea ice has enormous consequences for the exchange of heat, momentum, and gases across the air-sea interface (**Figure 8-10**). When sea ice is present, exchanges between the atmosphere and the ocean are reduced or blocked. Cracks in the sea ice, called **leads,** may permit exchanges between the atmosphere and the ocean. Open expanses of water encircled by sea ice, called **polynyas,** similarly affect air-sea exchange. Sea ice also reflects a significant portion of sunlight back into outer space. The reflectivity of sea ice affects Earth's albedo. Sea ice, ice-

bergs, snow, glaciers, reflective white clouds, sea foam, and other highly reflective materials have high albedos (**Table 8.2**). By contrast, seawater has a low albedo, around 2%. On average, Earth's albedo is about 31% but changes significantly depending on location and time of year. During winter, when the extent of sea ice is maximal, the reflection of sunlight by sea ice acts as a positive feedback mechanism to maintain the cold state of polar regions (**Figure 8-11**). Conditions that produce greater amounts of sea ice, such as ice ages, result in a climate that warms more slowly. Sea ice is also an important habitat for marine organisms and has a significant effect on the productivity of polar food webs (see Chapter 14).

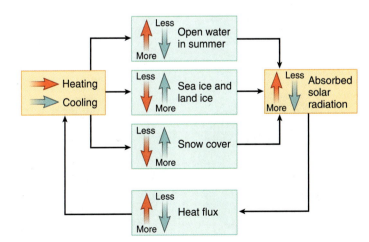

■ FIGURE 8-11 The ice-albedo positive feedback loop. As climate cools, reflective snow cover and sea ice grow in extent. In turn, Earth's albedo increases, leading to even more cooling. Melting of ice caps and sea ice can have the opposite affect. Decreases in ice extent accelerate warming by lowering Earth's albedo.

The Coriolis Effect

An important yet frequently misunderstood concept in oceanography and meteorology is the Coriolis effect. Named after French mathematician Gaspard Gustave Coriolis (1792–1843), who first derived its mathematics, the details of the Coriolis effect have been known for centuries. The difficulty in understanding the Coriolis effect stems in part because we live on the surface of a rotating sphere. From our perspective, the Earth appears to be flat and motionless. We even think of sunrises and sunsets as the Sun moving, not Earth. This "perspective" is what scientists call a fixed frame of reference. When driving from one city to the next, you have little concern that your destination is moving along with Earth's rotation while you are driving. That's because you, too, are moving along with Earth's rotation. Points on Earth's surface are fixed relative to each other. Thus, under the circumstances of our daily lives, the rotation of the Earth can be ignored. However, for objects moving freely across the surface of our planet—such as ocean currents and weather systems—Earth's rotation must be accounted for. In these instances, a rotating frame of reference is the preferred one. To account for Earth's rotation, oceanographers and meteorologists introduced an artificial force called the Coriolis effect. The artificial force is necessary to compensate for motions on a rotating sphere as viewed from a fixed frame of reference.

In its simplest form, the Coriolis effect is *the apparent deflection of moving objects to the right in the Northern Hemisphere and to the left in the Southern Hemisphere* (**Figure 8-12**). Air and water in motion are subject to this effect and exhibit deflections to the right or left as they travel across the face of the Earth. Even a baseball thrown over a distance of 30.5 m (100 feet) in 2 seconds in Yankee Stadium will experience a deflection of about 0.75 cm (about ¹/₄ inch) due to the Coriolis effect. However, water draining from a sink or bathtub is only minimally affected by the Coriolis effect. Other factors (such as the shape and levelness of the sink, and friction) determine the direction of rotation of water on small scales, where the Coriolis effect is, in most cases, negligible.

To gain an intuitive idea of the Coriolis effect, consider a person standing at the North or South Pole (**Figure 8-13**). This person rotates at a rate of one revolution per day but has no other motion than spinning. Now consider a person standing along the equator (**Figure 8-14**). This person moves eastward at a rate of 464 m s⁻¹ (1037 miles per hour) to traverse the entire circumference of Earth in a day. The rotational component of the Coriolis effect, or **angular velocity,** corresponds to the "rate of spin" for a given

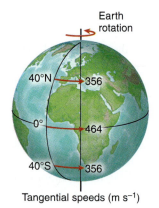

Tangential speeds (m s⁻¹)

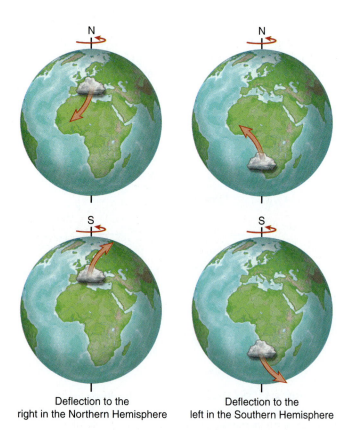

Deflection to the right in the Northern Hemisphere

Deflection to the left in the Southern Hemisphere

■ **FIGURE 8-12** The Coriolis effect makes objects in the Northern Hemisphere appear to deflect to the right. Moving objects in the Southern Hemisphere appear to move to the left.

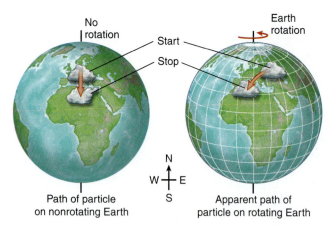

No rotation

Start

Stop

Earth rotation

N
W   E
S

Path of particle on nonrotating Earth

Apparent path of particle on rotating Earth

■ **FIGURE 8-13** If you view the Earth as a rotating sphere from outer space, a person (or object) on the equator would appear to be moving at 464 m s⁻¹ (1037 miles per hour), the linear (or tangential) speed of Earth's rotation. At latitudes closer to the poles, Earth's eastward velocity is slower. For an object like a cloud that starts at a point on the equator and drifts northward, its eastward velocity will be faster than points beneath it and the cloud will appear to move towards the right. The same reasoning applies for any objects moving about the face of the Earth. If their eastward velocity differs from points over which they are moving, they will appear to move right in the Northern Hemisphere or left in the Southern Hemisphere.

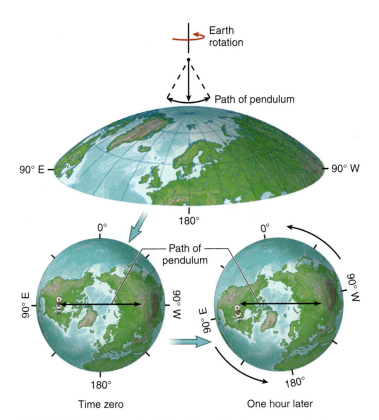

■ FIGURE 8-14 The Coriolis effect also has a rotational component that accounts for differences in Earth's rotational (or angular velocity). Consider a pendulum swinging between 90E and 90W at noon along the North or South pole. To a person standing next to it, the pendulum appears to rotate clockwise throughout the day (at a rate of 360 degrees/24 hours or 15 degrees per hour). This apparent motion results from the Earth's rotation beneath the pendulum, which swings constantly in the same direction, as viewed from outer space. Along the equator, an east-west swinging pendulum continues to swing east-west throughout the day. The rotational component at the equator is zero.

latitude. The eastward movement, or **linear velocity,** corresponds to the rate of motion tangential to Earth's surface (along the same plane as Earth's surface.) For a person standing halfway between the equator and the pole, the angular and linear velocities are half the rates observed at the pole and equator, respectively. So both of these components are important when calculating the magnitude of the Coriolis effect for objects that move freely over Earth's surface. The mathematics of these calculations is well beyond the scope of our treatment here, and further discussion runs the risk of confusing rather than enlightening the reader. For an excellent treatment of this subject, including the dangers of explaining the Coriolis effect in terms of turntables and merry-go-rounds, consult the references on our website. For sake of completeness, we summarize a few additional points in **Table 8.3.**

Global Atmospheric Circulation

General global atmospheric patterns can be identified by long-term averages of large data sets obtained over many decades from ships, moorings, satellites, and land-based observations. Because long-term averages smooth out short-term temporal and spatial variations in winds, the patterns they reveal are not representative of the "snapshot" views obtained from satellite images. Nonetheless, they prove useful for understanding air-sea interactions.

The Cause of Winds

Pressure differences in the atmosphere and ocean give rise to winds and currents, respectively. Fluids, such as air and water, move when

---

TABLE 8.3 Summary of the Coriolis Effect and Its Mathematics

1. The Coriolis effect accounts for the motion of a rotating sphere (i.e., the rotational effects and sphericity of the Earth) to explain and quantify the motion of objects from the perspective of a fixed frame of reference (i.e., on the surface of the Earth).

2. From our fixed frame of reference, moving objects such as winds and currents deflect to the right of their trajectories in the Northern Hemisphere and to the left of their trajectories in the Southern Hemisphere (Figure 8-12).

3. The Coriolis effect increases with latitude. (Formally, it is proportional to the Coriolis parameter, f, which is given by $f = 2 \, \Omega \sin \phi$, where ω is the rotation rate of the Earth and ϕ is latitude.)

4. The Coriolis effect for horizontal motion is zero at the equator. (At the equator, $\phi = 0$ and thus $\sin \phi$ and $f = 0$.)

5. The Coriolis effect is proportional to the velocity of the parcel of seawater or air; faster-moving objects experience greater deflections due to the Coriolis effect.

6. The Coriolis effect is most important for large-scale motions, that is, over scales greater than tens of kilometers (tens of miles), such as the circulation of the ocean and atmosphere.

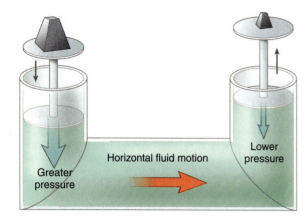

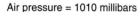

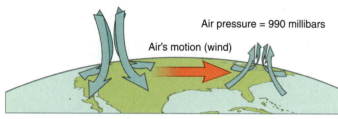

a.

b.

FIGURE 8-15 Horizontal gradients in pressure and their effects. (a) If atmospheric or ocean pressure is greater on one side than another, air or seawater will flow from the region of higher pressure to lower pressure. (b) Horizontal pressure gradients are responsible for the winds we observe in the atmosphere.

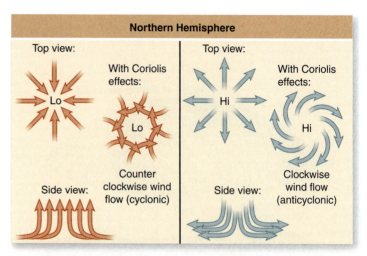

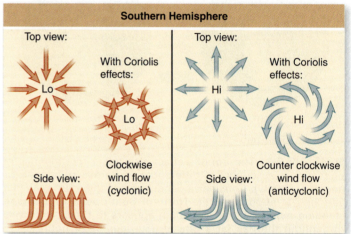

FIGURE 8-16 Low pressure centers draw air towards them. As the air moves, it rotates under the Coriolis effect in a cyclonic direction. Northern hemisphere cyclones rotate counterclockwise while southern hemisphere cyclones rotate clockwise. High pressure centers cause air to flow outwards. As it does so, the Coriolis effect causes it to rotate in an anticyclonic direction. In the northern hemisphere, anticyclones rotate clockwise while in the southern hemisphere they rotate counterclockwise.

pressure differences exist within the fluid (**Figure 8-15**). Differences in air pressure between two locations in the atmosphere give rise to **winds,** while differences in hydrostatic pressure in the ocean can create **currents.** Global meteorological data reveal patterns in air pressure that can be used to predict weather. Air masses with higher pressures are called **highs,** while air masses exhibiting lower pressure are called **lows.** Of course, these designations are relative. They designate **pressure centers,** locations where the highest or lowest air pressure is observed in a given region. Because air moves from high pressure centers to low pressure centers, meteorologists can predict the direction and magnitude of the winds. However, they must also account for the Coriolis effect. The Coriolis effect causes low pressure centers to rotate counterclockwise in the Northern Hemisphere and clockwise in the Southern Hemisphere (**Figure 8-16**). This pattern of circulation is called a **cyclone,** not to be confused with a tornado. The term *cyclone* refers to all low pressure systems, including tornadoes, hurricanes, and typhoons, all of which exhibit cyclonic circulation. Atmospheric high pressure centers rotate clockwise in the Northern Hemisphere and counterclockwise in the Southern Hemisphere. These systems are called anticyclones. Their circulation pattern is **anticyclonic.** Similar processes occur in the large-scale movements of ocean currents (see Chapter 9).

Major Patterns

Ferrel's three-cell model of atmospheric circulation envisions three large convection cells in each hemisphere that govern air flow in the global atmosphere (**Figure 8-17**). **Hadley cells** (named in honor of Hadley) form when warm air rises at the equator and sinks at 30°degrees N or S. **Polar cells** form from the sinking of cold air at the poles and the rising of air at 60°degrees N or S. Between the Hadley and polar cells, a third type of circulation cell occurs, the **Ferrel cell** (named in honor of Ferrel). Within the Hadley cell, trade winds blow toward the equator to replace air that rises due to heating by the tropical ocean. As the air moves, it deflects under the influence of the Coriolis effect, creating a corkscrew-type rotation that creates a spiral of air out of the tropics. In reality, the rising air may circle the Earth before it reaches the upper atmosphere. Eventually, cooling of this air causes its descent. The surface circulation of the Ferrel cell also moves easterly and poleward as the **westerlies.** Sailors traveling to America from Europe could conveniently sail southwest on the trade winds and return home toward the northeast on the westerlies. **Polar easterlies** form at high latitudes within the polar cell as air masses over the poles move equatorward and turn west under the influence of the Coriolis effect.

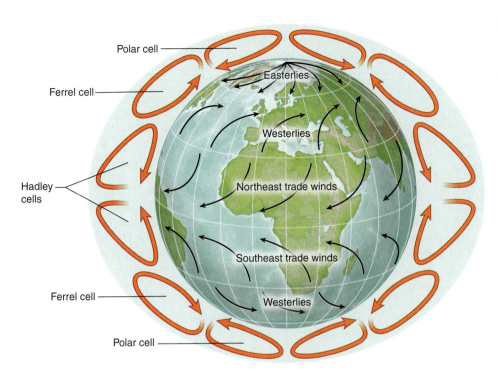

■ **FIGURE 8-17** Ferrel's three-cell model of atmospheric circulation. Ferrel expanded on Hadley's earlier model. In reality, the Ferrel and polar cells only persist intermittently in the atmosphere.

In actuality, the classic three-cell circulation pattern breaks down at high latitudes as weather systems overwhelm the large-scale circulation **(Figure 8-18)**. The easterly and poleward movement of air in the upper part of the Hadley cell feeds the **subtropical jet stream,** a fast-moving current of high-altitude air that stretches across the globe. The boundary where the poleward-mov-

ing westerlies encounter the equatorward-moving polar easterlies is called the **polar front.** At high altitudes, this convergence feeds the **polar jet stream.** This boundary separates the northerly cold, dry air masses from the southerly warm, moist air masses. Rossby waves also develop along the polar front, causing it to undulate in three to six long waves as it wraps around the poles **(Figure 8-19)**.

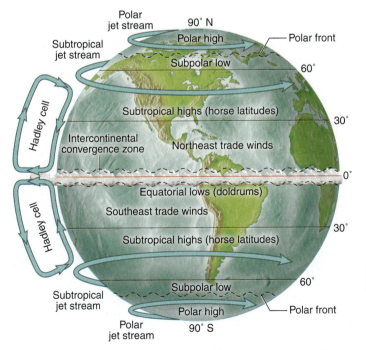

■ **FIGURE 8-18** Alternative model of atmospheric circulation. While the three-cell model provides a convenient starting point, many of its features appear only intermittently. The Hadley cells adequately describe low-latitude atmospheric circulation, but the Ferrel cells and Polar cells are interrupted by seasonal highs and lows and modified by Rossby waves.

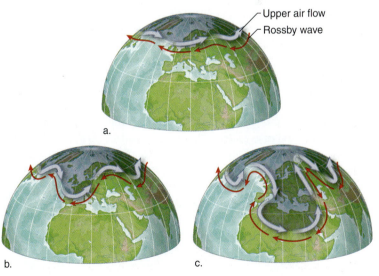

■ **FIGURE 8-19** (a) A Rossby wave propagates westward along the polar front. (b) Higher wind speeds and other atmospheric conditions may increase the number of Rossby waves and shorten their wavelengths. (c) The "crests" and "troughs" of Rossby waves may bring warm or cold conditions as the wave migrates across a given location.

Hurricanes and Global Warming: Is There a Connection?

Hurricanes and typhoons rank among the most deadly and destructive of all natural phenomena. In recent decades, economic and human losses associated with landfall of tropical storms and cyclones have increased worldwide. One hypothesis to explain these losses is that more people now live near the coastal zone, where hurricanes exert their greatest fury. While it is true that coastal populations have increased, is population density the only factor causing increased losses? Some scientists think that hurricanes may be occurring more frequently and with greater intensity, and that global warming is to blame. Does the scientific evidence support the hypothesis that global warming is causing more intense and frequent hurricanes?

Between 1944 and 1970, an average of 2.7 hurricanes appeared annually. From 1971 to 1994, the activity fell to an average of 1.5 hurricanes per year. From 1995 to 2000, the activity increased to 3.8 hurricanes per year, making it the most active six-year period of all time (1944–2000). Scientific analysis of active and quiet periods suggests that a multidecadal pattern of change in sea surface temperatures (SSTs) in the Atlantic may strongly influence the number of hurricanes formed in a season. The correlation between hurricane activity and the Atlantic multidecadal mode is so convincing that researchers predict hurricane activity will be higher than normal for the next 10 to 40 years!

While many scientists remain skeptical of a positive influence of global warming on hurricane activity, new studies suggest that the link to global warming is possible. Professor Kerry Emanuel of the Massachusetts Institute of Technology addressed these hypotheses in a paper published in *Nature* (August 4, 2005). His statistical analyses of data collected since the mid-1970s indicates that the number of tropical cyclones (hurricanes and typhoons) has fluctuated between about 80 and 100 per year for several decades but that no long-term increase in numbers is evident. However, Emanuel's analysis also included an estimate of the intensity of hurricanes over their lifetime. According to Emanuel, the intensity of Atlantic hurricanes has more than doubled since the mid-1970s; the intensity of Pacific typhoons has increased by 75%. At the same time, the lifetime of tropical storms—the amount of time from their formation to their dissipation—has grown over the past several decades. The destructiveness of hurricanes, as calculated by Emanuel, has shown a similar rising trend. While Emanuel attributes increased activity to increases in ocean warming, he acknowledges that SSTs alone cannot explain the patterns in his data.

Another scientist, Ken Trenberth of the National Center for Atmospheric Research, wonders if scientists are even "asking the right question." He argues that what is important to understand is *how* hurricanes are changing as a result of global warming. What is the role of water vapor in the trophosphere, wind shear, deep ocean circulation, and El Niño, all of which have been shown to exert an influence on the number, intensity, and duration of hurricanes? The way in which global warming affects these factors will determine the degree to which it influences hurricane activity and destructiveness. As Trenberth points out, the effects of human activity on hurricanes are clear. What is unclear is how this activity is changing hurricanes and whether these changes will override multidecadal patterns or enhance them.

Clearly, more data are needed. New satellite measurements of wind speeds over the ocean will help, but these data need to be "sea-truthed" with direct measurements from buoys at sea. Such studies are underway using data collected from the Bermuda Testbed Mooring and similar platforms. In that sense, oceanographers will play an increasingly important role in studying the nature of air-sea interactions and the role of the ocean in producing destructive hurricanes. Stay tuned as the scientific debate on this important problem continues.

FIGURE 8a Storm surge and flooding as a result of Hurricane Katrina devastated the Gulf Coast. Scientists continue to debate whether catastrophic hurricanes will become more frequent as a result of global warming.

■ **FIGURE 8-20** Satellite image of the equatorial Pacific. The ITCZ is clearly visible as clouds extending across the equator.

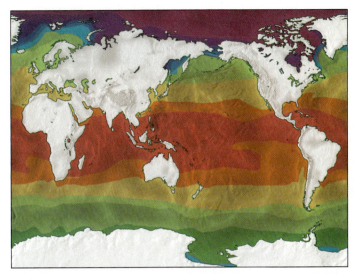

a. **January**

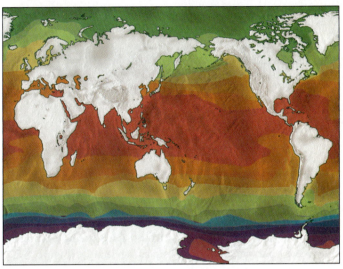

b. **July**

Surface Temperature (°C)

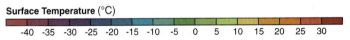

-40 -35 -30 -25 -20 -15 -10 -5   0   5   10   15   20   25   30

■ **FIGURE 8-21** Global distribution of sea surface temperatures in (a) January and (b) July based on shipboard measurements. Source: World Ocean Atlas 2001.

Rossby waves dominate atmospheric circulation at mid- latitudes and greatly influence the weather patterns across the United States. Interestingly, Rossby waves also occur in the ocean where they play a role in both physical and biological processes.

The convergence of trade winds at the equator creates a region called the **Intertropical Convergence Zone** (ITCZ). The ITCZ is characterized by low atmospheric pressure, calm winds, and stifling humidity, a region known to sailors as the **doldrums.** The humidity stems from high rates of evaporation in the equatorial ocean, which delivers enormous quantities of water vapor to the atmosphere. Cooling at higher altitudes causes the water vapor to condense as clouds and release its latent heat to the atmosphere. Formation of clouds within the ITCZ produces massive amounts of precipitation and surface waters in the ITCZ have a lower salinity than waters at higher latitudes. The ITCZ is typically visible in satellite images as a band of clouds centered near the equator (**Figure 8-20**). The heat energy is then transported to higher latitudes by the Hadley cell. In this way, the equatorial ocean acts as a kind of "firebox" for the rest of Earth's atmosphere. Descending air at 30° causes high-pressure and atmospheric light winds, a region known as the horse latitudes. Sailing ships carrying livestock are rumored to have thrown the animals overboard at these latitudes, presumably when they died of dehydration from voyages prolonged by a lack of winds.

The Physical Structure of the World Ocean

Heat exchange across the air-sea interface may warm or cool the surface of the ocean and change its temperature. At the same time, heat plays a role in evaporation and precipitation, processes that alter the salinity of surface waters. As a result, heat plays an important role in determining the density of surface waters in the world ocean (see Chapter 7). Exchanges of heat and matter, and the resultant changes in density determine the **physical structure** of the world ocean, the layering and movement of parcels of water and their distributions over temporal and spatial scales. Visualizing the ocean in this way

guides our thinking much in the same way that visualizing a rain forest—with its canopy, middle layers, and undergrowth—gives us a "model" of terrestrial systems.

Global Sea Surface Temperatures

One of the most common global views of the world ocean can be found in maps of **sea surface temperature** (SST). Monthly composites (averages) of SST for January and July illustrate the kinds of extremes found in the seasonal cycle (**Figure 8-21**). False color scales emphasize the differences in temperature at different locations using reds for warmer temperatures and blues for cooler temperatures. Comparison

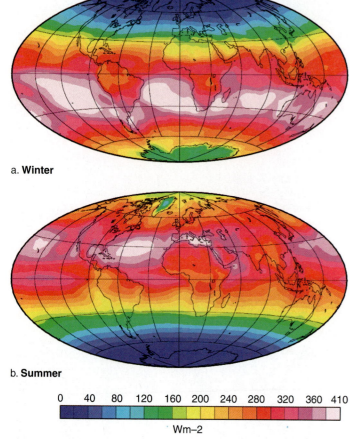

a. **Winter**

b. **Summer**

| 0 | 40 | 80 | 120 | 160 | 200 | 240 | 280 | 320 | 360 | 410 |

Wm–2

■ **FIGURE 8-22** Average of shortwave radiation (Watts per square meter) absorbed at the Earth's surface for (a) December–February (1985–1986) and (b) June–August (1985–1986). Compare this figure to Figure 8-21.

of seasonally averaged SST maps with the seasonal means for absorbed **shortwave solar radiation** reveal broad agreement (**Figure 8-22**). Though seemingly obvious, the correlation between shortwave solar radiation and SST indicates that solar radiation is the principal means by which the surface of the world ocean is heated.

Global Sea Surface Salinities

Evaporation and precipitation are the two principal factors that control **sea surface salinity** (SSS) in the world ocean. The salinity of a given parcel of water represents the balance between evaporation and precipitation. If evaporation exceeds precipitation, a saltier water mass will result. If precipitation exceeds evaporation, then the parcel of water will be less saline. Meteorologists express this quantity formally as precipitation (P) minus evaporation (E), so that:

If $P-E < 0$, then the ocean surface becomes saltier (salinity increases).

If $P-E > 0$, then the ocean surface becomes fresher (salinity decreases).

If $P-E = 0$, then the surface salinity remains the same.

The global seasonal averages of SSS and $P-E$ show broad agreement (**Figures 8-23, 8-24, and 8-25**). On local scales, wind

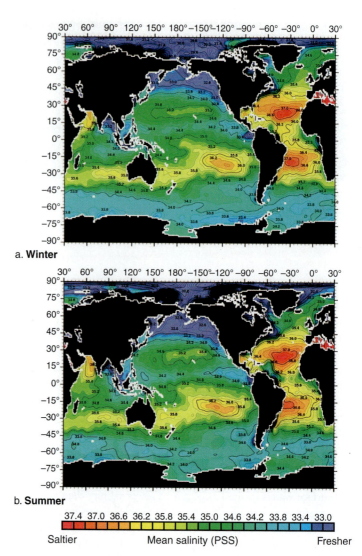

a. **Winter**

b. **Summer**

| 37.4 | 37.0 | 36.6 | 36.2 | 35.8 | 35.4 | 35.0 | 34.6 | 34.2 | 33.8 | 33.4 | 33.0 |

Saltier   Mean salinity (PSS)   Fresher

■ **FIGURE 8-23** Global distribution of surface salinity averaged for (a) Jan–March; and (b) July–September.

patterns, the outflow of freshwater from rivers, and the seasonal formation and decay of sea ice will influence SSS. Wind transport of water evaporated over the Atlantic Ocean and transported to the Pacific Ocean has been hypothesized to explain the greater mean value of salinity found in the Atlantic basin.

Meridional Sections of the World Ocean

Despite the tremendous advantages afforded by satellites, an understanding of the world ocean based solely on surface properties would be highly distorted. In view of the broad temporal and spatial scales now resolved by satellites, the need for measurements beneath the ocean has never been more critical. Concerted efforts to obtain a higher-resolution picture of the vertical structure of the world ocean have met with considerable success. Ship-based CTD and related depth-varying measurements of temperature, salinity, and other properties have now been assembled into large Internet-accessible databases. One such data set, the Levitus data set, boasts more than 7 million profiles. Though not as detailed as satellite measurements

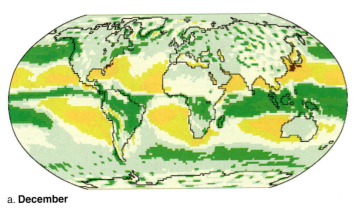

a. **December**

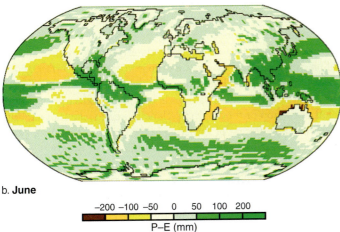

b. **June**

-200 -100 -50   0   50 100 200
P-E (mm)

■ **FIGURE 8-24** Global distribution of P-E in (a) December, and (b) June. Compare this figure with Figure 8-23.

of the sea surface, these data provide a view of the distribution of ocean properties that is impossible to obtain with satellites.

Vertical slices of ocean properties along meridians or lines of longitude are called **meridional sections.** Meridional sections of temperature and salinity obtained during the World Ocean Circulation Experiment provide a dramatic view of the layering of the world ocean (**Figure 8-26**). These sections reveal three fundamental features. The sharp vertical gradient in temperature—visible as the close spacing of color bands—reveals what is

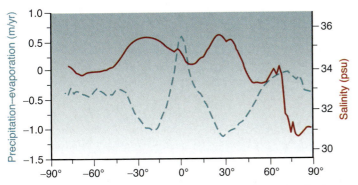

■ **FIGURE 8-25** P-E and surface salinity along a longitudinal section. Note the inverse relationship between P-E and salinity.

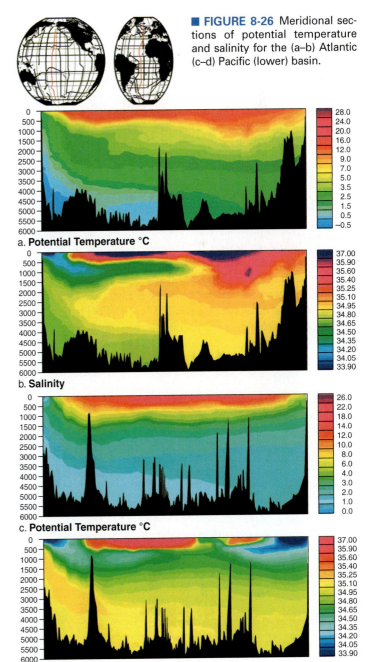

a. **Potential Temperature °C**

b. **Salinity**

c. **Potential Temperature °C**

d. **Salinity**

called the permanent **thermocline** (meaning "temperature slope or change"). The permanent thermocline separates warm surface waters from cold deep waters. A similar feature can be seen in the salinity sections, denoting the permanent **halocline** (meaning "salt slope or change"). Though not shown here, a meridional section of density—which results from the combined effects of temperature and salinity—would reveal the presence of a permanent **pycnocline** (meaning "density slope or change"). In most cases, the depths of the permanent thermocline and pycnocline are similar. This region marks the boundary between the upper ocean and the deep ocean. In general, surface processes dominate above the pycnocline, while deep-ocean processes dominate below the pycnocline. Of course, the word *permanent* should not be taken too literally. The depth of these features will vary between meridians and on seasonal scales.

■ **FIGURE 8-26** Meridional sections of potential temperature and salinity for the (a–b) Atlantic (c–d) Pacific (lower) basin.

Temporal and Spatial Variability in the Ocean-Atmosphere System

In addition to daily and seasonal changes, the ocean-atmosphere system exhibits longer-term patterns of change. **Climatological variability** refers to oceanic and atmospheric phenomena that change over time scales of years to millennia and space scales from local to global (see inside front cover). As a practical matter, oceanographers and atmospheric scientists find it useful to consider how these processes vary over interannual scales (year to year) or interdecadal scales (decade to decade) to assist in citizen preparedness and mitigation of potential weather hazards. At the same time, climate exhibits variability over spatial scales and in ways that appear to be connected. For example, El Niño events in the Pacific (see the section on El Niño in this chapter) generally promote mild hurricane seasons in the Atlantic. The apparent connections between ocean and atmospheric processes on global scales are called **global teleconnections.** Though little understood at present, global teleconnections represent one of the exciting areas of ocean-atmosphere research in the twenty-first century.

The Atlantic-Caribbean Hurricane Season

The period of time from June 1 to November 30 is officially known as **hurricane season.** It is during this time of year that oceanographic and meteorological conditions are favorable for the formation of hurricanes (**Figure 8-27**). Typically, hurricanes form over tropical waters where the humidity is high, winds are light, and sea surface temperatures exceed 26.5°C (80°F). The first stage of hurricane development comes during the convergence of sur-

face easterly winds, producing thunderstorms. Occasionally, these storms develop into a **tropical disturbance,** a loosely organized, self-sustaining group of thunderstorms.

Because they are regions of low pressure, tropical disturbances develop cyclonic rotation. At some distance from the center of the low pressure system, typically 400 km (250 miles), the converging winds begin to rise, a process called **ascent.** A rising spiral of converging air characterizes the circulation pattern of hurricanes. Moisture also plays a significant role in the formation of tropical cyclones. The ascent of air from the warm sea surface temperatures brings moisture-laden air aloft. The change in state of water vapor to a water droplet releases latent heat, which fuels an even greater ascent of air. Thick cumulus and cumulonimbus clouds are formed, the cloud type associated with thunderstorms. These towering clouds cause an even greater release of heat and greater condensation. This process, called **cumulonimbus convection,** is one reason tropical cyclones produce massive amounts of precipitation. The rise of water vapor from the sea surface and the release of latent heat aloft create a positive feedback that sustains and intensifies the circulation patterns in a cyclone. Convergence of winds aloft at the top of the tropical cyclone causes a sinking of air, called **descent.** The descent of air occurs in the center of the storm system and releases heat through adiabatic warming, that is, heat is released as descending air is placed under greater pressure (see Chapter 7). The warm, descending air eventually reaches the surface, creating a warm "core" in the center of the storm system called an **eye** (**Figure 8-28**). The adiabatic release of heat within the eye generally keeps the eye cloud-free. In the region immediately surrounding the eye, some 10–100 km (6–60 miles) in diameter, the intensely converging winds turn sharply upward, creating the most dangerous region of a hurricane, the **eye wall.** The highest wind speeds are produced along the right side of the eye wall, where the wind direction matches the direc-

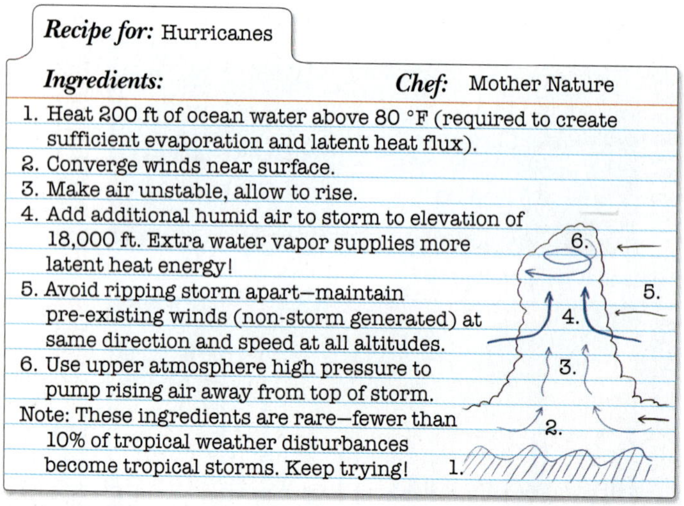

Recipe for: Hurricanes

Ingredients: **Chef:** Mother Nature

1. Heat 200 ft of ocean water above 80 °F (required to create sufficient evaporation and latent heat flux).
2. Converge winds near surface.
3. Make air unstable, allow to rise.
4. Add additional humid air to storm to elevation of 18,000 ft. Extra water vapor supplies more latent heat energy!
5. Avoid ripping storm apart—maintain pre-existing winds (non-storm generated) at same direction and speed at all altitudes.
6. Use upper atmosphere high pressure to pump rising air away from top of storm.

Note: These ingredients are rare—fewer than 10% of tropical weather disturbances become tropical storms. Keep trying!

■ **FIGURE 8-27** Stages of development of a hurricane.

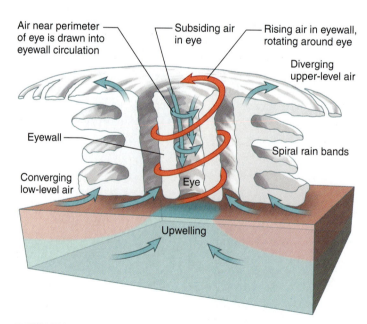

■ **FIGURE 8-28** Anatomy of a hurricane. A warm ocean (>80° F) supplies the energy that maintains a hurricane. At the same time, cyclonic rotation and low pressure at the sea surface beneath the eye causes upwelling of seawater. The most intense hurricanes occur when the surface mixed layer is both warm and deep.

Air near perimeter of eye is drawn into eyewall circulation

Subsiding air in eye

Rising air in eyewall, rotating around eye

Diverging upper-level air

Eyewall

Spiral rain bands

Converging low-level air

Eye

Upwelling

tion of forward motion of the storm. Hurricanes also exhibit spiral bands of clouds that spread outward from the eye in an anticyclonic flow. These spiral bands produce the greatest amount of precipitation in a hurricane, which is why they are often called **spiral rain bands.** These spinal bands produce the heaviest rains in a tropical cyclone.

A number of factors control the number and intensity of hurricanes in a given season. Climatological data on sea surface temperatures (SSTs) in the North Atlantic reveal periods of colder-than-normal SSTs and warmer-than-normal SSTs. **Cold regimes** from 1900 to 1925 and 1970 to 1994 were associated with fewer, less intense hurricanes. **Warm regimes** from 1926 to 1969 and 1996 to the present produced more numerous hurricanes with greater intensity. According to hurricane forecasters, SSTs in the North Atlantic have entered a multidecadal warm phase that may continue to produce more frequent and more intense hurricanes for the next 10 to 20 years or more.

Wind shear, a sort of tussle between upper- and lower-level winds, also plays a major role in hurricane activity. Strong wind shear disrupts the circulation of a tropical storm and inhibits its development. The warming of eastern Pacific equatorial waters during El Niño produces stronger-than-normal wind shear and weaker conditions for the development of North Atlantic hurricanes. On the other hand, La Niña conditions produce weak wind shear and promote North Atlantic hurricane activity. Thus, while people living on the western coast of the United States enjoy La Niña conditions, those along the eastern seaboard prefer the reduced hurricane activity brought by El Niño conditions.

The Pacific El Niño and La Niña

The periodic shifts in ocean and atmospheric properties in the equatorial Pacific have come to be known as **El Niño.** El Niño, roughly translated as "the small boy," specifically refers to the Child Jesus, who was believed to be responsible for bringing favorable warm currents to Peruvian fishermen around Christmastime each year. In modern times, the term *El Niño* is associated with the appearance of higher-than-average water temperatures in the eastern tropical Pacific. Unlike its benevolent historical use, modern use of the term *El Niño* suggests globally destructive weather patterns. El Niño's "partner," La Niña ("the small girl"), is a modern invention often used to describe lower-than-average temperature conditions in the eastern tropical Pacific. El Niño-La Niña describes the cycle of interannual fluctuations in eastern tropical Pacific sea surface temperatures and their associated oceanic and atmospheric processes. A related phenomenon, the **Southern Oscillation,** refers to the atmospheric processes that accompany El Niño and La Niña. Despite the distinct terminology, El Niño, La Niña, and the Southern Oscillation encompass a coupled ocean-atmosphere phenomenon. Many scientists use the term **ENSO** (El Niño-Southern Oscillation) to describe the coupled phenomenon, but we prefer to use El Niño to describe the entire phenomenon, as suggested by Philander (2004).

The 1982 El Niño brought sea surface temperatures of 7°C above normal along the coast of Peru, effectively shutting down the anchovy fishery. The western South Pacific experienced severe drought, bringing widespread fires to Australia and devastating famine to Indonesia and the Philippines. Globally, El Niño-related damage was esti-

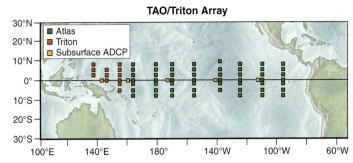

TAO/Triton Array

■ **FIGURE 8-29** The TAO-Triton array provides real-time data on meteorological and oceanographic conditions from 70 sites in the equatorial Pacific. This array provides critical information for the early detection of El Niño and La Niña.

mated at $8 billion. Perhaps most important, the 1982 El Niño event stimulated the deployment of an array of weather and oceanographic buoys across the equatorial Pacific, the TAO-Triton array (**Figure 8-29**). The completion of this array in 1994 enabled one of the most detailed studies of an El Niño to date in 1997, one of the most severe events on record. The data provided by the TAO-Triton array and other monitoring instruments revealed detailed information on the atmospheric and oceanic processes that give rise to El Niño.

A 100-year data set of sea surface temperatures near the Galapagos Islands reveals the patterns of El Niño and La Niña. By taking the average of the entire set of measurements, we can compute the mean temperature and represent it as a straight line. Temperature fluctuations above or below the mean, called **temperature anomalies,** represent El Niño warm-water or La Niña cold-water conditions, respectively (**Figure 8-30**). Officially, a 0.5-degree positive or negative anomaly must be sustained over a period of three months for an official El Niño or La Niña to be declared. Anything in between is now unofficially referred to as *La Nada* ("the nothing"). For the most part, the equatorial Pacific appears to be in a constant state of change between El Niño and La Niña. Some decades produce frequent and prolonged periods of El Niño. In other years, La Niña conditions persist. An alternative view of these events and their severity represents

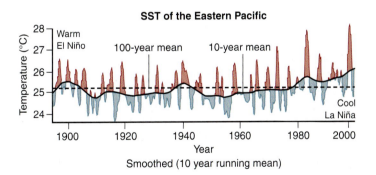

■ **FIGURE 8-30** A plot of sea surface temperatures in the equatorial Pacific since the late 1800s demonstrates the periodicity of El Niño and La Niña. Typically, the strength of an El Niño or La Niña is represented as the deviation, in SST from the 100-yr mean, the so-called temperature anomaly. Alternatively, some scientists use the average temperature over a ten-year period, the ten-year running mean. A ten-year running average appears to better represent the intensity and periodicity of the patterns.

the mean conditions as a moving 10-year average (**Figure 8-31**). In this instance, the otherwise solid line representing average conditions becomes a continually shifting one. While the frequency of El Niño and La Niña conditions remains the same, the intensity of warm- or cold-water periods appears much different. It is against this constantly shifting background of ocean and atmospheric climate that oceanographers and meteorologists must work to try to better understand the oscillations of El Niño and La Niña.

A basic understanding of El Niño starts with the atmospheric and oceanic processes that prevail in the equatorial Pacific under non- El Niño conditions. In the early 1900s, English physicist Sir Gilbert Walker (1868–1958) became interested the Indian Ocean monsoon (see Chapter 9) because widespread famine often occurred when the monsoon failed to bring rain. In 1923, Walker published a statistical analysis demonstrating that the difference in atmospheric pressure between the Pacific Ocean and the Indian continent periodically reversed. At certain times, atmospheric pressure over the Pacific Ocean was higher than atmospheric pressure over the Indian continent. At other times, the situation was reversed. In modern times, the atmospheric pressure difference is used to compute the Southern Oscillation Index (SOI), which describes the atmospheric pressure difference between Papeete, Tahiti and Darwin, Australia.

A positive SOI corresponds to normal or La Niña conditions while a negative SOI indicates El Niño conditions (**Figure 8-31**). Comparing the patterns of El Niño and La Niña in Figure 8-30 with the SOI reveals close agreement. Although unappreciated at the time, Walker's observations set the stage for an understanding of El Niño. In his honor, the pattern of circulation that results from the pressure differences came to be known as the **Walker circulation.** Typically, rising air in the ITCZ generates easterly trade winds and relatively higher air pressures in the eastern Pacific. Under these conditions, air pressure over Tahiti is higher than air pressure over Darwin, a positive state for the Southern Oscillation (**Figure 8-32**). At the same time, these conditions produce a flow

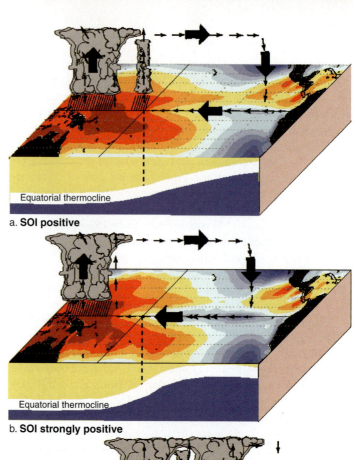

a. **SOI positive**

b. **SOI strongly positive**

c. **SOI negative**

■ **FIGURE 8-32** (a) The Walker circulation pattern consists of descending air over the South Pacific and ascending air over Indonesia and Australia. The Southern Oscillation Index rates the strength of the Walker circulation cell on the basis of differences in atmospheric pressure between Tahiti and Darwin, Australia. The greater the pressure difference, the greater the strength of the southeasterly trade winds. (b) If the pressure difference is exceptionally great, then the southeasterly trade winds produce La Niña conditions. (c) When the pressure difference weakens or reverses, El Niño conditions are produced.

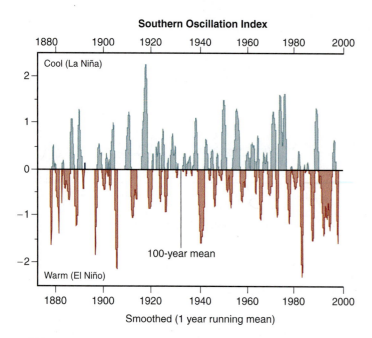

■ **FIGURE 8-31** The Southern Oscillation Index since 1880.

of surface currents toward the west and an upward slope in the tropical thermocline. The "piling up" of water in the western Pacific as a result of the easterly trade winds generates a horizontal pressure gradient that is balanced by the subsurface flow of water in the Equatorial Undercurrent (EUC), among the strongest currents in the world ocean. The warm waters of the western Pacific fuel the cumulonimbus convection that is important for the transfer of heat from the tropics to higher latitudes and is responsible for the seasonal monsoons over this region. In the eastern equatorial Pacific, the shallower thermocline permits upwelling of colder,

Robotic "Oceanographers" Witness Extreme Events

Oceanographers have gone to sea in ships for centuries to sample beneath the surface of the ocean. Ships provide a firsthand look at the ocean and allow oceanographers to conduct many types of experiments that cannot be performed otherwise. Nonetheless, extreme "ocean weather" conditions such as those produced by severe storms, hurricanes, and typhoons make ships a dismal platform for sampling. Extreme conditions may generate surface waves more than 10 m (33 feet) in height and currents as strong as those observed in the Gulf Stream (over 100 cm sec^{-1}; about 2 knots). Because extreme weather and episodic events have a tremendous impact on upper-ocean processes, it is essential that oceanographers and their instruments are deployed to observe, measure, and sample ocean properties when such events occur.

Ocean moorings outfitted with various oceanographic instruments and robotic sampling devices have been deployed in a few parts of the world ocean to attempt to observe oceanographic conditions during extreme events. The Bermuda Testbed Mooring (BTM see Chapter 7), sits in the track window of North Atlantic hurricanes and tropical storms. The BTM was the first mooring to "witness" oceanographic conditions before, during, and after a hurricane, successfully providing data during Hurricanes Felix (1995) and Fabian (2004), and Tropical Storm Harvey (2005). With these initial successes, oceanographers hope to deploy a greater number of moorings over an expanded region of the world ocean to observe extreme events.

Within the past few decades, autonomous unmanned vehicles (AUVs) and underwater gliders have been added to the oceanographer's toolbox (**Figure 8a**). AUVs and gliders also "brave the weather" and collect data when surface conditions might otherwise prevent observations. These instruments move freely through the water under their own propulsion, using buoyancy controls or wings to maintain their depth in the water column. Oceanographers can program these robotic platforms to move back and forth and up and down in the water column to collect data on temperature and salinity, and, in some cases, optical, chemical, and biological data. AUVs and gliders can even be commanded remotely via communication satellites to sample beneath the most intense portions of storms or hurricanes.

While ships will always be important for oceanographic research, moorings, AUVs, and gliders will allow collection of the massive volumes of data that oceanographers require to fully understand temporal and spatial variability in the ocean. Ultimately, data collected from these robotic samplers will be used to develop predictive models of "ocean weather" similar to weather predictions regularly featured on television news channels. When that occurs, you can be assured that robotic sampling platforms will represent the "front lines" of oceanographic data-gathering and help to provide the most complete picture of the upper ocean possible.

FIGURE 8a Robotic gliders descend into the depths to measure oceanographic conditions.

■ **FIGURE 8-33**
Temperature structure of an east-west section of the equator during (a) "normal" and (b) El Niño conditions. Relaxation or reversal of the trade winds during El Niño conditions permit the warm western equatorial water to slosh back towards the eastern Pacific.

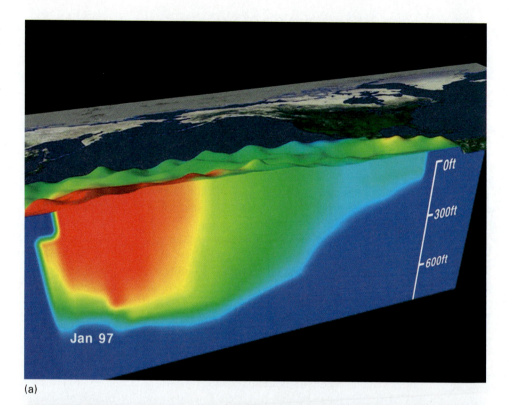

(a)

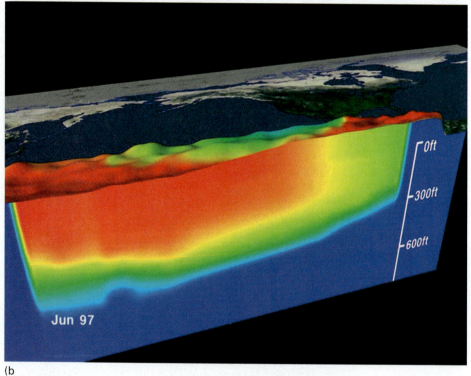

(b

nutrient-rich waters under the right conditions. Upwelling along the coast of Peru stimulates high biological productivity and a very productive fishery under normal conditions.

For reasons that are not well understood, these conditions reverse during an El Niño event. The normally strong trade winds weaken and may even blow in the opposite direction (i.e., toward the east). Atmospheric pressure shifts as well, generat-

ing higher pressure over Darwin than Tahiti, a negative state for the Southern Oscillation Index. Accompanying these changes is a flattening of the thermocline, which allows warm water from the western Pacific to "slosh" toward the eastern Pacific (**Figure 8-33**). As well, the flow of the EUC is considerably diminished or even absent. The deepening of the thermocline in the eastern Pacific effectively puts a lid on upwelling and severely limits biological productivity. At the same time, the eastward shift in warm surface waters fuels greater cumulonimbus convection and higher amounts of precipitation in the eastern Pacific. El Niño conditions often deliver powerful storms to the western coast of the United States and Central and South America. Regions dependent on monsoonal circulation for rainfall may experience severe drought. The shift in cumulonimbus convection also modifies the poleward transport of heat, which may alter the position and intensity of the subtropical jet stream. Anomalous wave patterns in the jet stream during El Niño conditions may partially explain the global tele-connections observed in other parts of the world.

La Niña conditions essentially represent "strong" normal conditions. A more positive Southern Oscillation results in much stronger trade winds and an even greater westward shift of warm surface waters. These conditions are accompanied by a much steeper tilt to the tropical thermocline, even to the point of exposing the colder subthermocline waters at the surface in the eastern Pacific. La Niña conditions can produce generally drier conditions along the western coast of the United States and central and South America while producing greater rainfall in the western Pacific.

Much of our progress in understanding and predicting these events will continue to come from long-term time series measurements, such as those provided by the TAO-Triton array, new satellite observations, and other observational and modeling programs. Nevertheless, scientists remain cautious. Philander (2004) cites an example of farmers and financiers in Zimbabwe who were advised to expect below-normal rainfall because of the intense 1997 El Niño. Taking scientists at their word, Zimbabwean bankers declined loans to farmers, fearing the inability of the farmers to pay back loans if their crops failed. As it turned out, Zimbabwe experienced normal rains that year, but crop production remained 20% below normal because farmers planted fewer seeds. Philander comments:

> "Our biggest challenge is to give appropriate weight to the inevitably uncertain scientific information when making policy decisions. This is too serious a matter to be left to scientists and economists. It is the joint responsibility of all of us because the policies we adopt reflect our values. . .Much can be learned from El Niño. We need to do so in a hurry before we succeed in changing him."

YOU Might Wonder

1. **Okay, I get it that water draining from sinks doesn't rotate clockwise in the Northern Hemisphere and counterclockwise in the Southern Hemisphere, but if the Coriolis effect is important, why doesn't it?**

   Two factors are important here. One is simply the length scale over which the Coriolis effect operates. The Coriolis effect in a sink is too weak, relative to other forces, to exert a noticeable effect on the rotation of the water. The major factor is the sink itself: the design of the basin and the opening determine how water rotates as it drains. Nonetheless, you are correct in thinking that the Coriolis effect is present.

2. **What's the deepest any human has dived in the ocean?**

   The world's record for "no limits" free diving (without an air supply) is 209 m (685.7 feet), set by Belgian diver Patrick Musimu on June 30, 2005. The women's free diving record is held by Tanya Streeter, who dove to 160 m (524.8 feet) on August 17, 2002. Musimu's success apparently stems from his technique of flooding the air spaces in his ears with water before a dive. By equalizing the pressure on his ear drums with water instead of air, he suffers less of the equalization problems divers have in descending to these depths. Scuba divers have now passed beyond 300 m (1000 feet) using specialized mixes of air and very long decompression times following the dive. On July 13, 2005, a 41-year-old French diver reached a depth of 330 m (1082.6 feet). Likely, by the time you read this, new records will have been set.

3. **If hurricanes are so dangerous, why don't we do something to stop them?**

   Well, in fact, attempts have been made to control hurricanes in the past. In the 1960s, Project Stormfury actually seeded hurricane rain bands with silver iodide particles that act as cloud condensation nuclei to "dry out" the hurricane. The experiments were largely unsuccessful and abandoned. More recently, scientists have been using computer models to replicate past hurricanes and, by altering the conditions one by one, look for a particular stimulus or factor that might alter the path or strength of the hurricane. They found that by ever so slightly altering the starting temperatures and winds, the path of a hurricane would be dramatically different. These modeling efforts help scientists to determine when and where hurricane intervention techniques might be effective. Seeding of clouds, spreading biodegradable oil on the water to reduce evaporation, or heating a portion of the ocean with a satellite microwave beam are among the techniques being proposed (Hoffman 2004).

Key Concepts

- The ocean and atmosphere exhibit the properties of fluids, albeit in a different fashion.
- Rates of evaporation and precipitation depend on the vapor pressure, which rises exponentially as temperature increases.
- Formation of clouds requires aerosols, some of which have biological origins.
- Wind stress imparts momentum on the sea surface and causes turbulence.
- Ice may restrict the exchange of energy, materials, and gases across the air-sea interface.
- Differences in atmospheric pressure centered around highs and lows cause winds.
- The Coriolis deflection alters the circulation patterns around high and low pressure systems.
- Seasonal variations in solar radiation cause changes in the physical structure of the water column, especially in temperate regions.
- Global sea surface temperatures correspond to latitudinal changes in incident solar radiation.
- Global sea surface salinities correspond to latitudinal differences in rates of evaporation and precipitation.
- Meridional sections of the ocean reveal the layering of water masses and the influence of polar environments on the structure of the deep ocean.
- Multidecadal variations in climate may affect weather on local and global scales.

Terms to Remember

Critical Thinking

1. Compare and contrast Earth's geophysical fluids in terms of their roles in heat, gas, and materials transports in the Earth system.

2. Trace the development of models of atmospheric circulation since the time of Hadley. Why is the three-cell circulation model an oversimplification?

3. Why might reliable, long-term forecasts of weather and hurricanes be impossible?

4. What is the relationship between temperature and equilibrium vapor pressure? Why is this relationship important for understanding processes that control the salinity of the ocean?

5. Compare the effects of wind stress on the upper ocean when it is stratified and unstratified. Describe how stratification of the water column might limit the effects of wind mixing.

6. How do seasonal changes in sea ice affect heat and gas exchange in the ocean?

7. Describe the wind circulation in a high pressure system and a low pressure system in the Northern Hemisphere. Why is knowledge of air pressure and the Coriolis effect important to an understanding of weather in the United States (or wherever you live)?

8. How are El Niño, La Niña, and hurricanes examples of climatological variability?

9. Describe the relationship between sea surface temperatures and the different stages of a hurricane. How does SST affect the formation, intensity, and duration of a hurricane?

10. Describe Walker circulation and how it changes during El Niño and La Niña. Discuss the "chicken-and-egg" problem of air-sea interactions during El Niño and La Niña. In other words, is the atmosphere driving the ocean, or is the ocean driving the atmosphere?

Explore Online

Visit www.mhhe.com/chamberlin1e for access to chapter quizzing, key term flash cards, video clips, interactive activities, and more. Further enhance your knowledge with web links to chapter-related material!

Exploration Activity 8-1

Exploring El Niño Through Scientific Investigation

Question for Inquiry

How do oceanographers monitor and predict the occurrences of El Niño?

Summary

The phenomenon of El Niño is just one example of multiannual and multidecadal variability in the ocean-atmosphere system. Nonetheless, its effects on human affairs give it prominence in scientific efforts to understand and predict its occurrence. In this activity, we explore oceanographic efforts to monitor El Niño using the TAO/Triton array in the equatorial Pacific.

Learning Objectives

By the end of this activity, you should be able to:

- Interpret color contour maps of sea surface temperature
- Describe the distribution, range and trends of oceano-graphic and meteorological data reported by the TAO/Triton buoy array
- Understand how the information obtained by the buoys is used to predict El Niño and La Niña events
- Appreciate the benefits and limitations of fixed-point sampling for understanding ocean processes

Materials and Preparation

For this activity, you will need access to the Internet and a printer. Alternatively, your instructor may provide a printout of figures sufficient for completing the activity. You may want to review the sections on global sea surface temperatures and El Niño in this chapter. Some of you undoubtedly are expert web surfers with little need for instructions. Others of you may be relatively new to browsing the Internet. We strongly encourage the "experts" to help the "newbies."

Part I: Familiarize Yourself With the TAO/Triton Buoy Array

The TAO/Triton Java applet at www.pmel.noaa.gov/toga-tao/java/taoinfo.html is a project managed by the National Oceanic and Atmospheric Administration's Pacific Marine Environmental Laboratories. The Project Home page is located at www.pmel.noaa.gov/tao/ and you should start here first.

From the TAO Project Home Page, view the TAO Story and answer the following questions:

1. What does TAO stand for?
2. What are the scientific goals of the TAO array?
3. How are TAO buoys deployed?
4. How do the data from the buoys get to the Internet?
5. What are the beneficial uses of the data?

Click on the tab marked Project Overview.

1. Describe the buoy array. What does it consist of? (See Technical Information—Moorings.)
2. What exactly do the buoys measure? (See Technical Information—Sensors.)
3. How often do the buoys collect data? (See Technical Information—Sampling.)
4. How do scientists know the data are reliable? (See Technical Information—Data Quality Control.)
5. What types of data are available for the buoy located at the intersection of the equator and the International Date Line? What depths are measured for sea temperature at this location? (See Technical Information—Data availability.)
6. How often are data updated on the TAO web pages? (See Technical Information—Data telemetry.)

Click on the El Niño/La Niña tab.

1. What is El Niño?
2. What is La Niña?
3. What are the global impacts of El Niño?
4. Choose one of the Frequently Asked Questions (FAQs) and summarize the answer here (be sure to include the question).
5. What are the benefits of El Niño prediction?
6. What are today's predictions and news releases concerning the possibility of an El Niño event occurring in the near future?

Part II: Familiarize Yourself With the TAO/Triton Java Applet

Go to the applet site here at www.pmel.noaa.gov/toga-tao/java/taoinfo.html

Please use this website http://www.pmel.noaa.gov/tao/jsdisplay/ or http://www.pmel.noaa.gov/tao/disdel/disdel.html to access the data and graphics. Regards, Nancy Soreide

Familiarize yourself with how the applet displays buoy locations and oceanographic data. When you have completed each task, put an "X" in the check box.

1. Roll your mouse cursor over one of the diamond symbols (i.e., buoy locations) displayed on the color contour map. Observe how information in the box marked Most Recent Data changes as you move from one buoy location to another.

2. Roll your cursor over a colored region away from one of the buoys. Observe how the latitude and longitude of your cursor are displayed at the lower left corner of the data display (above the Most Recent Data box).

3. Roll your cursor over the buoy located at the intersection of the equator and the International Data Line. Hold down the Shift key and click your mouse. You should see the TAO Buoy Summary Plot for 0, 180W.

4. Enlarge the image (the graphs depicting the buoy's data) by clicking on it. Observe the graphs for wind speed and direction, dynamic height (essentially, the height of the sea surface above a fixed depth in the water column—don't worry if you don't understand this; it's a fairly advanced topic) and water temperatures from the surface to 300 meters from November 2001 to November 2002 (and beyond). Observe that the colors on the temperature map correspond to specific temperatures.

5. Return to the applet. Find one of the square symbols on the map and roll your cursor over it. What is the difference between the square symbols and the diamond symbols? Shift-click on one of the square symbols. Observe the image summarizing, wind, dynamic height, and temperature data for this station.

6. Remain on this page and click on the link for the Glossary of TAO Terminology. Look up and make sure you understand the following terms: Mean, Anomaly, SST, WSPD, WDIR, T1-T10, Humidity, Isotherms, DYN (Dynamic Height), Heat, Transport, Ocean Measured, Ocean Derived.

Part III: Prepare a Weather Report on the Tropical Pacific

Return to the applet and prepare a weather report on current conditions in the tropical Pacific as indicated by the TAO/Triton array. Your report should take the format of a typical weather report as presented by a weatherperson on a TV or radio news program. Make it fun. Include the following:

- The latitude and longitude from the most northwestern Atlas buoy to the most southeastern Atlas buoy; this is the geographical region (the TAO region) for which you are reporting oceanographic and meteorological data.

- The dates over which your data were collected and are being reported (the five-day mean).

- An assessment of current sea surface temperatures within the TAO region, commenting on the range of temperatures observed (highest to lowest) where the highest and lowest temperatures occur (use lat-long data and/or directional information, i.e. in the center, at the southeast corner, at the upper northeastern quadrant, etc); and any north-south and east-west trends (does it get warmer or cooler as you travel from west to east, south to north?)

- A detailed report of data specific to one buoy each within the warmest, coolest and moderate temperature regions (that is, provide air temperature, sea surface temperature, 20-degree isotherm, dynamic height, heat content, wind speed, wind direction and humidity for three buoys, one in the warmest region, one in the coolest region and one in a region with temperatures between the warmest and coolest.

- A report on the annual variations in wind speed and direction, dynamic height and water column temperatures for one buoy in a hot region and one in a warm region.

- A summary of the current forecast for El Niño and its potential impacts where you live (i.e. above-normal air temperatures in the northeastern US; drought in the western US; heavy rainfall in Hawaii, etc).

- A statement about the limitations of benefits and limitations of weather buoys for understanding ocean processes (including statements about the spatial and temporal resolution of the array, its advantages and disadvantages compared to satellites, its costs and limitations on what it can and can't measure.

9 Surface and Deep Circulation

A current sweeps across a wreck in the Florida Keys.

Questions to Consider

1. Why do oceanographers consider the surface and deep circulation as interdependent parts of a single world ocean circulation?

2. How does Ekman transport create patterns of flow in the surface circulation of the world ocean?

3. What is geostrophic flow and how does it describe the general surface circulation patterns of the major gyres?

4. What causes upwelling and why is it important?

5. What are the differences between western and eastern boundary currents, and why are those differences important?

6. Why are models and observations important for studies of the world ocean circulation?

7. What are the general patterns of deep circulation, and why are they important?

8. What role does ocean circulation play in global climate?

9. What processes drive surface circulation and deep circulation?

World Ocean Circulation

In *20,000 Leagues Under the Sea,* Jules Verne's infamous Captain Nemo compared the movements of the world ocean to the human circulatory system: "Look at that sea! Who can say it isn't actually alive? . . . It has a pulse, arteries, sudden movements, and I side with the learned Maury who discovered it has a circulation as real as the circulation of blood in animals." Published in 1870, Verne's novel borrowed heavily from the oceanographic literature of the time. His musings on ocean circulation originate with Matthew Maury (1806–1873), who, in 1855, published *The Physical Geography of the Sea and Its Meteorology,* one of the first textbooks on modern oceanography.

The large-scale pattern of currents within the world ocean is called the **world ocean circulation.** Traditionally, ocean circulation has been divided into two components, the **surface circulation** and the **deep circulation.** However, oceanographers now view the division between surface and deep circulation as largely artificial. The flow of surface waters strongly influences the deep circulation and vice versa. Many oceanographers prefer the term *world ocean circulation* to describe the myriad of patterns and processes that influence the flow of water in the world ocean. Nonetheless, for the purposes of understanding world ocean circulation, it is convenient to treat the surface and deep circulation separately.

This chapter considers ocean physics from a dynamical perspective. **Dynamical physical oceanography** describes the ocean using physical laws and conservation equations in concert with a variety of atmospheric and oceanographic observations. A major goal of this branch of physical oceanography is to quantify flow patterns and property distributions in the world ocean. Dynamical oceanography has many practical applications as well, including ship routing, search and rescue operations, and marine pollutant trajectories. Our goal here is to explore the nature of world ocean circulation and to become familiar with the forces that govern its movements.

The Foundations of Dynamical Physical Oceanography

Since humans first ventured to sea in vessels, knowledge of currents has brought lucrative and often powerful advantages to the men and women who possessed it. Benjamin Franklin (1706–1790), who served as postmaster general of the United States, wanted to reduce the transit time for ships carrying mail from Europe to the United States. Using information provided by his cousin, Nantucket whaling ship Captain Timothy Folger (1706–1749), Franklin published one of the first detailed maps of the Gulf Stream around 1769 (**Figure 9-1**). In his pioneering book, Maury established a clear link between scientific observations and navigation of the sea. His global compilations of data on winds, currents, weather, and a number of other important oceanographic phenomena, combined with the appeal of his charming and sometimes emotional prose, produced a work that drew favor from a wide audience, including merchant seamen, naval officers, and the general public. By the first part of the nineteenth century, the major features of the circulation of the surface North Atlantic were relatively well established, although many details and important aspects remained to be worked out.

Despite evidence suggesting otherwise, many scientists in the early nineteenth century thought that the deep ocean was motionless. As early as 1798, American physicist Sir Benjamin Thompson, also known as Count Rumford (1753–1814), had proposed the concept of temperature-driven circulation in the ocean. Rumford's hypothesis was consistent with his discovery of **convection currents,** the rising and sinking of hot and cold water parcels in a liquid. Rumford based his proposal on deep water measurements taken by Captain Henry Ellis (1721–1806) of the British slave trader, *Earl of Halifax,* who reported in 1751 that water from 1189 m (3900 feet) was uniformly cold and good for a cool bath (see Chapter 6). According to Rumford, cold water sank in polar regions and flowed outward toward equatorial regions. A poleward surface flow completed the circuit. Rumford's idea represented the first conceptual model of thermohaline-driven deep circulation.

A number of scientists pursued these ideas during the 1800s, but it was data from the *Challenger* Expedition that provided the key evidence in support of thermohaline circulation. John Young Buchanan (1844–1925), a chemist and physicist aboard the *Challenger,* published in 1885 some of the first maps depicting the distribution of salinity at different depths and latitudes in the Atlantic Ocean. Buchanan also noted the intrusion of Antarctic Intermediate Water into the North Atlantic (see the section on water mass identification in this chapter). In 1895, working with *Challenger* data, British meteorologist Alexander Buchan (1829–1907) demonstrated the presence of deep water originating from the Antarctic in each of the three major world ocean basins. In 1911, the German Antarctic *Deutschland* expedition provided the first fully documented evidence for a southward flow of water from the North Atlantic wedged between intermediate and deep waters of the Antarctic. Based on these observations, Alfred Merz (1880–1925) and George Wüst (1890–1977) sketched in 1922 one of the earliest depictions of the circulation of deep waters in the Atlantic Ocean.

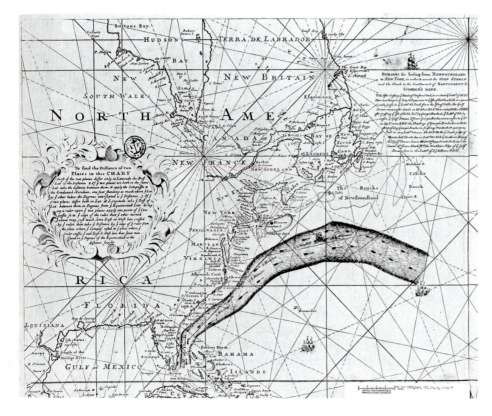

■ **FIGURE 9-1** Franklin and Folger's map of the Gulf Stream. Franklin also performed some of the first temperature measurements of the Gulf Stream.

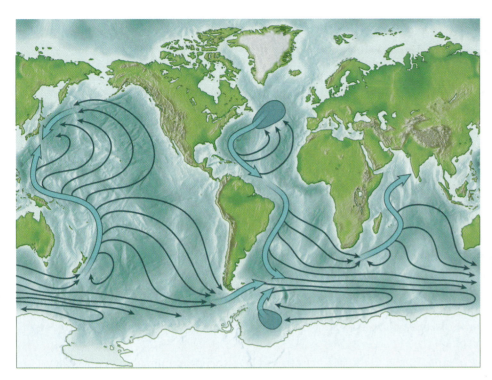

■ **FIGURE 9-2** Stommel's 1958 model of deep circulation. Subsurface flows are indicated by thin blue lines, while deep water flows are indicated by shaded lines. Open circles represent sites of sinking of deep water. Stommel's models were among the first to suggest a link between the surface and deep circulation.

In 1958, Henry "Hank" Stommel (1920–1992) proposed one of the first global models of abyssal circulation (**Figure 9-2**). Stommel has been called "the most original and important physical oceanographer of all time." His work on both surface and deep circulation brought significant advances to scientific understanding of ocean circulation. Stommel's abyssal circulation model suggested that the extra heat that is input into low-latitude regions was balanced through the slow upward transport of cooler waters from below. Moreover, he suggested that the sinking of dense water took place in only a few locations, designating deep water formation sites off Greenland and the Weddell Sea in the Antarctic. Stommel's model indicates that the final destination of water formed in the North Atlantic was the North Atlantic, even though this water may transit the Indian and Pacific Oceans before it returns. Several features of Stommel's model remain part of ocean general circulation models today.

The decade of the 1950s brought to physical oceanographers an invention that changed the course of oceanographic history: the computer. Out of this era emerged some of the first **ocean general circulation models,** or OGCMs. These models, many of which were first developed at the Geophysical Fluid Dynamics Laboratory at Princeton University, Princeton, New Jersey (the alma mater of co-author Professor Dickey), paved the way for modern numerical modeling of the ocean and atmosphere, and coupled ocean-atmosphere models. At the same time, the development of improved ocean observation capabilities provided much-needed data for initializing and testing OGCMs. Deep floats, drifters, current meter moorings, bottom pressure sensors, and satellite sensors, among other instrument platforms deployed in the twenty-first century, now provide the kinds of observational data required to fully understand the surface circulation in the world ocean.

An Ocean in Motion

The most obvious features of ocean circulation are the **oceanic gyres,** the near-circular movements of water that span ocean basins (**Figure 9-3**). **Subtropical gyres** include the subtropical regions, such as the North Atlantic, South Atlantic, North Pacific, South Pacific, and South Indian Ocean gyres. **Subpolar gyres** are centered in subpolar regions, such as those between Greenland and Norway and in the Gulf of Alaska. While Maury's charts provided the first definitive evidence of oceanic gyres, more than a century would pass before oceanographers would understand the causes of gyre circulation.

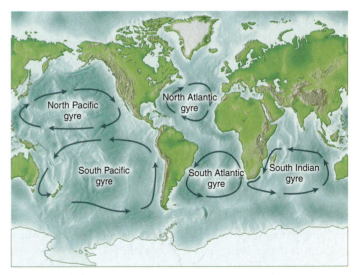

■ **FIGURE 9-3** The major subtropical oceanic gyres of the world ocean.

TABLE 9.1   Summary of Ekman's Theory

- Ideally (under no other influences), surface currents are directed 45° to the right of the direction of the wind in the Northern Hemisphere and 45° to the left of the direction of the wind in the Southern Hemisphere.

- The speed of wind-driven currents decreases with depth.

- Deeper currents move sequentially toward the right. (Think of a stack of note cards with each successive card moving farther to the right.)

- Plots of current vectors on a graph results in a spiral pattern, called the **Ekman spiral**. (Note that the "spiral" is the pattern obtained when velocity and direction are graphed as vectors; the Ekman spiral is not a whirlpool or eddy, a common misconception.)

- The **Ekman layer** depth is defined as the depth where the speed has decreased to a value of about 37% of its surface value. This is the depth where a current vector is pointing in the opposite direction (180°) to that of the surface current vector.

- The depth of the Ekman layer depends on the Coriolis parameter. For a given wind stress, higher latitudes have shallower Ekman layer depths, while lower latitudes have deeper Ekman layer depths (except for the equator).

- **Ekman transport**, defined as the summation of the flow (direction and velocity) from the surface to the Ekman layer depth, is directed 90° to the right of the surface wind stress in the Northern Hemisphere and 90° to the left of the surface wind stress in the Southern Hemisphere.

- The magnitude of the Ekman transport is proportional to the wind stress and inversely proportional to the Coriolis effect and seawater density.

Ekman Transport

In the late 1800s, a Norwegian explorer and zoologist, Fridtjof Nansen (1861–1930), became interested in the shipwreck of an American vessel, the *Jeannette*, whose wreckage had been carried from the New Siberian Islands off northern Russia to the southern tip of Greenland. Nansen postulated that a current must flow in a westerly direction across the North Pole. Interested in being the first man to reach the North Pole, Nansen further reasoned that a ship frozen into the ice on the Siberian side of the pole might float across or near the pole as a result of this current. Though many considered him mad, Nansen designed and built a very strong, round-bottom vessel named *Fram,* meaning "forward."

In 1893, Nansen led an expedition into the Arctic ice floes. In addition to making numerous important meteorological, oceanographic, and biological observations, Nansen observed that floating ice did not move in the same direction as the wind. Curiously, the ice moved approximately 20 to 30 degrees to the right of the prevailing wind direction. Upon his return, Nansen related these observations to V. Walfrid Ekman (1874–1954), a Swedish oceanographer. Ekman knew that sea ice is subject to movement not only by winds but also by currents, especially when a greater fraction of the ice lies beneath the ocean surface. From Nansen's observations, Ekman deduced that wind stress, friction, and the Coriolis effect alter the movement of currents in a predictable fashion. In 1905, Ekman published a paper describing a theory to explain Nansen's observations of surface currents. The key features of Ekman's theory are summarized in **Table 9.1.**

According to Ekman's theory, winds impart motion to the surface layer, which, as a result of friction, drags the layer beneath it, which in turn drags the layer beneath it, and so on. From this "dragging" of successive water layers, two important phenomena occur: the speed of wind-driven currents decreases with depth; and the direction of each successive layer continues to deflect to the right or left, depending on the hemisphere (**Figure 9-4**). Ekman transport is central to an understanding of the major surface currents of the world ocean.

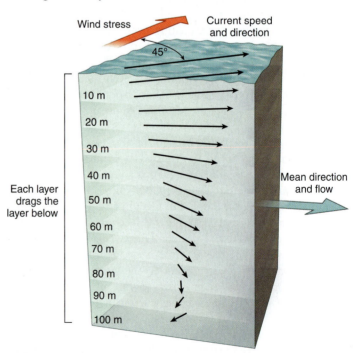

■ **FIGURE 9-4** The Ekman spiral and Ekman transport. See text for details

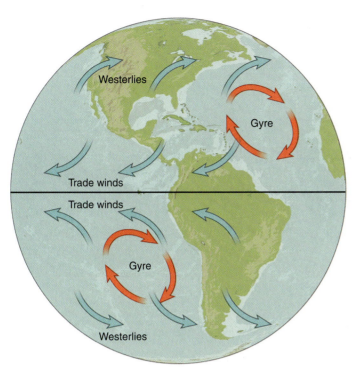

■ FIGURE 9-5 A simplified representation of the trade winds and the westerlies and their relationship to gyre circulation.

Geostrophic Flow: The Hill in the Gyre

As we learned in Chapter 8, the two most well-known atmospheric surface patterns are the trade winds, blowing in a westerly direction toward the equator, and the westerlies, blowing in an easterly direction away from the equator (**Figure 9-5**). If Ekman transport is directed 90° to the direction of the trade winds and westerlies, which way will the water go? Let's address this question step by step.

Ekman transport directs water toward the middle of the ocean basins. Of course, water does not "pile up" in the middle of the ocean, but this is a first step. Now imagine that the "pile" of water creates a sloping sea surface. Like water poured on a table, the sloping sea surface will generate a horizontal pressure gradient directed away from the "pile." (see Chapter 7). The "piling up" of water in the middle of gyres causes water to flow from the high pressure region in the middle

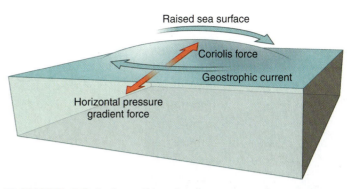

■ FIGURE 9-6 A three-dimensional representation of a geostrophic current in a Northern Hemisphere subtropical gyres, such as the North Atlantic gyre. Water directed toward the middle of the gyres by anticyclonic or clockwise winds is balanced by the Coriolis effect, resulting in a clockwise circulation. Winds drive currents indirectly by setting up the horizontal pressure gradient. Otherwise, geostrophic currents dominate the circulation patterns that we observe in ocean gyres.

to the low pressure region away from the center. However, this moving water is subject to the Coriolis effect. A parcel of water moved by a horizontal pressure gradient toward the north will tend toward the right and be moved eastward by the Coriolis effect. A tug-of-war exists between the horizontal pressure gradient and the acceleration of the mass due to the Coriolis effect. At some point, the pressure gradient force will exactly balance the Coriolis acceleration (technically, mass times Coriolis parameter times velocity) and the current will maintain a steady flow. This balance maintains current flows along lines of constant pressure (i.e., isobars) and generates the large-scale motions of the atmosphere and ocean, that is, the ocean gyres. These currents, called **geostrophic currents,** result from a balance between the horizontal pressure gradient created by Ekman transport (or similar forces) and the Coriolis acceleration (**Figure 9-6**). The term *geostrophic* means "earth-turning" and refers to a flow under the influence of the Coriolis effect (i.e., a rotating Earth).

An interesting consequence of geostrophic currents is an elevated sea surface in the center of the oceanic gyres. Detailed views of **sea surface topography** using satellite altimeters, such as TOPEX/Poseidon, revealed a sea surface marked by "hills" and "valleys," albeit modest ones (**Figure 9-7**). Across a given expanse of the ocean, the

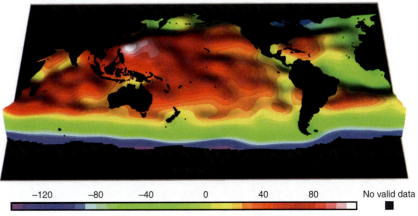

■ FIGURE 9-7 The "hills" and "valleys" of the sea surface as measured by TOPEX/Poseidon.

−120   −80   −40   0   40   80   No valid data

Ocean Dynamic Topography (cm) Oct 3–12, 1992

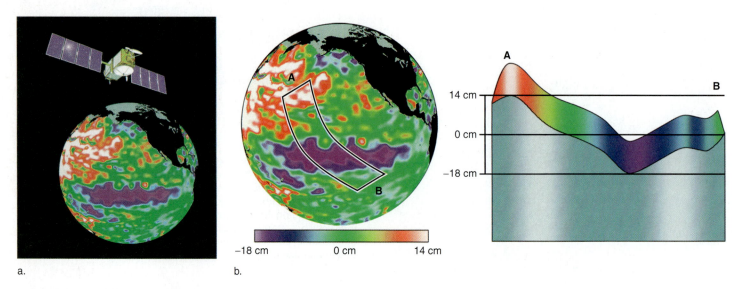

a.   b.

■ **FIGURE 9-8** (a) Satellite altimetry from Jason, launched in 2001. (b) Variations in sea surface height over a slice of the ocean.

change in sea surface elevation, or **sea surface height,** is rarely greater than 2 m (6 feet). Sea surface heights are too small to detect with human senses but well within the range of satellite sensors (**Figure 9-8**). When launched in 1992, TOPEX/Poseidon commenced a new era in the study of ocean circulation. Its 13-year record of continuous observations provided key data for understanding ocean circulation, heat transport, climate change, ocean tides, and El Niño/La Niña. TOPEX/Posedion was retired in January 2006.

Upwelling and Downwelling

Ekman transport also generates **upwelling,** the flow of subsurface waters to the surface, and **downwelling,** the movement of water downward in the water column. In the equatorial Pacific,

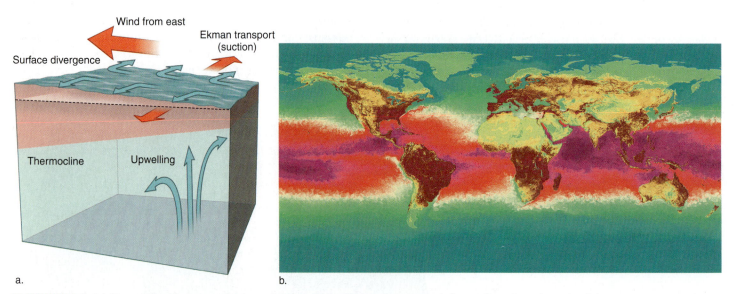

a.   b.

■ **FIGURE 9-9** (a) Generalized features of equatorial upwelling. Ekman transported directed north and south of the equator generates a surface divergence that draws deeper water towards the surface. (b) Satellite image of upwelling in the equatorial Pacific. Cooler waters are shown in shades of purple.

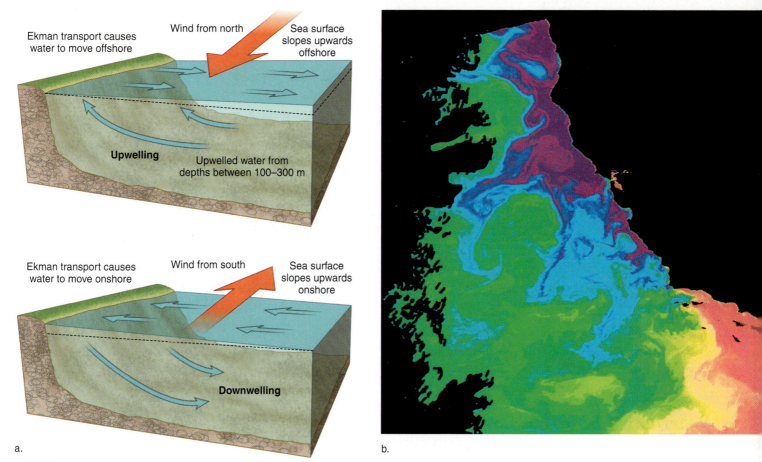

a.

b.

■ FIGURE 9-10 (a) Generalized features of coastal upwelling and downwelling in the Northern Hemisphere. Coastal upwelling occurs when Ekman transport is directed away from the coast. Fluctuations in wind stress and seafloor bathymetry influence the intensity and duration of upwelling. Downwelling occurs along coastlines when Ekman transport is directed toward the coast. (b) Coastal upwelling (shades of purple) appears as patches of cold water extending from the coast of California.

upwelling occurs as winds blowing westward in the Northern Hemisphere cause northward Ekman transport of surface waters and winds blowing in the same direction in the Southern Hemisphere cause southward Ekman transport. The resultant surface divergence leads to equatorial upwelling (**Figure 9-9**). Ekman transport also generates upwelling (and downwelling) along coastlines (**Figure 9-10**). Along west-facing coastlines, such as the California coast, northerly winds move surface waters

toward the west and cause upwelling of waters from depths of 100 to 300 m (300–1000 feet). These regions are among the most productive regions of the world ocean (see Chapter 14). Cyclonic low pressure systems can cause surface divergence and upwelling where Ekman transport is directed outward (**Figure 9-11**). In contrast, anticyclonic high pressure systems cause water to pile up or converge. Surface convergence in subtropical ocean gyres can deepen or depress the thermocline.

Northern Hemisphere

Ekman transport
Surface current
Wind

Surface divergence

Upwelling

Thermocline

Cyclonic wind

Surface convergence

Downwelling

Thermocline

Anticyclonic wind

■ FIGURE 9-11 Upwelling and downwelling also occur in association with low and high pressure systems, respectively. Divergence of surface waters cause upwelling and a depression of sea surface height, while convergence of surface waters cause downwelling and an elevation in sea surface height. Low pressure systems also draw the thermocline closer to the surface. Just the opposite occurs in high-pressure systems. These phenomena have important implications for ocean productivity.

General Patterns of Surface Circulation

Much of the early theoretical understanding of surface circulation was developed by Harald Sverdrup (1888–1957), a Norwegian-born oceanographer, Arctic explorer, and colleague of Vilhelm Bjerknes (see Chapter 9). Sverdrup examined the effects of a boundary, such as a continent, on ocean surface circulation. His theory allowed oceanographers to calculate surface flows directly from measurements of wind stress. In honor of his work, a unit of measurement of water transport was named for Sverdrup (1 Sverdrup [Sv] equals 1×10^6 $m^3 sec^{-1}$). The long-term averages of surface water transport provide a generalized description of the surface circulation of the world ocean (**Figure 9-12**). The surface circulation moves heat, dissolved and particulate substances, dissolved gases, and even organisms within and among the five major ocean basins. In this section, we compare and contrast the patterns of surface circulation in the ocean basins and explore the forces responsible for causing these patterns.

Western Boundary Currents

Currents located along the western sides of ocean basins are called **western boundary currents.** These currents arise from an intensification of gyre circulation along the western boundaries of basins in response to variations in the Coriolis effect with latitude (see the section on western intensification in this chapter). The best studied and most famous of western boundary currents is the Gulf Stream

in the North Atlantic subtropical gyre (**Figure 9-13**). Among ocean currents, the Gulf Stream is one of the fastest-moving currents, with peak velocities near 2.0 to 2.5 m s^{-1} (5 knots). The Gulf Stream transports warm waters from the Gulf of Mexico toward the north off the eastern seaboard of the United States. The Gulf Stream also generates unstable flows that separate from the main flow in the form of **Gulf Stream rings,** a type of mesoscale eddy (**Figure 9-14**). Rings that form on the landward side of the Gulf Stream rotate clockwise and have a warm, elevated center surrounded by a cold exterior. These rings are called **warm-core rings.** Rings that form on the seaward side of the Gulf Stream rotate counterclockwise and have a cold, depressed center surrounded by a warm exterior. These rings are called **cold-core rings.** Rings have diameters of about 100 to 300 km (60–180 miles), maximum rotational water speeds of over 100 cm sec^{-1} (3 feet per second), can extend down to depths of 1000 m (3300 feet) or more, and have lifetimes of a month to more than a year. Annually, the Gulf Stream produces an average of 20 warm core rings and 35 cold core rings.

The Pacific Ocean equivalent of the Gulf Stream is the **Kuroshio Current,** also known as the Japan or Black Current. Current speeds here also exceed 5 knots (about 2 miles per second). The Kuroshio arises from the same forces as the Gulf Stream and other western boundary currents. Unlike the Gulf Stream, however, the flow of the Kuroshio is complicated by the presence of the Japanese archipelago. The interaction of the Kuroshio with islands creates wakes and eddies, factors that may contribute to its considerable snake-like movements. Off the island of Honshu, the Kuroshio Current flows eastward as the **Kuroshio Extension,** a region representing one of the largest air-sea exchanges of heat in the world ocean.

Another notable western boundary current is the **Brazil Current.** Midway down the coast of Brazil, the Brazil Current

FIGURE 9-12 The surface circulation of the world ocean.

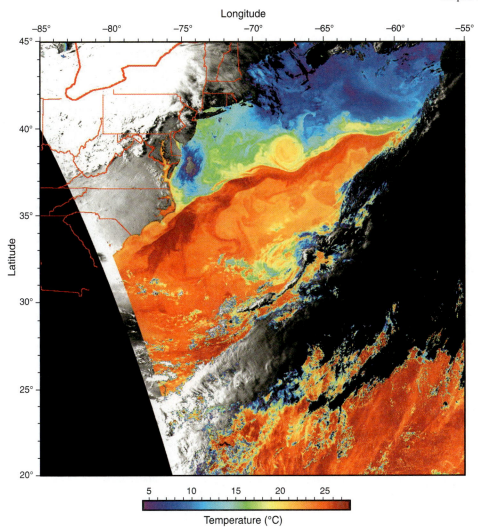

■ **FIGURE 9-13** The Gulf Stream and associated mesoscale features imaged by the NOAA-12 satellite.

runs head-on into the northward-flowing and colder Malvinas Current, also known as the Falkland Current. The **Brazil-Malvinas Confluence** has been called "one of the most energetic regions of the world ocean" and forms part of a highly productive large marine ecosystem (LME) known as the **Patagonian Shelf LME,** an internationally designated site.

Western Intensification

Western boundary currents such as the Gulf Stream, the Kuroshio, the Brazil Current, and others are faster and narrower than other currents. This difference results from **western intensification,** the increase in flows of western boundary currents caused by an increase

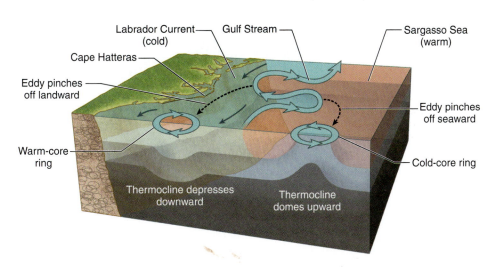

■ **FIGURE 9-14** Formation of cold- and warm-core rings. Warm-core rings form when a meander of the Gulf Stream pinches off northward and traps warmer water (relative to the Labrador Current). Cold rings form when a meander of the Gulf Stream pinches off southward and traps colder water (relative to the Sargasso Sea). Note their influence on the water column beneath them.

in the Coriolis effect at higher latitudes. Oceanographers long knew that faster currents occurred on the western side of ocean basins, but they did not know why. This problem caught the attention of Henry Stommel, who set out to explore why eastern and western boundary currents were different. He knew that Sverdrup's wind-driven circulation model published in 1947 mostly explained gyre circulation (albeit with one boundary) in terms of variations in winds with latitude (i.e., wind torque). Building on Sverdrup's model, Stommel included eastern and western boundaries and ran it assuming that the Coriolis effect was the same at all latitudes. The result was a symmetrical circulation pattern with no Gulf Stream. In the next run of the model, he included the variations in the Coriolis effect with latitude (**Figure 9-15**). This time, the model duplicated the observed western intensification. Stommel published his model in 1948. Today, powerful computers running complex numerical models are used to predict the path, meandering, and mesoscale eddies produced by western boundary currents. Nonetheless, Stommel's simple insight demonstrated the applicability of fundamental physical principles and mathematics to explaining and understanding the circulation of the world ocean.

Eastern Boundary Currents

Currents that flow along the eastern boundaries of ocean basins are called eastern boundary currents. The California Current, an eastern boundary current in the North Pacific subtropical gyre, exhibits their major features. The California Current begins off the coast of Vancouver Island, Canada, where the North Pacific Current splits northward into the Alaska Current and southward into the **California Current.** Characteristic of eastern boundary currents, the California Current is broad (about 1000 km or 600 miles) and generally slow-moving with maximum current speeds of about 35 cm s^{-1} (1 foot per second). Despite its quiescence, the flow of the California Current is punctuated by intense, energetic episodes of wind-induced upwelling that give rise to coastal eddies and **coastal jets,** narrow, meandering bands of cold, upwelled water (also called squirts and filaments). The cool California Current waters and coastal upwelling moderate California's climate, creating dry, Mediterranean-like conditions. These cold waters also produce considerable coastal fog.

■ **FIGURE 9-15** Stommel's model of western intensification. Building on Sverdrup's model, which reproduced gyre circulation in the presence of an eastern boundary (such as the European continent), Stommel was able to satisfactorily reproduce the major features of a Northern Hemisphere gyre. The key for Stommel was to include a factor that accounted for variations in the Coriolis effect with latitude.

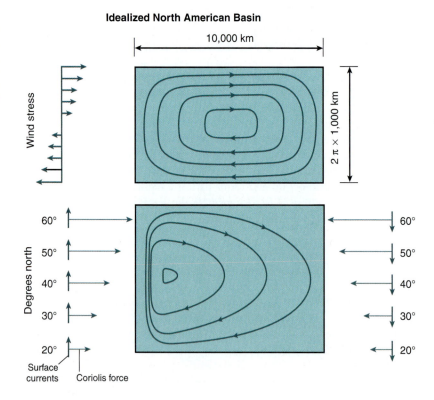

Idealized North American Basin

■ FIGURE 9-16 False color, satellite-derived SST image of the North Atlantic equatorial current, otherwise known as Hurricane Alley. Yellows and reds indicate warmer waters.

The **Benguela Current,** the eastern boundary current in the subtropical South Atlantic gyre, displays similar characteristics to the California Current. This region hosts the **Benguela upwelling system,** one of four highly productive eastern boundary upwelling systems in the world ocean. This upwelling system often exhibits El Niño–like features as warm water shifts the front southward and reduces upwelling. Though less intense than its Pacific counterpart, the **Benguela Niño** leads to similar reductions in fisheries productivity. The flow of the Benguela Current may also be interrupted by its interaction with the **Agulhas Retroflection,** a region off the southern tip of Africa. This prominent ocean feature sheds large anticyclonic eddies, some reaching nearly 320 km (200 miles) in diameter. These eddies transport large quantities of warm water from the Indian Ocean into the South Atlantic Ocean.

Equatorial Currents

All three major oceans feature a North Equatorial Current (NEC) that flows westward just north of the equator. To the south lies the North Equatorial Countercurrent, a weak, seasonally variable current that returns some of the equatorial flow eastward. These currents often interact with the westward-flowing South Equatorial Current (SEC) and the South Equatorial Countercurrent to deliver warm surface waters from southern oceans into northern ones. This cross-equatorial exchange plays an important role in the circulation of the equatorial Atlantic and may play a role in the formation of hurricanes. The equatorial Atlantic, sometimes called **Hurricane Alley,** supplies heat to developing tropical disturbances that on occasion grow into hurricanes (**Figure 9-16**). The westward-flowing components of the North and South Equatorial Currents in the equatorial Pacific are largely wind-driven and respond rapidly to seasonal shifts in the easterly trade winds. The eastward-flowing Pacific equatorial currents include the Equatorial Countercurrent (ECC) and the Equatorial Undercurrent (EUC; also known as the Cromwell Current.) The Equatorial Undercurrent, which extends for some 14,000 km, gradually rises from a depth of 200 m (650 feet) at its westward end to a depth of 40 m (130 feet) at its eastern end. The Equatorial Undercurrent is most intense and transports the largest volumes of water eastward during La Niña conditions. It is weak and almost disappears during El Niño conditions (see Chapter 8).

■ **FIGURE 9-17** Generalized circulation of the Indonesian Throughflow an important choke point in the world ocean. This region of the world ocean is under intense study for its possible role in climate change.

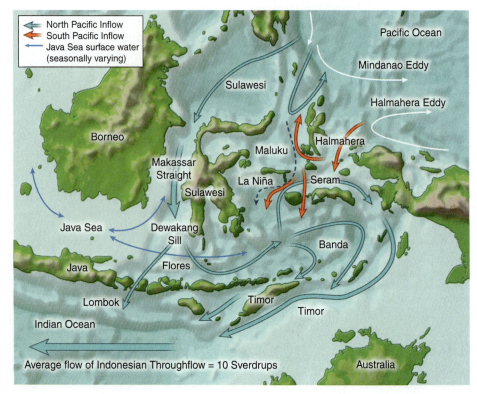

The Indonesian Throughflow

A southerly branch of the South Equatorial Current supplies the East Australian Current and the **Indonesian Throughflow,** a maze of currents traversing the Indonesian archipelago. The Indonesian Throughflow represents an important region for the exchange of waters between the Pacific and Indian basins. It also provides an important return route for Atlantic and Indian basin waters that flow into the Pacific via the Antarctic Circumpolar Current. As a result, oceanographers consider the Indonesian Throughflow to be an important **choke point,** a region whose flows exert rate-limiting control over the transport of heat and mass in the world ocean (**Figure 9-17**).

Monsoonal Circulation

The India-Eurasian continent restricts the size of the Indian basin. Thus, the current system here is geographically reduced in comparison to other basins. The prominent feature here is the monsoonal circulation in the Northern Hemisphere that reverses winds and alters currents over the seasonal cycle (**Figure 9-18**). **Monsoon** (from the Arabic word *mausim,* meaning "season") refers to a seasonal reversal of winds. In the Indian Ocean, sailors noted that persistent northeasterly winds during the boreal (Northern Hemisphere) winter switched to southwesterly during the austral (Southern Hemisphere) winter. Monsoons originate largely as a result of seasonal changes in atmospheric pressure over the land versus the ocean. The Northeast or **Winter Monsoon** occurs as the cooler winter air mass over Asia generates higher pressure than the atmosphere over the warm ocean, creating a flow of surface winds from the northeast, that is, from Asia toward the equator. As the season progresses and the Asian landmass heats up, this system

weakens and the stronger Southwest or **Summer Monsoon** takes hold, creating a reversal of winds. Arizona experiences a similar monsoonal circulation as heating of the Mohave Desert in summer generates low pressure that induces surface winds from the Gulf of California and the Gulf of Mexico. The seasonal reversal of winds affects currents in the Indian Ocean, largely in the Northern Hemisphere. Under the Northeast Monsoon, current patterns appear most similar to those of other ocean basins: the equatorial currents are present (NEC, SEC, ECC and EUC) in their usual configuration. However, during the Southwest Monsoon, the NEC reverses direction and merges with the ECC to form the Southwest Monsoon Current. Along the east coast of Africa in the Northern Hemisphere, the Somali Current, a western boundary current that normally flows northward, reverses direction during the Southwest Monsoon. The Somali Current is one of the fastest currents in the world ocean during the Southwest Monsoon.

The Antarctic Circumpolar Current

The eastward flowing **Antarctic Circumpolar Current** (ACC) is the only continuous current in the world ocean that is nearly unimpeded by land masses and the only one that makes a complete transit around the globe (**Figure 9-19**). The ACC traverses the southern end of all the major ocean basins. Like the Indonesian Throughflow, the **Drake Passage** is considered by some oceanographers to represent an important choke point in the world ocean. It's flow from the Pacific to the Atlantic through the Drake Passage runs counter to the prevailing polar easterlies and produces one of the most dangerous sea conditions in the world ocean. As a result, it may well be one of the most important currents in the world ocean, acting as a kind of "switchboard" that controls the exchange of water between ocean basins.

Northeast Monsoon (January)

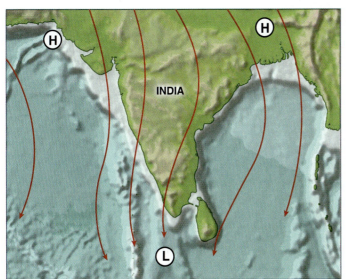

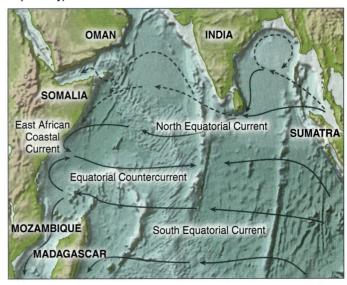

Southwest Monsoon (July)

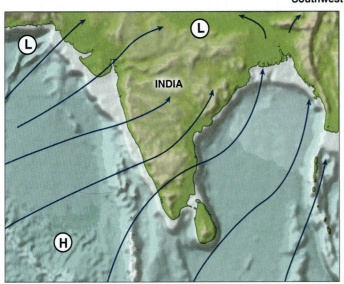

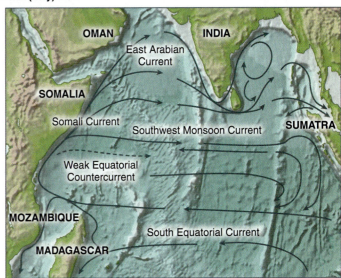

■ **FIGURE 9-18** Monsoonal circulation in the Indian Ocean. High pressure over India creates offshore flow and a pattern of currents similar to other oceans. During summer, the winds reverse, causing a reversal of currents. See text for additional details.

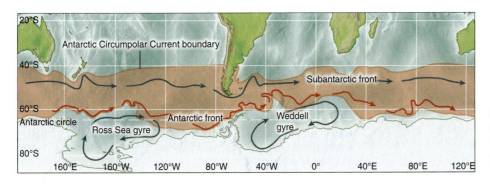

■ **FIGURE 9-19** The Antarctic Circumpolar Current (ACC). The ACC encircles Antarctica and acts as a conduit for the exchange of waters between the three major ocean basins, the Atlantic, Pacific, and Indian.

General Patterns of Deep Circulation

The deep circulation extends from roughly 1 km (about 0.5 miles) below the sea surface to the ocean bottom, about 4 km (2.5 miles) on average. Thus, the greater volume of the world ocean circulation is involved in the deep circulation. Near the poles, the deep circulation may start at the surface: deep convection of cold and dense water masses essentially mixes the water column completely from the surface to the bottom. Whereas the North Atlantic gyre completes a circuit in about two-and-a-half years, the average transit time for most deep water masses is somewhere between 600 and 1000 years. If we think of the surface circulation as a kind of speedy hare, rapidly distributing heat and materials from one part of the globe to another, the deep circulation represents a kind of slow-but-steady tortoise, a feature that has important consequences for long-term climate change. The deep circulation derives part of its flow from buoyancy changes in water masses that occur as the result of air-sea exchanges of heat and water, which alter the temperature and salinity and, hence, the density of surface waters. Mechanical mixing of deep waters across the permanent thermocline into the upper ocean also drives deep circulation.

T-S Diagrams

An understanding of deep circulation starts with the **temperature-salinity (T-S) diagram,** most often drawn as a theta-salinity (θ-S) diagram. Oceanographers prefer to use potential temperature (θ) because it more accurately represents the conditions of formation at the surface and takes into account adiabatic changes (see Chapter 7). The T-S method was first introduced by the Norwegian oceanographer Bjørn Helland-Hansen (1877–1957), who showed that deviations from a "normal" T-S plot in the North Atlantic represented the influence of water masses with different origins. The method is little changed since Helland-Hansen's time, and it provides a simple yet powerful tool for identifying water masses in the world ocean.

If we plot temperature (T) or potential temperature (θ) versus salinity (S) for a number of seawater samples, we generate a graph that looks like **Figure 9-20.** Often included on these graphs are lines that connect the combinations of T and S that produce equal densities. These lines are called **isopycnals.** Typically, isopycnals are expressed as a quantity known as **sigma-t** (officially, $\sigma_{s,t,p}$, or if potential temperature is used, σ_θ), a shorthand for expressing seawater density (ρ). The significant variations in density occur in the last four digits (when expressed in units of kg m^{-3}), so sigma-t simplifies expressions of density. Sigma-t is simply the density of seawater minus 1000 kg m^{-3}, or

$$\text{Sigma-t} = \rho - 1000 \text{ kg m}^{-3}$$

See our website for additional details.

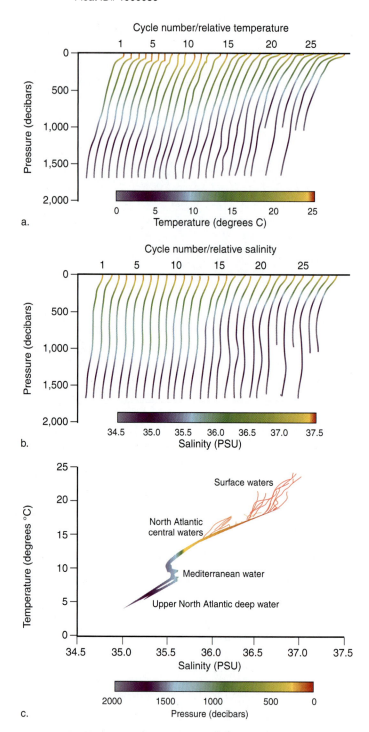

FIGURE 9-20 ARGO float 1900036 acquired data north of the Canary Islands in the North Atlantic (about 31–37° N, 25–27° W) on T (a) and S (b) that were plotted to produce a *T-S diagram* (c). Note the greater variability of surface waters (red) versus deeper waters (blue). The *T-S* diagram is one of the most widely used tools in physical oceanographic research. *T-S* diagrams help oceanographers to identify specific water masses and make inferences over processes that lead to their formation and subsequent evolution. They also help oceanography students to better understand density and its affect on the distribution of water masses in the world ocean.

TABLE 9.2   Major Categories of Water Masses Found in the World Ocean

1. **Surface (upper) waters**—waters that occur between the surface of the ocean and the permanent thermocline/pycnocline

2. **Central water masses**—water masses found at the surface in the central gyres of the regional oceans

3. **Intermediate water masses**—water masses that occur between the surface and the deep or bottom water masses

4. **Deep waters**—water masses that occur near the bottom (below intermediate water masses) of the major ocean basins and, during part of the year, may also extend from the surface to greater depths in polar or subpolar regions

5. **Bottom waters**—water masses that occur at the bottom of the major ocean basins and, at least for part of the year (usually winter), extend from the surface to the bottom in specific polar or subpolar regions (i.e., bottom water formation areas)

6. **Mode waters**—water masses characterized by weak stratification and a general homogeneity in temperature and density over a broad range of depths

Water Mass Identification

Parcels of water exhibiting a somewhat narrow range of temperature and salinity properties are called **water masses.** Water masses typically appear as a fairly well-defined segment of a *T-S* diagram. In contrast, **water types** consist of a single point on a *T-S* diagram, typically representing a specific site or condition of origin. Water masses are typically composed of a number of different water types with similar but not identical properties. The *T-S* diagram affords oceanographers the opportunity to give names to water masses as a kind of shorthand to indicate a specific set of properties. Names of water masses usually refer to the geographic location where the water mass originates, such as Labrador Sea Water, formed in the

Labrador Sea. In addition to the site of origin, oceanographers also identify a water mass according to one of six broad categories that describe the generalized layer (roughly, the depth range) in which it is found. These categories are listed in **Table 9.2.**

For example, Antarctic Bottom Water forms in the Antarctic and flows along the sea bottom (**Figure 9-21**). North Atlantic Deep Water forms in the North Atlantic and can be found in a layer just above the bottom. Sometimes, an additional description is necessary, especially for water masses with a complex history and formation process. Thus, we also have Upper North Atlantic Deep Water and Lower North Atlantic Deep Water to differentiate the characteristics of two water masses that make up North Atlantic Deep Water.

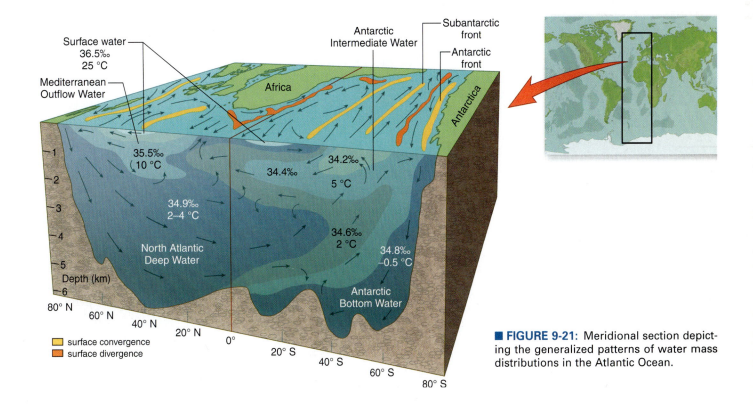

■ **FIGURE 9-21:** Meridional section depicting the generalized patterns of water mass distributions in the Atlantic Ocean.

TABLE 9.3   A Few Major Water Masses and Their Properties

Water mass	Acronym	Potential temperature (°C)	Salinity (psu)	Sigma-t (kg m⁻³)	Formation rates (Sv)	Source(s)
Atlantic Ocean						
North Atlantic Deep Water	NADW	1.8 to 4	> 34.9	> = 27.80	13–17	At least five sources in North Atlantic
Mediterranean Outflow Water	MOW	11 to 13	36.6 to 36.5	27.5 to 27.7	0.5	Mediterranean seawater plus North Atlantic Central Water
Pacific Ocean						
North Pacific Deep Water (also called Pacific Common Water)	NPDW	1.1 to 1.2	34.68 to 34.69	N/A	8	Antarctic Circumpolar Water
North Pacific Intermediate Water	NPIW	7	< 33.8 to 34.4	26.7 to 26.9	12	Kuroshio-Oyashio Interfrontal Zone
Indian Ocean						
Indian Central Water	ICW	N/A	34.5 to 35.4	N/A	N/A	N/A
Indian Deep Water	IDW	2	34.75 to > 35.8		4–10	North Atlantic Deep Water
Southern Ocean						
Antarctic Bottom Water	AABW	−0.30 to 0.04	34.66 to 34.71	> 28.27	8–12	Weddell Sea plus Indian Pacific sector
Subantarctic Mode Water	SAMW	4 to 15	34.2 to 35.8	26.5 to 27.1	25	North of the polar front
Antarctic Intermediate Water	AAIW	3 to 5	34.2 to 34.4	27.2 to 27.3	14	Uncertain: Subantarctic Mode Water plus Antarctic Surface Water postulated
Circumpolar Deep Water (Upper and Lower)	CDW	0 to 2	34.68 to 34.72	27.4 to 28.2	8–13	North Atlantic Deep Water
Arctic Ocean						
Arctic Surface Water	ASW	−1.9 to −1.5	28 to 33.5	N/A	N/A	North Atlantic surface water and freshwater runoff into Arctic basin
Arctic Bottom Water	ABW	−0.95 to −0.8	34.95	N/A	N/A	

Note: N/A = not available
Table based on multiple sources. See website.

Table 9.3 lists a few of the major water masses found in the world ocean. Note the narrow range of properties for a given water mass and the differences in the density of different water masses. For example, the potential temperature of **North Atlantic Deep Water** (NADW) ranges from 1.8° to 4°C (35.2–39.2° F), while its salinity is greater than 34.9. By comparison, **Antarctic Bottom Water** (AABW) is colder, −0.3° to 0.04° C (31.46–32.1° F) and slightly fresher, from 34.66 to 34.71. Despite its lower salinity, AABW is the among the densest water masses in the world ocean. Nevertheless, NADW is more abundant than AABW and influences deep circulation over a greater region of the world ocean. The site of formation of NADW is thought by some oceanographers to represent an Achilles' heel for climate change. These oceanographers postulate that shut down of NADW formation causes abrupt climate change, like that seen in the movie *Day After Tomorrow* (albeit nowhere near as fast). Other ocean-ographers, while acknowledging an important role for NADW, believe that water masses in the Southern Ocean govern long-term climate. You may recall from our earlier discussion that surface waters from the Antarctic are fed into the South Atlantic gyre via the Benguela Current and eventually into the North Atlantic. By this reasoning, it is postulated that upwelling in the Southern Ocean may control the flow of surface waters available for the formation of NADW. As much as 15 Sv of deep water may be upwelled to the surface in the Drake Passage. This scenario emphasizes a stronger role for winds and wind-induced mixing in the control of ocean circulation and climate.

Antarctic Intermediate Water (AAIW), which forms in the region of the **Antarctic Convergence,** may also be important for climate. This region, also known as the Antarctic Polar Front Zone, encircles the Antarctic continent where warm surface waters from the north meet cold waters from the south (see Figure 9-19). It extends northward to about 30° N in the Atlantic, where it separates the deeper NADW from the less dense subtropical surface waters at a depth between 800 and 1000 m (2625–3281 feet). AAIW occupies a large expanse of the Pacific Ocean to about 20° N. It can also be found in the Indian Ocean. This water is notable for its relatively low salinity, a property it acquires from melting sea ice at the Antarctic Convergence. Some evidence exists that this water mass has been changing as a result of global warming.

Whereas Atlantic and Southern Ocean water masses have dynamic and "youthful" histories, **North Pacific Deep Water** (NPDW) represents a more "mature" type of water mass. It has been called the "oldest and most isolated water mass in the world ocean," owing to its longevity and minimal interaction with surface waters. NPDW occupies the greatest volume of all water masses in the ocean and represents the return flow of bottom-hugging Circumpolar Deep Water (CDW), which moves northward from the Antarctic. Unlike in the Atlantic Ocean, little interaction occurs between the abyssal and surface waters of the North Pacific.

The highly saline and warm **Mediterranean Outflow Water** (MOW) flows out of the Straits of Gibraltar. Under the influence of the Coriolis effect, it is deflected northward and mixes with North Atlantic Water in the Gulf of Cadiz in Portugal. MOW reaches neutral buoyancy at a depth of about 1000 m (3281 feet) and can be found in a narrow layer throughout most of the North Atlantic Basin. The circu-lation pattern of MOW also gives rise to Mediterranean eddies called **meddies.** These meddies transport heat, salts, and even pollutants to the Azores and other regions of the central North Atlantic.

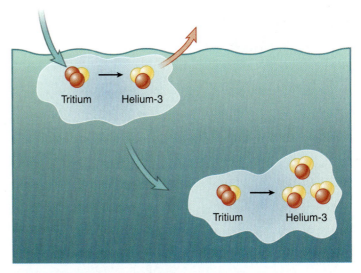

■ **FIGURE 9-22** Radioactive elements such as tritium can be used to trace water masses as they descend and flow in the deep cir-culation. The decay of tritium to helium-3 occurs naturally, but at the surface, helium-3 is released to the atmosphere. When a water mass descends and loses contact with the atmosphere, the con-centration of helium-3 increases. Knowledge of the decay rate of tritium to helium-3 allows oceanographers to estimate the "age" of the water mass, that is, the time since its submergence.

In addition to common properties such as temperature and salinity, oceanographers characterize water masses based on a wide variety of other chemical compounds. These chemical **tracers** can be used not only to identify and track water masses but also to infer water mass ages (**Figure 9-22**). Radioactive compounds introduced by human activities decay predictably. Carbon-14, a byproduct of nuclear testing, can be measured in water masses to determine how long a particular water mass has been isolated from the atmosphere. The concentration of refrigerants such as chlorofluorocarbons (CFCs), introduced into the atmosphere in the 1950s, can also be measured in deep waters to determine the approximate time when that water mass sank and lost contact with the atmosphere. Water masses in which no CFCs can be found must be older than 50 to 60 years. Other location-specific tracers, especially those associated with hydrothermal vents, can be used to tag the movements of water masses within specific regions of the ocean.

The Circulation Puzzle: What Drives Deep Circulation?

While density differences give rise to the downward flux (sinking) of water masses at high latitudes, it is the upward flux of these waters across the thermocline over the expanse of the ocean that generates the work needed to drive the deep circulation. This upward mixing arises primarily as a result of wind-induced upwelling and turbulence and, possibly, tidal interactions with seafloor features. The idea of deep circulation being "pulled" by the stirring action of winds and

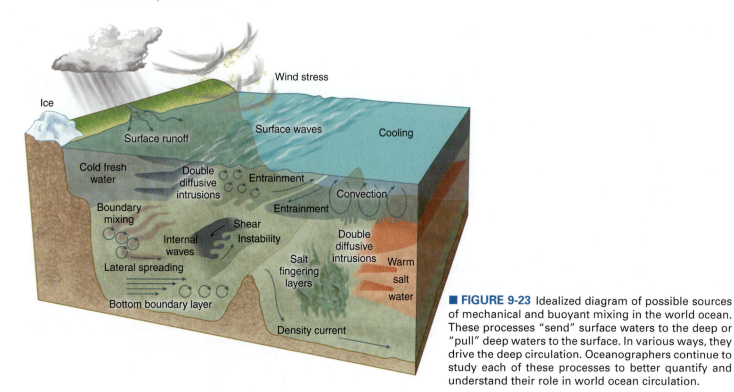

■ **FIGURE 9-23** Idealized diagram of possible sources of mechanical and buoyant mixing in the world ocean. These processes "send" surface waters to the deep or "pull" deep waters to the surface. In various ways, they drive the deep circulation. Oceanographers continue to study each of these processes to better quantify and understand their role in world ocean circulation.

tides rather than "pushed" by density differences is a relatively new one, although Stommel and others provided early clues.

For purely physical reasons, convection alone cannot explain the return of water masses to the surface. In fact, without some kind of mechanical mixing, the entire ocean would stagnate into a giant pool of cold, salty water in a few thousand years! This startling scenario is a consequence of **Sandström's theorem,** which states that deep convection is only possible when the heat source is lower than the cooling source. This effect is best understood by considering what would happen to a pot of water heated from below (like the atmosphere) or above (like the ocean). Heating from below reduces the density of the fluid, causing it to rise and mix with the overlying fluid. Heating from above imposes stratification on the pot of water and reduced mixing of its water layers. To overcome this effect in the ocean, vigorous mixing is required to carry warm fluids downward or cold fluids upward. Some oceanographers have gone so far as to conclude that the rate at which heat and mass are transported via deep circulation depends on the rate at which energy is supplied through mechanical mixing (**Figure 9-23**). In 1998, Walter Munk and Carl Wunsch proposed that tides might provide the energy needed to mix the ocean sufficiently. According to Munk and Wunsch, the world ocean requires roughly 2 terawatts (1 TW equals 1 billion kilowatts) of energy to maintain the observed separation of deep and surface waters. Buoyancy changes and geothermal heating were found to be minor, but winds and tides could supply 1 TW of energy each, according to their calculations.

Oceanographers continue to debate whether deep circulation is "pushed" by buoyancy changes, "pulled" by upwelling, or driven by some combination of the two processes. Nevertheless, new research efforts are being directed toward better quantifying the sources of mechanical mixing in the ocean. Improved observational capability, allowing greater temporal and spatial coverage of ocean properties, will stand at the forefront of efforts to resolve this challenge.

World Ocean Circulation and the Great Global Conveyor

By the late 1960s, it became clear that existing observational data were inadequate to confirm or refute the predictions of circulation models. As a result, Stommel organized several researchers to address this problem as part of the Geochemical Ocean Sections Study (GEOSECS), which occurred during the **International Decade of the Ocean,** 1970–1979. The GEOSECS program (1968–1978) provided the first global look at ocean geochemistry using modern analytical techniques. It also represented one of the first large-scale, multimillion-dollar oceanographic programs to be undertaken by the National Science Foundation. GEOSECS served as a model for the major programs that followed, including the World Ocean Circulation Experiment (WOCE) and the Joint Global Ocean Flux Study (JGOFs) in the 1980s and 1990s.

Wallace Broecker of Lamont-Doherty Earth Observatory of Columbia University was a participant in GEOSECS and set out to analyze nutrient and other chemical tracers to identify water masses and their transports. With Tsung-Hung Peng of Oak Ridge National Laboratory, Broecker published the book *Tracers in the Sea* (1982), a comprehensive treatment of modern geochemistry that included many GEOSECS findings. Broecker's ideas concerning the movement and transport of water masses grew over the next decade through collaborations with Arnold Gordon, a physical oceanographer and colleague at

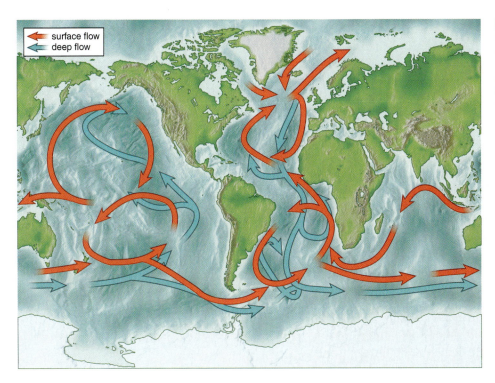

■ **FIGURE 9-24** The Great Ocean Conveyor Belt, a highly idealized representation of world ocean circulation. Though now "famous," this diagram presents an over-simplified view of world ocean circulation. Its strength lies in illustrating the connections between surface and deep circulation.

Lamont-Doherty Earth Observatory. At nearly the same time, they published descriptions and illustrations of the formation and transport of NADW throughout the world ocean: Gordon's in a scientific journal, *Journal of Geophysical Research* (1986); and Broecker's in a science magazine, *Natural History* (1987). Broecker's depiction became the now-famous **Great Ocean Conveyor Belt** of thermohaline circulation (**Figure 9-24**).

The origins of the ocean conveyor belt are discussed by Broecker in a 1991 paper in the journal *Oceanography*. The diagram was intended to illustrate the general concept of thermohaline circulation to a general readership. An attractive feature of the virtual three-dimensional diagram was that it showed not only deep circulation but also the return paths of water to the surface of the ocean. Perhaps because of the simplicity and aesthetic quality of the diagram, it took on a life of its own and has been widely used, oftentimes out of context. The diagram suggests the many linkages of the deep and surface circulation and rather beautifully illustrates the connectivity of the world ocean. However, as noted by Broecker, the diagram fails to capture several important points and can lead to erroneous interpretations and oversimplifications by those unfamiliar with its intended use, especially in discussions of climate change.

While no diagram can adequately represent ocean currents, the more detailed models and schematics of Schmitz (1995) better illustrate the general patterns of deep circulation in the world ocean (**Figure 9-25**). In particular, Schmitz separates the circulation of bottom, deep, and intermediate waters and indicates the exchanges between them. The role of intermediate waters has been emphasized in recent years, particularly as these waters may act as a kind of go-between in the exchange of waters between the deep and surface circulation. This portrayal of intermediate waters comes into play when we consider the mechanisms that control the deep circulation, especially those

factors that return deep waters to the surface. Schmitz's illustrations should give you an idea of the complexity of ocean circulation and the dangers of oversimplification.

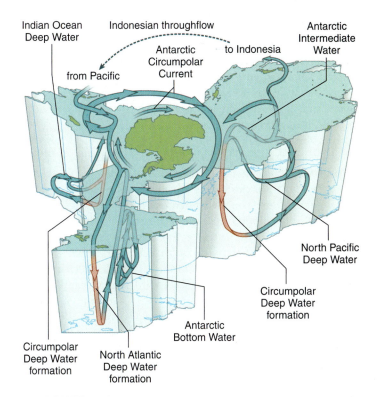

■ **FIGURE 9-25** The world ocean circulation showing the coupled surface and deep circulation. Note that waters from all three major basins—Atlantic, Pacific, and Indian—intermingle in the Southern Ocean.

Ekman Spirals: Putting a Theory to the Test

Science is like a tennis match. On one side of the net is the "theory player" and on the other side is the "observation player." You need both players to advance knowledge of the natural world. One of the more interesting "oceanographic" tennis matches has focused on V. Wilfrid Ekman's elegant theory that mathematically predicted Ekman spirals (see the discussion in this chapter). The back-and-forth "volley" between observationalists and theorists began in the early 1900s. Norwegian Fritjof Nansen, the only oceanographer to win a Nobel Prize (specifically, the Nobel Peace Prize), observed a high-latitude ice floe that was directed roughly 20° to 40° to the right of the prevailing wind direction. Nansen's observations were described to his Norwegian colleague, V. Walfrid Ekman, who quickly developed an elegant analytical theory to explain this phenomenon, which he published in 1905.

Long after Ekman's theory was published, oceanographers sought to put the theory to the test—the tennis ball flew to the observationalists' side of the net! After a century of searching, only a few direct observations supported the existence of the Ekman spirals. Why? First, all theories make assumptions, some modest and some major, that serve to idealize the problem of interest and make it more mathematically tractable. For example, several sources of ocean motion are excluded from Ekman's theory. These include surface and internal gravity waves, tides, upwelling and downwelling motions, and large-scale, pressure-driven currents. These factors are not typically negligible. The theory also assumes a homogeneous ocean, whereas stratification is almost always present to some degree in the upper ocean. Ekman assumed that winds were constant and uniform in direction, which rarely is the case. So it is not surprising that the relatively weak or slow currents predicted by Ekman's theory might be swamped by other motions.

Another problem has been direct and simultaneous measurements of currents at many depths in the ocean. Until fairly recently, it was difficult to separate motions of waves from motions of currents. The advent of acoustic Doppler current profilers (ADCPs) and some very well-designed mechanical current meters finally provided the spatial resolution required to observe ocean currents throughout the upper ocean. In 1995, ADCP observations of an Ekman layer in the California Current were reported by Teresa Chereskin. Subsequently, the Ocean Physics

Laboratory at the University of California, Santa Barbara began using an ADCP deployed from the Bermuda Testbed Mooring (BTM) and observed a *rotating* Ekman spiral (see our website for an animation of these data). The BTM observations occurred during and after the passage of Hurricane Fabian in September 2004 (**Figure 9a**). The strength of wind forcing by Hurricane Fabian apparently led to dominance of the Ekman currents over competing ocean motions, resulting in "beautiful" rotating Ekman spirals. The convergence of observations and theory in this case scores on both sides of the net!

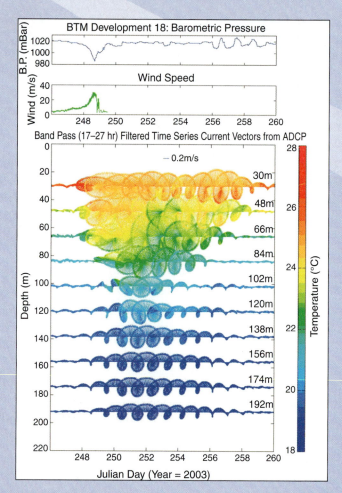

FIGURE 9a Ekman spiral as observed by the Bermuda Testbed Mooring.

YOU Might Wonder

1. **If you put a message in a bottle and toss it overboard at sea, will it really wash up on a beach somewhere?**
Unfortunately, just about anything thrown overboard or lost at sea ends up on a beach somewhere (if it doesn't sink to the bottom of the ocean). Nonetheless, drift bottles have been an important tool in studies of ocean currents. One drift bottle dropped in the Mississippi River ended up on a beach in England in just ten months! One famous story of a message in a bottle recounts a young boy in the Azores who found a message in a bottle from a New York radio station offering a reward to the finder. A month after writing the address on the message, the boy received a check for $1000. The bottle was thrown overboard by the station's owner, who thought it would make a great way to celebrate the station's twenty-fifth anniversary. Numerous stories have been told of bottles containing last wills and testaments, messages to loved ones, messages asking for a hand in marriage, and even messages asking for recipes.

2. **Is it true that oceanographers have used rubber duckies to track ocean currents?**
True, but not intentionally. Oftentimes, ships at sea caught in rough weather will accidentally lose the contents of their shipping containers. Rubber duckies and sneakers are among the best examples. By tracking the path of these floating objects from their source to the beach and the time it took them to reach the beach, oceanographers have been able to piece together a picture of surface currents in a few regions of the ocean. While such experiments are not ideal, they do provide insights into the mean flow patterns of the upper ocean.

3. **Some currents have obvious names, but where did names such as Agulhas and Benguela come from?**
Like many currents, the Agulhas takes its name from a nearby geographic feature. Portuguese sailors named it after the Cabo das Agulhas, or Cape of Needles, off the coast of Africa. The Benguela Current is named after a city founded in 1617 by the Portuguese.

Key Concepts

- Though it is commonly studied as surface circulation and deep circulation, oceanographers recognize a single, interdependent circulation, the world ocean circulation.
- Ocean currents are primary driven by winds and buoyancy and perhaps tides.
- Modeling of upper ocean circulation has evolved from the relatively simple Ekman theory to sophisticated numerical models that include many additional processes and are applied to ocean basins and the global ocean.
- Upwelling occurs where Ekman transport causes surface divergence; downwelling occurs where surface waters converge.
- New observations and increased computing power are expected to enable useful predictions of ocean circulation in the near future.

Terms to Remember

Agulhas Retroflection, 175

Antarctic Bottom
Water, 181

Antarctic Circumpolar
Current, 176

Antarctic Convergence, 181

Benguela Current, 175

Benguela Niño, 175

Benguela upwelling
system, 175

bottom waters, 179

Brazil Current, 172

Brazil-Malvinas
Confluence, 173

California Current, 174

central water masses, 179

choke point, 176

coastal jets, 174

cold-core rings, 172

convection currents, 166

deep circulation, 165

deep waters, 179

downwelling, 170

Drake Passage, 176

dynamical physical ocean-
ography, 165

Ekman layer, 168

Ekman spiral, 168

Ekman transport, 168

geostrophic currents, 169

Great Ocean Conveyor
Belt, 183

Critical Thinking

1. How has scientific understanding of ocean circulation progressed with the development of technology for observing and measuring ocean currents?

2. What role do mathematical models of ocean circulation play in scientific understanding of ocean currents? How do models and observations work together to further our knowledge of ocean circulation?

3. What is meant by the life cycle of an upwelling event or Gulf Stream ring, and what factors influence the life cycle of these features?

4. Compare and contrast the surface circulation and the deep circulation. How do these two components of the world ocean circulation interact and influence each other?

5. Why are *T-S* diagrams such a powerful tool for understanding the formation and global distribution of water masses?

6. In what ways is the ocean conveyor belt a "cartoon" of ocean circulation? How does this illustration better help us to understand the role of the ocean in climate change?

7. How are ocean "choke points" like freeway interchanges?

8. Compare and contrast the surface circulation of two ocean basins.

9. How do the concepts of density and buoyancy apply to the deep circulation?

10. Why is the study of world ocean circulation important to humankind?

Explore Online

 Visit www.mhhe.com/chamberlin1e for access to chapter quizzing, key term flash cards, video clips, interactive activities, and more. Further enhance your knowledge with web links to chapter-related material!

Exploration Activity 9-1

Exploring Ocean Circulation Through Scientific Investigation

Question for Inquiry

How do measurements of ocean physical properties help oceanographers better understand the nature of the world ocean?

Summary

It's no accident that much of our focus in Chapters 7 and 8 concerned heat energy and its distribution in the world ocean. Heat drives the motions of the ocean, especially the deep ocean. The distribution of heat also affects the distribution and productivity of ocean life. Here we explore oceanographers' valiant efforts to map and understand the distribution of heat and other properties in the world ocean.

Learning Objectives

By the end of this activity, you should be able to:

* Acquire data that includes temporal and spatial distributions in each of the three major ocean basins
* Analyze and characterize the distribution of seawater temperature in different types of graphical representations of data
* Interpret spatial and temporal distributions of temperature and salinity within the context of what you know about the seasonal cycle of solar radiation and the distribution of water masses in the world ocean
* Articulate an opinion on the importance of synoptic efforts to map temperature distributions in the world ocean over large temporal and spatial scales

Materials and Preparation

For this activity, you will need access to the Internet and a printer. Alternatively, your instructor may provide a printout of figures sufficient for completing the activity. You may want to review the sections on the Argo system and the seasonal thermocline in Chapter 7. You may also wish to check again the sections in Chapter 8 on the physical structure of the world ocean. Some of you undoubtedly are expert web surfers with little need for instructions. Others of you may be relatively new to browsing the Internet. We strongly encourage the "experts" to help the "newbies."

Part I: Exploring Argo

The Argo system represents one of the most ambitious attempts to map the distribution of ocean properties ever attempted. A principle goal of this effort is to better understand the dynamics of ocean circulation and property distributions as they affect climate. Despite more than 3000 profiling drifters deployed by the end of 2006, the world ocean remains undersampled. Nonetheless, the volume of data being collected and reported in near-real time on the Internet is impressive. In this part, you will learn how to access the data sets and interpret them.

√ Explore the Argo website at www.argo.ucsd.edu/. Make notes on the major goals and milestones of the mission. Review especially the sections About Argo and the FAQs. These pages contain valuable information. Check out the Picture Gallery for a closeup view of the instruments, how they are deployed, and how they operate. There is a good animation on this page as well.

Part II: Exploring Argo Profilers

Finding the data from the UCSD site can be tricky, so go directly to Coriolis, www.coriolis.eu.org/, the European site that archives data and makes it available to scientists and the general public. Feel free to learn more about the project. Otherwise, click on Data Service and look for the Argo>All Profilers link. Click it. A page of numbers will appear. Each number corresponds to active individual profilers.

Go back to the Data Service main page and click on Geographic Map. Use your mouse to select a portion of the map. (Click and drag on the map to create a square. The map will zoom to that region.) In the zoom view, you will see the individual float numbers. These correspond to the table you just looked at on the All Profilers page. You can zoom closer or pull out to view the floats inhabiting a particular region of the world ocean. This will be useful when you choose individual floats in each basin for your data analysis.

Return to the All Profilers page. Each Argo profiler (floats) delivers data on various ocean properties. The particular type of data transmitted by a profiler depends on its configuration but most of them measure temperature and salinity. The website organizes the profilers (floats) by the ocean basin in which they operate. Click on any one of them and take notes on the types of data and information associated with a float. You will see a heading here for Float cycles. If the float cycle has a + (plus) mark next to it, click on the + mark to expand the page and view the dates associated with different float cycles (the time from descent to ascent and data transmission; note that the float may descend and ascend several times in one cycle before transmitting data). This information will be useful when choosing your floats below. Click also on the + mark next to Map to view the location and trajectory of the float. Continue to Part III.

Part III: Acquiring Data from the Argo Profilers

Your task is to report on the spatial and temporal variability at one location in each ocean basin (Atlantic, Pacific, and Indian). To do this, you will need to find a float that is currently operating in each ocean basin and that has data available over a year or longer. Note that the Coriolis site reports dates in the European format, that is, day/month/year.

1. Use the Geographic Map to identify several floats in any region of your choosing in each of the three major ocean basins. Write them down. You should have three lists, one for each basin.
2. Return to the All Profilers page and find the floats that you wrote down. Click on the float number and review the Float Cycles data to determine if the float has been in operation for at least a year. If not, choose another one. Keep exploring until you find three floats, one in each basin.
3 For each float, do the following:
   a. Print the map of its location. (Click on the + sign next to map.)
   b. Expand the waterfall profile by clicking the + mark. At the top left of the expanded view, you will see two symbols. The first allows you to open the

waterfall profile image as a separate window so you can save it by right-clicking (PC) or CRTL-clicking (Mac). The second symbol allows you to print. Do either one or both (consult your instructor).
   c. Repeat for Overlayed Profiles, TS Diagram, and Section Plots. Save or print the plots.
   d. Click on Vertical Profiles and save the image.
   e. You should now have a waterfall profile, overlayed profile, TS diagram, section plot, and vertical profiles for profilers in the Atlantic, Pacific, and Indian basins

Part IV: Analyzing Data from the Argo Profilers

One of the challenges of a large data set is to make sense of the numbers. It often requires years of experience to look at a graph and "know" if it makes sense or if it is unusual. If an oceanographer does spot an unusual graph, their first task is to verify the data by checking to see if nearby instruments (if available) recorded similar data. If the data are unique, then there was possibly an instrument malfunction. Otherwise, if other instruments recorded similar data, then an oceanographer begins to explore possible natural causes. This is often the most exciting time of a project because this is where discoveries are made. It is with that sense of discovery that we hope you approach your data analysis.

1. Start with the vertical profiles and label the thermocline and halocline in the temperature and salinity images, respectively. It's possible to have multiple thermoclines or haloclines so if you think your data exhibit multiple layers, then indicate so. You may also observe inversions (cold water on top of warm water) or places where the salinity goes to zero. Note these on your graphs. Include any other notes or observations on the data that you think are relevant. Circle any features you don't understand and put a question mark by them. These are *real* data so you never know what you might find!
2. Create a table of data for each date where a vertical profile is available and indicate the range of temperature and salinity values observed.

3. Open the waterfall profiles and compare the shape of the profiles. Record observations on trend of the thermocline and halocline over the course of the months. In other words, are the depths of the thermocline and halocline constant? Do they deepen? Shallow? Record any other unusual patterns that might appear, especially intrusions of other water masses (often appearing as a spike in temperature and/or salinity) Also note anything you didn't notice in the individual profiles.

4. Examine the overlay plots. For each depth interval, record the range of values of temperature and salinity. For example, if the lowest surface temperature is 12 °C and the lowest is 7 °C then the range of surface temperatures is 5 °C.

5. Record observations on the overlay profiles. Are they generally the same? Do their shape vary by a few degrees or several degrees? Do the same "bumps and wiggles" appear in all of the profiles or are they different? Record their similarities and differences.

6. Open your section charts. These "maps" provide a three-dimensional view of your profiler's data. For example, a temperature section indicates temperature (as a color), depth (the vertical axis), and time (increasing from left to right on the horizontal axis). Looking at the sections, describe how temperature and salinity changed over time. Note the location (climate zone and hemisphere) and season for each cycle shown on the section.

7. Examine the TS diagram and mark the inflection points, the regions where the plot changes rapidly.

Part V: Interpreting and Reporting Argo Profiler Data

What does it all mean? In truth, that's what *you* have to figure out. Interpreting data of these type relies on inference and, to some degree, speculation. The important thing is to connect your ideas with existing knowledge. Does your interpretation make sense?

1. Prepare separate summaries of your data for each profiler. Your summary should start with a statement that describes the location and dates of data collection. You should then summarize your notes and observations.

Support your observations with references to your figures (which you should number) and any tables. Write up your data as if you were writing a results section in a scientific paper.

2. Write a discussion section that includes the following:
   a. An interpretation of the effects of the seasonal cycle of solar radiation on the vertical structure of temperature in the water column at each location.
   b. A comparison and interpretation of the seasonal cycle between locations (For example, if one location is located in a polar zone and the other in a temperate zone, use that knowledge to interpret their differences. If all of your profiles are located in tropical waters, explain their similarities or differences.)
   c. An interpretation of water masses present at each location (see Tables 9-2 and 9-3 and Figures 9-20 and 9-21).
   d. A general statement on the limitations of your limited data set (What additional observations or data analyses would you need to test your inferences and ideas?)
   e. A statement of how this activity has affected your view of physical oceanography and the work of physical oceanographers to obtain synoptic data on the world ocean over a broad range of temporal and spatial scales.
   f. Use the Rubric for Analyzing and Presenting Scientific Data and Information and discussions with classmates and your instructor to refine your work (See Exploration Activity 1-1.)

10 Ocean Waves

A surfer blasts down the face of an extreme wave at Shark Park, one of California's newest extreme wave surfing locations.

Questions to Consider

1. What does it mean to say that waves are solar-powered?

2. What kinds of disturbing forces generate waves, and how do they impart their energy to the ocean?

3. What is the difference between surface waves and internal waves?

4. How does the motion of particles within a traveling wave affect the wave's interactions with the seafloor?

5. What causes waves to "bend" or refract, and what effect does wave refraction have on coastlines?

6. What is the fate of a wave from its moment of formation to the point it crashes on a beach?

7. Why are extreme waves an important area of study for oceanographers?

In Search of the Perfect Wave

Waves hold a special fascination for us. Perhaps it's the power with which they hit the shore and explode in a fury of sound. Perhaps it's the regularity of their patterns and rhythms that holds our attention. Or maybe waves fascinate us because of their many forms: a moving train of hills and valleys stretching across the ocean; concentric ripples expanding across a pond; or the slosh of water back and forth in a bathtub. Perhaps it's just as simple as the search for a "perfect" wave, as documented by filmmaker Bruce Brown in the 1966 surfing classic, *The Endless Summer*.

Fundamentally, a study of waves provides insights into the coupling between earth, atmosphere, and ocean processes. The greatest exchanges of matter and properties across the air-sea interface occur when large waves are breaking and turbulent mixing is most intense. The exchange of biologically important gases such as carbon dioxide and oxygen is affected by waves. Marine organisms, particularly those living in the surf zone, must be adapted for the tremendous forces exerted by breaking waves or they will perish. Knowledge of waves also has many important practical and aesthetic applications. Waves can affect the safety and integrity of human-made structures along the shoreline and at sea, so knowledge of their properties is vital to ocean engineering. Oceangoing vessels must be capable of withstanding the force of waves or risk capsizing. Captains of vessels require considerable knowledge and experience with waves to ensure the safety of their cargo, crew, and passengers. Of course, waves can simply be enjoyed, either as a casual observer or an active participant in surfing or other forms of wave-related recreation. And it is not unusual to see marine mammals surfing or riding the bow waves of ships, so perhaps even marine life enjoys waves as well.

The Development of Wave Theory

The earliest studies of waves were carried out by the Pacific Ocean-dwelling people of Oceania, in particular, the Marshallese. From the vantage posts of coastal hills and mountains and during extensive travels in oceangoing canoes, they acquired a working knowledge of waves and other oceanographic phenomena. Little is known regarding the history of the development of their ideas, but colonization of distant island archipelagoes at least 2,000 years ago suggests that their knowledge existed then. Evidence of their understanding of waves appears in the form of stick charts, or **mattang,** which were used to teach young sailors how to read patterns in the ocean swell (**Figure 10-1**). This knowledge enabled them to navigate across broad expanses of the open ocean using a technique called **wayfaring.** Unlike traditional forms of navigation, wayfaring relied on the skills of the navigator, who could discern the patterns and interactions of ocean swell in the vicinity of islands well beyond the horizon. By following these patterns, the navigator could successfully guide the canoe between islands separated by hundreds of miles.

Sir Isaac Newton (1642–1727) attempted the first mathematical description of the motions of waves in his book *Principia*, published in 1687. Newton even proposed that the motion of water particles in a wave was circular, but more than a century would pass before these ideas would be developed further. Between 1775 and 1825, Pierre Simon, Marquis de Laplace (1749–1827), introduced a description of waves in terms of the displacements of individual water particles, an important consideration in the development of wave theory. By 1802, Czech mathematician Franz Josef Gerstner (1756–1832) developed the first primitive wave theory to describe the circular motion of water particles. Gerstner noted that particles traveling at the crest of the wave moved up and forward, while those traveling at the trough of the wave moved down and backward, relative to the direction of the wave. The motions of the individual water particles traced a circle whose height at the surface was equal to the height of the wave.

Gerstner's work caught the attention of Ernst Weber (1795–1878), a professor of anatomy, and his brother, Wilhelm Weber (1804–1891), a physicist known for his work on electricity. Together, they built the first **wave tank,** a very long and narrow rectangular vessel for studying waves (**Figure 10-2**). These first wave experimentalists confirmed Gerstner's circular motions. Suffice it to say that many notable scientists have contributed to the development of wave theory and various mathematical descriptions of waves, including Joseph Louis-Lagrange (1736–1813), John Scott Russell (1808–1882), George Biddell Airy (1801–1892), George Stokes (1819–1903), and Lord Rayleigh (1842–1919) (see our website for a summary of their contributions).

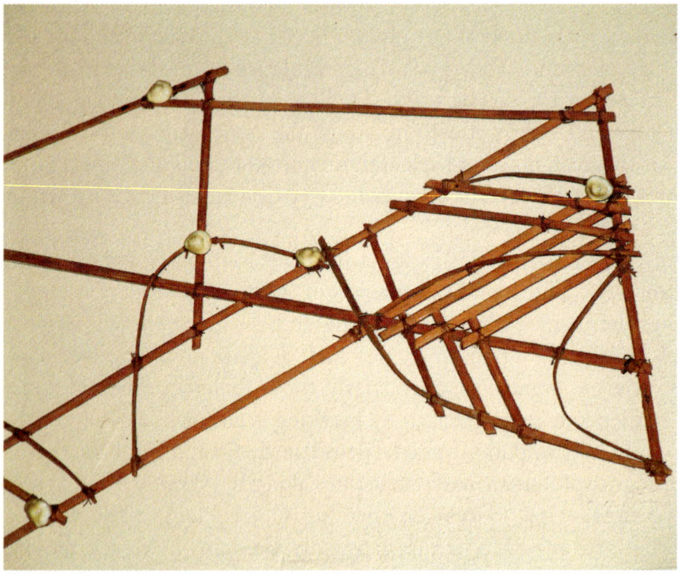

■ **FIGURE 10-1** A stick chart or mattang used by the Polynesians to teach a specially chosen young person how to use patterns in waves to navigate from island to island. Each shell represents an island. The arrangement of sticks indicate different wave patterns that might appear in the vicinity of islands for a given period and direction of swell.

■ **FIGURE 10-2** Wave tanks continue to provide valuable information on the nature of waves. This wave tank at the Cabrillo Aquarium, in San Pedro, California, features a mechanical paddle that produces a regular train of waves. The well-behaved waves produced in wave tanks allow oceanographers to test ideas concerning the nature of waves and their interaction with the shore or man-made objects. They also help laypersons better understand and appreciate waves.

■ **FIGURE 10-3** Buoys such as this one in Prince William Sound in Alaska regularly measure and transmit data on waves and other oceanographic and meteorological information.

World War II brought a different set of needs to the development of wave theory. Before attacking enemy-held beach positions, commanders wanted to know the type of sea conditions they might expect. Walter Munk, then a meteorologist with the Army Air Corps, and Harald Sverdrup, then with the University of California Division of War Research at Scripps Institution of Oceanography, began work on surf prediction. Their success in forecasting conditions at the Allied invasion of North Africa stimulated the development of **surf forecasting,** the prediction of wave conditions at a given location and time. Munk continued to develop this field. In the early 1960s, he deployed wave-measuring instruments at six stations across the Pacific from New Zealand to Alaska and found that wave arrivals in California could be traced to storms in the Indian Ocean several days earlier. Describing his work, Munk concluded that "predictions of surfing conditions at Honolulu" are possible but "from what I know of the surfing profession, I do not suggest this as a lucrative enterprise." Of course, surf forecasting and wave prediction have become integral tools for recreational and economic interests in modern times.

Technology continues to drive studies of waves. Wave heights, wind speeds, and other data are now routinely measured using satellites and buoys (**Figure 10-3**). These data are incorporated into wave models to forecast wave heights and other important ocean weather information. Other tools, such as time-lapse photography, seafloor pressure sensors, digital tide gauges, recording surface floats, and a number of others, are routinely deployed to understand the characteristics and behavior of ocean waves. A major driving force behind studies of waves is their application to coastal engineering problems and vessel safety at sea.

What Is a Wave?

In its simplest scientific form, a **wave** is an expression of the movement or progression of energy through a medium. Note that the medium itself does not move appreciably; rather, energy passes through the medium. The flick of an outstretched rope attached to a wall does not move the rope, but a wave passes down its length. Ocean waves are similar: energy moves through the water but the water does not move (other than the circular motion of individual water particles). Waves that result from the movement of energy through a medium are called **progressive waves,** a category that includes ocean waves, seismic waves, and sound waves. Progressive waves result from a **disturbing force,** any force that transmits energy to the water column or sea surface, such as the wind or an undersea earthquake. Waves produced from episodic forces, such as winds and earthquakes, result in **free waves,** waves that travel without any further influence from the disturbing force. **Forced waves,** such as tides, are continuously under the influence of their disturbing force, the gravitational forces of the Moon and the Sun. Forces that restore the sea surface, **restoring forces,** drain the wave of its energy. For very small waves, such as capillary waves, the surface tension of water acts as a restoring force. However, gravity acts as the restoring force for most free-traveling waves in the ocean. Water parcels, displaced upward or downward by the passing wave, are returned to their neutrally buoyant state by gravity (see Chapter 7). Thus, most waves in the ocean (and atmosphere) are broadly classified as **gravity waves.**

Anatomy of a Wave

As energy is transmitted through a fluid as a wave, the particles in the fluid move in a circular orbit called an **orbital.** The motion of a ball floating on the sea surface confirms this theory. The ball rises and moves slightly forward as the crest of the wave passes and falls, and moves slightly backward as the trough passes. Other than motions associated with currents (and Stoke's drift; see our website), the ball returns to the same location where it started. If we trace the position of the ball (or an individual wave particle) during the passing of a wave, the resultant shape resembles a sinusoid or sine wave

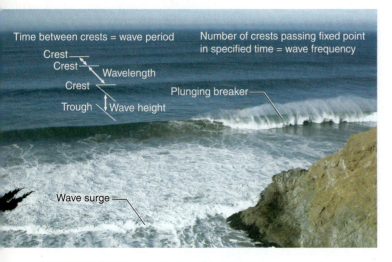

■ FIGURE 10-4 Parts of a typical ocean wave.

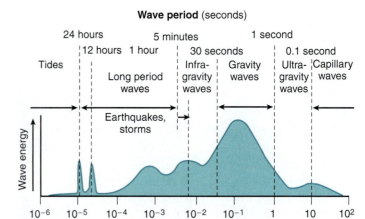

■ FIGURE 10-6 Types of ocean waves and their typical frequencies, periods, and energies.

(**Figure 10-4**). The peak of such a wave is called the **crest,** while the valley is the **trough.** The crest and trough represent displacements from the **still water level,** the "resting" or equilibrium state of a sea surface in the absence of a disturbing force. The horizontal distance between two crests is called the **wavelength,** and the vertical distance between the crest and the trough is known as the **wave height.**

Energy in Motion

Because waves represent energy in motion, it is useful to define their time-dependent variables (**Figure 10-5**). The **wave period** represents the time (usually in seconds) it takes for a single wave—from crest to crest—to travel past a given point. Alternatively, the

wave period may be thought of as the time for a crest to travel a distance of one wavelength. Unlike other properties of waves, the period of a wave does not change once the wave is formed. A related time-dependent variable is **wave frequency,** the number of crests that pass a fixed point in a specific interval of time, usually in number per second (or cycles per second).

Classification of Waves

Waves may be classified according to their period, wavelength, frequency, disturbing force, restoring force, energy, and other wave characteristics. **Table 10.1** and **Figure 10-6** represent two such classifications. For the most part, wind-generated waves and swell occupy our attention in this chapter. We will explore tides in Chapter 11.

The time between cars depends upon their speed and the distance between them.

$T = \dfrac{L}{S}$

50 mph   50 mph

The number of cars passing by every minute depends upon their speed and the distance between them.

$\dfrac{1}{T} = \dfrac{S}{L}$

50 mph   50 mph

The time between wave crests depends upon their speed and the distance between them.

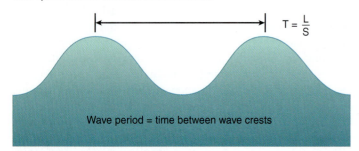

$T = \dfrac{L}{S}$

Wave period = time between wave crests

The number of wave crests passing by every minute depends upon their speed and the distance between them.

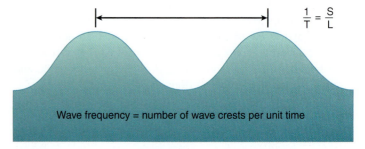

$\dfrac{1}{T} = \dfrac{S}{L}$

Wave frequency = number of wave crests per unit time

■ FIGURE 10-5 Concept of wave period and wave frequency.

TABLE 10.1 Categories of Waves Based on Their Disturbing Forces

Wave category	Characteristic	Cause(s)
Ripples (capillary waves), wind waves, and swell	Very short to moderate periods (seconds to minutes) and a wide range of wavelengths (centimeters to 1000s of meters)	Winds at the air-sea interface
Internal gravity waves	Short to long periods (up to 24 hours) and a wide range of wavelengths (centimeters to kilometers); vertical amplitudes greater than surface gravity waves	Current shear, surface disturbances, flow over topography, and other factors
Tsunami (see also Chapter 7)	Very long period (10–100 minutes) and very long wavelengths (100–1000 km)	Seismic activity, underwater land slides, meteorites, and other factors
	Gyroscopic gravity waves and Rossby waves Barotropic Rossby waves (independent of depth) propagate westward at speeds on the order of may arise from flow instabilities and 50 m s−1 and can cross the ocean in a few days; baroclinic Rossby waves (varying with depth) propagate at a few centimeters per second with periods on the order of a year or more and can take a decade or more to cross an ocean basin; baroclinic internal Rossby waves have larger vertical amplitudes than surface barotropic Rossby waves	Very long period and very long wave lengths; Variations in winds and atmospheric pressure and influenced by the Coriolis effect; Rossby waves eastward- and poleward-propagating Kelvin waves
Equatorial and coastal Kelvin waves (see also Chapter 11)	Similar in characteristics to Kelvin waves except that equatorial Kelvin waves propagate eastward along the equator; coastal Kelvin waves propagate northward in the Northern Hemisphere and southward in the Southern Hemisphere; surface Kelvin waves (barotropic) can travel 200 m s−1, internal (baroclinic) Kelvin waves travel from 0.5 to 3.0 m s−1; internal (baroclinic) Kelvin waves have larger vertical amplitudes than surface (barotropic) Kelvin waves	Equatorial Kelvin waves may be caused by wind burst events; coastal Kelvin waves are produced as equatorial Kelvin waves encounter land masses; the Coriolis effect influences Kelvin waves
Tides	Very long and predictable periods (hours-day) and wavelengths (1000s km); dominant periods are 12.0–12.4 hours for semidiurnal tides and 23.9–25.8 hours for diurnal tides; internal tides also occur with larger vertical amplitudes	Varying gravitational forces between Earth, Moon, and Sun

Wave Theory

The mathematical description of waves allows us to predict their behavior under a given set of circumstances and forecast their arrival on distant beaches. Surf forecasters make regular use of wave theory to predict when and where ocean waves will have their greatest impact on the shore. Your knowledge of wave theory will guide your understanding of the sections that follow.

Wave Speed

The speed of an individual wave (C) may be expressed dimensionally as a function of the wavelength (L) and the wave period (T), or:

$$C = L/T \qquad \text{Equation 10.1}$$

According to this equation, waves with different wavelengths and periods will travel at different speeds. Because L and T are proportional, increases in wave speed must be accompanied by increases in wavelength, wave period, or both. Remember, the period does not change once a wave is formed.

Deep- and Shallow-Water Waves

More formally, the speed of an individual wave crest (phase speed) wave can be predicted from the equation:

$$C = \sqrt{\frac{gL}{2\pi}} \tanh \frac{2\pi h}{L}$$ Equation 10.2

where g is the acceleration of gravity (9.8 m s^{-2} or 32 feet sec^{-1}), L is the wavelength, π is the constant pi (3.14), and h is the water depth. The hyperbolic tangent function, tanh, may be unfamiliar to you, but it simply describes a curve that rises linearly and levels off, in this case, as a function of $\sqrt{h}$. This relationship between wave speed, wavelength, and water depth illustrates the principal difference between shallow and deep-water waves (**Figure 10-7**).

If $2\pi h/L$ is large or h is greater than $L/2$, then tanh $2\pi h/L$ is approximately equal to 1. In that case, the wave speed equation reduces to

$$C_{deep} = \sqrt{\frac{gL}{2\pi}} \text{ or } C_{deep} = 1.25\sqrt{L}$$ Equation 10.3

or, by substitution and rearrangement,

$$C_{deep} = 1.56\ T.$$ Equation 10.4

Note that wave speed, in this case, only depends on wavelength or wave period, with longer waves traveling faster than shorter waves. Depth is not a factor here. Waves that satisfy these criteria—where $2\pi h/L$ is large or where h is greater than $L/2$—are defined as **deep-water waves.**

If values of $2\pi h/L$ are small, then tanh $2\pi h/L$ is approximately equal to $2\pi h/L$, and the wave speed equation reduces to

$$C_{shallow} = \sqrt{gh}$$ Equation 10.5

Waves where values of $2\pi h/L$ are small or where h is less than $L/20$ are defined as **shallow-water waves.** The speed of shallow-water waves depends solely on water depth and not on wave period or wavelength.

Surface waves that fall between the deep- and shallow-water approximations are called **intermediate waves.** For these waves, the exact hyperbolic tangent equation must be used to find the wave speed. Examples of deep-, intermediate-, and shallow-water calculations are shown in **Figure 10-8**. One word of caution applies to the terminology used here. Despite what the name implies, shallow-water waves can occur in the deep ocean. In fact, a tsunami with a wavelength on the order of hundreds of kilometers (1 km equals 0.62 miles) and a wave period approaching hours meets the criteria of a shallow-water wave (i.e., h is less than $L/20$).

Water Particle Motions

For deep-water waves at the surface, the diameter of a wave orbital is determined by the height of the wave. However, the diameter of wave orbitals becomes progressively smaller with depth (**Figure 10-9**). At some depth, the motion of water particles disappears altogether and the effects of wave energy are not felt. As it turns out, the depth at which orbital motion ceases is equal to one-half the wavelength of the wave. For shallow-water waves, the influence of the ocean bottom causes wave orbitals to

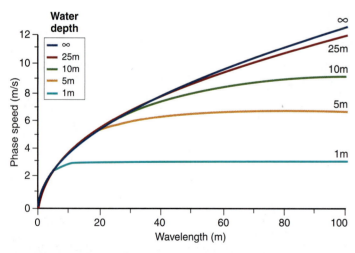

■ **FIGURE 10-7** A family of curves based on Equation 10.2, illustrating the mathematical relationship between wave speed, wavelength, and water depth. In shallow waters, wave speed for nearly all wavelengths is governed solely by water depth. In deeper waters, wave speeds generally increase with increasing wavelength.

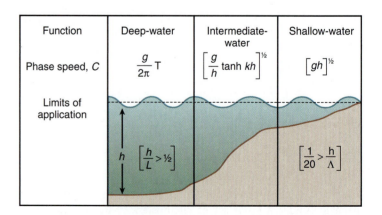

■ **FIGURE 10-8** Mathematical formulations and limits of applicability for deep, intermediate, and shallow-water waves.

take the form of an ellipse rather than a circle. This effect can be observed watching the back-and-forth motions of seaweeds on a shallow bottom.

Wave Energy

Wave energy is distributed between the potential energy that results from displacement of water particles from their resting state and the kinetic energy that results from the circular motion of water particles in a wave. The equation for the total energy (E) in a wave is

$$E = 1/8 \, \rho \, g \, H^2 \qquad \text{Equation 10.6}$$

where ρ is water density, H is wave height, and g is the gravitational acceleration, 9.8 m s^{-2} (32 feet sec^{-1}). This equation tells us that energy increases as the square of wave height. Thus, a doubling of wave height produces a wave with four times the energy. For example, a wave with a height of 6 m (20 feet) contains four times as much energy as a wave with a height of 3 m (10 feet).

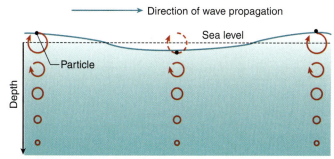

a. Deep-water wave

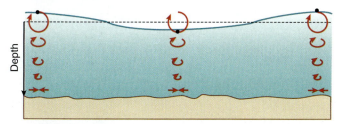

b. Intermediate-water wave

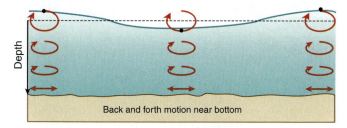

c. Shallow-water wave

■ **FIGURE 10-9** Wave orbitals for (a) deep-water; (b) intermediate-water; and (c) shallow-water waves.

Analyzing Natural Waves

If you have spent any time watching the surface of the ocean or a lake, you will have noticed that wave heights and wavelengths vary seemingly at random. A wave you follow with your eye may seem to disappear while another wave may suddenly loom. As emphasized by oceanographer Willard Bascom (1916–2000) in his best-selling classic *Waves and Beaches,* "concepts of wave period and wavelength lose their meaning under natural conditions."

Wave Interference

The combination of many different waves can be thought of as the **superposition,** or "adding up," of a series of waves. When the crests or troughs of two waves overlap exactly, their superposition produces a **constructive interference** (**Figure 10-10**). Where two sinusoidal (sine or cosine) waves are perfectly offset, their superposition results in a subtraction of their surface displacements, called **destructive interference.** The resultant profile of the sea surface represents the combined effects of constructive, destructive, and intermediate interference of waves. When the waves are "in phase," they produce a wave that is twice the height of the individual waves. When they are "out of phase," the resultant wave has zero height.

The Wave Energy Spectrum

Modeling of wave fields using combinations of sine waves yields good approximations to the sea state under natural conditions, but this approach can be unnecessarily cumbersome. A convenient way to analyze a wave profile involves dividing it up into

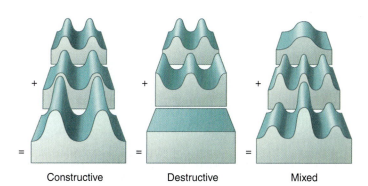

Constructive   Destructive   Mixed

■ **FIGURE 10-10** Constructive interference results when the crests and/or troughs of two waves coincide exactly. The resultant wave is larger than either of the individual waves. Destructive interference occurs when the crests and/or troughs of two waves are exactly out of phase. The resultant wave is smaller than either of the individual waves. In reality, the sea surface represents the sum of contributions of many individual waves.

frequencies and amplitudes using a powerful mathematical tool known as Fourier analysis. Oceanographers apply Fourier analysis to decompose complex waves into the sine waves of which they are comprised. Fourier analysis is also highly useful for analyzing time-series data, such as changes in ocean properties over time as measured by a remote platform such as a buoy or satellite. Though well-beyond our treatment here, awareness of Fourier analysis can help you appreciate representations of scientific information that you might find on NOAA wave buoy sites and other sources of real-time wave information.

As we have learned, the energy in a wave is proportional to the square of its height. Knowing the height and frequency of waves, which are relatively easy to measure, oceanographers can plot a **wave energy spectrum** for a given location and set of conditions (**Figure 10-11**). Wave energy spectra provide a convenient way for understanding the evolution of waves as the wind blows. Bigger and more energetic waves are formed as the factors that affect waves continue to act upon the sea surface, as we shall see.

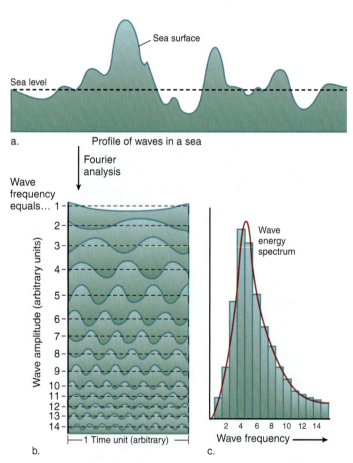

FIGURE 10-11 The wave energy spectrum. The sea surface at any given time and location (a) represents the superposition of many different sine waves (b). A wave energy spectrum (c) is produced by plotting wave frequency versus wave energy, calculated from measurements of wave height (twice the amplitude). In general, lower frequency waves (or longer period waves) contain more energy than higher frequency waves (or shorter period waves) (data from Wells, 1997).

Wave Generation by Winds

The development of waves from small ripples to the large waves desired by surfers requires a combination of factors that contribute to the transfer of energy from the atmosphere via winds to the surface of the ocean (**Figure 10-12**). The amount of energy transmitted to the upper ocean depends on the balance of factors that enhance the transfer of wind energy to the sea surface (such as wind stress) and factors that dissipate wind energy (such as gravity). While some of the absorbed wind energy causes surface currents, most of the energy transferred from the wind to the sea surface results in waves. The speed of the wind, the length of time it blows, and the distance over which it blows, called the **fetch,** ultimately determine the size of waves. In combination, these three factors can generate waves of considerable size, exceeding 20 to 30 m (65–98 feet) or more. Hurricanes, while exhibiting the highest wind speeds, were once thought to lack sufficient fetch and duration to produce large waves. However, buoy measurements of waves produced by Hurricane Ivan in 2005 revealed waves with heights in excess of 27 m (90 feet), so

FIGURE 10-12 Sustained winds over large expanses of the ocean such as those produced by storm systems generate most of the waves observed along the shore.

hurricanes may produce large waves under some circumstances. In general, storm systems operating in the open ocean with high sustained wind speeds of great duration and long fetch generate some of the largest and most powerful waves found in the world ocean. Winds blowing unhindered around Antarctica are famous for the large waves they produce.

From Calm to Fully Developed Sea

The development of waves from a calm ocean to ocean swell generally takes place over time in a series of stages that depend on wind speed, duration, and fetch. A general model of wave development is presented in **Figure 10-13**. At first, winds blowing over the sea surface generate **capillary waves** that exist as small wrinkles or cat's-paws on a calm fluid surface. These waves are only a few millimeters in height with wavelengths of only a few centimeters. Because they appear in even the slightest of winds, they may be the most abundant waves in the ocean. These waves provide a rough surface against which the wind can push. Thus, the amount of wind energy transferred to the sea surface increases. Additional wind energy produces ripples, the smallest type of gravity wave. As the wind continues to blow, larger waves form, the familiar and chop wind-waves. These waves increase in height and present an even larger surface area by which wind energy may be transferred to the ocean surface. However, wind-to-wave interactions are far from simple. The motion of waves on the sea surface causes fluctuations in the velocity and pressure of the overlying air in a region known as the **critical layer** (**Figure 10-14**). Under certain conditions, the fluctuations become synchronized with the waves in a way that increases the transfer of wind energy to waves, a phenomenon called **wind-wave resonance.** Thus, wave-induced air flow at the air-sea interface increases wave height beyond what otherwise might be expected assuming a linear transfer of wind energy to waves.

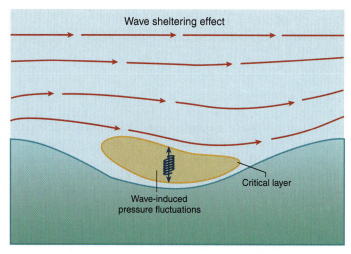

■ FIGURE 10-14 Wave height is not a simple linear function of wind speed. Complex interactions between waves and the overlying air may enhance wave growth. Like a spring, waves on the sea surface create fluctuations that compress and expand the overlying air in a region known as the critical layer. Synchronous (i.e., resonant) interactions increase the efficiency of transfer of energy from the wind to waves.

This theory, first proposed in 1957, was confirmed in 1995 using the research platform FLIP (**Figure 10-15**).

At a certain point, the height of the wave reaches a limit for a given wind speed beyond which the wave no longer maintains its

■ FIGURE 10-15 Confirmation of a wave theory. The theory of wind-wave resonance was first developed in 1957 by John Miles, an oceanography professor at Scripps Institution of Oceanography. In 1995, oceanographers confirmed Mile's theory using the research platform, FLIP (Floating Instrument Platform). A manned spar buoy at a length of 355 feet, FLIP can be towed to sea and flipped vertically by opening ballast tanks in its hull. Because most of its length remains submerged, FLIP is very stable, making it an ideal platform for oceanographic studies. The oceanographers who confirmed Mile's theory used a long mast extended away from the ship to make careful measurements of wind and wave properties. Their results were published in *Nature* in March 2003.

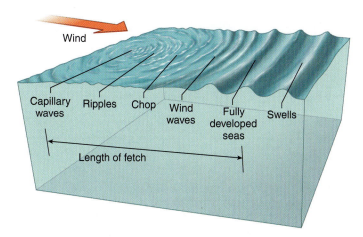

■ FIGURE 10-13 Conceptual model of wave development from ripples to ocean swell.

form. This limit occurs when the **wave steepness** exceeds a value of about one-seventh (0.142). Wave steepness may be defined as:

$$S = H/L \tag{10.7}$$

where S is wave steepness, H is wave height, and L is wave period.

The limit on wave steepness means that a wave with a wavelength of 7 m (or feet) can be no higher than 1 m (or foot) before breaking, at which point the crest topples and forms a whitecap (**Figure 10-16**). The spilling of wave crests as whitecaps generates turbulence that provides additional energy to the wave.

As the wind continues to transfer energy into the sea, the wave energy spectrum changes. The peak of wave energy increases and shifts toward a narrow range of lower frequencies. For a constant wind blowing offshore, the height, period, and energy of waves increase with greater distance from the shore (**Figure 10-17**). Eventually, the rate at which energy is being put into waves equals the rate at which energy is being dissipated, and the sea reaches a steady state called a **fully developed sea.** The time to reach a fully developed sea at a given wind speed depends on fetch and wind duration. Light winds produce a fully developed sea in a few hours, whereas high winds must blow for two to three days before a fully developed sea state is reached. The meteorological and oceanographic conditions of a fully developed sea for a given wind speed are given by the **Beaufort Wind Force Scale** (**Table 10.2**). The Beaufort scale also includes predictions of the **significant wave height,** the average height of the highest one-third of all waves. Significant wave heights are especially useful for wave forecasters and coastal engineers.

Wave Dispersion and Spreading

When waves travel beyond the reach of the wind, they change in behavior. A careful look at the equations we have just discussed reveals important consequences of the propagation of deep-water waves away from the storm centers. The fastest waves, the waves with the longer periods and wavelengths, will travel more quickly and separate from the slower waves, which take longer to cover the same distance. The separation of deep-water waves according to

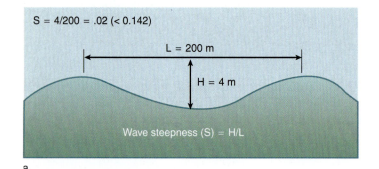

a.

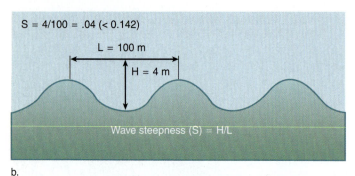

b.

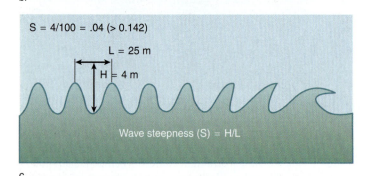

c.

■ **FIGURE 10-16** Wave steepness is defined as the ratio of wave height to wavelength, as shown in examples (a) and (b). Waves whose steepness exceeds 1:7 or 0.142 (c) become unstable and break.

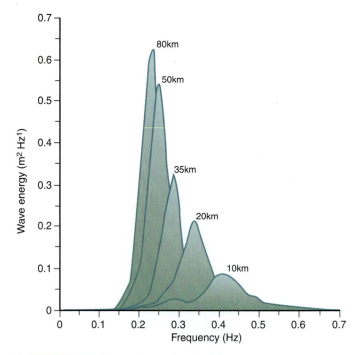

■ **FIGURE 10-17** Wave spectra at different distances (increasing fetch) from a constant wind source. As fetch increases, wave energy increases and wave frequency decreases. Put another way, a longer fetch produces waves with a longer period.

TABLE 10.2   The Beaufort Wind Force Scale and Associated Ocean and Land Conditions

Beaufort Wind Force	Wind speed 10 m (33 feet) above sea or ground surface Knots (miles per hour)	World Meteorological Organization classification	Description	
			Ocean	Land
0	< 1 (1)	Calm	Sea surface glassy	Smoke rises vertically
1	1–3 (1–3)	Light air	capillary waves	Smoke drifts in direction of wind
2	4–6 (4–7)	Light breeze	Small wavelets; no breaking	Wind felt on face; leaves rustle; wind vanes move
3	7–10 (8–12)	Gentle breeze	Larger wavelets; some wave breaking; scattered whitecaps	Leaves and small twigs constantly moving; flags slightly extended
4	11–16 (13–18)	Moderate breeze	Wave heights of 1 m (3.3 feet); numerous whitecaps	Dust, leaves and loose paper lifted; small tree branches move
5	17–21 (19–24)	Fresh breeze	Wave heights of 2 m (6.6 feet); numerous whitecaps; some sea spray	Small trees begin to sway
6	22–27 (25–31)	Strong breeze	Wave heights 3 m whitecaps (9.9 feet); common; more sea spray	Larger tree branches moving; whistling sound in overhead wires
7	28–33 (32–38)	Near gale	Wave heights 4 m (13.1 feet); sea heaps up; streaks of white foam coming from breakers	Whole trees moving; wind resistance felt while walking
8	34–40 (39–46)	Gale	Wave heights 5.5 m (18 feet) with greater length; crests of waves blown off into heavy foam in streaks spray;	Greater whole tree motion; greater wind resistance while walking
9	41–47 (47–54)	Strong gale	Wave heights 7 m (23 feet); sea swell obvious; dense streaks of foam and spray; reduced visibility	Slight structural damage; slate blows off roofs
10	48–55 (55–63)	Storm	Wave heights 9 m (29.5 feet); sea white with foam; rolling seas; lowered visibility	Considerable structural damage; trees uprooted
11	56–63 (64–72)	Violent storm	Wave heights 11.5 m (37.7 feet); patches of foam cover sea surface; low visibility	Widespread damage
12	64 or more (73 or more)	Hurricane	Wave heights exceeding 14 m (46 feet) air filled with foam; sea completely white with driving spray; greatly reduced visibility	Potentially catastrophic damage; roofs blown off; walls toppled

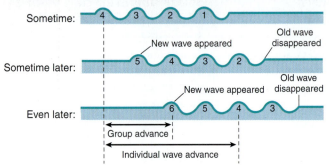

■ FIGURE 10-18 The phenomenon of dispersion can be observed by dropping a pebble in a still pond. The ripples created by the pebble become more distant from each other as they progress from the site of initial disturbance. The ripples with longer wavelengths move out faster than the waves with shorter wavelengths and they disperse. However, these front-running waves soon disappear and a new wave appears at the back of the group of ripples. If you watch carefully, this new wave will travel through the group until it becomes the front wave. Once again, this wave disappears and another wave forms to the rear.

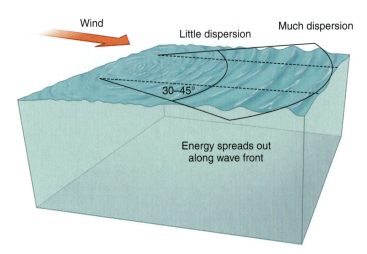

■ FIGURE 10-19 Dispersion and spreading of a wavefront as it propagates outwards from its source.

their wave speed is called **dispersion** (**Figure 10-18**). In practical terms, this means that waves with the longest period or wavelength will reach the shore first, followed by swells of progressively shorter wavelengths and wave periods. Because waves sort out or disperse according to wave speed, waves will tend to travel together in groups with similar periods and wavelengths. These **wave trains** exhibit properties that are different from individual waves. The speeds of individual waves in a wave train (their phase speeds) exceed the speed of the group, called the **group speed.** In fact, the phase speed is twice the group speed, or, put another way, the group speed is half the speed of an individual wave. Thus, although the longest waves move the fastest from a storm center, they arrive two times later than expected from their individual wave speeds alone. This knowledge of group speed is vital when predicting the time of impact of waves on a shoreline. Note that shallow-water waves, whose speed depends only on water depth and not on wave period or wavelength, are non-dispersive.

Most swells travel outward from a storm center within an angle of 35° to 40° from the direction of the wind. Near the storm center,

there is little dispersion of waves. However, farther from the storm center, greater separation of waves with differing wavelengths (and periods) occurs. At the same time, the wave front tends to spread outward (**Figure 10-19**). Because the total energy is conserved, the amount of energy along a given segment of the swell decreases as the wave spreads outward. The decrease in wave energy as the result of widening of the wave front is called **spreading loss.**

Wave Interactions with the Seafloor

When a deep-water wave encounters water depths that begin to interfere with the orbital motions of its water particles, the wave is said to "feel bottom." When this occurs, the deep-water wave begins its transformation to a shallow-water wave. Recall that wave speed then becomes dependent only on water depth, that is, $C_{shallow} = \sqrt{gh}$

Friction of the wave with the seafloor also slows down the wave and decreases its wavelength. (Remember, wave period does not change). Wave orbitals at the bottom of the wave are prevented from completing their circular motions. Water particles move in a more elliptical or even back-and-forth horizontal motion. At the same time, the height and steepness of the wave suddenly increase as the energy in the wave is confined to an increasingly shallow water column. Eventually, the speed of water particles in the crest exceeds the speed of the wave, and the wave breaks (**Figure 10-20**). Put another way, the wave becomes overly steep, and the water particles in the crest of the wave are thrown forward, that is,

$$S = H/L > 1/7.$$

Seafloor bathymetry exerts a strong influence on when and where waves break on a shoreline. Nearshore ridges or canyons,

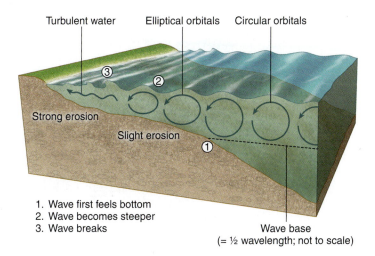

Turbulent water   Elliptical orbitals   Circular orbitals

Strong erosion

Slight erosion

1. Wave first feels bottom
2. Wave becomes steeper
3. Wave breaks

Wave base
(= ½ wavelength; not to scale)

■ **FIGURE 10-20** Sequence of events as a wave approaches the shore.

sand bars, nearshore troughs, and other variations in the shape of the sea bottom will alter the speed, shape, and direction of a wave (see **Table 10.3** and sections on wave refraction, reflection, and diffraction in this chapter). For some waves, the massive lip created by a plunging breaker creates a spectacular waterfall, the tube so highly sought after by surfers. Other waves simply curl in a more turbulent fashion, a spilling breaker, or topple, a collapsing breaker. On very steep beaches, the wave face rushes up on shore without breaking, a surging breaker.

Wave Refraction and Focusing

The most common observation of waves approaching a shoreline is **wave refraction,** the bending of a wave front. Refraction occurs in response to changes in speed along the wave front as different parts of the front encounter different water depths. For example, a wave approaching from an angle other than parallel to the shoreline will slow down where the wave front first reaches shallow water (i.e., where $C_{shallow} = \sqrt{gh}$; a shallower depth, h, gives a slower speed). The part of the wave front that has not yet felt bottom will continue at full speed. The net result of one part slowing and the other part maintaining its speed is a bending of the wave. Wave refraction is nearly always observed along a shoreline because waves rarely approach a shoreline straight on and because wave crests tend to align with **isobaths,** contours of constant depth. Wave refraction may also be observed along headlands and in bays as the leading edge of the wave front "feels" the bottom near the headland or along the edge of the bay. Notice how wave fronts often match the shape of a bay or headland. Wave refraction also has consequences for the distribution of wave energy along a wave front. Where the wave is refracted

TABLE 10.3   Characteristics of Breaking Waves

Plunging breakers
- Wave form is arching, convex back, and concave front
- When crest curls over, it plunges downward, pushing surfer forward and downward
- Usually associated with groundswell
- Can be produced by wind swell on steep beaches

Spilling breakers
- Foamy and turbulent at crest
- Top of wave moves faster than wave as a whole
- Occur on gently sloping shorelines, often well away from shore
- Seen on beaches during storms as waves that are short and steep

Collapsing breakers
- Similar to plunging breakers, but front face collapses and crest does not curl over
- Produced on beaches with moderate slopes and winds

Surging breakers
- Occur on the steepest-sloping beaches
- Produced by long period and long wavelength waves
- Long, low waves with front faces remaining generally intact as waves move up the beach

Plunging breakers

Spilling breakers

FIGURE 10-21 Wave refraction in an embayment.

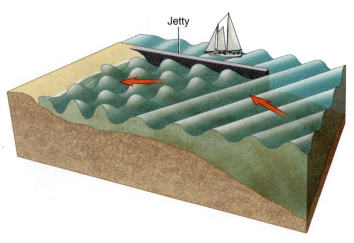

FIGURE 10-22 Anatomy of the Wedge. When a southerly swell approaches, wave fronts reflect off the jetty. The oncoming waves interact with the reflected waves to produce through constructive interference a "wedge" of water roughly double the original wave height. Typically, the Wedge is most active during summer, when Southern Hemisphere storms generate large southerly swell. Though exhilarating, surfing at the Wedge can be quite dangerous.

toward a headland, the wave energy will be concentrated and more focused. In an embayment, expansion of the wave front distributes its energy along a greater length of the wave, and, as a result, these waves tend to be more gentle (**Figure 10-21**). Most of the world-class surf spots, the places where the largest, barrel-shaped waves occur, are the result of **wave focusing,** the convergence of wave energy on a region of the nearshore. Where the depth contours (isobaths) are concave with respect to the direction of the wave, the wave front converges. Where the contours are convex, the wave front will disperse. Typically, long-wavelength waves with higher speeds build more rapidly and break more intensely than short-wavelength waves. Similarly, seafloor slopes that are gradual will build waves more slowly and generate less intense waves than steeply sloping seafloors, where the wave energy is suddenly concentrated.

Wave Reflection

Surface waves also reflect off of steep slopes or vertical boundaries, such as shoreline cliffs or human-made structures such as seawalls. While some energy is lost as waves hit these objects, significant energy reflects backward. A classic **surface wave reflection** occurs near the entrance to Newport Beach Harbor in California at a place called the Wedge (**Figure 10-22**). Here, a long jetty protrudes in such a way that waves are often reflected off the jetty and back toward the beach. As a result, the reflected waves superimpose on top of the incoming (nonreflected) wave, creating a rough doubling of wave height. The Wedge was featured in *The Endless Summer* and continues to be a favorite spot for bodysurfers, bodyboarders, and kneeboarders. Wave reflection may also occur head on, such as along a seawall. In this situation, the reflected wave returns in the opposite direction of the incoming waves, resulting in a series of **standing waves** beyond the seawall.

Wave Diffraction

Wave diffraction, the transfer of wave energy laterally along the crest of a wave, can also alter the direction of the wave front. Diffraction most typically is seen where waves are blocked by an island or human-made barrier (**Figure 10-23**). As a wave approaches an island or barrier, part of the wave is cut off, creating a shadow zone. However, some of the energy in the passing wave is transmitted into the shadow zone, generating waves where they might not be expected. Diffraction of waves around the point of an island or at the narrow entrance of a harbor can negatively impact boats otherwise protected in the lee of the island or the safety of the harbor.

FIGURE 10-23 A jetty at a harbor entrance blocks incoming waves and causes the wave front to diffract around the rocky point.

Applications of Wave Theory

There are many practical reasons for studying waves. Planning of ship routes and avoidance of problematic wave regions rely on knowledge of wave conditions and the forces that produce them. Evacuation strategies for populations living in coastal zones subject to tsunamis, hurricanes, typhoons, storm surges, and other extreme wave-producing events require an understanding of waves. Beach erosion and beach nourishment projects dominate the political landscape of many coastal cities as a growing demand for public beaches and beachfront hotels interferes with the natural processes that replenish beaches. Barrier islands such as those lying off the Carolinas of the United States are vulnerable to a combination of rising sea level and accelerated shoreline erosion associated with extreme weather, more intense waves, and climate change. On a more positive side, an understanding of waves enables predictions of prime surfing conditions, safe swimming conditions, or calm boating conditions. Wave energy also potentially represents a harvestable source of alternative energy for generation of electricity.

Surface Waves and Air-Sea Exchanges

From an oceanographic perspective, surface waves play an important role in a variety of physical, chemical, and biological processes. Surface waves generate surface currents through the transfer of momentum from the atmosphere to the ocean. Breaking surface waves and whitecaps contribute to mixing of near-surface waters in the form of turbulent eddies. In fact, the greatest exchanges of heat, matter, and momentum occur across the air-sea interface when large waves are breaking and turbulent mixing is most intense. The exchange of biologically important gases such as carbon dioxide and oxygen is enhanced by breaking waves and the bubbles they produce. The effects of extreme waves produced by hurricanes and typhoons are little known, but cooling of surface waters and enhancement of phytoplankton growth in the path of hurricanes may result in part from these waves.

Internal Waves and Ocean Mixing

Along the density boundaries between the fluid layers of the ocean can be found immense, slow, and lumbering waves known as **internal waves.** The height of internal waves ranges into the hundreds of feet. Wave speeds occur in meters per minute as opposed to meters per second. Like ocean swells, these waves may travel great distances with little loss of energy. Though we cannot directly view these waves at the surface, they are readily apparent to oceanographic instruments and occasionally visible as slicks on the surface or alternating light and dark bands of color (**Figure 10-24**). Internal waves can propagate and send energy over great depths and over long distances in the ocean. Thus, they

■ FIGURE 10-24 Internal waves in the Sulu Sea as observed on April 8, 2003 by the Moderate Resolution Imaging Spectroradiometer (MODIS) aboard the Aqua satellite.

affect properties and mixing far from their points of initiation. Internal waves may bring nutrients and biota into the upper ocean and exert a positive effect on ocean food webs. As internal waves approach coastlines, they often break and cause significant mixing and resuspension of sediments, affecting sediment transport and deposition along the shelf as well as disturbing habitats of bottom-dwelling organisms. In areas where toxic materials are buried, internal waves can interact with the bottom and cause reexposure and transport of harmful substances. Submarines and other submerged vehicles can be moved over great depth ranges by internal waves: some losses of submarines at sea have been attributed to encounters with large-amplitude internal waves. Suggestions have even been made that it may be possible to harness the energy of internal gravity waves.

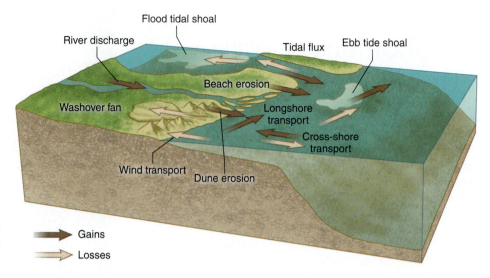

■ FIGURE 10-25 Nearshore sedimentary processes. Waves, winds, tidal currents, river discharge, and runoff are among the many processes that affect the gain and loss of sediments on a beach.

→ Gains

→ Losses

Nearshore Sediment Transport

Wave-generated movements of sediments along coastlines are considerable, approaching 1 million cubic meters (about 1.3 million cubic yards) per year in some regions. Interruption of this process by natural or human-made barriers (sea stacks, rocky promenades, groins, jetties, breakwaters, etc.) may starve coastlines of their sand and cause erosion of beaches and sea cliffs. Along developed coastlines, coastal erosion puts homes and other structures at risk from storm surge, as witnessed along the Gulf Coast in the wake of Hurricane Katrina in 2005. Tourism suffers as well when formerly expansive beaches are reduced to a fraction of their original size. An understanding of nearshore sedimentary processes that govern the transport of sediments is essential to maintaining the integrity and economic value of coastlines (**Figure 10-25**).

Any coastal region with unconsolidated sediments may be classified as a **beach (Figure 10-26).** While we often think of beaches as long stretches of golden sand, beaches range in type from fine sediments on a mud flat to loose rocks on an exposed coast. Typically, a beach can be divided into four segments. The backshore and foreshore represent the exposed portions of the beach. The nearshore and offshore represent the submerged portions of the beach. The backshore experiences wave action only during strong storms and may exhibit one or more **berms,** a terrace of sand or other sediments deposited by storm waves. The region between the berm crest and the level of the mean low tide is called the **foreshore.** This region includes the beach face and, on some beaches, a more gently sloping to flat portion called the **low-tide terrace.** The beach face marks the uppermost limit of normal wave action. The low-tide terrace, subject to the back-and-forth motion of waves, includes the **swash zone.** Wave swash consists of alternating periods of **run-up,** the advance of water from a wave, and **backwash,** the retreat of water from a wave. Typically, wave swash results from the bore-like advance of water from waves that have broken in the surf zone.

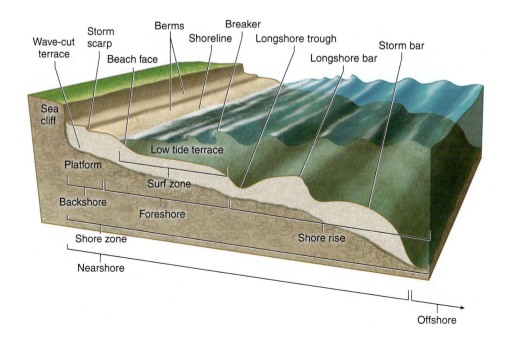

■ FIGURE 10-26 Anatomy of a beach.

Waves supply energy for movement of sediments on a beach and affect the **beach profile,** the changes in shape or elevation of a beach when viewed from the side (like your own profile). A back-and-forth movement of sediments across the beach face is called **cross-shore transport.** Cross-shore transport results in changes to the beach slope as sand is piled up or removed from the beach face. Sediments are moved forward by the run-up of waves and carried backward by the returning backwash. The difference between run-up and backwash determines the net transport of sediments toward or away from the shore. Because the energy of wave run-up normally exceeds the energy of backwash, there will always be a tendency for sediments to pile up on the beach face. At some point, however, the slope of the beach becomes steep enough such that gravity enhances the intensity of the backwash. When the forward and backward movements of sediment are equal, then the beach is said to be in equilibrium (**Figure 10-27**).

At the same time, sediments move parallel to the beach face, a process called **longshore transport.** Waves striking the beach at an angle generate a current of water called a **longshore current** (**Figure 10-28**). Both wave height and angle of approach influence the speed of the longshore current: higher waves and steeper angles increase its velocity. If you have ever gone swimming in moderate surf and found yourself at a considerable distance from your entry point, the speed and strength of the longshore current may be duly appreciated! The movement of sediments by longshore transport may result in a net loss or gain of sediments on a given stretch of beach. When the longshore current is interrupted by a breakwater or jetty, sediments may accumulate on the upstream side. If the buildup of sediments occurs in an inlet or harbor, regular dredging must be performed. Dams along rivers (one of the sources of sediments to a beach) or interruption of the longshore current by coastal structures may starve a beach of its sediments.

Artificial importation of sand to a beach, or **beach nourishment,** may be necessary to combat beach erosion. Concerns for the effects of beach erosion have led to increasing efforts to develop **sediment budgets** for beaches, an accounting of the volumes of sediments supplied to or lost from a particular beach. Sediment budgets often embody the concept of a **littoral cell,** a self-contained section of a coast whose beach sediments are controlled by identifiable local mechanisms of supply and loss (**Figure 10-29**). The coast of southern California provides an ideal example of a series of littoral cells whose sediments are largely supplied by rivers and removed by submarine canyons. Other processes may create subcells and interact to supply or remove sediments to the major littoral cell. This is particularly true for locations where inlets, barrier islands, and embayments alter the flow patterns of sediments, such as along the coast of North Carolina and Washington.

The Surf Zone

In an attempt to predict the intensity of breaking waves, some researchers have developed an index of wave intensity called the **vortex ratio,** simply the ratio of the length of the barrel of a plunging wave to its width. A comparison of wave breaks at different locations (**Table 10.4**) reveals that the lower the ratio, the more "intense" the breaking wave. Waves with vortex ratios greater than

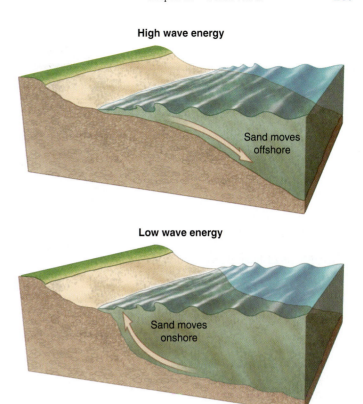

■ **FIGURE 10-27** The profile of a beach changes in response to wave-generated cross-shore transport. At any given moment, the beach profile represents the sum of processes that move sediments onto the beach and those that take sediments away. When wave energy is high, sediments are carried offshore away from the beach face and "stored" as a sand bar. Low-energy waves move sediments from sand bars onto the beach face causing a buildup of sand. In general, high-energy waves generate a flat foreshore with a narrow berm. Low-energy waves create a steep foreshore and a wide berm. The exact shape of the beach profile depends on wave conditions and the grain size of sediments on the beach. Periodic disturbances, like storms, temporarily disrupt the equilibrium profile.

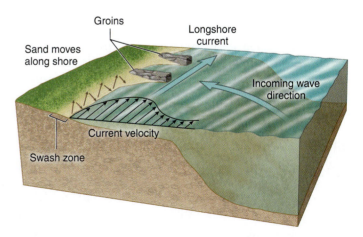

■ **FIGURE 10-28** The wave-generated longshore current moves sediments down the beach, i.e., parallel to the shore. The strength of the longshore currents varies with distance from the shore, reaching a maximum velocity near the middle part of the surf zone. Where man-made structures interfere with longshore transport, sediments will accumulate on the upstream side and be depleted on the downstream side.

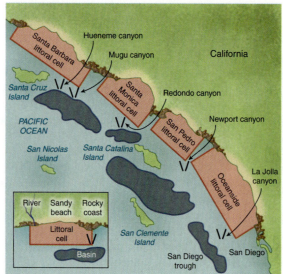

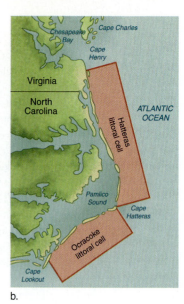

■ **FIGURE 10-29** Examples of littoral cells in (a) Southern California; (b) Outer Banks

a.   b.

3.0 lose their tubes and, hence, their desirability as prime surfing waves (although the skill level and type of surfer will also influence their desirability). The development of a classification system for surfable waves applies to the construction of artificial surfing reefs, an area of active and growing interest.

In some locations of the world ocean, waves reach sizes of epic proportions. These **extreme waves** are generated in a number of different ways, but nearly all of them catch our attention for their awe-inspiring power and their costly and occasionally deadly impacts. Matt Warshaw's *Encyclopedia of Surfing* defines a "big" wave as one in excess of 4.6 m (15 feet). Nonetheless, renewed interest in big wave surfing and the use of personal watercraft for tow-in surfing in the 1990s have led to what might be called the era of **extreme wave surfing.** Wind-generated ocean waves in excess of 9 to 15 m (30–50 feet) have now been ridden at popular extreme wave locations such as Mavericks (Half Moon Bay, California), Jaws (Maui, Hawaii), North Shore (Oahu, Hawaii), Todos Santos (Baja, Mexico), Pico Alto (Lima, Peru), and Dungeons (Cape Town, South Africa). What these locations feature in common is a steeply sloped seafloor bathymetry that focuses wave energy on an undersea shoal. When a long-period ocean swell arrives from the right direction, extreme waves may be produced. Some locations may produce even higher waves on rare occasions. Surf forecasters estimate that under perfect conditions, waves in excess of 30 m (100 feet) may be possible at Cortes Bank, a seamount off the coast of San Diego, California.

Rogue Waves

At any given time, the open ocean wave field is complex and consists of waves with many different periods, wavelengths, and speeds. A statistical probability exists, however, that **rogue waves** will appear simply by chance constructive superposition of multiple waves. Rogue waves can be quite frightening as they often occur unexpectedly and without warning. Since the 1980s, more than 200 cargo ships longer than 200 m (650 feet) have been lost at sea due to rogue waves, nearly one every two weeks. The largest rogue wave, measuring 34 m (112 feet) was observed on February 7, 1933, aboard the Navy vessel *USS Ramapo* in the Pacific Ocean. In 2005, students aboard the Semester at Sea Program's M/V *Explorer* experienced a 15 m (50 foot) rogue wave that broke windows on the bridge of the ship and damaged the ship's engines. All hands made it home safely, but their oceanography lesson was one that they would never forget. A three-week satellite study of wave heights in the world ocean revealed 10 waves nearing 30 m (100 feet). Thus, rogue waves may not be as rare as once thought. Such waves probably went largely unnoticed by oceanographers until new technologies provided a means for observing them.

TABLE   10.4   A Comparison of Vortex Ratios at Various Surfing Spots Around the Globe

Relative intensity of wave	Vortex ratio	Representative location
Extreme	1.6–1.9	Pipeline, Hawaii
Very high	1.91–2.2	Padanag Padang, Indonesia
High	2.21–2.5	Kirra Point, Australia
Medium high	2.51–2.8	Bells Beach, Australia
Medium	2.81–3.1	Manu Bay, New Zealand

Seiche

A type of standing wave that occasionally reaches "extreme" proportions (up to 6 m or 20 feet) in enclosed or semi-enclosed waterways is a **seiche** (pronounced *say'sh*). The seesaw motion of water in a bathtub when disturbed at one end gives you a general idea of the motion of a seiche. First described for alpine lakes in Switzerland, seiches occur as oscillations of the surface in response to a sudden change in atmospheric pressure or earthquakes that excite the **resonance frequency** of a body of water, including swimming pools. The 1994 Northridge, California, earthquake generated seiches in swimming pools across the Los Angeles basin and caused their water to spill out. In lakes, harbors, and bays, seiches can cause considerable damage and even loss of life. A seiche produced by the low pressure eye of a hurricane passing over Lake Okeechobee, Florida, on September 17, 1928, claimed the lives of 1,836 people, making it the second-deadliest natural disaster in US history.

Storm Surge

Storm surge is a type of extreme wave that occurs when high winds and low atmospheric pressure combine with high tides to cause coastal flooding. In 1970, an estimated 300,000 people were killed as a result of storm surge in the Bay of Bengal. This region is often called "the storm surge capital of the world" because of the frequency with which storm surge occurs there. Storm surge generated by Hurricane Katrina in August 2005 caused widespread destruction along the panhandle of Florida and the Gulf coasts of Alabama, Mississippi, and Louisiana, making it the costliest natural disaster in the continental United States to date (**Table 10.5**).

Storm surge results from several hurricane-related phenomena that generate above-normal sea level in a broad region surrounding the eye of a hurricane. Low pressure associated with the eye of a hurricane creates a doming of water like that which occurs

TABLE 10.5   Deadliest and Costliest Hurricane and Storm Surge Events in the United States and Territories Since 1900

Rank	Hurricane	Intensity/ Pressure	Date	Area affected	Surge height	Deaths and costs (in 2005 dollars)
Deadliest	Galveston Hurricane of 1900	Category 4; 936 mb (27.64 in)	September 1900	Galveston Island and parts of the Texas coast	2.4–4.6 m (8–15 feet)	6,000–12,000 deaths; $540 million– $1.1 billion
Costliest	Hurricane Katrina	Category 4; third-lowest central pressure; 920 mb (27.17 in)	August 2005	Florida panhandle and Gulf coast states of Alabama, Mississippi, and Louisiana	7.6–9.1 m (25–30 feet)	600–1200 deaths; $100 billion
Second deadliest	San Felipe– Okeechobee Hurricane	Category 4; 929 mb (27.43 in)	September 1928	Leeward Islands to Florida, especially shoreline surrounding Lake Okeechobee	2.7–3.7 m (9–12 feet)	1,836 deaths in Florida; 312 in Puerto Rico; $800 million
Second costliest	Hurricane Andrew	Category 4; fourth-lowest central pressure; 922 mb (27.23 in)	August 1992	South Florida and the Upper Keys; Louisiana	Up to 5.2 m (17 feet) in Florida; 2.4 m (8 feet) in Louisiana	23 deaths in Florida; $36 billion
Third deadliest	Florida Keys Labor Day Hurricane	Category 5; second-lowest 892 mb (26.35 in); central; pressure surpassed only by Hurricane, Gilbert 1988, at 888 mb (26.22 in)	September 1935	Middle Florida Keys and Florida Gulf Coast	Up to 6 m (20 feet)	408 deaths; $82 million
Third costliest	Hurricane Charley	Category 4; 941 mb (27.79 in)	August 2004	Gulf coast of Florida	Less than 2.1 m (7 feet)	10 deaths; $15–$17 billion

beneath low pressure systems in the atmosphere. Winds influence the slope of the sea surface and cause Ekman transport toward or away from the coast. Waves, tides, seafloor bathymetry, and the presence or absence of barriers, such as islands, wetlands, dunes, and other coastal features, will also affect the height and intensity of storm surge.

Storm surge advances onshore like a rapidly rising tide. During Hurricane Katrina, trapped residents reported both steady and sudden increases in storm surge at rates of up to a few meters (several feet) per hour. Run-up estimates varied with location, with the highest storm surge occurring east of the hurricane eye where it made landfall near Buras-Triumph, Louisiana. Aerial observations revealed a 1.6 km (about 1 mile) wide swath of destruction, suggesting storm surge penetrated at least this distance from the shoreline. Storm surge along the Biloxi River at Wortham, Mississippi, reached 8 m (26 feet) during the hurricane. Evacuations and emergency preparations undoubtedly saved thousands of lives but did little to prevent the widespread destruction of some 233,000 km² (about 90,000 square miles) of coastal cities, towns, and countryside. The devastation of Hurricane Katrina along the Louisiana–Mississippi–Alabama coast in many ways resembled the devastation of Banda Aceh, Indonesia, and other coastal communities following the 2004 Indian Ocean tsunami.

Tsunami

Without question, the most dangerous form of extreme wave is the tsunami. Since 1990, 11 destructive tsunami have struck coastlines, with a death toll numbering in the hundreds of thou-

sands (**Figure 10-30**). Before 2004, tsunami were often mistakenly called tidal waves, largely because the sudden increase in sea level associated with tsunami resembled an incoming tide. Unlike tides, however, tsunami are generated by vertical displacements of the seafloor caused by earthquakes, undersea landslides, or volcanic eruptions. Ironically, the word *tsunami* (the singular and plural form) translates to "harbor wave" in Japanese, an expression used to describe tsunami damage that fishermen witnessed when returning home after an apparent calm day at sea. These early observations underscore one of the principal features of tsunami, mainly, that the waves they generate in deep water may be only a few meters in height. Because their energy is distributed throughout the water column, however, their "heights" when approaching coastlines may be considerable.

Oceanographers define three stages in the life cycle of a tsunami in the ocean. The first stage, **tsunami generation,** involves forces on the seafloor that cause impulsive disturbances of the water column (**Figure 10-31**). Vertical displacement of the seafloor by earthquake faulting or undersea landslides may be the principal factor in determining the height of a tsunami. As witnessed in the 2004 tsunami, however, the speed, direction, and extent of faulting along the seafloor also play a role in the intensity and direction of the tsunami.

The second stage, **tsunami propagation,** involves interactions of the tsunami with the seafloor as it travels outward from its source (**Figure 10-32**). Because tsunami have very long wavelengths, on the order of hundreds of kilometers, they display speeds and movements typical of shallow-water waves (although some of their properties resemble deep-water waves). Tsunami can travel several hundred kilometers per hour in the open

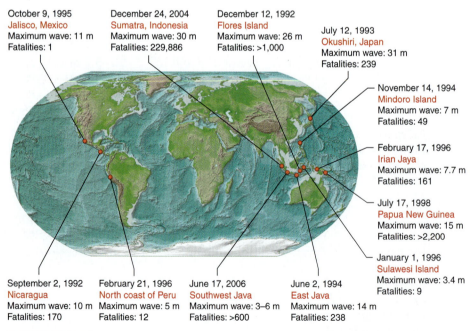

October 9, 1995
Jalisco, Mexico
Maximum wave: 11 m
Fatalities: 1

December 24, 2004
Sumatra, Indonesia
Maximum wave: 30 m
Fatalities: 229,886

December 12, 1992
Flores Island
Maximum wave: 26 m
Fatalities: >1,000

July 12, 1993
Okushiri, Japan
Maximum wave: 31 m
Fatalities: 239

November 14, 1994
Mindoro Island
Maximum wave: 7 m
Fatalities: 49

February 17, 1996
Irian Jaya
Maximum wave: 7.7 m
Fatalities: 161

July 17, 1998
Papua New Guinea
Maximum wave: 15 m
Fatalities: >2,200

January 1, 1996
Sulawesi Island
Maximum wave: 3.4 m
Fatalities: 9

September 2, 1992
Nicaragua
Maximum wave: 10 m
Fatalities: 170

February 21, 1996
North coast of Peru
Maximum wave: 5 m
Fatalities: 12

June 17, 2006
Southwest Java
Maximum wave: 3–6 m
Fatalities: >600

June 2, 1994
East Java
Maximum wave: 14 m
Fatalities: 238

■ FIGURE 10-30 Location of tsunami since 1990. Prior to the 2004 Indian Ocean tsunami, the Papua New Guinea tsunami caused the greatest number of fatalities.

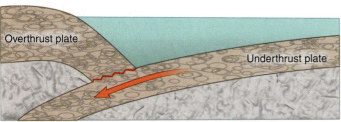

Before earthquake

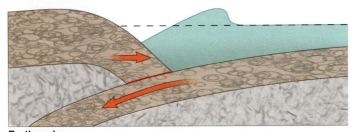

Earthquake

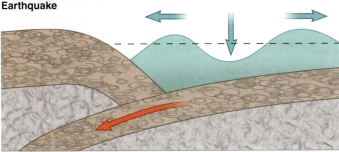

Tsunami wave propagates in all directions

■ **FIGURE 10-31** Vertical displacements of the sea floor are one force that generate tsunami. The greater the distance of displacement (up or down), the greater the energy transmitted to the overlying water column and the more powerful the tsunami.

ocean, according to $C_{shallow} = \sqrt{gh}$. Interactions with mid-ocean ridges, seamounts, and islands may cause **tsunami scattering,** the refraction (bending) and reflection of tsunami in response to seafloor features (see Spotlight 10.1). Thus, seafloor bathymetry plays a major role in the propagation of tsunami. As tsunami enter shallow coastal areas and harbors, their phase speed slows down (as small h gives small c) and the water "piles up," causing large amplitude waves.

The least understood stage is **tsunami impact,** the behavior of a tsunami in the vicinity of a coastline. Two key measurements of tsunami impact are **tsunami run-up,** the height above sea level to which a tsunami reaches, and **inundation,** the distance from shore inland to which a tsunami reaches (**Figure 10-33**). Contrary to popular belief, tsunami do not break on shore like conventional ocean waves. The power of a tsunami lies in its considerable length and the enormous energy it contains. Tsunami advance on shore as a powerful wall of water that may continue for miles inland. Though their heights at the coastline may be modest, on the order of several to a few tens of meters, their momentum carries them far inland. Thus, while the "height" or more properly, the run-up, of a tsunami may be 30 m (100 feet), the actual face of the wave may be closer to 3 to 6 m (10–20 feet) when observed approaching the beach.

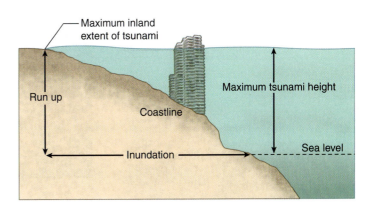

■ **FIGURE 10-33** Definitions of runup and inundation.

■ **FIGURE 10-32** Propagation of a tsunami varies as a function of sea floor bathymetry because tsunami propagate as a shallow-water wave. The expanding "rings" of a tsunami waves take on different shapes as the wave fronts are scattered by islands, continental shelves, shelves and banks, and other sea floor features.

The 2004 Indian Ocean Tsunami, Part 2: The Wave

The 2004 Indian Ocean tsunami sealed the term *tsunami* in the global vocabulary forever. Within hours, professional and amateur still and video images of the tsunami were disseminated to news organizations and websites. At the same time, a global array of scientific instruments quickly delivered observations and data. For the first time ever, a tsunami was captured by Earth-orbiting satellites, including Jason-1 and TOPEX/Poseidon, which carry altimeters that measure sea surface heights using microwave radar. The speed of dissemination, the amount of data, and the diverse types of information available on this event were unprecedented.

As discussed in Spotlight 3-1, the magnitude 9.3 (Mw) Sumatra-Andaman megathrust earthquake fractured the seafloor along a 1,207 to 1,287 km (750-800 mile) section. The amount of energy delivered to the overlying water column, 4.2×10^{15} J, represented less than 0.5% of the total energy released by the earthquake, roughly equivalent to the total energy contained within a large thunderstorm. Nonetheless, this "small" amount of energy was sufficient to generate a powerful wave that literally swept the world. Tide gauges around the world ocean recorded the passage of the tsunami over the next several days as the waves propagated throughout the world ocean. On Cocos Island, approximately 1,700 km (1,056 miles) from the tsunami, sea-level perturbations of about half a meter (1.6 feet) were recorded, while other stations of equal distance observed sea-level changes up to nearly 3 m (10 feet). These data suggest that the direction of tsunami propagation from the tsunamigenic source region is an important factor in the impact of a tsunami and that oceanic ridges and continental ridges act as "wave guides" for tsunami (**Figure 10a**). Because tsunami are shallow-water waves, their speed of propagation and factors affecting their refraction and reflection depend on water depth. Thus, bathymetric features along the seafloor act to focus tsunami wave energy and influence the direction of propagation.

The earthquake may also have generated "multiple" tsunami waves from different locations along the rupture zone. In some cases, a wave crest moved outward from the source; in others, a wave trough may have formed the leading edge of the tsunami. The Maldives observed an increase in sea level at the onset of the tsunami, while Banda Aceh and Phuket experienced a retreat of sea level when the tsunami hit. This observation dispels the notion that a rapid retreat of sea level on a shoreline always precedes an impending tsunami.

If there is any silver lining to this catastrophe, it is that the world now recognizes the destructive power of a tsunami. Very few places along the world's shorelines are immune from the possibility of tsunami. Thus, even if you don't live near the beach, an ounce of knowledge may save your life. Consider the story of eighth-grade British schoolgirl, Tilly Smith, who had just studied tsunami the semester before her visit in Thailand. Recognizing that a tsunami was imminent on that fateful morning, she warned her family, and they fled to safety. If more people had known to run to high ground when the sea acts "strange," more lives might have been saved.

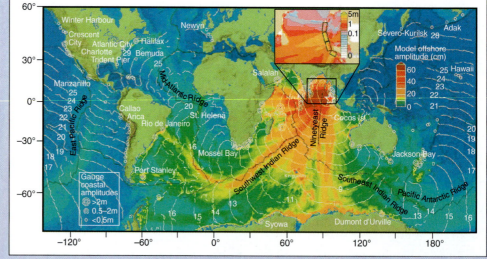

FIGURE 10-1a Propagation of the 2004 Indian Ocean tsunami throughout the world ocean. The inset shows four distinct tsunamigenic "regions" propagating waves. Contours represent time of arrival of the leading edge. Interaction with the seafloor is evident as a curling (slowing) of the wavefronts in the vicinity of oceanic ridges.

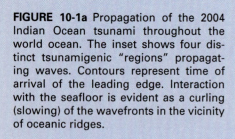

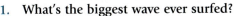

YOU Might Wonder

1. **What's the biggest wave ever surfed?**
   As of 2004, the biggest wave was 21 m (70 feet), surfed by Pete Cabrinha at Peahi, Hawaii (also known as "Jaws"). For that effort, he was awarded $70,000 by Billabong, a sponsor of big wave contests. Cabrinha's record is not likely to stand for long because Billabong has offered a $200,000 prize to any professional surfer who rides a 30 m (100 foot), wave (and lives to prove it). In recent years, extreme wave surfing and big wave contests have attracted a lot of attention and money. Who says surfing doesn't pay!

2. **Has anyone ever surfed a tsunami?**
   Contrary to popular belief, tsunami are not very good waves for surfing. They come onshore at very high speeds with a face that is more like a tidal bore than a surfable tube. Nonetheless, legend has it that in the 1860s, a Hawaiian farmer was swept out to sea during a tsunami but managed to climb atop the dislodged door of his house to surf the tsunami waves that followed the initial wave. Perhaps more believable is the story of Felipe Pomar and Pitty Block, who paddled out at the island of Punta Hermosa following a magnitude 7.2 earthquake in Lima, Peru, on October 3, 1974. The two were dragged a mile offshore as a tsunami struck but managed to surf a 3 m (10 foot) secondary wave that brought them to shore. Pomar repeated the feat some years later along with several other surfers who rode 6 m (20 foot) tsunami waves off the coast of Hawaii.

3. **What are some other unusual surfing feats that have been accomplished?**
   No doubt there are lots of stories that fall under this category. In Houston, surfers ride a boat into the Houston Ship Channel to surf the wake from large tankers and freight ships that ply these waters. In some cases, they may surf the same wave for several minutes. Some rank the "wraparound" wave at Fort Point, San Francisco, California, as among the top-five unusual surf spots. Under just the right conditions, the wave refracts around the point by nearly 90 degrees before entering the cove. Surfers have also trekked to Antarctica and other exotic surfing spots just to catch a wave. And if the isolation of Antarctica doesn't suit you, perhaps you can get in on the next attempt to achieve the most surfers on a surfboard. In July 2005, 60 surfers clamored aboard a 12 by 3 m (40 by 10 foot) giant surfboard in Huntington Beach, California, and "surfed" a wave to set the new world's record, easily surpassing the previous record of 47 people.

Key Concepts

- Waves represent the transmission of energy through a medium.
- Waves are created by forces that disturb sea level and removed by forces that restore sea level.
- Wind speed, duration, and fetch determine the size of swell in a fully developed sea.
- Waves approaching a shoreline will "feel" bottom and slow down, but their period will remain the same.
- Breaking waves concentrate vorticity in the crest of the wave.
- Wave refraction is the bending of shallow waves to conform with the sea bottom.
- Internal waves occur at the interface between fluids with different densities.
- Tsunami are caused by disruptions of the seafloor by earthquakes, undersea landslides, or other such events.

Terms to Remember

backwash, 206
beach, 206
beach nourishment, 207
beach profile, 207
Beaufort Wind Force Scale, 200
berms, 206
capillary waves, 199
constructive interference, 197

crest, 194
critical layer, 199
cross-shore transport, 207
deep-water waves, 196
destructive interference, 197
dispersion, 202
disturbing force, 193
extreme waves, 208

extreme wave surfing, 208
fetch, 198
forced waves, 193
foreshore, 206
free waves, 193
fully developed sea, 200
gravity waves, 193
group speed, 202

intermediate waves, 196
internal waves, 205
inundation, 211
isobaths, 203
littoral cell, 207
longshore current, 207
longshore transport, 207
low-tide terrace, 206

Critical Thinking

1. Compare and contrast the ocean swell that results from a tropical cyclone with 129 m/hr (80 mph) winds and an Antarctic storm system with a 1000 km (621 mile) fetch and 80 m/hr (50 mph) winds. Use as much quantitative information as provided in the text to estimate wave heights, wavelengths, wave periods, and speeds.

2. Summarize the formation of a wave in the open ocean. What factors will influence the size and speed of the resultant wave?

3. Summarize the formation of a wave in an enclosed bay, such as Puget Sound or Chesapeake Bay. What factors will influence the size and speed of the resultant wave?

4. Describe the transition of a deep-water wave to a shallow-water wave in terms of wave characteristics and wave energy.

5. Why is a tsunami a shallow-water wave? Compute the speed of a tsunami in a water column whose depth is 1000 m (3,281 feet).

6. List the ways in which waves impact the lives of humans living and recreating in coastal regions of the world ocean. How is an understanding of waves important to these activities?

7. Why is an understanding of extreme waves vital to the safety and survival of persons living in coastal regions and who travel at sea on ships?

Explore Online

 Visit www.mhhe.com/chamberlin1e for access to chapter quizzing, key term flash cards, video clips, interactive activities, and more. Further enhance your knowledge with web links to chapter-related material!

Exploration Activity 10-1

Exploring Surf Forecasting Through Concept Mapping and Data Investigation

Question for Inquiry

How do oceanographers monitor and forecast wave heights and swell periods along coastlines?

Summary

Though we often associate big waves with great surfing, waves can also cause property damage and loss of life. Surf forecasting, the ability to predict the timing and severity of surf along a coastline, enables homeowners, beachgoers, and city managers to prepare for episodes of high surf, protect structures and warn the public. In this activity, we explore some of the tools that oceanographers use to predict surf along a coastline. With a little practice, you, too, can predict surf.

Learning Objectives

By the end of this activity, you should be able to:

- Identify the types of data needed to forecast wave conditions on a given shore
- Acquire and interpret data on meteorological and oceanographic conditions
- Assemble the appropriate data to make a forecast for a particular beach
- Test your predictions against wave forecast sites and available surf-cams

Materials and Preparation

For this activity, you will need access to the Internet and a printer. Alternatively, your instructor may provide a printout of figures sufficient for completing the activity. You may want to review the sections on winds and wind stress in Chapter 8. You may also want to review the section on Concept Mapping in the Exploration Activity for Chapter 4. For best results, Parts II-IV should be completed on the same day to insure consistency.

Part I: Create a Concept Map for Surf Forecasting

Nearly all of the information presented in this chapter relates to making predictions of surf. Create a concept map with a box for Surf Forecasting in the center. Create additional boxes and links that relate to surf forecasting. For example, you may create a box called deep-water waves with an arrow labeled "predicted by" connecting it to Surf Forecasting. Your concept map should grow quite large as you include equations and processes that affect waves and the ability to predict them. Review your concept map with your classmates or instructor. Use the Rubric for Physical and Conceptual Models to refine your work. (See Exploration Activity 2-1.)

Part II: Exploring Wind Data

Somewhere on your concept map, you should have placed winds because wave formation depends on winds. We begin by exploring available wind data. Be forewarned: wind data are not generally produced for non-specialists' use! The myriad of symbols used and the lack of global coverage (owing to limitations in the coverage by satellites and weather buoys) can make wind charts confusing at first. Fortunately, we can gain a basic appreciation of winds with a few simple tools.

The most common way of representing wind speed and direction on a map is with a wind barb, a symbol resembling a flag (in a very abstract way!), such as ⟋. This wind barb has three bars, two long and one short. Each long bar represents 10 knots and each short barb equals 5 knots (or the equivalent speed in meters per second or miles per hour). So what is the wind speed represented by this barb? (10 + 10 + 5 = 25 knots). The direction of the line (from the barbed end to the unbarbed end) indicates the direction from which the wind is blowing (the direction the wind is coming from) based on compass points with North at the top of the page. In this example, the wind direction would be northeast.

1. A convenient place to start to learn about weather systems and winds is NOAA's Geostationary Satellite (GOES) Server website, http://www.goes.noaa.gov/. This site provides a bird's eye view of weather systems across the globe and offers access to wind data. Go to the GOES website and browse the visible, infrared (IR), and water vapor images. Click on one or two of the loops. Spend some time observing the movements of clouds and water vapor. Record your observations. (Don't forget to record the time and date of the image and the location you are viewing.)
2. Click on the GOES full disk linked in the column on the left side of the page. View both the GOES East and GOES West Visible and IR. What general features discussed in Chapter 8 appear in these images? Record your observations.

3. Click on the High Density Winds link. Click on Northern Hemisphere Infrared. You will see a set of wind barbs overlaid on the GOES infrared image. Spend a few minutes deciphering the winds. Click through the different time cycles (00Z, 03Z, 06Z, etc.) What do you observe? Are any hurricanes present? Any large storm systems? Where are the highest winds located? Do they follow a generally straight pattern or are they curved or rotating? Explore the different images and record your observations. Print one or two of them to use in Part III.

4. Explore the Severe Storm Sectors and Daily Significant Event imagery and report on any severe or significant events, especially hurricanes. This site is an excellent resource for learning more about global weather.

5. Go to the Coastwatch site (http://coastwatch.noaa.gov/cw_index.html). Click on Data Products. This site provides satellite data on sea surface temperature, ocean color, and winds. Click on Ocean Surface Winds and click on the most recent imagery for both type of wind data. Find a region that interests you and click on its PNG file. Record your observations. Find a region with high winds (refer to your data from the GOES satellites) and record your wind observations. What is the difference between the two types of wind observations?

6. NASA's WINDS home page (http://wins.jpl.nasa.gov/) is another good resource for learning about winds, especially those recorded during hurricanes. Search the site and find wind data for a hurricane. Record your observations. Note especially the changes in wind direction associated with a hurricane. Print one or two of the hurricane wind data images for use in Part III below.

7. Based on your wind observations, where are the most intense winds at this time? Which of those systems do you think will produce the most intense waves?

Part III: Calculating Fetch

As you learned in this chapter, fetch is the distance over which the wind blows at a constant speed and direction. Calculations of fetch are based on uniform direction and speed. If wind speed or direction changes, then a new fetch must be calculated. An excellent tutorial on fetch and its calculation from wind data is available at the University Corporation for Atmospheric Research's Meteorology Education and Training website, http://meted.ucar.edu/topics_marine.php. Click on Wave Cycle I: Generation. Select the print version and view the topics on fetch.

Our analysis will be very similar to theirs. Use the printouts of wind data from Part II to complete the following:

1. Locate regions of your wind data (the wind barbs) where wind speed and direction remain constant. Use a ruler to draw a straight line along these regions. Perform this analysis on the GOES wind data and the hurricane wind data. Label each line with a different number or letter.

2. Determine the latitude and longitude of the start and end points for your fetch lines. Create a table with this information. Go to this website, http://jan.ucc.nau.edu/~cvm/latlongdist.html (or Google for a similar one) to calculate the linear distance of your fetch lines. Enter the distance in kilometers in your table along with the wind speed.

3. Create another column in your table and use the wave nomogram on the UCAR site to predict wave heights for a given wind speed and fetch.

4. Where did you observe the longest fetch? How does hurricane fetch (as determined in this analysis) compare to the fetch of larger storm systems? Why are determinations of fetch using the type of analysis we employed not always suitable for hurricanes? (Hint: see the UCAR website.) Under what conditions might a hurricane produce a larger than apparent fetch, i.e., a large dynamic fetch?

Part IV: Combining Winds and Swell

Fortunately, predictions of wave heights can be automated using computer models (but had we told you this earlier, you might not have appreciated winds and fetch as much as you do now!). Wave heights and wave periods can also be measured directly using moored instruments. In this part, we explore both types of data.

1. Visit the Fleet Numerical Meteorology and Oceanography Center (FNMOC) website at www.fnmoc.navy.mil/PUBLIC/index.html. Maintained by the Navy, this site provides abundant data products on meteorological and oceanographic conditions, especially those that impact operational aspects of naval vessels. Most importantly, FNMOC provides nowcast (current conditions) and forecast models of future conditions. Click on the Meteorology link and select WXMAP. Choose one of the regions where you measured fetch in Part III. Click on the NGP link. Select the FNMOC Wave Height Data with Ocean Surface Winds (Note that we have shortened their title here.). You may select the ALL link to the left or start

with the 000 hour forecast.

2. Take a few moments to study the 000 hour forecast for your region of interest. Notice that wave heights are represented as color contours. (The key to the colors appears at the top right of the image.) The contours are also labeled. Superimposed on the contours are wind barbs. Locate the highest winds. Do they correspond to the largest waves? Follow the forecast model for each successive time step (out to 144 hours). Observe and record the changes in wind intensity and wave height predicted by the model. View the loop as well to gain a sense of how the model predicts changes over several days.

3. Compare your wave predictions with the FNMOC model. Do they agree? What wave-generating factor(s) were missing from your analysis in Part III?

4. Now visit the National Data Buoy Center (NDBC) website at http://www.ndbc.noaa.gov/. Use the interactive map to click on one of the regions where buoys can be found. Ideally, you will find buoys in one of the regions you analyzed in Parts II–IV.

5. Select one of the NDBC Moored Buoys and click on it. Take a few moments to become familiar with the type of data provided by these buoys. Note that observations are updated hourly and that data are available for the previous 24 hours. Find the data for wind speed and wave height. How do these data compare to the table of data you created? Why do you think they are similar or dissimilar?

6. Click on the graph symbol next to Wind Speed under Current Conditions. You will see a graph of wind speed over the past several days. Describe the trends in wind speed over the days shown. Was wind speed steady? Rising? Falling? Up and down? How did you account for wind duration in your nomogram wave prediction? (Hint: you didn't!) Why is wind duration an important part of predicting wave heights? Can it explain any discrepancies between your predictions and the FNMOC model predictions?

7. Perform the same analysis for Wave Height. Describe the trend over the previous several days. Does the trend in wave height match the trend in wind speed? Why or why not?

8. The NDBC site also provides data on wave direction, wave steepness, and wave period. Record their trends. Why might these data be important for surf forecasting? Be sure to review their definitions and descriptions.

9. Find the plot of wave energy versus frequency and period. Record the frequency (or period) where wave energy is maximum. What information does this graph provide?

10. If you are interested in learning more or if you desire an alternate source of the above information, check out the Coastal Data Information Program operated by Scripps Institution of Oceanography (http://cdip.ucsd.edu/) or Oceanweather, Inc. (www.oceanweather.com/; see current marine data).

Part V: Exploring and comparing surf forecasts

For persons simply interested in surf conditions at their local beach, a number of websites provide data on current and future conditions. Some of these even feature surf cams that provide live video feeds of local beaches (usually on a subscription or fee basis). Use the list of sites below to complete the following:

1. Explore at least three websites that provide forecasts of wave conditions in textual or graphic form. Try to find forecasts for regions you selected above. For each site, make a table that includes forecast data for surf height, wave period, and other important observations. Compare the wave models and forecasts. In what aspects are the similar and in what aspects are they different? How do your predictions compare?

2. Write a two-page summary of the steps involved in surf forecasting that includes an analysis and opinion of the strengths and weaknesses of available data. Choose your favorite beach and propose a set of observations that might improve your ability to predict surf. Finally, include a brief statement of the importance of surf forecasting in general and for your favorite beaches (even if you live far inland!)

Useful websites for surf forecasting:

National Weather Service Marine Forecasts
   www.weather.gov/om/marine/home.htm
Coastal Data Information Program at Scripps Institution of Oceanography
   http://cdip.ucsd.edu/
Surfline
   www.surfline.com
Wetsand
   www.wetsand.com/
Stormsurf
   www.stormsurf.com/

11 Ocean Tides and Sea Level

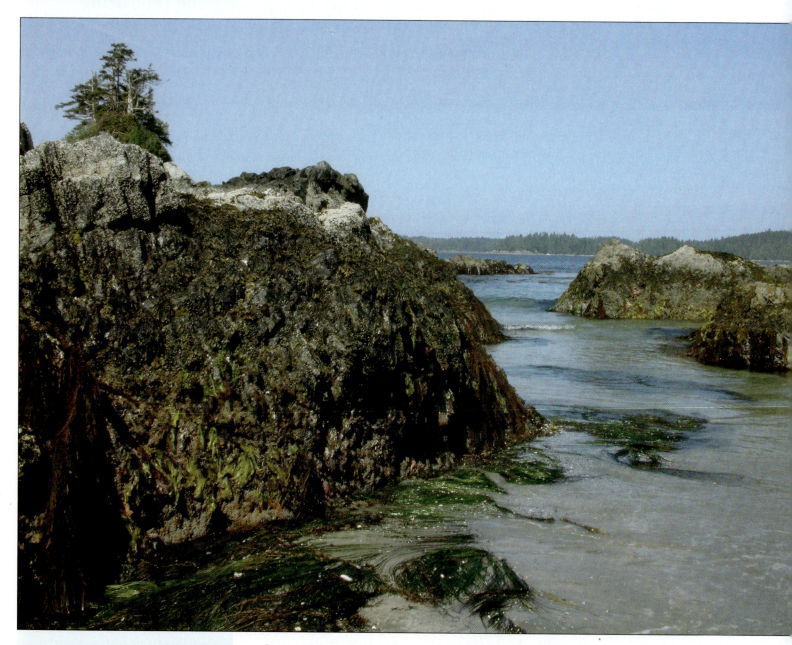

Ocean tides appear as the periodic advance and retreat of sea level. Low tide exposes organisms to terrestrial-like conditions while high tide submerges them in an aquatic environment.

Questions to Consider

1. How have physicists contributed to scientific understanding of tides?

2. How do the relative positions of the Sun and the Moon affect the heights and pattern of the tides observed in the world ocean?

3. Why are both the equilibrium and the dynamic models of the tides useful for understanding the patterns of the tides?

4. What are the tidal patterns observed at your favorite shoreline, and how do they change over days, months, and years?

5. What temporal and spatial factors alter sea level?

6. Why is an understanding of tides and sea level important to the safety and welfare of humans?

The Rhythm of the Sea

On the South Pacific island of Palau, there is a saying "we are born on one tide, we die on another." Time here is not measured by the tick of a clock. Rather, the days are marked by **tides,** the daily rise and fall of sea level. When sea level is at its greatest height, the tide is said to be high. When sea level is at its lowest extent, the tide is said to be low. Tides are a planetary phenomenon. Astronomers define tides as the distortions of land and sea produced by the gravitational attraction of the Moon and the Sun on every part of Earth. Tide-producing forces, though quite small, give rise to horizontal movements of water, called **tidal currents.** It is the tidal currents that bring water from one location to another (a horizontal movement) and that cause the rise and fall of sea level (perceived as a vertical movement).

Tides play a major role in geological, physical, chemical, and biological processes in the world ocean. The daily alternating submergence and exposure of rocks on the shore, combined with the action of the waves and heat from the Sun, accelerate weathering and erosion along the coast. Internal tides interact with the seafloor and supply energy for mixing of surface and deep waters. Tides along coastlines enrich the land, which, in turn, nourishes the sea. Many marine organisms synchronize their behavior to the tides.

Navigation, shipping, travel by sea, food-gathering, shell collecting, and fishing have long been associated with the tides. In modern times, tides are being used to turn turbines and provide electrical energy to coastal communities. Tides have long played a role in warfare, influencing when and where to place troops on a beach. Sea level is an important factor in the consideration of tides. Rising sea levels exacerbate the action of the tides, especially in low-lying coastal areas and oceanic islands. Scientific understanding of the factors that alter sea level provides insights into the role of global warming in the accelerating rise of sea level in the twenty-first century.

A Brief History

Knowledge of tides and their patterns dates back to at least 2300 BC. In the ruins of coastal cities along the Gulf of Cambay in India, archaeologists uncovered evidence of tidal docks, structures that allow boats to enter at high tide and, by means of a gate, trap the water and keep the boats afloat when the tide recedes. Because tides may rise and fall in excess of 10 m (30 feet) in this region, they likely played an important role in the culture and commerce of these ancient people. Indian religious texts from 300–400 BC suggest a link between the tides and the phases of the Moon, a highly advanced claim for its time. In contrast, civilizations surrounding the Mediterranean, where tidal ranges are on the order of a few feet, gave little more than a passing interest in tides. The Greek philosopher Posidonius (135–51 BC) made observations near Cadiz in southern Spain and formulated one of the first understandings of the relationship between tides and the positions of the Sun and the Moon. Near 1 AD, the Chinese observed the relationship between phases of the Moon and the size of the tides on the Qiantang River near Hang-Chou. Nevertheless, many centuries would pass before a more scientific approach to understanding and predicting the nature of tides would appear.

Several notable scholars advanced theories of the tides, including Galileo, but the first complete and fundamentally correct explanation was published in 1687 by Sir Isaac Newton (1642–1727) in his *Principia*. Newton introduced the law of gravitation, which states that the gravitational attraction between two planetary bodies is the product of their masses divided by the square of their distance times the gravitational constant. Offering a few pages on tides, Newton's work set the stage for others, notably Dutch mathematician Daniel Bernoulli (1700–1782), who with others developed the **equilibrium theory of the tides** (**Figure 11-1**). Differences in gravity at different points on Earth's surface cause vertical and horizontal forces, but the vertical forces are much too small to generate tides. A friend and colleague of Bernoulli's, Swiss-born mathematician Leonard Euler (1707–1783), demonstrated that the tide-causing forces result from the horizontal forces acting along the surface of Earth (i.e., tangential to the surface). These horizontal forces cause horizontal motions of water toward points directly beneath and on the opposite side of the Moon and Sun, the **tidal bulges.** The equilibrium model of the tides remains popular in modern textbook descriptions of tides.

While useful for explaining the basic principles of the tides, the equilibrium theory is insufficient for practical application on a rotating Earth with ocean basins and continents. Recognizing the limitations of a static model, Pierre Simon, Marquis de Laplace (1749–1827), introduced the **dynamical theory of the tides** in various publications between 1775 and 1825. In essence, Laplace's theory states that tides can be viewed as a wave caused by the rhythmic pulses of horizontal tidal forces (**Figure 11-2**). This tidal wave has the same period as these forces and propagates as a wave in a rotary fashion in ocean basins. The major advancement of dynamical theory is that it takes into account the shape and depth of ocean basins and accounts for the Coriolis effect. Modern tidal predictions are based on the dynamic model.

Another significant advance in the nineteenth century was the application of harmonics to tidal prediction, the representation of tidal constituents as sine curves. Harmonic analysis of tides was first applied by Sir William Thomson, otherwise known as Lord Kelvin (1824–1907), who built one of the earliest tide-predicting machines. Using a system of gears and shafts, the contribution of each of ten tidal constituents could be represented (**Figure 11-3**). By mechanically linking the rotations of each gear, the predicted tide could be recorded on a paper-and-pen recording device (i.e., a chart recorder). Of course, the speed of rotation exceeded that of the natural cycles so that a set of tide table predictions could be acquired and published. In 1867, the Coastal Survey of the United States began to produce tide tables using Thomson's machine. Until the advent of digitally predicted tides using computer simulations in 1966, all tides were predicted in this manner.

Equilibrium model

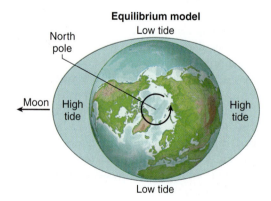

■ **FIGURE 11-1** Newton's Law of Gravitation provided the basis for Bernoulli's equilibrium model of the tides. In this model, the Earth is treated as a sphere with an ocean of uniform depth. The horizontal component of gravitational forces creates two tidal bulges. Earth's rotation beneath these bulges generates the tides. See section "What Causes Tides?" in this chapter for details.

Dynamic model

■ **FIGURE 11-2** Laplace introduced the dynamical description of the tides, in which tides move as rotating waves in an ocean basin, diagrammatically illustrated here. See section "Dynamic Models of Tides" in this chapter for details.

surface pass beneath the tidal bulges. Locations directly beneath the tidal bulges will experience a higher sea level, called **high tide.** Locations between the tidal bulges will experience a lower sea level, called **low tide.** Because tidal forces give rise to bulges on both sides of Earth (for reasons discussed in the section on tidal forces), the rotation of Earth under these bulges will cause (in theory) two high tides and two low tides each day, a pattern called a **semidiurnal tide** (*semidiurnal* meaning "twice daily") (**Figure 11-4**). Tides along the eastern coast of the US generally follow a semidiurnal pattern. Some locations along the western coast of North America experience a mismatch in the height of successive high and low tides. This daily tidal pattern of unequal high and low tides is called a **mixed semidiurnal tide** (or simply a mixed tide). A third tidal pattern is also observed, one that seems at odds with the equilibrium model. In some locations, notably the Gulf of Mexico, only one high tide and one low tide occur each day, a pattern called a **diurnal tide.** This tidal pattern results from the fact that the Moon and the Sun and the bulges they produce are not always directly above the equator, as we shall see.

■ **FIGURE 11-3** Tide prediction machines used mechanical gears and chart recorders to simulate the natural components of tides. Because the overall tidal cycle is composed of several individually varying components, several different-size gears are required to simulate the tides.

During the nineteenth century, another important component of tide measurements was introduced. At the suggestion of German geophysicist Alexander von Humboldt (1769–1859), Antarctic explorer James Clark Ross (1800–1862) established a tidal datum in Tasmania. The **tidal datum** is a fixed point on the shore from which **tide heights** may be measured. Establishment of a permanent, fixed location from which to measure tides enabled accurate measurements of tides and comparison of tide heights and patterns between locations.

What Are Tides?

We most commonly observe tides along the coast, but tides occur in the open ocean as well. The height of tides, however, varies over temporal scales of days to millennia (see inside front cover). Burntcoat Head in the Bay of Fundy experiences one of the largest tidal ranges in the world, up to 16.1 m (53 feet). On Ellesmere Island in Canada, the tidal range is a mere 0.1 m (4 inches). Tidal patterns vary spatially. Along the US coastline, some locations experience two tides a day, while others experience only one.

Daily Patterns in Tides

The simplest explanation of tidal patterns follows from the equilibrium model. As Earth rotates on its axis, locations on its

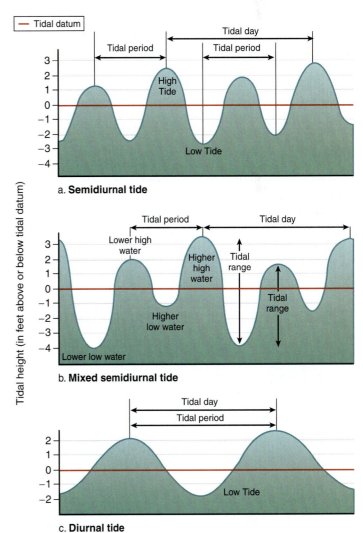

■ **FIGURE 11-4** Daily tidal patterns. (a) semidiurnal (b) mixed semidiurnal (c) diurnal tidal patterns.

Some useful definitions in the description of tidal patterns refer to the time periods involved and the range of the tides. The time between successive high tides or low tides is the **tidal period,** about 12 hours and 25 minutes where semidiurnal and mixed tides occur. A **tidal day** refers to one complete revolution of Earth beneath the tidal bulges, or about 24 hours and 50 minutes. For diurnal tides, the tidal period and tidal day are equal.

The **tidal range** is determined by subtracting the heights for the high and low tides in a tidal period, or

tidal range = tide height (high tide) − height (low tide)

Remember, when subtracting a negative number, which is often the case for low tides, the tide heights are added to obtain the tidal range. For example, a high tide of 1 m (3 feet) and a low tide of −0.6 m (−2 feet) would have a tidal range of 1.6 m (5 feet), that is, 1 − (− 0.6) = 1.6.

The Lunar Month

To understand variations in the height and time of the tides, we have to take into account the combined effects of the Moon and the Sun, each of which creates a tidal bulge. The twice daily or semidiurnal contribution of the moon is designated as the M_2

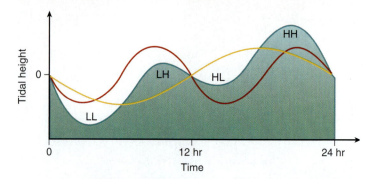

■ **FIGURE 11-5** The actual tide can be represented as the sum of the tidal constituents. In this simplified example, the combination of the M2 lunar semi diurnal constituent (blue) and O1 lunar diurnal constituent (red) generate a mixed semidiurnal tide (black). LL = low low tide, LH, = low high tide, HL = high low tide, HH = high high tide.

tide, and the semidiurnal component of the tide caused by the Sun is called the S_2 **tide.** M stands for Moon and S stands for Sun and the 2 indicates the twice-daily nature of these **tidal constituents.** Table 11.1 lists the nine major tidal constituents. Each of these constituents can be represented mathematically and summed to yield a prediction of the daily tide (**Figure 11-5**).

Two factors alter the timing of the M_2 and S_2 tides: the orbit of the Moon around the Earth; and the orbit of Earth around the Sun,

TABLE 11.1   The Nine Major Tidal Constituents and Their Characteristics

Tidal constituent	Symbol	Period (hours)	Description
		Semidiurnal tides *(twice daily)*	
Principal lunar semidiurnal	M_2	12.42	Lunar contribution to tidal bulges; represents Earth's rotation with respect to Moon
Principal solar semidiurnal	S_2	12	Solar contribution to tidal bulges; represents Earth's rotation with respect to Sun
Lunar elliptic semidiurnal	N_2	12.66	Accounts for variation in Earth-Moon distance
Lunisolar semidiurnal	K_2	11.97	Modulates the amplitude and frequency of the M_2 and S_2 tides
		Diurnal tides *(once daily)*	
Principal lunar diurnal	O_1	25.82	Together with K_1, accounts for variations in Moon's declination
Principal solar diurnal	P_1	24.07	Together with K_1, accounts for variations in Sun's declination
Lunisolar diurnal	K_1	23.93	Together with O_1, accounts for Moon and Sun declination
		Long period	
Lunar fortnightly	M_f	327.86	Accounts for variation in lunar declination
Lunar monthly	M_m	661.3	Accounts for variation in rate of change of lunar speed and distance

Source: *NOAA Tidal Glossary.*

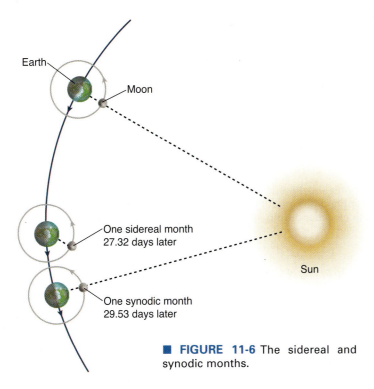

■ **FIGURE 11-6** The sidereal and synodic months.

respectively. Let's consider the Moon first. The orbit of the Moon around Earth takes approximately 27.3 days, the **sidereal month** (**Figure 11-6**). The Moon orbits in the same direction as Earth's rotation (eastward) so that after 24 hours (the earth day), the Moon has advanced about one-twenty-seventh from its original position, that is, $1/27 \times 24$ hours or about 53 minutes. Thus, it takes 53 minutes longer each day for the Moon to reach the same position in the sky (to an observer on Earth). That means that the M_2 tide advances 53 minutes

each day. In other words, if the Moon alone caused the tides, the time of the high tide and the low tide would be 53 minutes later each day (which is a good approximation of the daily advance of the tides). The orbit of Earth around the Sun also plays a role in lunar tides. Each day Earth advances in its orbit around the Sun by about 0.985 degrees per day (i.e., 360 degrees/365.24 days). By the end of the 27-day orbit of the Moon, Earth has moved about 26.5 degrees forward. An additional two days are required for the Moon to reach the same configuration relative to Earth, the Moon, and the Sun. Thus, a complete lunar tidal cycle takes 29.5 days, the **synodic month**.

Spring and Neap Tides

The path of the Moon on its orbit around Earth creates an additional pattern in the tides, one clearly visible in the synodic month. For a couple weeks during the month, the tidal range reaches a maximum. This pattern is called the **spring tide** because the tide appears to "spring up" like a jack-in-the-box. During alternate weeks, the tidal range is minimal. This pattern is called the **neap tide** (**Figure 11-7**).

The phases of the Moon provide the most convenient framework for understanding these monthly tidal patterns. When the moon is completely illuminated in the night sky, it is said to be full. A **full moon** phase occurs when the side of the Moon facing Earth is fully illuminated by the Sun. At this time, the Moon-Earth-Sun appear to be in a straight line as viewed from above or below. When the Moon is between Earth and the Sun, the side of the Moon facing away from Earth is illuminated, but we cannot see it. To us, the Moon is completely dark. This phase of the moon is called the **new**

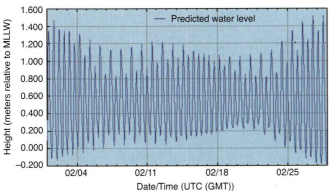

a. **Buzzards Bay, MA**

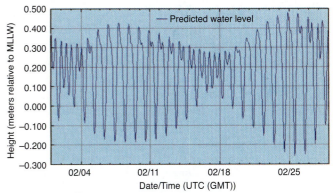

b. **Galveston, TX**

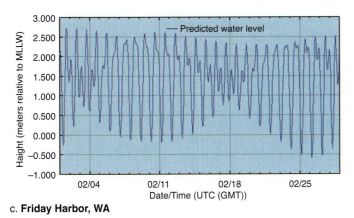

c. **Friday Harbor, WA**

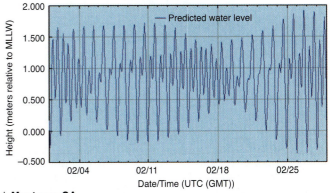

d. **Monterey, CA**

■ **FIGURE 11-7** Tides predicted for the month of February, 2006, for (a) Buzzards Bay, MA (b) Galveston, TX (c) Friday Harbor, WA (d) Monterey, CA. Periods of maximal tidal range are called spring tides while periods of minimal tidal range are called neap tides.

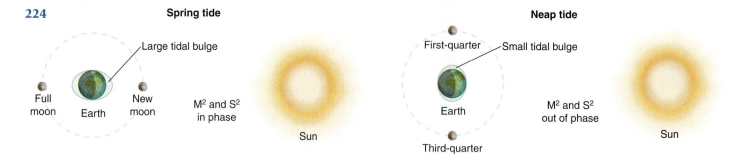

Spring tide

Large tidal bulge

Full moon   Earth   New moon

M2 and S2 in phase

Sun

Neap tide

First-quarter   Small tidal bulge

Earth

Third-quarter

M2 and S2 out of phase

Sun

■ **FIGURE 11-8** The position of the Moon relative to the Sun determines whether the M2 and S2 tides are in phase, creating spring tides, or out of phases, generating neap tides. Two spring and two neap tides occur during each tidal cycle.

moon. When the Moon is at the halfway point in its orbit between the new-to-full moon or full-to-new moon, that is, when the Moon is at a 90-degree angle relative to the Earth-Sun line, only half of the Moon is illuminated. These positions are called the **quarter moons,** representing the first quarter (after the new moon) or third quarter (after the full moon) of the Moon's orbit. In between these positions, the Moon may appear crescent-shaped (concave) or gibbous-shaped (convex) and may be growing in intensity (waxing) or shrinking in intensity (waning).

The positions of the Moon cause changes in the location of the tidal bulges in the M_2 and S_2 semidiurnal tides (**Figure 11-8**). When the Moon is aligned with Earth and the Sun, the two tidal constituents will be "in phase" and additive. The resultant tidal bulges will be much greater than either bulge alone. These are the spring tides and they occur during the full and new moon, that is, when the Moon, Earth, and Sun are aligned. When the Moon, Earth, and Sun are not aligned, the two tidal constituents will be "out-of-phase" and subtractive. The resultant tides will exhibit less of a dynamic range. These are the neap tides and they occur in conjunction with the first and third quarter phases of the moon. Note that the relative positions of the Moon place it in line twice a month, approximately every 14.5 days. Similarly, the Moon is at a 90-degree angle to the Earth-Sun line twice a month. Thus, as expected, there are two spring tides and two neap tides ever 29.5 days, that is, every synodic month.

Declination

Unfortunately, the positions of the Moon and Sun relative to Earth are insufficient to explain mixed semidiurnal or diurnal tides. For that, we need to explore a few additional characteristics of their orbits. First, Earth's plane of orbit around the Sun, called the **ecliptic,** varies relative to the equator. Earth's 23.5-degree **axial tilt,** the angle between Earth's axis of rotation and the ecliptic, is the reason for the seasons on Earth (see Chapter 7). The seasonal variation of the ecliptic is observed as changes in the path of the Sun across the sky. The angle between the orbital plane of the Sun (and Moon) and the equator is called **declination** (**Figure 11-9**). When we observe the Sun move higher in the sky from winter to summer or lower in the sky from summer to winter, we witness the declination of the Sun with respect to the equator. Similarly, the Moon's orbital plane is tilted relative to the ecliptic (at an angle of about 5 degrees). With respect to tides, these astronomical factors cause the maximum displacement of M_2 and S_2 tidal bulges to occur at locations above or below the equator. A person standing at one location on Earth will experience a different tide height as Earth rotates beneath different parts of the tidal bulges. Similarly, different locations on Earth will experience different tide heights due to declination. Some locations will miss the second bulge altogether. The net result is an inequality in the height of the tides at a given location or even an absence of a high or low tide. Thus, declination accounts for the observation of mixed tides and diurnal tides at different locations in the world ocean (**Figure 11-10**).

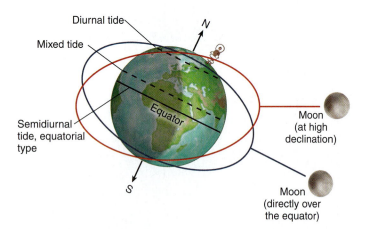

Diurnal tide

Mixed tide

N

Semidiurnal tide, equatorial type

Equator

S

Moon (at high declination)

Moon (directly over the equator)

■ **FIGURE 11-9** The concept of declination. Declination is similar to latitude: celestial objects at the equator have zero declination while those above the equator have positive declination (to +90 degrees at the north pole) and those below the equator have negative declination (to -90 degrees at the south pole). The declination of the Moon determines the position of the tidal bulges relative to the equator.

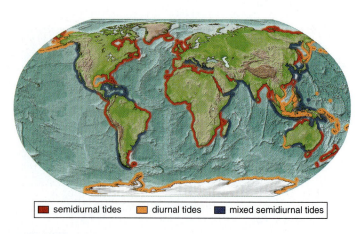

| ■ semidiurnal tides | ■ diurnal tides | ■ mixed semidiurnal tides |

■ **FIGURE 11-10** Types of tides found in different locations of the world ocean.

Tidal Fronts: Where the Action Is!

If you have ever stood on a jetty or pier or even crossed an inlet or bay in a ferry or similar craft, you may have noticed places where the water suddenly changes color. This color change often marks the boundary between two bodies of water with different properties: the dark-colored "bay" waters and the semitransparent "ocean" waters. The boundary between two water masses is called a front. Fronts can occur virtually anywhere in the world ocean where two bodies of water with different properties meet. Tidal fronts occur where tidally mixed waters meet stratified waters, generally where coastal waters encounter oceanic waters. Tidal fronts exhibit a wide range of spatial and temporal variability, from meters to kilometers and hours to years, respectively. While the hydrodynamics of tidal fronts are complex, the biological processes associated with them have enormous implications for coastal ecology. Since the "discovery" of tidal fronts in the 1970s, oceanographers have focused a great deal of research on understanding their dynamics and their productivity, especially for fisheries.

The most prominent feature of tidal fronts is the abundance of phytoplankton along their stratified edge, observed as increases in chlorophyll concentrations along the frontal boundary. As tidal flow increases during spring tides, the area of the tidal front expands and so do the phytoplankton. Because tides are predictable, the position of the tidal front can be modeled with some accuracy, allowing oceanographers to predict the extent of chlorophyll associated with the front. One of the significant findings during the 1970s was that tidal fronts remain productive throughout the summer months, when other regions of the temperate ocean tend to exhibit low rates of productivity. Since that time, numerous researchers have shown that many species of fishes migrate to spawn in the vicinity of tidal fronts and time their spawning so that their larvae drift within the most productive waters, that is, the waters that are the best suited for feeding. More than a dozen spawning sites have been identified for Atlantic cod. There is even some evidence that site selection by different groups of the cod has led to genetic differences between them. Hydrographic isolation reduces mixing of gene pools and permits genetic divergence between breeding stocks with preferred spawning sites.

Satellite images reveal that tidal fronts form in very nearly the same location and at the same time from year to year. The stability of these features has no doubt led to their exploitation by migrating species, such as fishes, marine mammals, and seabirds. Basking sharks have been observed to follow zooplankton stocks associated with tidal fronts. Sperm whales and bottlenose dolphins exhibit an affinity for frontal features, which has been interpreted as feeding-related. The "masters" of tidal front feeding must certainly be seabirds, which flock to these regions by the hundreds and thousands. The development of technology for recording water temperatures during the feeding cycles of diving birds supports their association with fronts.

The occurrence of tidal fronts has two major implications for productivity in ocean waters. First, they provide a temporally "continuous" source of food for a wide range of marine organisms. Both pelagic (free-swimming) and benthic (bottom-dwelling) organisms benefit from the high productivity of phytoplankton in tidal fronts. Spawning fishes and invertebrates that release planktonic larvae find a reliable food supply in tidal fronts. Second, tidal fronts create spatially heterogeneous or patchy distributions in planktonic and pelagic food webs. Feeding aggregations are frequently associated with tidal fronts and other frontal features so that high densities of marine organisms occur in limited areas. In some cases, feeding aggregations are associated with the release of eggs and larvae into tidal fronts, such as occurs on the Great Barrier Reef during the synchronized mass spawning of corals in November.

Unfortunately, the predictability of tidal fronts makes them especially vulnerable to human activities. Knowledge of the timing and location of tidal fronts often means the difference between a bountiful catch and poor one for fishers. The outfitting of fishing vessels with global positioning systems, satellite downlinks, oceanographic sensors, and "fish finders" has led to an increased ability to find and exploit tidal fronts.

FIGURE 11a Whales and other organisms often take advantage of the dense aggregations of plankton found in association with tidal fronts.

Tides in the Coastal Ocean

While the tidal cycle in coastal waters is generally synchronous with the nearby ocean tidal cycle, the ranges and phases vary regionally. Seafloor bathymetry and the shape of the inlets and passages through which water must move play important roles. The incoming tidal current as the tide is rising toward high tide is called the flood current or **flood tide.** The outgoing current is called the ebb current or **ebb tide.** The period between the flood and the ebb, when the current speed falls to zero and the direction of the current changes, is called the **slack tide.** Tidal currents can be quite large as large volumes of water are constrained to move through relatively small cross-sectional areas. Tidal currents in the Nakwakto Rapids, a narrow inlet on the Pacific coast of Canada, have been clocked at speeds of 29 km hr^{-1} (18 miles per hour). Such currents are clearly hazardous to navigation and recreational activities. In other parts of the world, tidal currents propagate as a **tidal bore,** a wall of water that moves into a bay or river as the leading edge of a rising tide. Tidal bores form when the tidal wave front moves faster than the speed of a shallow-water wave. This is somewhat analogous to a sonic boom that forms when the pressure wave from an aircraft is faster than the speed of sound. Most tidal bores are less than 1 m (3 feet) in height, but some can reach up to 5 m (16 feet). Tidal bores may move upstream at speeds close to 6 m sec^{-1} (13 miles per hour), too fast to be outrun by humans but perfect for tidal bore surfing and rafting.

What Causes Tides?

The equilibrium model of the tides, while not suitable for a description of the real tides, nonetheless helps to explain the major patterns of the tides. A number of popular misconceptions exist concerning the equilibrium model and the causes of tides. We begin with a brief summary of the major aspects of tides and tide-causing forces:

1. Tides are caused by gravity; both Earth's gravity and the gravitational attraction of the Moon and Sun are important.
2. Other planetary bodies in our solar system make very small (practically negligible) contributions to tides.
3. Tide-producing forces, called **tidal forces,** result from *differences* in gravitational attraction at various points on Earth's surface.
4. Tidal forces have a vertical component and a horizontal component; only the horizontal component causes movements of water.
5. The gravitational attraction of the Moon (or Sun) does not pull water toward the Moon (or Sun); rather, it causes the water to move horizontally as a tidal current.
6. The vertical pull of the Moon overhead is to reduce the weight of an object by about 0.000035%, a very tiny amount; for comparison, a meal might change your weight by 1%; put another way, an overhead Moon would have the same effect on the weight of a ship as if a gull took flight from its perch on the ship.
7. Tides would occur even if Earth, the Moon, and the Sun were not moving (an admittedly impossible situation); the motion of these bodies has no bearing on tides.

8. Explanations of tides invoking centrifugal and centripetal forces are an unnecessary complication when discussing tides; tides are a gravitational phenomena that do not require equations involving planetary motions.
9. The height of tides is referenced from Earth's surface (the geoid): Earth bulges at the equator (by about 23 km or 14 miles) because of Earth's rotation about its axis, but this has no bearing on tides other than changing the shape of the geoid (and the tidal reference point).

Tidal Forces

Both the Moon and the Sun exert a gravitational force on Earth and cause tides. For the sake of simplicity, we will consider the Moon only. Imagine a straight line drawn through the center of Earth and the Moon. An object nearest the Moon would experience a greater gravitational attraction from the Moon than a point on the side opposite the Moon. Of course, the object would remain fixed to Earth because the gravitational attraction of Earth is much greater than the Moon's on Earth's surface. However, the small difference in gravitational attraction at different points on Earth's surface relative to Earth's gravitational attraction is sufficient to move water, as we shall see.

Newton very wisely deduced that the gravitational attraction between objects could be formulated as the distance to their centers, or their center of mass. For solid bodies, such as Earth and the Moon, the average gravitational force between them (F_{ME}) can be expressed in terms of Newton's law of gravitation:

$$F_{ME} = (G \times M_E \times M_M)/d^2$$

where G is the gravitational constant (6.6×10^{-11} Newtons m^2 kg^{-2}), M_E is the mass of Earth, M_M is the mass of the Moon, and d is the distance between their centers of mass.

To calculate the tidal forces (δF, or delta force), we simply take the mathematical difference (by subtraction) between the average

TABLE 11.2 Useful Astronomical Constants for Computing Gravitational Attraction and Tidal Forces

Earth mass = 5.98×10^{24} kg

Moon mass = 7.35×10^{22} kg

Sun mass = about 1.99×10^{30} kg

Distance Earth to Moon = 384,400,000 m

Distance Earth to Sun = 149,597,890,000 m

Mean radius Earth = 6.37×10^6 m

Mean radius Moon = 1.74×10^6 m

Mean radius Sun = 7.36×10^{22} m

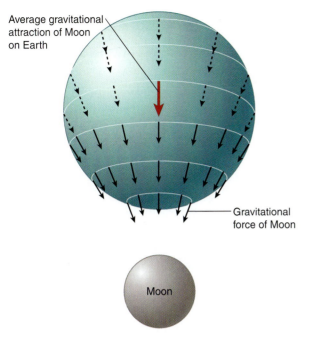

Average gravitational attraction of Moon on Earth

Gravitational force of Moon

Moon

■ **FIGURE 11-11** Force vectors for the Moon's gravitational attraction on points along Earth's surface (small arrows) and the average gravitational attraction on Earth's center of mass (bold arrow). Note that gravitational forces are larger for points closer to the Moon and smaller for points furthest from the Moon.

gravitational force (F_{ME}) and the actual gravitational force for a water parcel on Earth's surface (F_{MW}), or:

$$\delta F = ((G \times M_E \times M_M)/d^2) - ((G \times M_W \times M_M)/(d-r)^2)$$

where M_W is the mass of the water parcel and r is the distance between Earth's center and the water parcel on the surface. Sparing you the details (see our web site), this simplifies to:

$$\delta F = (GM_M)/d^3$$

Thus, we can see that the tidal forces vary approximately as the cube of the distance between the attracting bodies. You may wish to compute the gravitational forces of the Moon and Sun on Earth and the tidal forces from the Moon and the Sun using the astronomical constants provided in **Table 11.2.** If you do so, you will see that the gravitational attraction of the Sun is about 179 times that of the Moon (we orbit the Sun, after all). However, the tidal force of the Sun is only about 46% of the tidal force of the Moon. Thus, the Moon has about twice the effect on tides as the Sun.

The tidal forces can also be represented visually as vectors. If we draw a vector that represents the average gravitational force of the Moon (the bold red arrow in **Figure 11-11**) and also draw vectors to represent the Moon's gravity at various points on Earth's surface (the small arrows), we can subtract them. The resultant difference vectors represent the tidal forces (**Figure 11-12**). However, we should not be misled by Figure 11-12. It does not represent the actual forces, just their differences. Note that the direction of the tidal forces is opposite on the side away from the Moon. The result is correct mathematically but may seem counterintuitive.

The movement of water as tides comes from horizontal tidal forces only. That is, the ocean moves in response to tidal forces directed horizontally (i.e., tangentially) to the surface of Earth, as

Tidal forces = actual force − average force

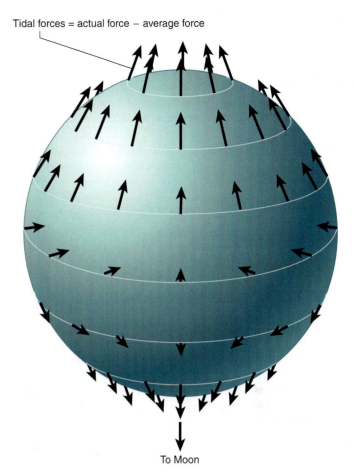

To Moon

■ **FIGURE 11-12** The difference between the actual gravitational attraction and the average gravitational attraction yields the tidal forces. The tidal forces are much smaller than the actual forces of gravitational attraction. As well, the tidal forces caused by the Moon are about twice the tidal forces caused by the Sun. Note that the tidal forces point both towards and away from the attracting body (because they represent the difference in gravity at different points on Earth's surface).

shown in **Figure 11-13.** Vertical gravitational forces do not play a role in generating tides because Earth compensates for this component of the force. The vertical component causes **earth tides,** vertical movements of Earth's crust in response to gravitational forces. Directly beneath the Moon and on the opposite side of Earth are no horizontal forces. As a result, there are no movements of water here. Water in the bulges is not being pulled toward the Moon.

Horizontal tidal forces

Moon

Earth

■ **FIGURE 11-13** The horizontal component of the tidal forces moves water and generates tides.

The water rises at the bulges because it is moved there by horizontal forces. For similar reasons, points halfway between the moonward and moonless sides have no horizontal forces and no movements of water, even though the ocean reaches its lowest level here. Thus, points directly beneath and opposite the Earth-Moon line (the major axis) and points halfway between this line (the minor axis) experience zero tidal forces. Halfway between the major and minor axis, the horizontal component of the tidal forces reaches its maximum value. Tidal currents at these locations will be greatest.

An understanding of tidal forces as horizontal forces that appear because of the differences in gravity at different points on Earth's surface is central to a proper understanding of tides. When viewed in this way, misconceptions about the bulges disappear. The tidal bulge nearest the Moon is not the result of water being "pulled" toward the Moon. The tidal bulge opposite the Moon is not the result of water being "flung" from Earth's surface (the centrifugal argument), nor is it the result of being "left behind" (another commonly stated misconception). The tides are not caused by gravity *per se* but by differences in gravity. Differential gravity is not sufficient to cause tides in a bathtub or even a small lake. The vertical component of gravity, the "pulling force," may very slightly decrease the weight of water along the Earth-Moon line and very slightly increase the weight of water where directed inward (perpendicular to the Earth-Moon line). However, the force plays no role in generating the tidal bulges.

Dynamic Models of Tides

Thus far, we have considered factors that cause tides using a model based on an Earth uniformly covered by an ocean. However, we know from experience that tidal bulges do not pass over the continents twice each day! The dynamic model of tides takes into account the flow of tidal currents around land masses, the changing depth of the seafloor, the Coriolis effect, meteorological conditions, and other factors that affect the tides. Dynamic models of tides produce better agreement with actual ocean observations and allow us to accurately predict tidal levels.

As in the equilibrium model, the tidal forces supply energy to the ocean. In this case, however, the tidal forces set up a natural resonance, a periodic oscillation of waves in an ocean basin. The amplitude and periodicity of that oscillation will depend on the geometry of the basins, each basin having its own natural resonance. As a result, dynamic models treat tides as if they were a long-wavelength wave, a **tidal wave.** You should not confuse a tidal wave with a tsunami. Both are long-wavelength waves, but their disturbing forces are quite different (see Chapter 10). Because their lengths are much greater than the average depth of the ocean (i.e., depth less than L/20), tidal waves (and tsunami) move like shallow-water waves. Even so, their maximum velocity is much slower than Earth's rotational velocity over most of the world ocean (between 60°N and 60°S). Thus, there is a **tidal lag,** a delay between the passage of the Moon overhead and the high tide. At the equator and lower latitudes, the tidal lag may be up to six hours.

Tidal waves propagate through the ocean basins in the form of a **standing wave,** a "fixed" wave pattern that results from the constructive interference of reflected waves. The wave pattern appears to stand still (although we know that waves are moving back and forth across

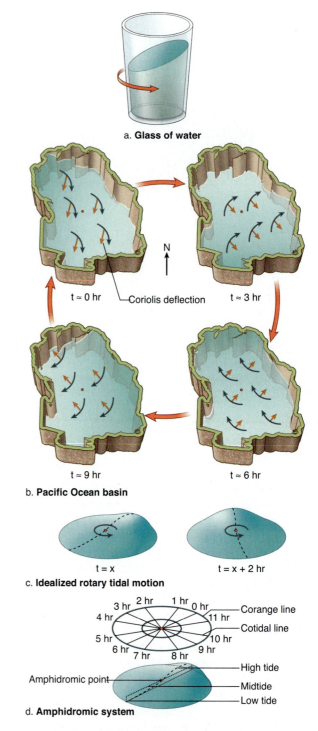

a. **Glass of water**

b. **Pacific Ocean basin**

t ≈ 0 hr   Coriolis deflection   t ≈ 3 hr

t ≈ 9 hr   t ≈ 6 hr

N

c. **Idealized rotary tidal motion**

t = x   t = x + 2 hr

d. **Amphidromic system**

3 hr 2 hr 1 hr 0 hr
4 hr   11 hr — Corange line
5 hr   10 hr — Cotidal line
6 hr 7 hr 8 hr 9 hr

Amphidromic point
High tide
Midtide
Low tide

■ **FIGURE 11-14** (a) Standing waves resemble the rolling motion of water in a glass when you swirl it. (b) Similarly, tidal forces generate a standing wave that rotates around an ocean basin. One complete revolution takes 12 hours and 25 minutes for the M2 tidal constituent. The passage of the standing wave creates a horizontal pressure gradient such that water flows "down hill" under the influence of the Coriolis effect. (c) The net result is a wave that rotates within an ocean basin and produces the tides. (d) The point around which the rotary wave rotates is called the amphidromic point.

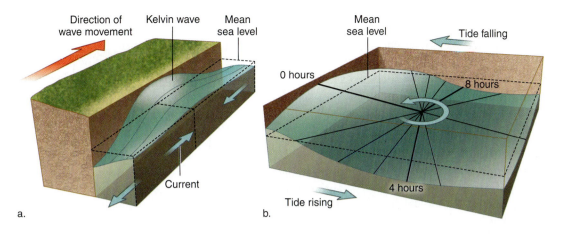

■ **FIGURE 11-15** Propagation of a Kelvin wave around an amphidromic point in a basin. (a) The Kelvin wave will have its high point along the boundary of the basin, i.e., the coastline. As the wave rotates, the tide at a given point will rise or fall as the wave "sweeps" around the basin. (b) The entire rotation of a Kelvin wave takes 12 hours and 25 minutes. Modified from Pugh.

the basin to produce this pattern). A wave reflected from a boundary, such as a continent, will return along the same or a similar path as the original wave and produce a standing wave pattern. In the simplest case, a single standing wave in an enclosed basin, the wave will look like a seesaw: it will appear to oscillate around a point called a **node.** The vertical amplitude around the node will be zero, while at the two ends, the amplitude of the wave will be greatest. However, the water will move horizontally—back and forth—at the node. These horizontal currents are the important ones where tides are concerned.

Anytime water moves as a current over long distances, the Coriolis effect becomes important. In ocean basins, the Coriolis effect gives rise to a wave that rotates about a point equivalent to the node of a standing wave (**Figure 11-14**). This "nodal" point is called an **amphidromic point.** The rotating wave, called a **Kelvin wave,** generally moves counterclockwise in the Northern Hemisphere and clockwise in the Southern Hemisphere. Along the equator, Kelvin waves propagate eastward (and are called equatorial Kelvin waves). Coastal Kelvin waves move northward along western boundaries in the Northern Hemisphere and southward along western boundaries in the Southern Hemisphere. Not all Kelvin waves are tidal waves, but they represent the most common type of tidal wave. Other types of waves may also be produced by tides.

The propagation of a Kelvin wave around an amphidromic point reveals the way in which tides progress along a coastline (**Figure 11-15**). In practice, the high point of the Kelvin wave along the coast is followed as it moves around the basin. Thus, the propagation of a Kelvin wave can be divided into 12 segments or phases. Each phase is represented by a **co-tidal line,** the line along which the phase of the tide is the same. Put another way, co-tidal lines represent the location of the high tide at a particular time. Numbers on co-tidal lines indicate tidal lag based on passage of the Moon overhead at a specified location, usually the Greenwich meridian. **Co-range lines** indicate places where the amplitude of the tide is the same. These lines typically intersect at right angles to co-tidal lines (giving the overall appearance of a map covered in spider webs). A Kelvin wave viewed from the side reveals its increase in amplitude from its amphidromic point to the coastline. Co-range lines indicate the sweep of that portion of the Kelvin wave whose amplitude is the same. Numbers on co-range lines indicate the amplitude of the tide in meters or feet. When viewed as color contours, the characteristics of tidal waves in various parts of the world ocean become more apparent (**Figure 11-16**).

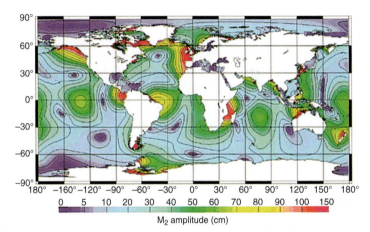

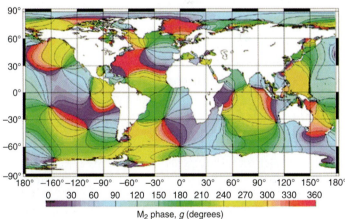

■ **FIGURE 11-16** M2 tide heights and M2 phases across the world ocean, viewed as color contours. The upper figure represents the amplitude of the M2 tide in the world ocean. Lines of constant amplitude (co-range lines) are shaded by different colors from purple for zero amplitude to red for the highest amplitudes. Thus, purple regions generally indicate amphidromic points while red regions indicate places where the range of tides is greatest. In the lower figure, the phase of the M2 tide is shown. Each line represents 1/12th of the tidal cycle and is color coded to show the sweep of the tidal wave from its starting point (purple) to its completion (red). Using such maps, we can gain better insights into the nature of tides around the world and obtain much better predictions of tidal periods and heights.

Sea Level and Tide Heights

Tidal predictions necessarily involve an understanding of **sea level,** the average level of the sea. At any moment, the observed sea level may vary according to a number of factors, including tides, storm surge, and various other meteorological and oceanographic factors. In theory, true sea level, the surface of the sea in the absence of any disturbing forces, can be approximated by the **geoid,** the equipotential (equal magnitude) surface of Earth's gravitational field (determined mathematically, using a least-squares fit of the gravity field to mean sea level). By definition, all points on the geoid form a surface so that each small elemental area is perpendicular to Earth's gravitational field. Thus, the geoid defines average or **mean sea level.** Your first thought might be that mean sea level would be flat. In fact, the surface of the geoid varies considerably and exhibits a number of bumps and depressions (**Figure 11-17**). This "lumpy" surface comes from the uneven distribution of matter in the interior of Earth. Because gravity depends on mass, variations in the distribution of mass will cause variations in the geoid. Undersea features such as ridges, trenches, and seamounts also change the gravitational field locally. The end result is a geoid—and mean sea level—that varies up to several tens of meters over the expanse of the world ocean.

While simple in theory, accurate measurements of tide heights are difficult to obtain because of the absence of a fixed reference point from which to measure tide height. The simplest method involves establishment of a tidal datum in the form of a **tide gauge benchmark**—literally, a permanent marker, usually metal, fixed to a rock. A method used in satellite determinations of sea level (see Spotlight 11-1) defines a **reference ellipsoid,** the geometric form that most closely conforms to the major and minor axes of Earth's ellipse. Sea surface heights are then reported relative to this ellipsoid. On an operational level, tide heights are measured relative to the **mean lower low water** (MLLW). NOAA's Center for Operational Oceanographic Products and Services (CO-OPS) is the agency responsible for collecting and distributing tidal observations and predictions. CO-OPS uses a 19-year average of mean lower low water to establish tide heights, a time period called the **National Tidal Datum Epoch.** The 19-year average includes the relevant astronomical cycles, in particular the 18.6-year lunar nodal cycle. A number of technologies may provide tide information, including pressure sensors deployed on the sea bottom. Bottom pressure sensors are now routinely deployed to provide warning of catastrophic changes in sea surface height due to tsunamis. NOAA's Deep-Ocean Assessment and Reporting of Tsunamis (DART) program deploys bottom sensors near buoys linked to satellites to provide real-time data on sea-level variations (**Figure 11-18**). If sudden or larger-than-normal increases are observed, a tsunami warning system is triggered.

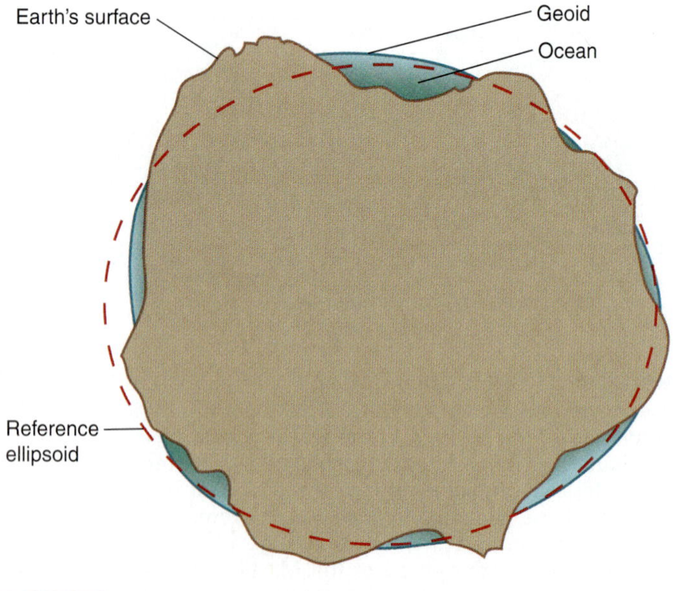

■ **FIGURE 11-17** A model of the Earth illustrating the geoid relative to the surface of the Earth and a perfect ellipsoid.

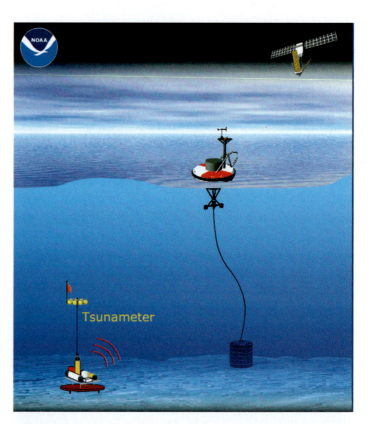

■ **FIGURE 11-18** NOAA's DART buoys track the daily fluctuations of sea level using pressure sensors deployed on the sea floor. The DART system acts as an early warning system for tsunami by detecting unusual fluctuations in sea level. A greater number of DART buoys over an expanded area, including the Atlantic, are being deployed to prevent a catastrophe like the 2004 Indian Ocean tsunami.

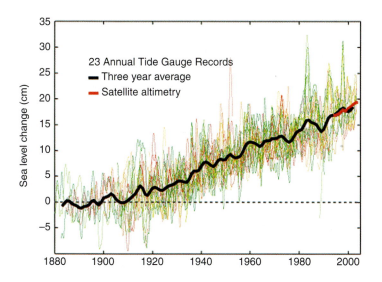

■ FIGURE 11-19 Sea level rise since the 1880s.

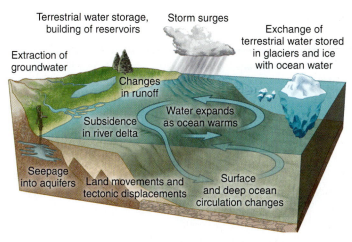

■ FIGURE 11-20 Principal factors causing sea level change. Melting of land-bound ice and thermal expansion of seawater account for the largest percentage of increase in sea level.

Sea-Level Rise and Global Warming

Concerns over rising sea levels as a result of global warming and damage caused by storm surge in low-lying areas have stimulated new interest in accurate measurements of tides and sea level. The surge-induced flooding of New Orleans and the sinking of Venice, Italy, are two examples of the problems that sea level rise may cause. Costs to adapt to a modest 50-cm (about 20 inch) increase in sea level have been estimated in the hundreds of billions of dollars in the United States. While human activities likely contribute to sea-level rise, at least part of the observed rise stems from natural causes. At the present time, Earth is experiencing an **interglacial,** a period between ice ages, which result primarily from changes in incident solar radiation over time periods that correspond to the Milankovitch cycle (see Chapter 5). The last ice age reached its peak about 20,000 years ago, when sea level was about 20 m (393 feet) lower compared to present-day levels. The latest interglacial period, the Holocene epoch, began about 10,000 years ago. Thus, sea level continues to rise in modern times as a result of glacial melting over the past 10,000 years (**Figure 11-19**).

A rise in sea level may occur by **eustatic change,** the melting of glaciers or land-ice causing an increase in the mass of the world ocean; **steric change,** thermal expansion of seawater due to warming (without a change in mass); **glacial isostatic adjustment** (GIA), isostatic rebound of the crust (uplift or subsidence) as the weight of glaciers diminishes or disappears; tectonic- or geologically driven isostatic adjustments, such as stress along faults, weathering, sediment loading, and so on; and other factors, such as winds, waves, storms, atmospheric pressure, and geostrophic flow, that may cause regional increases in sea level (**Figure 11-20**).

Observations of sea-level rise based on satellite measurements since 1992 indicate that sea level is rising at a rate of approxi-

mately 2.8 mm yr^{-1} (0.11 inches per year, 3.1 mm yr^{-1} or 0.12 inches per year if GIA is removed). However, measurements from tide gauges over the past several decades suggest a more moderate rise in sea level, from 1 to 2 mm yr^{-1} (0.04 to 0.08 inches per year). Calculations of the volume of melt from continental ice and thermal expansion of seawater indicate a rise of only about 0.5 to 0.7 mm yr^{-1} (0.02 to 0.03 inches per year) over the past several decades. The difference between measured and indirect estimates of sea-level rise has generated considerable debate among scientists and even been given a name, the **attribution problem.** Satellite measurements represent only a very short period of time relative to the more than 100-year record of tide gauge measurements. On the other hand, satellite measurements provide global-scale data on sea level, while tide gauges are limited to locations on the coastlines of continents and islands. Satellite measurements reveal that some regions of the world ocean exhibit a tenfold-higher sea level than the global mean. Some scientists have suggested that the limited coverage of tide gauge stations biases the global mean based on tide gauges and that correcting for this bias brings satellite and tide gauge stations into agreement. However, this interpretation attributes the observed rise in sea level almost exclusively to thermal expansion. Evidence from the Antarctic ice shelves reveals both melting and growth of glaciers, which may have no net effect on sea level (see Spotlight 7-1). Because the Arctic ice cap is mostly sea ice (which is already displacing its volume of seawater in accordance with Archimedes' principle), no effect of melting in the Arctic on sea level is expected. Continued measurements from tide gauge stations and satellites combined with other types of data and modeling will help better constrain the amount of sea-level rise in different locations around the globe. These studies, combined with more accurate estimates of the sources of sea-level rise, may soon yield an answer to the attribution problem. Until then, scientists will continue to seek a better understanding of sea-level rise and its causes, and they will continue to debate their findings in the spirit with which modern science is practiced.

Gauging Sea Level from Space

Coastlines and islands make convenient locations for installing tide gauges to measure sea level. But what about the rest of the world ocean? Until the invention of instruments for measuring sea level from space, sea-level measurements over most of the world ocean were impossible. Fortunately, an armada of Earth-orbiting satellites is now acquiring continuous information on sea level across the entire expanse of the world ocean. Satellite-based altimeters work a lot like echosounders (see Chapter 4). However, instead of sound, they send and receive pulses of microwaves that bounce off the ocean surface. By measuring the time interval between the transmitted radar pulse and its reflected signal (Δt) and by having a precise value for the speed of the radar pulse (c_R), the distance (D) between the satellite and the surface of the ocean can be computed. Of course, a number of factors complicate these measurements. The continuous motion of the sea surface from waves and tides has to be taken into account. Water vapor in the atmosphere can interfere with the signal. And determination of the exact position of the satellite is critical. Data pro-cessing and analysis are perhaps the most challenging aspect of satellite altimetry. Averaging and filtering the data to isolate the scales of interest require sophisticated mathematics and powerful computers. If you consider that the distance between the satellite sensor and the sea surface is roughly 1400 km (870 miles) and the accuracy of sea surface height must be on the order of 2 to 3 cm (0.8–1.2 inches), then you have some idea of the precision required.

The sea surface height measurements begun by TOPEX/ Poseidon in 1992 and now carried out by Jason provide a consistent and continuous record of sea level in the world ocean. In many ways, these measurements are diagnostic of the "state" of the world ocean, providing information on upper-ocean heat storage, El Niño/La Niña, and climate variability. Tracking of El Niño/La Niña has been especially successful, enabling oceanographers to witness firsthand the development of these phenomena (**Figure 11b**). Satellite-based altimeters launched by the European Space Agency (e.g., ERS-1 and ERS-2) work in concert with the Jason satellite to develop a synoptic and accurate picture of the sea surface. For people concerned about the global rise in sea level over the past several decades, it is a remarkable and timely achievement.

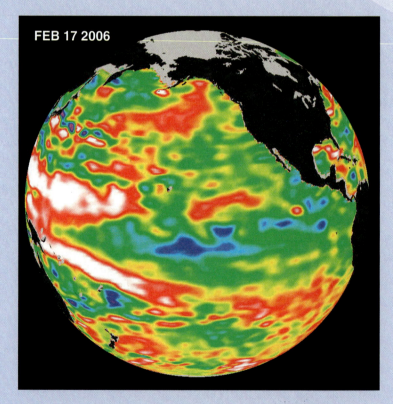

FIGURE 11b La Niña in the equatorial Pacific? Jason tracked a pool of colder-than-normal water in the equatorial Pacific in early 2006.

YOU Might Wonder

1. **Is it true that tides can cause earthquakes?**
   Scientists long dismissed the idea that earth tides, the periodic rise and fall of Earth's crust under the gravitational influence of the Moon and Sun, were sufficiently strong to trigger earthquakes. There simply was no evidence of an association between tides and earthquakes. Yet new studies show that earth tides may cause earthquakes in some types of faults. In 2004, a group of Japanese scientists demonstrated that the most common type of earthquake, the strike-slip earthquake, may be triggered by tides if the direction of the tidal force acts in the same direction as the tectonic stress on the fault. Rather than focus on the intensity of the tidal forces, these scientists examined the tidal stress azimuth, the directional component of tidal stresses, and found that when it aligns with the tectonic stress azimuth, a higher percentage of earthquakes was observed. More work remains to determine if this effect is universal across the globe, but the short answer is that, under certain circumstances, tides may trigger earthquakes!

2. **Are tidal waves caused by tides?**
   The term *tidal wave* is an unfortunate synonym for a tsunami. Tides and tsunami have entirely different causes (see Chapter 10). The term *tidal wave* has its origins in the similarity of an approaching tsunami to a rapidly rising tide. Tsunami were often mistaken as "unusual" tides until public understanding of their causes and behavior corrected the misconception. Still, the term *tidal wave* remains in popular use (try Googling it).

3. **Why do many marine organisms time their spawning to the tides, and how do they know what tide it is?**
   Synchronous spawning has two principal advantages. First, it maximizes the chances for fertilization when eggs and sperm are released at the same time or maximizes the opportunity for mating when males and females arrive "on location" at the same time. Second, the enormous quantities of sperm and eggs produced overwhelm predators that feed on the gametes to the point of satiation, at which point the survivorship of remaining gametes increases. Similarly, spawning aggregations of squid and grunion may overwhelm their predators, a kind of safety in numbers. Scientists are not certain how organisms "know" the precise time for spawning. Bottom-dwelling species such as corals may respond to hydrostatic pressure, or they may simply react to temperature changes that occur when strong tidal currents flow over a reef. Lunar light has also been postulated as a stimulus for spawning. For some species, multiple triggers may be required. Sensitivity to a stimulus may require preconditioning, such as a consecutive number of days where seawater temperatures exceed a particular value or similar clues. Synchronous spawning remains a fascinating and intriguing behavior for scientific study.

Key Concepts

- Tidal patterns vary across the world ocean and include daily-, monthly-, and decadal-scale cycles.
- Tidal patterns result from the time-varying, individual contributions of the tidal constituents.
- Tide-producing forces, called tidal forces, result from *differences* in gravitational attraction at various points on Earth's surface. Only the horizontal component of the tidal forces is important to ocean tides.
- The Moon exerts about twice the effect on tides as the Sun. Nonetheless, the gravitational pull of the Sun is about 179 times that of the Moon.
- Sea level represents the average level of the sea with variations due to waves removed.
- Observed sea level may change as a result of tides, storm surge, and meteorological and oceanographic factors.
- Sea level appears to be rising at about 1.5 mm (0.06 inches) per year.

Terms to Remember

amphidromic point, 229

attribution problem, 231

axial tilt, 224

co-range line, 229

co-tidal line, 229

declination, 224

diurnal tide, 221

dynamical theory of the tides, 220

earth tides, 227

ebb tide, 226

ecliptic, 224

equilibrium theory of the tides, 220

eustatic change, 231

flood tide, 226

full moon, 223

geoid, 230

glacial isostatic adjustment, 231

high tide, 221

interglacial, 231

Kelvin wave, 229

low tide, 221

M_2 tide, 222

mean lower low water, 230

mean sea level, 230

mixed semidiurnal tide, 221

National Tidal Datum Epoch, 230

neap tide, 223

new moon, 223

node, 229

quarter moons, 224

reference ellipsoid, 230

S_2 tide, 222

sea level, 230

semidiurnal tide, 221

sidereal month, 223

slack tide, 226

spring tide, 223

standing wave, 228

steric change, 231

synodic month, 223

tidal bore, 226

tidal bulges, 220

tidal constituents, 222

tidal currents, 219

tidal datum, 221

tidal day, 222

tidal forces, 226

tidal lag, 228

tidal period, 222

tidal range, 222

tidal wave, 228

tide gauge benchmark, 230

tide heights, 221

tides, 219

Critical Thinking

1. What patterns of tides would Earth experience if no Moon were present?
2. How has astronomy contributed to our understanding of the multidecadal patterns of tides? Defend the notion that a study of ocean tides is a study of solar system astronomy.
3. What are the limitations of the equilibrium model of tides? How does the dynamic model of tides overcome those limitations?
4. What is tidal prediction? How accurate is it? What factors make tidal predictions inaccurate at a given time or location?
5. Why is sea level so difficult to measure? Design an apparatus to measure the "sea level" of water in your bathtub. What complications might arise in trying to determine your bathtub sea level?
6. Compare and contrast the life of a marine animal in a diurnal tidal pattern and a mixed semidiurnal pattern. How might the physical, chemical, and biological environments of the animal differ under the two tidal patterns? Include both daily and annual cycles in your discussion. Use tide predictions from actual locations to support your answers.

Explore Online

 Visit www.mhhe.com/chamberlin1e for access to chapter quizzing, key term flash cards, video clips, interactive activities, and more. Further enhance your knowledge with web links to chapter-related material!

Exploration Activity 11-1

Exploring Sea Level Through Concept Mapping and Data Investigation

Question for Inquiry

What are the effects of global warming on sea level?

Summary

Sea level rise and fall, as described in the last section of this chapter, involves a complex number of geological, physical, and chemical factors over a wide range of temporal and spatial scales. Historically, tide gauges have provided long-term information on sea level fluctuations but these data have been limited to coastlines. The TOPEX and Jason satellites have now enabled oceanographers to measure sea level over broad geographic scales but these data span only a few decades. In this activity, we explore both of these types of measurements with the goal of developing a deeper understanding and appreciation for addressing the impacts of global warming on sea level.

Learning Objectives

By the end of this activity, you should be able to:

- Analyze and interpret tide data from tide gauge stations and ocean buoys
- Analyze and interpret satellite data on sea surface height
- Analyze and interpret long-term sea level trends at various locations in the world ocean
- Summarize the strengths and limitations of current sea level measurement technology
- State an opinion over the severity of sea level rise for human concerns

Materials and Preparation

For this activity, you will need access to the Internet and a printer. Alternatively, your instructor may provide a printout of figures sufficient for completing the activity. You may want to review the sections on tide patterns and sea level in this chapter. You may also want to review the section on Concept Mapping in the Exploration Activity for Chapter 4.

Part I: Create a Concept Map for Sea Level

Figure 11-20 presents some of the factors that influence sea level in the world ocean. Take a few moments to review this figure and the accompanying text. You may also want to check out Wikipedia's excellent summary on sea level rise (http://en.wikipedia.org/wiki/Sea_level_rise) and the section of the 2001 report on sea level rise produced by the Intergovernmental Panel on Climate Change (www.grida.no/climate/ipcc_tar/wg1/408.htm/) From these references, create a concept map with a box for Sea Level in the center. Create additional boxes and links that relate to factors that alter sea level and ways in which it is measured. For example, you may create a box called "glaciers melting" with an arrow labeled "cause a rise in" connected to Sea Level. Make sure to revise your concept map as you learn more in the activities that follow. Use the Rubric for Physical and Conceptual Models to refine your work.

Part II: Exploring Tide Gauge Data

NOAA's CO-OPS provides a wealth of data on tides, currents, and sea level on their website, http://tidesandcurrents.noaa.gov/. Because much of your historical understanding of sea level rise is based on tide gauges, we start by learning how to access and analyze tide data.

1. Go to the CO-OPS website and browse the About Us section. What are the primary goals of CO-OPS and what measurements and data products do they provide? Why is their mission important to the general public? Review the pages on Measuring Tides in the Education section of the site. Provide a brief summary.
2. From the home page, click on Tides Online. Report on any high water advisories or storm tide alerts. Storm

surge is the #1 killer during hurricanes so these alerts provide a very important service.

3. Find the link for State Maps and go there. You will see a clickable map of the United States. Choose one station each from the western US, the eastern US, the Gulf coast, and an island state or US territory for a total of four stations. Click on each station and print or save the tide chart. You may also want to print or save the data table associated with each station.

4. For each chart, create a table that includes the following:
   a. The time and height of the predicted low and high tides (record all that occur)
   b. The time and height of the observed low and high tides
   c. The difference between predicted and observed low and high tides (otherwise known as the residuals)
   d. The tidal range for a given tidal period (height of high tide minus height of next low tide)
   e. Label each table with the type of tidal pattern you observe (see Figure 11-4)

5. Compare and contrast your four stations in terms of tidal ranges. Where do the greatest tidal ranges occur and where do the smallest occur? Explain these variations in terms of what you learned about the monthly tidal cycle and declination in this chapter.

6. Now visit the Sea Level Trends section of their website at http://tidesandcurrents.noaa.gov/sltrends/sltrends.html. Find the same four stations (or sites nearby) and print or save 1) the mean sea level trend; 2) the seasonal sea level trend; 3) the interannual variation of mean sea level for all data; and 4) the interannual variation of mean sea level since 1980. Explain the difference between each type of graph (in other words, describe how each graph differs in its representation of sea level data).

7. Record your observations of sea level trends in each of the four charts from each of the four locations. Is sea level rising/falling over time? Why is sea level higher sometimes than other times? Why is it important to remove seasonal effects? What kinds of phenomena or processes may be responsible for the observed variations in sea level trends.

8. Find the mean sea level anomalies map on the same page as the mean sea level data. Print or save the map. What is shown on this map? Record your observations of sea level anomalies along the coastlines of the US, noting especially regions where sea level is higher or lower than average. Use the drop down menu to view maps from each year previous in the current month. Record your observations. Now view the animation of mean sea level anomalies and record your observations.

9. Find the link for global stations and record observations for Narvik, Norway and Mera, Japan. Record the mean sea level trends for these stations. Why is sea level rapidly falling in Norway and rapidly rising in Japan? What is happening at these locations that affects sea level?

10. Based on your observations, provide an assessment of sea level trends globally. What magnitude and trend of sea level is supported by your data?

Part III: Exploring Satellite Sea Surface Height Data

Although tide gauge data have provided excellent insights into the nature of sea level variations at specific locations over the past 100+, satellite measurements have "opened our eyes" to the extent of sea level variability over broad spatial scales. In this part, you will explore satellite measurements of sea surface height based on radar altimetry (see Figure 9-8).

1. Visit the University of Colorado Sea Level Change website at http://sealevel.colorado.edu/. This site is a excellent source of scientific references and websites pertaining to sea level research, a great topic for further inquiry.

2. Click on the time series link. Both global and regional data are available (regions shown as colored sections on a map). You will see a term on this page that may not be familiar to you: inverted barometer effect. This term compensates for atmospheric effects on sea level due to air pressure (which are relatively minor but important to take into account). What do you think happens to sea level when atmospheric pressure is higher? Check your answer in the online glossary of meteorology, http://amsglossary.allenpress.com/glossary.

3. Click on the image (jpg) for global mean sea level that corrects for seasonal and barometric factors. Describe the various parts of the graph. What are the axes? What is the difference between red circles and green squares? What is 60-day smoothing? How do the circles, squares, and 60-day smoothing line vary in relation to one another? What does the straight black line indicate? Record the rate of sea level rise provided on the graph.

4. Click on the maps section. Describe the image with the inverted barometer applied. What does this map show? In what regions of the world ocean is sea level falling? In what regions of the world ocean is it rising? Locate your tide gauge stations from Part II. Does the satellite map agree with the trends you observed in your tide gauge data?

Part IV: Exploring the Applications of Sea Level Research

The websites you visited provide only a glimpse of the kinds of information oceanographers are using to determine changes in sea level on local and global scales. Far from a scientific exercise, sea level research has enormous practical applications.

Low-lying coastlines and islands stand to become permanently inundated if the current rise in sea level continues or accelerates. Sea level research provides information in the preparation and mitigation of future impacts on coastlines.

1. Visit NASA's Ocean Surface Topography from Space website at http://sealevel.jpl.nasa.gov/. This site features a wealth of information on the TOPEX/Poseidon and Jason missions, the tracking of El Nino/La Nina, and the links between sea level and climate change. Spend some time browsing the site. Review carefully the sections on sea level and the section by Dr. Lee-Lueng Fu on the links between ocean circulation, sea level, and climate change.

2. Write a two-page paper on the importance of sea level research to human affairs. Include a brief summary of the sea level data you observed and answer these questions: What appear to be the trends of sea level over the decades? How does sea level vary from location to location? Explain the difference between daily tides and sea level trends. How are these measurements related? In what ways are measurements of tides and sea level important to human affairs? How might climate change affect sea level and present a danger to human property and lives?

12

Ocean Life and Its Evolution

A mob of starfish wrestle for a cool moist crack in the rocks. The evolution of biochemical, physiological, and behavioral adaptations for acquiring water is just one of the many examples of survival in the world ocean.

Questions to Consider

1. Why is the theory of evolution considered to be as revolutionary and all-encompassing as the theory of plate tectonics?

2. How are the tools of molecular biology helping oceanographers to better understand the diversity of marine life in the world ocean?

3. What forms of life exist in the world ocean?

4. Why is the classification of life in the world ocean a difficult task?

5. What adaptations do organisms exhibit that help them survive and reproduce in the world ocean?

From Microbes to Mammals

The world ocean represents the single largest habitat for life on Earth. Moreover, as far as we know, it is the single largest habitat for life *anywhere* in the universe. Yet, from our terrestrial perspective, the world ocean seems a forbidding and hostile environment for life. Crushing pressures, near-freezing temperatures, complete absence of light, and a scarcity of food characterize most of the world ocean. The challenges of survival in the world ocean are considerable, yet a rich assemblage of life persists. The world ocean houses a startling variety of microbes whose nature and abundance we are just beginning to appreciate and understand. It is also home to the largest animal that has ever existed on Earth—the blue whale. Where did ocean life come from and why is it so diverse?

The diversity of life found in the world ocean today represents the species that have survived and evolved over geologic time. The survivors have one trait in common: they are living and reproducing within the range of conditions that exist in the world ocean today. By all counts, they are a select group. Scientists estimate that 99.9% of all species that have ever existed on our planet are now extinct. Yet the survivors did not persist by mere chance. They were favored through **natural selection,** an interaction between organisms and their environment that produces differential survival and reproduction. While chance plays a role in producing a favorable characteristic through small alterations in an organism's genetic make-up, chance plays less of a role in natural selection, which produces differential survival and successful reproduction of the best-suited individuals. **Evolution,** the inheritable changes of a population of organisms over many generations, involves both replication and interaction. Replication transmits variations from parent to progeny, and interaction determines which of those variations survive. In this sense, evolution is a two-step process. The evolution of life on Earth is intimately tied with the evolution of the geological, physical, chemical, and biological processes throughout Earth's history. Accordingly, the mutual interaction of life and the geology and chemistry of Earth are an inescapable part of the evolution of life. Scientific understanding of the coevolution of organisms and the surrounding environment, the science of **geobiology,** has just begun. This exciting, new interdisciplinary science promises to revolutionize scientific thinking about the history of life on Earth.

The Foundations of Evolutionary Theory

In 1814, Prussian naturalist, explorer, and oceanographer Alexander von Humboldt (1769–1859) proposed an affinity between different species of plants and their geography. The idea that the distribution of organisms was influenced by their physical surroundings was a new one. Humboldt's ideas laid the foundations of modern **biogeography,** the study of the distribution of organisms in relation to their environment. British naturalist Edward Forbes (1815–1854) was among the first scientists to study the distribution of plants and animals within the sea. His 18-month study of the Aegean Sea, part of the Mediterranean Sea between Greece and Turkey, published in 1843, led him to conclude that geological events explained the distribution of organisms. Forbes, like many naturalists of his time, believed in the "immutability" of life: every species present had always been present, and all species, born of a single set of parents, had migrated from their centers of origin to their present locations. Where species distributions were not continuous, Forbes hypothesized that a geological upheaval had produced a barrier between them. This hypothesis is not supported by modern evidence. Despite evidence to the contrary, Forbes also proposed that the deep sea was lifeless, the "azoic" zone as he called it. (Perhaps Forbes was not aware of deep-sea collections clearly indicating life on the deep seafloor.)

During this period, British naturalists Charles Darwin (1809–1882) and Alfred Wallace (1823–1913) were developing their ideas on the origins and extinctions of species by the process of natural selection. Central to their ideas was the growing acceptance of geologic time and uniformitarianism. First proposed by James Hutton (see Chapter 2) and later championed by Oxford University geologist Charles Lyell (1797–1875) in *Principles of Geology,* uniformitarianism emphasized gradual change over long periods of time. Ironically, a key piece of evidence in support of evolutionary theory came from an outspoken opponent of uniformitarianism, French naturalist Georges Cuvier (1769–1832). According to Cuvier, Earth had undergone several catastrophic upheavals in the past, such that some species survived and others went extinct. Inadvertently, Cuvier had planted the "revolutionary" idea that some species did not survive through geologic time.

Charles Darwin studied carefully Lyell's *Principles of Geology* (1830) during his epic five-year voyage aboard the *HMS Beagle* (during which time he was terribly seasick). In the Cape Verde Islands, Darwin observed alternating layers of basalt and calcareous sediments. This observation convinced him that Earth changed gradually through geologic time, building up by volcanoes and grinding down by erosion. At various stopovers, Darwin made extensive observations on plant and animal species. He was particularly intrigued by three species of Galapagos Island mockingbirds, each inhabiting a different island and bearing a resemblance to ones he had seen in South America. Where did these birds come from and why was each of them distinctly different? Darwin could not be sure, but his observations led him to pursue these questions with other species. In 1839, Darwin published *Voyage of the Beagle* (as it is now known). In this book,

Darwin presented extensive observations on barnacles, an animal he studied for eight years (**Figure 12-1**).

Darwin's study of barnacles demonstrated that slight differences in body parts allowed different species to inhabit different environments. For Darwin, the incremental differences in the shells of different barnacle species were proof of evolution and descent from a common ancestor. What's more, Darwin observed that even within a species, there was considerable variety from

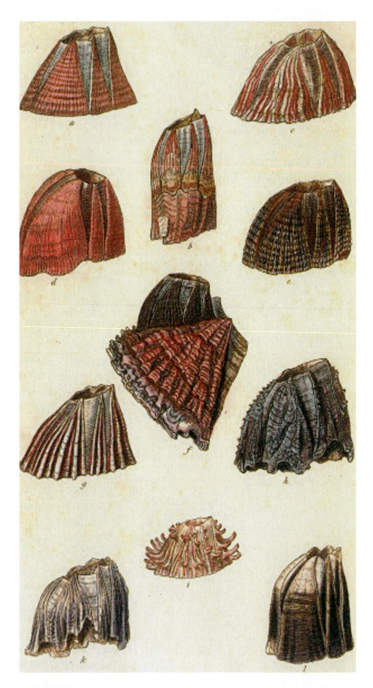

■ **FIGURE 12-1** Darwin's barnacles. Through careful analysis of the shape of the individual components of the shells of barnacles, Darwin discovered and described new species and formulated ideas concerning the nature of evolution and natural selection. Although Darwin is best known for his studies of Galapagos finches, his eight-year study of barnacles established his credentials as a scientist. For his work, Darwin was awarded the Royal Medal of the Royal Society in 1854.

one individual to the next, an important principle of evolutionary theory. He surmised that environmental conditions acted against the survival of *some* individuals of a species. For example, a "tall" barnacle might be more easily dislodged from a rock in a high-energy environment than a "short" barnacle. This "disfavoring" of certain individuals of a species reduced the likelihood that they would survive and reproduce. Those individuals best adapted for an environment continued to produce progeny. Over many generations, variability within species led to evolution of new species. These ideas reached a feverish pitch in 1858, when Darwin's colleague, Alfred Wallace (1823–1913), on a collecting expedition in Indonesia, penned a quick letter to Darwin outlining some new thoughts on the adaptedness of species. Alarmed that he was about to be scooped, Darwin allowed his and Wallace's ideas to be presented to London's Linnean Society in July 1858 (without Wallace's permission, although he was told later). Darwin was pressed to complete his work, and in 1859, he published *On the Origin of Species by Natural Selection.* Like Wegener's theory of plate tectonics for geology, Darwin's theory of natural selection explained a wide number of seemingly different biological phenomena. Both Wallace and Darwin are credited with developing the first theories regarding the evolution of life on Earth by natural selection (see **Table 12.1**).

Nonetheless, acceptance of Darwin and Wallace's ideas was not immediate. While plant and animal breeders of Darwin's time accepted artificial selection to produce superior breeds (more fruit, faster growth, longer wool, etc.), the leap from improving a breed to creating new species was difficult to accept. A mechanism was needed for passing on new characteristics from one generation to the next in a way that allowed the accumulation of changes over time. Discovery of the principles of heredity by Austrian monk and "father" of genetics Gregor Mendel (1822 –1884) paved the way for understanding how natural selection operates. Mendel's work on flower color in pea plants and subsequent work by geneticists in the early 1900s revealed that "novel" characteristics could be passed on from one generation to the next in a discrete and predictable fashion. In the 1930s, three scientists contributed major works that established a modern theory of evolution incorporating gene theory and population genetics, the **modern synthesis,** also known as neodarwinism. British geneticist Sir Ronald Fisher (1890–1962) established the role of genes in *The Genetical Theory of Natural Selection* (1930). Another British geneticist, J.B.S. Haldane (1892–1964), wrote *The Causes of Evolution* (1932), in which he applied Mendelian genetics to explain evolution by natural selection. The third founder of the modern synthesis, American geneticist Sewall Wright (1889–1988), contributed to theoretical population genetics and the role of mutations in natural selection. In the decades that followed, dozens of prominent scientists, including Thedodosius Dobzhansky (1900–1975), Ernst Mayr (1904–2005), George Gaylord Simpson (1902–1984), and John Maynard Smith (1920–2004), among others, advanced scientific knowledge of evolutionary theory. Within 100 years of Darwin's and Wallace's publications, the science of modern evolutionary biology was firmly established.

An enormous body of evidence accumulated over nearly 150 years supports evolution as a fact. Nonetheless, the mechanisms by which natural selection operates continue to be debated vigorously. As one example (and as an example of how science never completely shuts the door on a theory), Cuvier's catastrophe theory has gained greater support in modern times. Scientists now recognize the possible role of meteorites (and other catastrophic events) in producing **mass extinctions,** the sudden disappearance of a majority of living species (**Figure 12-2**).

At the same time, scientists recognize through the fossil record long periods of geologic time during which species persist unchanged and short periods during which speciation is rapid. This "fits-and-starts" style of evolution, championed in 1972

TABLE 12.1   Comparison of Darwin's Five Theories, Species Versus the Modern Synthesis

Darwin's theories	Modern synthesis
1. Species change through time.	1. Characteristics of organisms are determined by their genes, sequences of DNA (deoxyribose nucleic acid) that code for proteins and govern cellular functions.
2. All organisms descend from a common ancestor.	2. Genes are the units of heredity; they are passed on from one generation to the next.
3. Evolution is gradual.	3. Mutations in the DNA of genes create variability among individuals.
4. Species diversify.	4. Several mechanisms, including but not limited to natural selection, may alter the frequency of genes and their variants.
5. Natural selection increases the frequency of individuals best suited for an environment and decreases the frequency of ill-suited individuals.	5. New species arise through the gradual accumulation of gene mutations.

Source: After talkorigins.org

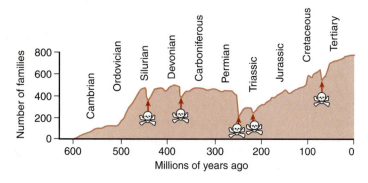

■ **FIGURE 12-2** At least five mass extinctions are recorded in the fossil record. There is some evidence that the disappearance of dinosaurs and two-thirds of all marine animal species at the end of Cretaceous was caused by a meteorite impact in the Gulf of Mexico. Some scientists fear that a sixth mass extinction is underway because of global-scale environmental change and rapid loss of biodiversity caused by humans.

by Niles Eldredge and Stephen J. Gould (1941–2002), is called **punctuated equilibrium.** Controversial to this day, punctuated equilibrium remains a valid hypothesis for explaining some aspects of the evolutionary record.

The Tree of Life

Darwin's discovery that living organisms change through time in response to natural selection provided evidence and a mechanism for the **theory of common descent,** the evolution of organisms from a common ancestor. While Darwin's theory of evolution gained widespread acceptance, it wasn't until the advent of **molecular phylogeny**—the mapping of evolutionary relationships based on the genetics and biochemistry of organisms—that the theory of common descent gained firm footing. Today, a wide body of evidence supports a common ancestor for life on Earth.

Classification and Systematics

The evolution of life also provides a basis for arranging related organisms into groups, a process called **classification.** By studying the phylogenetic relationships of organisms, biologists can also study the patterns of change and the evolutionary history of life on Earth, a field of study called **systematics.** Classification is just one of the many tools systematists use to study the history of life.

The system of classification most widely known, the hierarchical classification of organisms, was first introduced in 1735 by Swedish botanist Carl Linnaeus (1707–1778). Linnaeus proposed an arrangement of organisms from the most general to the most specific (**Table 12.2**). At the most specific level, Linnaeus gave every organism a two-part name. In this scheme, every species of organism is given two names, a genus name and a species name, such as *Homo sapiens.* Using the characteristics of organisms, scientists can construct a **phylogenetic tree,** a representation of the "branching" or diversification of species of organisms from their common ancestor.

TABLE 12.2   Examples of the Hierarchical Classification of Organisms

Organism	Taxonomic level	Example
Prochlorococcus marinus, the smallest known photoautotroph	Domain	Bacteria
	Kingdom	Eubacteria
	Phylum	Cyanobacteria
	Class	Prochorophytes
	Order	Chroococcales
	Family	Prochlorococcaceae
	Genus	*Prochlorococcus*
	Species	*marinus*
Blue whale, the largest animal on Earth	Domain	Eukarya
	Kingdom	Animalia
	Phylum	Chordata
	Class	Mammalia
	Order	Cetacea
	Family	Balaenidae
	Genus	*Balaenoptera*
	Species	*musculus*

In modern times, evolutionary scientists prefer an alternative system of classification called **cladistics,** essentially, a study of the genealogy of organisms. Cladistics seeks to arrange organisms according to their similarity (or lack of similarity) and the presence of ancestral versus derived features. Cladistics can be based on morphological features, biochemical characteristics, genomic features, or other characteristics of a species. One of the advantages of cladistics is that it allows scientists to use mathematical tools to study the relationships between organisms (**Figure 12-3**). Cladistics is a means for testing hypotheses of evolutionary relationships and exploring them in a quantitative manner.

Molecular Approaches to Classification

The latter decades of the twentieth century experienced rapid growth in the application of techniques borrowed from molecular biology and genetics for studying the diversity of the world ocean. **Ocean genomics,** the study of the genetic material of marine organisms, has quickly progressed as a tool for exploring the diversity and phylogenetic relationships of marine organisms. At the same time, ocean genomics is providing clues about the adaptations and ecology of marine organisms, especially microbes. Genomics allows oceanographers to determine the diversity of microbial spe-

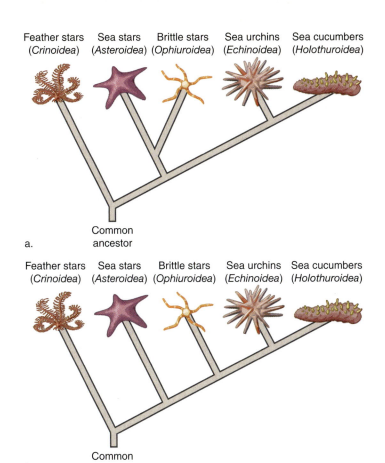

FIGURE 12-3 Cladistic analysis based on morphological descriptions (a) or genetic information (b) yield different phylogenetic trees for echinoderms, a group of organisms that includes sea stars. Disagreements over the ancestry of organisms force scientists to look for additional clues to settle the matter. The end result is a much improved understanding of organisms and their evolution.

cies present. Typically, identification of a microbial species requires that a scientist grow them in a laboratory culture. Unfortunately, some 90% of all marine bacteria cannot be cultured by current techniques. Ocean genomics provides a way of determining the number of genes in the microbes inhabiting a particular region of the ocean, a result that can be applied to estimating species diversity. By collecting seawater samples, extracting the bulk genetic materials, and applying techniques such as **whole-genome shotgun sequencing,** a means for rapidly determining the sequence of nucleotides, the gene sequences can be established (**Figure 12-4**). A study in the Sargasso Sea off Bermuda revealed at least 1.2 million previously unknown gene sequences, estimated to represent at least 1,800 new species of marine microbes. Such studies suggest that an enormous diversity of life-forms exists in the world ocean.

Life in Three Domains

All of life may be divided into three **domains:** Archaea, Bacteria, and Eukarya (**Figure 12-5**). Though not universally adopted, the three-domain system gives prominence to the **Archaea,** an ancient and important group of microbes found throughout the world ocean (**Table 12.3**). Most Archaea inhabit regions of Earth known for their extreme conditions. Thermophiles prefer the hot and acidic conditions that scientists believe were widespread at Earth's formation. The

FIGURE 12-4 Techniques borrowed from molecular biology are now routinely used aboard oceanographic vessels for identifying and characterizing populations of marine microbes. The large quantities of biological material needed to conduct molecular analyses require oceanographers to collect hundreds of liters of water.

methanogens produce methane from carbon dioxide and hydrogen, and have been implicated in maintaining Earth's anoxic, methane-rich atmosphere from 3.5 to 2.5 billion years ago (see Chapter 2). The **Bacteria** introduced oxygen into Earth's atmosphere through the evolution of oxygenic photosynthesis (**Table 12.4**). They also became active participants in the global cycling of elements and have been called "nature's recyclers" for their role in the decomposition of organic matter. The Bacteria inhabit the widest range of habitats, from deep within Earth and beneath the seafloor to the upper atmosphere. The domain **Eukarya** includes humans and the organisms most familiar to us, including single-celled eukarya, plants, animals, and fungi (**Table 12.5**). The cells of Eukarya exhibit visible internal structures called **organelles.** A cell's nucleus, which contains the DNA, is an example of an organelle. The eukaryotic cell type has been hypothesized to have evolved from a joining together of **prokaryotes,** cells lacking organelles (e.g., Bacteria and Archaea), a process called **endosymbiosis.**

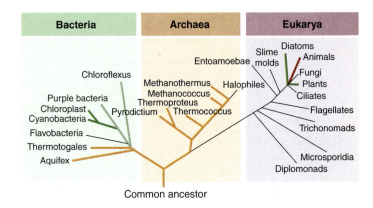

FIGURE 12-5 The three domains of life, based on molecular phylogeny.

TABLE 12.3 Major Groups of Marine Archaea

Oceanographers are just beginning to appreciate the metabolic diversity of the Archaea and their ecological significance in ocean food webs.

Major groups	Description
Crenarchaeota	Low-temperature pelagic forms found throughout water column in open ocean; psychrophiles found in polar regions
Euryarchaeota	Includes thermophiles, thermoacidophiles methanogens, and halophiles; found in hydrothermal vent and subseafloor habitats

TABLE 12.4 Major Groups of Marine Bacteria

Until the advent of molecular techniques for studying marine bacteria, most species went unnoticed and unstudied. While isolating and growing marine bacteria species continues to challenge oceanographers, inferences about their metabolic pathways and ecological role can be obtained by studies of their biochemistry and molecular biology.

Major groups	Description
Acidobacterium	Inhabit marine snow and sediments
Actinobacterium	*Kocuria marina*, a halotolerant species is found in marine sediments
Aquificales	Found on deep-sea hydrothermal vents
Chlamydia	Pathogenic; infects marine mammals
Cyanobacteria	Both coccoid and filamentous forms occur. Filamentous forms are important in nitrogen fixation. Coccoid forms contribute greatest percentage of photosynthetic carbon fixation in world ocean
Cytophagales	Abundant; free-living and attached to marine snow
Fibrobacter	Free-living; some species associated with deep chlorophyll maximum
Flexistipes	Halophilic, iron-reducing (FeIII) obligate anaerobe found in marine sediments
Green Non-Sulfur	Inhabit marine sediments and hypersaline environments
Green Sulfur	An extremely low-light adapted species found at 100 m (328 feet) in the Black Sea may carry out the slowest photosynthesis on Earth; also found on hydrothermal vents, where they have been speculated to photosynthesize on light generated by vent reactions
Low Guanine-Cytosine	Inhabit marine biofilms on the surface of submerged objects and organisms
Gram-Positive Nitrospira	Nitrite-oxidizing bacteria important in the nitrogen cycle of marine environments and saltwater aquariums
Planctomycetes	Inhabit marine snow and phytodetritus
Proteobacteria	The most abundant and diverse group in the ocean
Spirochaetes	Free-living and parasitic on a variety of organisms, including sponges, mollusks, and marine birds
Thermotogales	Inhabit hydrothermal vents, oil reservoirs, and marine sediments
Thermus/Deinococcus	Inhabit hydrothermal vents
Verrucomicrobia	Inhabit sulfur-rich muds and marine sediments

TABLE 12.5   Major Groups of Marine Eukarya

Number of species is based on described species, as reported in the *World Atlas of Biodiversity* (2002) and other sources. Millions of species remain undescribed. Classification of the Eukarya has undergone major revisions in recent years. Based on ultrastructural characteristics and molecular phylogenies, the International Society of Evolutionary Protistology (ISEP) has published a new classification scheme (Adl et al., 2005) in which the Eukarya consist of six major groups: the Amoebozoa (naked and testate amoeboid forms); the Opisthokonta (including the fungi and all animals); the Rhizaria (including the foraminifera, radiolarian, and other groups); the Archaeplastida (including the green "algae" and other groups); the Chromalveolata (including the cryptomonads, diatoms, dinoflagellates, and other groups); and the Excavata (euglenoids and other groups). The ISEP system will likely prove controversial, but its emphasis on the "microbial" dominance of life on Earth bears importance to studies of life in the ocean.

Major groups	Approximate number of marine species	Description
Single-celled Eukarya (Protists)	35,500	Single-celled multicellular algae (seaweeds); a catchall group for organisms that do not easily fit into other groups
Fungi	1000s	Once thought fairly rare, marine fungi play an important role in the decomposition of cellulose and other detrital plant material
Angiosperms	255	Seagrasses, mangroves, salt marsh plants
Porifera	10,000	Sponges
Cnidaria	10,000	Corals, jellyfish, sea anemones
Ctenophora	100	Comb-jellies
Platyhelminthes	15,000	Flatworms, mostly benthic
Dicyemida	50	Internal parasites on the kidneys of benthic cephalopods
Nemertea	750	Ribbon worms
Mollusca	100,000	Bivalves, gastropods, cephalopods, chitons, tusk shells
Annelida	12,000	Polychaete worms, feather-dusters
Echiura	150	Spoon worms
Sipuncula	150	Peanut worms
Arthropoda	38,000	Crustaceans
Nematoda	12,000	Roundworms
Nematomorpha	250	Horsehair worms
Priapulida	10	Penis worms
Loricifera	10	Brush-head worms
Phoronida	20	Horseshoe worms
Kinorhynchs	150	Mud dragons
Brachiopoda	350	Lamp shells
Bryozoa	4000	Moss animals
Chaetognatha	100	Arrow worms
Hemichordata	100	Blood worms
Echinodermata	7000	Sea urchins, sea stars, brittle stars
Chordata	17,000	Urochordates (sea squirts, pelagic chordates) and vertebrates

Ocean Biodiversity

Marine organisms encompass a vast number of species, from microbes to marine mammals. A satisfactory introduction to marine life requires many semesters of study (e.g., Marine Microbiology, Algology, Invertebrate Zoology, Vertebrate Zoology, Marine Mammalogy, Marine Ecology, etc.). Here, we consider marine life according to their three predominant modes of living. The three groups—the benthos (sessile), plankton (drifting), and nekton (swimming)—include representatives from nearly every phylum of organisms. These "lifestyles" provide useful insights into how organisms solve the problem of living in the marine environment. Be aware that even this scheme is overly simplified: many animals belong to two categories and some defy this classification scheme altogether. Nonetheless, this approach illustrates the broadest of adaptations of marine organisms for surviving and reproducing in the various habitats of the world ocean.

The Benthos

Animals that live on, within, or attached to the seafloor are called **benthos** or benthic organisms. The benthic environ-ment is defined largely by the type of substrate of which it is composed. Benthic environments composed of sediments (silts, muds, oozes, sands, etc.) house a **soft-bottom benthos,** organisms adapted for life on, within, or between sediment grains. Cobbles, boulders, and solid rock, and the variety of morphologies associated with rocky environments (e.g., caves, shelves, walls, crevices, etc.), create a habitat for the **hard-bottom benthos.** Benthic organisms may also be classified according to where they live. Organisms that live upon the substrate are called **epifauna.** Those that live within the substrate are called **infauna.** The infauna typically inhabit soft bottoms where they may burrow into or move freely within the sediments. Infauna may also include organisms that can penetrate solid substrates, such as rock-boring clams. Epifauna may occur on either soft or hard bottoms where they live or travel upon the surface of sediments or rock. Some very tiny organisms may even live between the grains of sediments. These **interstitial organisms** can be found from the seashore to the abyss.

Examples of epifauna include many of the organisms you see along rocky shores in tide pools, such as sea stars, sea anemones, crabs, mussels, and barnacles (**Figure 12-6**). Relatives of these animals can be found practically everywhere, from submarine canyons to hydrothermal vents. Much of what we know about

a.

c.

b.

d.

■ **FIGURE 12-6** Various types of benthic organisms: (a) basket sponge, (b) brain coral, (c) Christmas tree worm, (d) mud crab.

deep-sea epifauna comes from studies using ROVs, sleds mounted with cameras, and submersible observations. Infauna are more difficult to observe directly, though they often leave telltale signs. Clams, ghost shrimp, numerous types of worm, and some surprisingly large crustaceans, such as deep-sea benthic amphipods, can be found in the mud. Infauna are most often recognized by jets of debris or water coming from holes or by fecal deposits left on the surface.

For practical reasons, benthic organisms are also classified according to their size. Oftentimes, the only way to obtain organisms for study is to grab a sample of the mud in which they live. To separate the animals from the mud, oceanographers employ sieves with various mesh sizes that capture (or permit to pass) organisms of certain sizes. The size categories of benthos can be found in **Table 12.6.**

Of these size classes, the meiofauna represent numerically one of the most diverse and abundant groups of organisms on Earth. At least 20 animal phyla are represented by meiofauna, five of which belong exclusively to this size class. Examples include tiny mollusks, numerous small crustaceans, and many types of small worms, including nematodes and flatworms. Meiofauna play a vital role in the health and vitality of beaches, potentially eliminating harmful human parasites and helping to decompose organic matter. They are also an important food source for juvenile fishes.

The Plankton

Animals whose movements are governed by wind, currents, or tides are called plankton. (**Figure 12-7**). The **bacterioplankton,** the marine bacteria that inhabit the water column, represent the most numerous and diverse of all the plankton. Their role as primary producers and as a food source for marine protists has gained widespread attention since the 1970s. Marine viruses, though technically not alive, have gained attention for their role

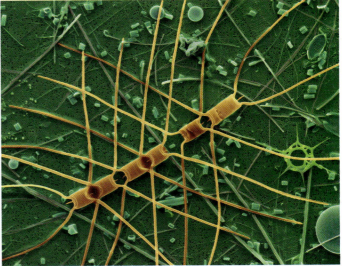
a.

b.

c.

TABLE	**12.6**	**Major Size Classes of Benthos** *
Macrofauna		Animals retained on a mesh larger than 500 μ m
Meiofauna		Animals that pass through a 500-μm mesh but are retained on a 62-μm mesh
Microfauna		Animals that pass through a 62-μm mesh

*The size ranges of benthos may differ among various scientists and authors. The sizes presented here are the ones defined by the International Association of Meiobenthologists, an organization dedicated to the study of meiofauna.

■ **FIGURE 12-7** Various types of plankton: (a) Antarctic diatoms and a silicoflagellate, (b) krill, (c) jellyfish.

in controlling marine bacteria. Bacterial lysis by **virioplankton** may be an important source of dissolved organic matter in the world ocean (see Chapter 6).

The **phytoplankton** represent the single-celled photosynthetic drifters. Though the term *phytoplankton* literally means "plant drifter," phytoplankton are not, technically speaking, plants. Their numbers consist of two groups: the photosynthetic bacteria (Domain Bacteria) and the photosynthetic "protists" (Domain Eukarya). Because of their importance to marine food webs, biogeochemical cycles, and climate, we explore the phytoplankton in greater detail in Chapter 13.

The **zooplankton** represent the animal drifters. Zooplankton inhabit the three dimensions of the water column in the world ocean. Approximately 5000 species of "true" zooplankton are known. Like the benthos, zooplankton can be classified ataxonomically, according to their habitat or size. However, zooplankton may also be categorized according to their body type ("soft" or "hard") and their length of stay in the plankton. The classification of zooplankton by body type is often overlooked, but increasing recognition of the importance of gelatinous forms in the sea merit greater awareness of this distinction. Animals whose bodies are gelatinous or saclike and whose form depends on the buoyancy of water to keep it intact are known as gelatinous zooplankton or **jellyplankton.** The most well-known jellyplankton include the jellyfish and ctenophores, but many other animals, including siphonophores, the pelagic invertebrate chordates (salps, dolioloids, and larvaceans), and the soft-bodied larvae of benthic forms, may be abundant as well. Until the advent of scuba and ROVs, the abundance and diversity of jellyplankton was poorly known, owing to the tendency of their bodies to fall apart in plankton nets.

The hard-body forms represent the "classic" zooplankton and include the shelled protozoan protists (foraminifera and radioloria), a few pelagic polychaetes, the crustaceans (copepods, krill, amphipods, ostracods, mysids, decapods), the chaetognaths, and most of the planktonic larvae of benthic animals. These zooplankton are the ones most readily caught in nets and the ones that provide a reflective surface for acoustic instruments used to track their abundance and motions.

Another important distinction in zooplankton concerns whether they are permanent residents of the zooplankton, the **holoplankton,** or temporary residents, the **meroplankton.** The meroplankton include the planktonic larval forms of benthic invertebrates and larval fishes, sometimes called the **ichthyoplankton.** Some meroplankton feed on smaller plankton (planktotrophic), and some rely on yolk sacs and are nonfeeding (lecithotrophic). Many of the meroplankton are capable of absorbing and utilizing dissolved organic matter, a nutritional mode that allows them to disperse over great distances. The planktotrophic larva of an Antarctic sea urchin has been shown to exhibit the highest growth efficiency of any animal on Earth.

As with benthic organisms, zooplankton size classification readily lends itself to the most common method of collection, the

TABLE 12.7 Major Size Classes of Zooplankton

Smaller size classes than indicated here exist for phytoplankton, bacterioplankton, and virioplankton. Larger size classes have been defined for nekton.

Nanozooplankton	2.0–20 µm
Microzooplankton	20–200 µm
Mesozooplankton	0.2–20 mm
Macrozooplankton	2–20 cm
Megazooplankton	20–200 cm

Source: After Sieburth et al., 1978.

plankton net (see Chapter 13). A listing of the major size classes of zooplankton can be found in **Table 12.7.**

Representatives of the nanozooplankton include the single-celled flagellates, tintinnids, and other ciliates. Microzooplankton include a wide range of zooplankton species from the larger protists (foraminifera and radiolaria) to small copepods, especially copepodites (the immature life stages of a copepod). Mesozooplankton probably represent the largest category of zooplankton in abundance and diversity, and include most copepods, krill, chaetognaths, and gelatinous forms, such as the hydromedusae and ctenophores. Macrozooplankton refer to the larger zooplankton, especially gelatinous forms, such as the scyphomedusae and salps. Since colonial salps as long as 46 m (50 yards) have been observed, the upper limit on this size category should be considered only an approximation. The assignment of any particular species to a size category may vary according to its life stage, growth conditions, or other factors, so generalizations concerning species and their sizes are understandably broad.

One goal of the size classification scheme was to consolidate taxonomic information and nutritional roles for organisms. Reducing the number of components (i.e., by lumping organisms into groups) facilitates efforts to model and understand complex systems, such as plankton. Size classifications have been used with some success in that regard. The size of organisms corresponds to their growth and metabolic rates: smaller organisms tend to grow faster and have higher metabolic rates than larger organisms. Similarly, the distribution of biomass within an ecosystem exhibits a size-dependent relationship: smaller organisms tend to have a greater biomass than larger organisms. The relationship between size and other properties of an organism (growth rates, metabolic rates, generation time, biomass, etc.) is known as an **allometric relationship.** Allometry has a well-established history in theoretical biology and ecology, and has proven useful in biological oceanography as well.

The Nekton

Nekton are animals that swim through the ocean, independent of the movement of currents (**Figure 12-8**). These are the animals with which humans have the closest affinity and the animals on which much of the world depends for food. The largest nekton, the whales, once provided the world's supply of oil for lighting until nearly hunted to extinction. Now they inspire millions to learn more about the world ocean. The ability to move allows nekton to seek out ideal places to live and reproduce. Their movements also help them to avoid predators. Nekton employ a wide variety of responses to keep from being eaten, including locomotory escape (i.e., running away), mechanical defense (spines), chemical defenses (poisons), cryptic defenses (looking like a piece of seaweed), and schooling (swimming synchronously). Despite the freedom of movement, nekton are constrained by the limits of their physiology and their tolerance to temperature, salinity, and pressure. To some extent, they are also limited by the distribution of their prey. Some nekton, such as humpback and California gray whales, may migrate far from their feeding grounds to give birth. Many fishes, including sharks, tuna, and salmon, also migrate. By maintaining **homeothermy,** a constant or near-constant body temperature, these migratory nekton are able to survive a wider range of environmental conditions than might otherwise be possible.

One invertebrate taxon—the cephalopods—and four representatives among the chordate taxa make up most of the recognized nekton. Some authors consider krill and some types of pelagic crabs to be types of nekton, so the distinctions are not always clear-cut. Unlike the benthos and zooplankton, the nekton are mostly visible and relatively easy to identify, so classifications based on size, biochemistry, residency time, and other characteristics are not necessary. The major groups of nekton are indicated in **Table 12.8.**

The nektonic cephalapods include the squids, cuttlefish, chambered nautilus, and a few pelagic octopuses. In many ways, the cephalopods are among the most highly adapted nonvertebrate

a.

b.

c.

■ **FIGURE 12-8** Various types of nekton: (a) cuttlefish, (b) yellow-banded sweetlips, (c) whale shark.

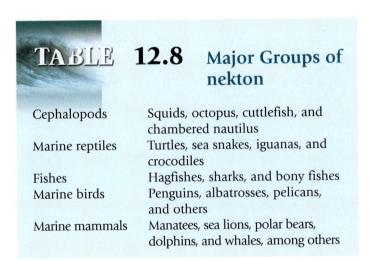

TABLE 12.8	Major Groups of nekton
Cephalopods	Squids, octopus, cuttlefish, and chambered nautilus
Marine reptiles	Turtles, sea snakes, iguanas, and crocodiles
Fishes	Hagfishes, sharks, and bony fishes
Marine birds	Penguins, albatrosses, pelicans, and others
Marine mammals	Manatees, sea lions, polar bears, dolphins, and whales, among others

The Evolution of Fishes

If someone were to call you a lobe-finned fish, you might knock them up side their head and return the insult. However, technically speaking, you are a lobe-finned fish! As John Maisely points out in his 1996 book, Discovering Fossil Fishes:

> We are fishes whether we like it or not…. Our anatomy bears the indelible traces of our fish ancestry: …the incompressible notochord, our spinal cord with all its myelin-coated nerves, our internal and external bony skeletons, our highly modified pectoral and pelvic fins…the point is made.

Approximately 370 million years ago, a relative of the lobe-finned fishes, called a **tetrapod**—with fin bones possibly related to the upper-arm bone (humerus) of vertebrates—pushed its way up onto land. From those humble beginnings arose the evolution and diversification of all land vertebrates and—in a reversal of fortunes, so to speak—the aquatic mammals. With more than 50,000 species encompassed by the Osteichthyes, you might say that the first tetrapod step was one giant leap for vertebrate-kind. Even without the terrestrial offshoots, the success of the Osteichthyes in the ocean is impressive. One reason might be the evolution of their jaws.

If you have ever watched a dog catch a Frisbee or observed a cat carrying a kitten, you may begin to appreciate the importance of the evolution of the jaw. The jaw has been called one of the greatest innovations in vertebrate evolution and rightly so. For fishes, the jaws expanded considerably their ability to capture and consume prey. Jaws could also be used to defend against predators, making the heavy armor of early fishes less important. With less armor to restrict their movements, fishes could swim faster, farther, and in more ways than ever before. Although the jawless fishes—which arose about 500 million years ago—preceded the first jawed fishes by about 50 million years, they were soon "put out of business" by the newcomers. By the end of the **Age of Fishes**, as the Devonian is known, only the hagfish and lampreys remained. All other jawless fishes went extinct. Of course, many jawed fishes did not survive to modern times either. As is often the case in evolution, small changes (intrinsic factors) confer great advantages but only under the right conditions (extrinsic factors). Geological, physical, chemical, and biological changes through time have allowed some species of jawed fishes to survive while driving others to extinction. What remains in modern times are those species that were adapted to survive those changes.

Among the survivors of the Devonian are a group of fishes that have changed relatively little over their some 400 million year history. The cartilaginous fishes, which include the sharks, ratfishes, and rays, evolved a body plan that enabled them to exploit a variety of habitats (**Figure 12a**). Serendipity may have played a part: the presence of abundant food resources, including fishes, cephalopods,

and invertebrates, kept them well-fed. While ancestral sharks differed considerably from modern-day sharks, they provided the basic blueprint from which evolved the highly efficient and effective top predator that roams the ocean today. Much of what we know about the earliest sharks comes from fossils found in Kansas and Ohio, where a broad sea once extended southwest from the St. Lawrence Seaway to Arkansas. Rich deposits and sediments have yielded hundreds of thousands of sharks' teeth and even the well-preserved almost 2 m (6 foot) long Devonian shark. Some specimens still contain the remains of whole fish in their fossil bellies.

Like the cartilaginous fishes, the bony fishes have an evolutionary history that dates some 400 million years. Nonetheless, the lineages are murky as interpretations based on fossils tend to generate controversy. Bony fish had diverged by the middle of the Devonian into two distinct groups: the **lobe-finned fishes** (Sarcopterygii) and the **ray-finned fishes** (Actinopterygii). The lobe-finned fishes include the freshwater lungfishes, found in Africa, Australia, and South America, and the coelacanths, a "fossil fish" discovered among the living in South Africa in 1938. The ray-finned fishes include just about every fish you can think of, with nearly 24,000 species inhabiting nearly every conceivable aquatic habitat. Among these, 23,681 species are teleosts, the fish with symmetrical tails, whose diversity surpasses that of all other vertebrates combined.

A common misconception is that sharks and their relatives represent a "primitive" form of fish, whereas the "true" or bony fishes are more advanced. However, both sharks and bony fishes appear in the fossil record in late Silurian (about 400 mya). Their evolutionary history and their "solution" to the problem of living in an aquatic environment are quite different. Most scientists agree that the cartilaginous fishes and the bony fishes represent two independent designs for being a fish. Both have enjoyed enormous success, and both continue to flourish in the modern ocean.

FIGURE 12a Sharks exhibit a number of specialized anatomical, sensory, and physiological adaptations that make them highly efficient predators in an aqueous habitat.

animals in the ocean. Some scientists maintain that "cephalopods functionally are fish." They range in size from the pygmy squid (*Idiosepius*), at a few inches, to the giant squid, whose mantle is on the order of several meters but whose measurement from fin to tentacle may exceed 17 m (55 feet). The cephalopods include one of the most poisonous animals on Earth, the blue-ringed octopus (*Hapalochlaena*), and one of the strangest, a squid without tentacles that evokes visions of Count Dracula, otherwise known as the vampire squid (*Vampyroteuthis*).

One word appears to describe the driving force behind the evolution of cephalopods: predation. Throughout the Mesozoic, fishes and reptiles competed with and preyed upon cephalopods, much as they do today (except that the plesiosaurs are now extinct). As a result, cephalopods have evolved some of the most sophisticated morphologies and behaviors of any animal. Their body is an arsenal of sensors, weapons, disguises, and propulsion systems designed to avoid being eaten. They also live life on the fly: they grow very rapidly, store very little food as energy reserves, and have very short life spans, less than a year in many species. Their bodies, behaviors, and life history have evolved to react and adapt very quickly.

The phylogenetics of fishes has undergone considerable change in recent decades and still remains a subject of considerable debate. According to the California Academy of Sciences Catalog, there are at least 26,000 species of fishes that have valid scientific descriptions. At least 15,080 species inhabit the world ocean. Of these, 77.5% live in coastal waters (11,700 species). Open-ocean pelagic species make up only 2% (338 species). About 1300 species of abyssal fishes (9%) and 1664 species of deep-sea benthic fishes (11%) are known. At the broadest level, fishes can be divided into two groups: those lacking jaws (Agnatha), and those with jaws (Gnathostomata). Within the first group, the Agnatha, taxonomists treat the lampreys and hagfish as distinct groups. Some scientists reject the hagfishes as vertebrates altogether and treat them as a separate taxa of chordates. Within the Gnathostomata, three major groups can be found (one more than you will find in prior treatments of fish taxonomy). Scientists now recognize the cartilaginous fishes, including the sharks, rays, and ratfishes; the lobe-finned fishes (Sarcopterygii), which include the coelacanths and lungfishes; and the ray-finned fishes (Actinopterygii), making up the majority of fishes on our planet with nearly 24,000 species.

Though Aristotle originally classified dolphins as fish (a classification shared by Sri Lankan Buddhist fishermen until recent times), British naturalist John Ray in 1693 recognized their proper placement among mammals. Unlike other animals, mammals posses three middle-ear bones, hair, and mammary glands. The marine mammals are no exception. In toothed whales and dolphins, the hair is present at birth and lost. Marine mammals, like reptiles and birds, returned to the sea from a terrestrial ancestor. The bearlike ancestry of polar bears and the weasel-like ancestry of sea otters are fairly obvious. Seals, sea lions, and walruses, all of which belong to the taxon Pinnipedia (from the Latin *pinna*,

meaning "feather," and *pedis*, meaning "foot," hence, "featherfoot"), have origins that remain disputed between latrine (otter) and ursid (bear) or other carnivore affinities. On the other hand, scientists generally agree that manatees and dugongs evolved from an elephantlike (proboscidean) ancestor. The evolution of cetaceans involves a less obvious transition from a terrestrial form to an aquatic form, but paleontological and genomic data now support their origin from a group of even-toed ungulates called ariodactyls, the same group that gave rise to the deer and hippopotamus (see **Spotlight 12-2**).

The seals, sea lions, and walruses make their home on land and sea, though there is little question of their prowess in water. Pinnipeds exhibit a number of exceptional adaptations for both a terrestrial and aquatic existence. While their mobility on land is limited, their swimming and diving achievements rival those of whales and dolphins. They can see and hear equally as well on land and sea, and can tolerate extremes of temperature in both. An estimated 25 million pinnipeds distributed among 33 species inhabit the world ocean. Nearly 90% of them are phocids, largely because of the Antarctic crabeater seal, *Lobodon carcinophagus*, which has an estimated population of about 15 million. At the other extreme, the Caribbean monk seal, last seen in 1952, was formally listed as extinct in 1996 and is no longer included in species counts. For the most part, the distribution of pinnipeds is global in extent, although seals comprise most of the pinnipeds found in polar regions.

Of the 83 or so modern species of whales, only 11 are baleen whales, the **Mysticetes**. The toothed whales, the **Odontocetes**, comprise the largest taxa of cetaceans and exhibit a greater diversity of habitat (freshwater, marine, shallow, deep) and behavior. The toothed whales, such as bottlenose dolphins, killer whales, and sperm whales, have simple, peglike teeth for holding fish and squid (versus cutting and tearing) before swallowing. The baleen whales, like the gray whale and blue whale, use baleen, a fibrous, hairlike material, to strain benthic amphipods from the mud or krill from the plankton, respectively. Because feeding on nekton requires greater agility and prey-locating adaptations than feeding on plankton, the toothed whales have developed a highly sophisticated sonar system for finding food, a behavior known as echolocation. Some evidence exists that sperm whales and spotted dolphins may use sound to stun or disorient their prey. Many toothed whales employ sounds in social behaviors and communication, as do the baleen whales. Toothed whales tend to be more streamlined and faster, frequently travel in pods (groups of cetaceans, often related), and exhibit a number of diverse behaviors and lifestyles. On the other hand, the baleen whales have evolved a more capacious, symmetrical skull designed to capture and process large volumes of water. Baleen whales often have throat grooves that expand like pleats to catch enormous volumes of water. Mysticetes tend to be larger, move more slowly over great distances, and exhibit a diverse number of behaviors and lifestyles. And while it is not yet established whether they use echolocation, the mysticetes produce elaborate sounds and repeatable songs the nature of which we are only beginning to understand.

Ocean Habitats

In many ways, oceanographers are just beginning to explore and recognize the diverse habitats that persist in time and space in the world ocean. Temporal and spatial variability in the geology, physics, chemistry, and biology of the world ocean (see inside front cover) creates conditions that allow many different species to coexist. As we have already learned, the upper ocean responds to seasonal changes in sunlight that alter the physical structure of the water column. Episodic events, such as storms, hurricanes, and upwelling, also generate variability in the upper ocean over temporal scales. Over longer scales, interannual and multidecadal climate change influence the upper and deep circulation. Spatially, the ocean varies vertically *and* horizontally. Conditions at the ocean surface are far different than conditions on the deep seafloor. Similarly, conditions in the equatorial ocean differ from those in the polar ocean. Although classification of ocean habitats has traditionally been confined to spatial descriptions, oceanographers have begun to identify temporal states of the ocean as well. We'll explore the different temporal states of the ocean in our study of food webs in Chapter 14. Here, we examine ocean habitats in terms of their spatial scales of change.

Ocean Life Zones

The most well-known classification system for ocean habitats is based on the vertical structure of the water column, the **ocean life zones** (**Figure 12-9**). The modern version of ocean life zones is derived from Joel Hedgpeth's seminal paper, "Classification of Marine Environments," published in 1957. According to

Hedgpeth, the surface of the ocean is called the **epipelagic zone** (from the Greek *pelagikos,* meaning "of the sea"). This region is also known as the **euphotic zone,** which extends to the depth of the 1% light level (at noon), typically the region in which net photosynthesis may occur. Because light penetration varies, the depth of this zone is flexible but typically reaches 100 m (330 feet) in regions of high productivity and perhaps as deep as 200 m (about 650 feet) in the clearest waters. It also represents the region most affected by seasonal mixing and changes in the depth of the seasonal thermocline.

The next zone is popularly called the **twilight zone** (partly in honor of Rod Serling's television series of the same name) and officially known as the **mesopelagic zone.** This region, which extends from 200 to 1000 m (650–3300 feet), is characterized by extremely dim illumination during the daylight hours. As a rule of thumb, this zone represents the region of intermediate and central water masses, and is a transition zone between warm surface waters and cold deep waters. The mesopelagic zone includes the **oxygen minimum zone,** a zone of reduced dissolved oxygen brought about by bacterial and animal decomposition of organic materials (see Chapter 6). Some animals are known to avoid the oxygen minimum zone, so its presence acts as a boundary for the habitat of certain organisms. Several fascinating biological phenomena occur in this zone. The **diel vertical migration** of animals upward or downward in response to or in avoidance of light acts like a **biological elevator** to transport materials up and down (largely as fecal pellets) in the water column on a daily basis. Both positive taxis (attraction to light) and negative taxis (avoidance of light) have been observed, the reasons for which are a rich source of scientific discussion. These vertical migrations give rise to the **deep scattering layer** (see the section on vertical migration in this chapter). Another biological characteristic of this zone is

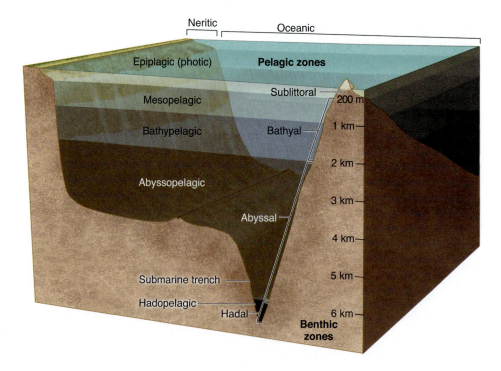

■ FIGURE 12-9 Ocean life zones for pelagic and benthic habitats.

the apparent dominance of **pelagic crenarchaeota,** a group of Archaebacteria (see **Table 12.3**). Bioluminescence, biologically produced light, is widespread in this zone.

The **bathypelagic zone,** at depths from about 1000 to 4000 m (3280–13,200 feet) marks a region of generally constant physical and chemical characteristics (other than pressure, which reaches 400 atmospheres at the bottom of this layer). Seawater temperatures range from 2° to 5° C (36–41° F), and oxygen concentrations are generally high due to reduced biological activity. Biogeochemical transformations take a different form in this zone and appear to be uncoupled from surface processes, such as primary production and particle flux. Merely the faintest of blue-green surface light is detected here, but bioluminescence is common. Carnivory appears to dominate this zone, as organic remains, even in the form of fecal pellets, are few.

The **abyssopelagic** zone extends from about 4000 m (2.5 miles) to the bottom (nominally to 6000 m or 3.7 miles), except where submarine trenches occur, where the **hadopelagic** zone has been defined. Here, we find the coldest waters, the highest pressures, and the scarcest food of any zone. Given the difficulty of studying these regions, it should not be surprising that our knowledge of their characteristics is limited.

Ocean Biomes

While the classification of marine environments in the vertical domain helps to define the range of conditions that an organism may encounter with depth, it represents only a partial description. On the broadest scale, the upper ocean can be divided into **biomes,** regions characterized by a particular type of vegetation, in this case, phytoplankton (**Figure 12-10**). The field of ocean biogeography strives to understand these spatial patterns in the distribution of organisms. Thus, biogeographic provinces represent distinct geographical regions where a particular organism or assemblage of organisms may be found.

Alan Longhurst, a pioneer in ocean biogeography, has proposed a scheme to partition the ocean into ecological provinces based on physical features as they affect the growth and dynamics of phytoplankton. The coupling between physical and biological processes addresses temporal and spatial variability and represents, to some degree, a systems approach for understanding variability in the ocean. As Longhurst writes: "What is attempted here is a framework for a regional ecology of the ocean that brings together our knowledge of relevant physical features of regional oceanography with what we know of the response of planktonic algae to seasonal forcing by physical processes." He also notes that oceanographers "should be able to do better" than merely divide the pelagos into coastal waters and oceanic waters.

Delineation of biomes comes primarily from satellite determinations of chlorophyll concentration. The data provided by these instruments reveal major spatial and seasonal climatological trends. The SeaWIFs climatological data reveal average higher concentrations of chlorophyll coincident with the coastal, polar, and trade biomes, and generally lower concentrations associated with the westerlies biomes. The coastal biome above the continen-

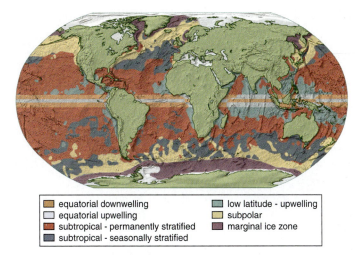

■ equatorial downwelling		■ low latitude - upwelling	
■ equatorial upwelling		■ subpolar	
■ subtropical - permanently stratified		■ marginal ice zone	
■ subtropical - seasonally stratified			

■ **FIGURE 12-10** Ocean biomes, horizontal "life" zones of the world ocean.

tal shelf is characterized by complex interactions among processes that govern mixed layer depth and primary productivity. The polar biome includes regions where sea ice is present and acts as a boundary for ice-edge upwelling and the melting of which creates a surface brackish layer that constrains mixing. The trades biome refers to equatorial and near-equatorial regions where the mixed layer depth is driven by geostrophic adjustment to global-scale winds. The westerlies biome includes central gyres and temperate regions where the mixed layer depth is controlled by local winds and seasonal changes in irradiance. The definition of these biomes is not meant to be comprehensive nor explicit. Some oceanographers disagree with them altogether and have adopted alternative schemes. The primary goal of any scheme for classification of the marine environment is to discover general patterns in the world ocean that allow us to better understand and predict the evolution, growth, and dynamics of marine organisms.

Adaptations of Ocean Life

Marine organisms exhibit adaptations to the marine environment that range from the commonplace to the bizarre. Some of these adaptations are shared among multiple groups of organisms. Some exert global-scale effects on the distribution of properties in the world ocean. To provide a few general examples of the adaptations of marine organisms, we examine four phenomena exhibited by marine organisms throughout the world ocean.

Photoadaptation

Photoadaptation in its broadest sense refers to the response of organisms to varying conditions of light. Increases in skin pigments in humans exposed to intense sunlight (i.e., suntanning)

■ FIGURE 12-11 Corals and other photoautotrophs have evolved to grow optimally under very specific conditions of light. Some species require intense sunlight to achieve maximum rates of photosynthesis. Others flourish where light is dim. "Tuning" of the photosynthetic machinery to different intensities and spectra of light contributes to the high diversity of photoautotrophs observed on coral reefs and in the water column of stratified oceans.

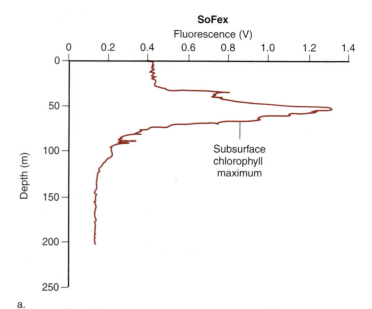

a.

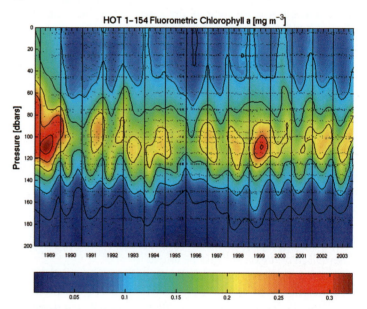

■ FIGURE 12-12 (a) The subsurface chlorophyll maximum appears as an increase in chlorophyll at depth, followed by a sharp decline. In many instances, cell numbers remain constant. This observation indicates photoadaptation—increases in the cellular concentration of chlorophyll, rather than increases in the population of phytoplankton. (b) A time series of the chlorophyll maximum off the coast of Hawaii indicates annual variability in the depth and concentration of chlorophyll.

may be considered a form of photoadaptation. The presence of sun-adapted and shade-adapted corals in the well-lit and dim regions of a coral reef, respectively, illustrates photoadaptation at the ecological level (**Figure 12-11**). Photoadaptation of phytoplankton affects their productivity and the distribution of their pigments in the water column.

The most common type of photoadaptation in phytoplankton occurs in response to decreases or increases in the intensity of available light. As light intensity diminishes, phytoplankton may increase the cellular concentrations of chlorophyll *a*, their major light-harvesting pigment. By increasing their cellular chlorophyll concentrations, phytoplankton maximize their ability to absorb the available light. In the world ocean, this phenomenon may be recognized as a **subsurface chlorophyll maximum,** typically at the base of the mixed layer in a stable water column during spring and summer (**Figure 12-12**). Simultaneous measurements of cell numbers and chlorophyll often reveal that chlorophyll increases with depth while cell numbers remain nearly constant. Photoadaptation of phytoplankton explains this observation. Phytoplankton may

also reduce their internal concentrations of chlorophyll when light intensities exceed their photosynthetic capacity.

Phytoplankton also exhibit photoadaptive responses to changes in the color of light penetrating the water column. Because different wavelengths of light are selectively absorbed, largely blue and green light are available in the deeper regions of the euphotic zone. Where phytoplankton concentrations in the surface waters are high, most blue light may be absorbed (by chlorophyll *a*) such that only green light is available. To accommodate this shift in the spectrum of available light, phytoplankton manufacture accessory pigments, such as chlorophyll *b* or fucoxanthin, pigments whose maximal absorption occurs in the longer wavelengths. Increases in the abundance of accessory pigments are commonly observed near the base of the euphotic zone.

The photoadaptive response of phytoplankton has important implications for the distribution of species within the water column. Much like a tropical rain forest, the "canopy" species will occupy the surface waters, while "understory" species will inhabit more dimly lit regions. An understanding of the adaptations of phytoplankton for particular light environments explains, in part, the evolution of diversity among phytoplankton. It also becomes important when considering the activities of phytoplankton in the global carbon cycle.

Vertical Migration

Each night at dusk begins the largest animal migration in the world, a phenomenon that attracts none but the most dedicated observers. The diel vertical migration of marine organisms, the 24-hour cycle of their upward and downward movements, has been known since the time of the *Challenger* Expedition. However, a complete understanding of its causes and significance remains elusive. Many taxa of marine organisms migrate, including dino-flagellates, numerous species of zooplankton (copepods, krill, shrimp), cephalapods, siphonophores, and many fishes, especially lanternfishes (**Figure 12-13**). Those that migrate upward at night and downward at dawn undergo nocturnal migration, the most common form of migration observed in the world ocean. Some species ascend and descend twice a day, a pattern called twilight migration. Still others migrate downward at dusk and upward at dawn, a form of reverse migration. Migration is not always to the surface: species living at depth may ascend to a particular depth and back over a 24-hour cycle. Travel distances also vary from several tens of meters to hundreds of meters. Some shrimp and krill may travel up to 800 m (2625 feet) at speeds from 100 to 200 m hr^{-1} (328–656 feet hr^{-1}).

Though long known, diel vertical migration gained prominent attention as wartime application of echosounders gained in popularity. Sonar operators were confronted with a mysterious deep scattering layer, a layer where sound reflection increases sharply, which appeared like a "false bottom" on their sonar plots. This mysterious layer rose at night and confounded the search for submarines. Oceanography came to the rescue in the early 1940s when Martin Johnson, a marine zoologist at Scripps Institution of Oceanography, correctly surmised that the deep scattering layer was produced by the diel vertical migration of marine organisms. Acoustic techniques are now the methodology of choice for studying the size distribution and migration patterns of the deep scattering layer (**Figure 12-14**).

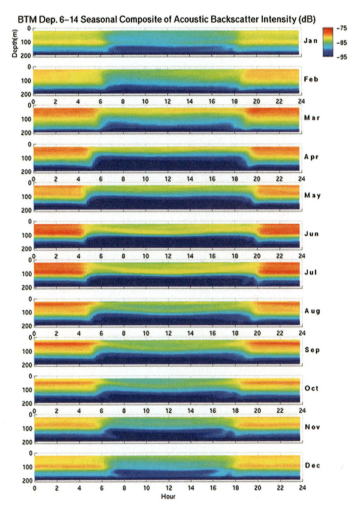

FIGURE 12-14 A seasonal composite of acoustic backscatter reveals in unprecedented detail the daily and seasonal patterns of vertical migraters. The weakest signal (blues and greens) occurs during daylight hours when the organisms are absent; the strongest backscatter (yellows and reds) is observed at night when the organisms migrate towards the surface. Note the effects of seasonal changes on the patterns of migration. Vertical migraters remain at the surface for less time during the long days of summer yet their numbers are greatest in spring and summer.

FIGURE 12-13 Lanternfish dwell in the dark midwater regions by day and migrate upwards to feed at night. This group of fishes (myctophids) serves as an important prey item for large predatory fish. Accounting for up to 90% of the biomass of all deepwater fishes, lanternfish contribute significantly to the deep scattering layer.

Vertical migration presumably enables organisms to feed in more productive surface waters under the cover of night while avoiding visual predators. However, studies of this phenomenon have identified several possible advantages as well as disadvantages, none of which can be eliminated satisfactorily. In addition, multiple cues for vertical migration may exist, and initiation of migration and its duration may depend on the stage of development and the physiological state of the organism along with other factors.

On a global scale, the impact of this migration is staggering. Perhaps as many as a billion tons of organisms migrate each day. As a result, significant quantities of carbon and other materials are removed from surface waters (through feeding) and transported to deeper waters (through downward migration and defecation). The packaging, vertical transport, and release of carbon comprises a major component of the carbon cycle (see Chapter 14). Because of its implications for the global carbon cycle and climate change, research efforts are under way to quantify the fluxes due to the biological pump and better understand factors that contribute to its temporal and spatial variability.

Aggregation

Clumping, crowding, schooling, or the coming together of a number of individuals is termed **aggregation.** The major implication of aggregation is that the distribution of organisms is not continuous in space. Aggregation may be passive, the result of geological, physical, or chemical factors that cause organisms to be clumped or to grow or reproduce more rapidly in a certain location; or it may be active, where the behavioral responses of organisms bring them together. Food-finding, reproduction, and settlement (among benthic forms) act as major stimuli for active aggregation. In such cases, aggregation increases the probability of obtaining food, successfully mating, or settling in a suitable habitat.

The nonuniform distribution of plankton is called **patchiness.** Among phytoplankton, physical and chemical conditions that promote or inhibit growth cause differential rates of growth that lead to higher or lower concentrations at a given location. Zooplankton patchiness may be a response to phytoplankton patchiness, or it may result from active behaviors, such as vertical migration, escape from predators, or mate-finding, that draw them together or apart. Physical processes may aggregate or disperse organisms through the action of winds, tides, waves, and any of a number of other physical phenomena. Interactions of currents with seafloor features and islands may generate flows that concentrate plankton. A related phenomenon found in stable water columns is **thin layers,** high concentrations of plankton confined to a narrow depth interval. Thin layers may occur along pycnoclines as the result of accumulation of marine snow or as a result of *in situ* growth at a particular depth where phytoplankton have achieved neutral buoyancy. They may persist for hours to weeks, depending on conditions in the water column. Research on thin layers is aimed at determining their origins and their importance for food webs and carbon flux.

Mating aggregations are widespread. Sea hares, a hermaphroditic snail-like mollusk, may number 50 to 100 individuals in a tide pool in what has been called a "sea hare orgy." Hundreds of thousands of horseshoe crabs come together each year to spawn. Numerous organisms undergo seasonal vertical migration associated with mating. Palolo worms swarm to the surface annually to mate, much to the delight of American Samoans who celebrate a Palolo-caviar feast around this behavior. The bioluminescent firefly squid put on a spectacular show during annual mating aggregations off the coast of Japan. California grunion amass in the thousands on beaches from southern California to Baja during spring and summer mating runs (**Figure 12-15**).

Aggregations of fishes in which individuals swim and orient themselves synchronously are called **schools.** Schooling in fishes serves a number of purposes, including feeding, mating, protection, and, possibly, other social interactions. Feeding aggregation in fishes such as herring often attracts multiple predators. Tuna feeding on herring may attract dolphins and other marine mammals as well as marine birds. These intense aggregations of fishes, mammals, and birds, swirling and diving in a massive food feast, can be spectacular to watch. Feeding aggregations may also be associated with krill, which typically move as swarms. Swarms of krill attract fish and mammal predators.

Larval settling behavior may cause aggregations of individuals. Among benthic invertebrates, planktonic larvae typically aggregate on or near the discarded shells or calcareous "scars" of

■ **FIGURE 12-15** Spawning of grunion on the shores of southern California from February through August in the days following the full and new moons is a spectacle not to be missed. Visit www. grunion.org for more details.

adults, which act to stimulate their settling behavior. Even where settlement does not aggregate individuals, the force of waves and waterborne objects may dislodge clumps of individuals, leading to patchiness in their distribution.

Numerous other behaviors and interactions may lead to aggregation of marine organisms, and it is highly likely that patchiness and aggregation are general features of their spatial distribution. That said, efforts to quantify rate processes in the ocean will be subject to the vagaries of such nonuniform distributions. Sampling in regions with concentrations of individuals may lead to overestimates when extrapolated to the whole ocean. Similarly, rate processes determined in "dilute" regions of the ocean may underestimate the importance of episodic aggregations, passive or active. Patchiness in the world ocean confounds our efforts to accurately determine the flux of carbon and other elements. Concerted efforts are under way to develop better sampling techniques aimed at improving the temporal and spatial resolution of properties and processes in the world ocean.

Bioluminescence

Bioluminescence, the production of visible light by organisms, has numerous examples in the world ocean. Marine bacteria, dinoflagellates, worms, crustaceans, starfish, sea cucumbers, octopus, squids, sea squirts, and fishes are among the several hundreds of marine organisms that display bioluminescence. In all of these organisms, the bioluminescent reaction involves a pair of molecules, luciferin and luciferase, that participate in an oxygen-utilizing, light-emitting reaction. In some organisms, an excited-state luciferin emits photons of light in particular wavelengths, most commonly blue (440–479 nm). Alternatively, the excited-state luciferin may pass its energy to a fluorescent protein that emits its own characteristic wavelength of light, such as the green glow of some jellyfishes and relatives (about 505 nm) or the near-infrared glow of the black dragonfishes (about 705 nm). At least five basic luciferin-luciferase systems have been identified, each with different luciferins and cofactors that participate in the reactions (**Figure 12-16**). Most organisms have developed their own bioluminescent systems and are called self-luminous species, but others, notably certain squids and fishes, have evolved specialized organs for promoting the growth of luminescent bacteria, or **luminous symbionts.** Bacterial light organs differ structurally from the self-luminous light organs of other organisms, called **photophores.** They are also less numerous; most symbiotic species, such as the anglerfishes, have one or only a few light organs. Self-luminous species may have hundreds of elaborate photophores under nervous control or a single-celled light organ, called a photocyte. Still other organisms use secretory bioluminescence, releasing a luminescent secretion. Most strikingly, bioluminescence may have evolved independently at least 30 different times among organisms (**Figure 12-17**). Thus, the selective pressures for bioluminescence and its ecological advantage to organisms must be considerable.

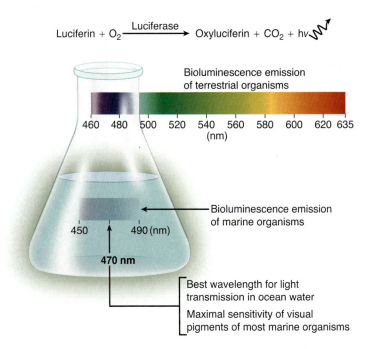

Luciferin + O_2 $\xrightarrow{\text{Luciferase}}$ Oxyluciferin + CO_2 + hv

Bioluminescence emission of terrestrial organisms

460 480 500 520 540 560 580 600 620 635
(nm)

450   490 (nm)

470 nm

Bioluminescence emission of marine organisms

Best wavelength for light transmission in ocean water

Maximal sensitivity of visual pigments of most marine organisms

■ **FIGURE 12-16** A chemical reaction between luciferin and luciferase produces bioluminescence in a wide number of marine and terrestrial species. Bioluminescence represents a form of convergent evolution, where a wide number of species evolve a similar strategy for "solving" a particular problem presented by their environment.

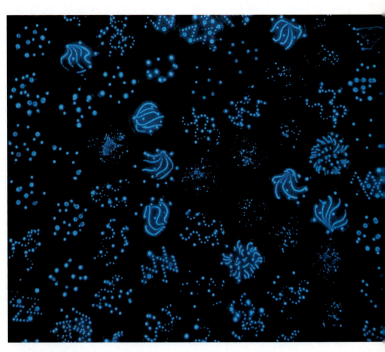

■ **FIGURE 12-17** Living art, a "painting" of bioluminescent bacteria created by students and faculty at Montana State University–Bozeman.

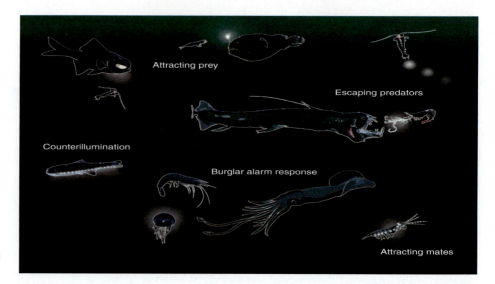

■ **FIGURE 12-18** Hypothesized functions for bioluminescence in marine organisms.

Hypothesized functions for bioluminescence fall into four broad categories: 1) defending against predators; 2) attracting prey; 3) camouflage; and 4) intraspecific communication associated with mating and schooling (**Figure 12-18**). One of the simplest bioluminescent organisms, the dinoflagellates, generates a luminescent response to handling by predators. This flash elicits a startle response from predators, such as copepods, that have been shown to alter their swimming pattern when flashed. The **burglar alarm defense** is especially prominent among organisms that secrete bioluminescence, such as jellyfish, marine worms, shrimp, and even some fishes. In these animals, the luminescent secretion may stick to the predator, making it even more effective for attracting the predator's predators. Bioluminescent secretions may also act as a "smokescreen" behind which an animal may escape. Some organisms shed a body part, such as the arm of a brittle star, which continues to flash after amputation, perhaps as a decoy. Use of bioluminescence to attract prey has been raised to an art form among deep-sea fishes such as the anglerfish and viperfish. Their dangling, glowing appendages act as lures to bring unsuspecting prey closer to the fishes' jaws. Luminous camouflage occurs among organisms living in the twilight zone, where a faint blue glow emanates from above. Here, an organism may apply the principles of **counterillumination,** matching the photophores along the underside of its body to the intensity and color of the water column above it. In such a way, these organisms blend in to the water column and become "invisible." Nearly all mesopelagic fishes employ counterilluminating photophores along the ventral surfaces of their body. In a completely dark environment, bioluminescence may help to attract potential mates. The patterns of glows and flashes among bioluminescent organisms can be quite complex, suggesting a form of communication. Specialized photophores other than those that function for counterillumination can be found in fishes, which, in many cases, exhibit differences between the sexes. Scuba observations of luminescent ostracods, tiny, free-swimming crustaceans, reveal a sexual dialogue on a par with that of fireflies. Communication using bioluminescence is

suggested by divers' observations on the Humboldt squid, an animal that may reach 1.2 to 1.8m (4–6 feet) in size. Flashes of light associated with schooling and perhaps even coordinated attacks on fish prey (and divers!) suggest that these animals employ bioluminescence in social behaviors other than mating.

Biodiversity and Global Warming

The polar oceans hardly seem the places to find diverse and bountiful marine life. With water temperatures below 0° C (32° F), a solar radiation cycle that varies from 24 hours of light to 24 hours of dark, and extremes of weather that generate immense waves and sudden freezing of the sea surface, the polar oceans may well be the most hostile environments on Earth. Nonetheless, polar oceans rank among the most productive marine environments on Earth, one that supports the feeding activities of numerous marine birds and mammal species, and one well-known for its abundant fishes. Yet, global warming and an unprecedented melting of the Arctic and portions of the Antarctic threaten the productivity and biodiversity of these extreme environments. As oceanographers bolster their efforts to explore the polar oceans, the rich variety of organisms that inhabit these waters has both surprised researchers and raised their concerns.

NOAA's "Hidden Ocean" expedition in summer 2005 brought together 45 US, Canadian, Chinese, and Russian scientists to explore the Canada Basin in the Arctic Ocean. A major emphasis of this effort was to acquire digital visual information from underwater still and video cameras, high-definition camcorders mounted on an ROV, and microscopic images. As a result, scientists acquired more than 5000 still macroscopic images, 3000 microscopic images, and 8 terabytes of digital video of marine organisms from the surface to the seafloor 3.2 km (2 miles) deep. More than a dozen new species were discovered, among them, a new species of jellyfish, and a new species of comb jelly (ctenophore). ROV explorations on the seafloor uncovered a dense community of benthic organisms whose abundance appeared at odds with the available food sources. Analyses of photographs and samples along with quantitative data collected during the expedition will continue for years. Despite their successes, oceanographers are concerned that many of the organisms they observed may be threatened as the extent of sea ice diminishes. Those organisms that live under multiyear ice—sea ice that persists for more than one year—may be especially at risk.

Similar threats face Antarctic organisms, particularly in regions where the melting of ice shelves has exposed previously "hidden" habitats. The Larsen B ice shelf along the eastern coast of the Antarctic Peninsula has been breaking apart over the past several decades and collapsed in 2002 (**Figure 12-19**). The disintegration of the ice shelf provides a unique opportunity to study the paleoceanographic history beneath the ice shelf and its relationship to climate change. Studies led by Eugene Domack at Hamilton College in Clinton, New York, suggest that the ice shelf has existed for at least 400 years and possibly since the end of the last ice age 12,000 years ago. In March 2005, Domack returned to this region to continue bottom mapping and sample collection. During an ROV survey of the seafloor, Domack's team discovered 0.6 to 0.9 m (2–3 foot) mounds covered by white bacterial mats and surrounded by large (several inches across) clams. The "Hershey Kiss"–shaped mounds strongly resemble the hydrogen-

sulfide-based, cold-seep communities found in Monterey Bay and the Gulf of Mexico (see Chapter 3), although the source of their energy is uncertain. These findings represent the first observations of chemosynthetic communities in the Antarctic. However, melting of the ice shelf has released large quantities of sediments to the seafloor. Scientists note evidence for deterioration of the bacterial mats and a reduction in their extent due to burial by sediments. As noted by the researchers, the "window of opportunity" to study these unique ecosystems may pass before we have a chance to uncover their mysteries.

In view of these threats, oceanographers have begun an ambitious program to determine the bioviersity of life in the world ocean. In 1997, Jesse Ausubel, program director of the nonprofit philanthropic Alfred P. Sloan Foundation, funded several workshops to help oceanographers to "think outside the box," to push the limits of their creativity and imaginations. He challenged some of the most accomplished oceanographers to answer: 1) What lived in the ocean in the past? 2) What lives in the ocean in the present? and 3) What might live in the ocean in the future? Because knowledge of ocean species is fundamental to an understanding of the ocean, oceanographers were asked to formulate ideas concerning approaches and technologies that could be applied to accomplish an accounting of all of the species of the world ocean. Today, more than 1000 scientists from 70 nations are engaged in a project called the Census of Marine Life, a 10-year initiative to assess and explain the diversity, distribution, and abundance of marine life in the oceans—past, present, and future. It is anticipated that new pharmaceuticals and industrial compounds will be among the potential uses of the estimated thousands of presently undescribed species that will be found during the Census of Marine Life and similar research programs (**Figure 12-20**). Underlying all of these issues is concern for humankind's very existence on planet Earth; our existence is likely highly dependent on the stewardship and maintenance of a livable and sustainable environment and its diverse life-forms.

■ **FIGURE 12-19** The collapse and melting of the Larsen B ice shelf has provided new opportunities to study the seafloor beneath these floating shelves of ice. At the same time, they threaten communities of organisms that may have persisted for at least 10,000 years.

■ **FIGURE 12-20** Oceanographers brave freezing waters in search of new life forms.

Life Beneath the Seafloor

The discovery of hydrothermal vents and chemosynthetic ecosystems on the seafloor off the Galapagos Islands in 1977 transformed the way scientists view life on Earth. Since that time, chemosynthetic communities have been found associated with methane cold seeps, icy gas hydrates, seafloor extrusions of asphalt, and hydrogen-sulfide-rich sediments, where the world's largest bacterium was discovered. Remarkable as each of these discoveries has been, perhaps the most astounding realization of the twenty-first century is that life abounds *within* Earth and *beneath* the seafloor. Subsurface microbial communities, recognized by scientists working in oil fields since the 1920s, have more recently been found living in igneous rock several kilometers beneath Earth's surface. The Ocean Drilling Program provided the first evidence for marine subsurface microbial communities in the 1970s. Microbial activity was detected in sediment cores at depths of more than 800 m (2625 feet) in sediments that were tens of millions of years old.

Direct observations of "erupted" subsurface microbes came in 1991, when researchers working aboard Alvin (the same submersible that gave witness to the first hydrothermal vents) on the Juan de Fuca Ridge observed what looked like a "blizzard" of white bacteria following volcanic activity on the seafloor (**Figure 12b**). Subsequent genomic studies of microbes from "snow-blower" vents revealed viable species distinct from those found in the water column. Subsurface microbial communities flourished beneath the seafloor.

Microbiological and genomic studies of sediment cores, vent fluids, and samples from seafloor, subseafloor, and subsurface terrestrial environments have expanded considerably the boundaries of life as we know it. In 2003, scientists at the University of Massachusetts discovered a hyperthermophilic microbe (Archaea) from a hydrothermal vent capable of tolerating temperatures as high as 121° C (250° F). Microbes have been cultured from Arctic and Antarctic permafrost where temperatures fall below − 10° C (14° F). The ability of microbes to "survive" almost goes beyond belief. In 2000, microbes from a 250 million-year-old salt deposit were "brought back to

life," the oldest known living microbes ever recovered. This ability, called cryptobiosis or "hidden life," enables microbes to survive under adverse conditions. It is precisely this ability that makes studies of microbial activity difficult under natural conditions. Just because a particular microbe is present does not mean that it is metabolically active.

Studies of microbes in extreme environments, such as hydrothermal vents and cold seeps, reveal subtle genetic differences that become more pronounced the greater the geographic distance between the populations. The "biogeographical" distribution of microbes challenges commonly held ideas that microbes are distributed uniformly across the globe. These observations suggest that natural selection acts on microbes in much the same way that it does for higher organisms. In essence, microbes are not everywhere but disperse and diversify under geographic isolation with limited gene flow between populations. As noted by one team of researchers, if this result holds true, the diversity of microbes on our planet may far exceed current expectations.

FIGURE 12b A "blizzard" of bacteria from a deep ocean hydrothermal vent. Oceanographers now believe that these bacteria reside beneath the seafloor and "emerge" during periods of extreme fluid flows, submarine earthquakes or undersea volcanic eruptions.

1. **I've learned that scientific theories provide a basis for predicting natural events. What kinds of things does the theory of evolution predict?**

   In the broadest sense, evolutionary theory predicts that organisms incapable of surviving and reproducing when their environment changes will go extinct. It also predicts features of speciation, predator-prey dynamics, and rates of mutation. Evolutionary theory also allows scientists to make retrodictions, "predictions" about information not yet discovered. Darwin "retrodicted" that human ancestors arose from Africa, a result later verified. Like the theory of plate tectonics, the theory of evolution explains a broad number of disparate events and phenomena. It explains the existence of fossils, it explains the predominance of marsupials in Australia, it explains the genetic similarities between humans and apes, and it even explains the effects of humans on size distributions of commercial fishes. As our understanding of the mechanisms that influence the evolution of organisms grows, so, too, will our ability to predict the effects of human activities on the evolution of ecosystems and biodiversity.

2. **How can organisms tolerate the extreme high pressures of the abyss?**

   By maintaining an equal pressure inside their bodies and out, organisms are not affected by the high pressures. Consider humans living on the surface of our planet under the weight of the atmosphere. The reason we can "tolerate" 14.7 pounds per square inch on the outside of our bodies is because the same pressure is exerted outward inside our bodies.

3. **I was really hoping to learn more about sharks or whales and dolphins, at least. Don't oceanographers study these animals?**

   That depends on how you look at it. Biological oceanographers generally view the ocean as an ecosystem and focus on its interactions with other earth systems, such as the lithosphere and atmosphere. The pathways of energy and material flows and factors that influence them are a major focus. To the extent that the behaviors of organisms and the role they play in food webs are important to the structure and functioning of ecosystems, then, yes, oceanographers study these animals. Studies of the biology of marine organisms fall more in the domain of marine biology. Though the distinctions are by no means clear-cut, oceanographers take a systems approach, whereas marine biologists focus more on the organisms and their ecological roles.

Key Concepts

- Species evolve through geologic time as a result of natural selection.
- Evolution is a two-part process: mutations create variation within species, and natural selection acts upon the variable species to remove those least adapted for a particular suite of environmental conditions.
- Evolution may be punctuated by events that lead to rapid speciation and mass extinction.
- Classification of organisms into related groups is a tool of systematics, a reconstruction of the history of life on Earth.

- Molecular techniques have revolutionized the study of systematics and have led to the discovery of thousands of new species in the world ocean.
- Ocean life may be broadly classified into three lifestyles: benthic, planktonic, and nektonic.
- The world ocean may be divided into vertical habitats called ocean life zones and horizontal habitats called biomes.
- Organisms exhibit a wide variety of adaptations that enable them to survive under a given set of geological, physical, chemical, and biological conditions.

Terms to Remember

abyssopelagic, 253

Age of Fishes, 250

aggregation, 256

allometric relationship, 248

Archaea, 243

Bacteria, 243

bacterioplankton, 247

bathypelagic zone, 253

benthos, 246

biogeography, 240

biological elevator, 252

bioluminescence, 257

biomes, 253

burglar alarm defense, 258

cladistics, 242

classification, 242

counterillumination, 258

deep scattering layer, 252

diel vertical migration, 252

domains, 243

endosymbiosis, 243

epifauna, 246

epipelagic zone, 252

Eukarya, 243

euphotic zone, 252

evolution, 239

geobiology, 239

hadopelagic, 253

hard-bottom benthos, 246

holoplankton, 248

homeothermy, 249

ichthyoplankton, 248

infauna, 246

interstitial organisms, 246

jellyplankton, 248

lobe-finned fishes, 250

luminous symbionts, 257

mass extinctions, 241

meroplankton, 248

mesopelagic zone, 252

modern synthesis, 241

molecular phylogeny, 242

Mysticetes, 251

natural selection, 239

nekton, 249

ocean genomics, 242

ocean life zones, 252

Odontocetes, 251

organelles, 243

oxygen minimum zone, 252

patchiness, 256

pelagic crenarchaeota, 253

photoadaptation, 253

photophores, 257

phylogenetic tree, 242

phytoplankton, 248

prokaryotes, 243

punctuated equilibrium, 242

ray-finned fishes, 250

schools, 256

soft-bottom benthos, 246

subsurface chlorophyll maximum, 254

systematics, 242

tetrapod, 250

theory of common descent, 242

thin layers, 256

twilight zone, 252

virioplankton, 248

whole-genome shotgun sequencing, 243

zooplankton, 248

Critical Thinking

1. What is the relationship between geologic processes and extinction of a species? Provide an example.
2. How might the biological activities of one species lead to the extinction of another species? Provide an example.
3. Why is it useful for oceanographers to divide the ocean into vertical and horizontal life zones?
4. Debate this statement (for or against): The earth system would do just fine without humans.
5. How might our understanding of evolution help us design better fishing methods that do not drive a species toward smaller sizes?
6. Compare and contrast the three major "lifestyles" of marine life. Why do some organisms maintain more than one lifestyle? Provide an example.
7. How is the ocean carbon cycle affected by vertical migration? What factors accelerate the sedimentation of organic matter to the deep sea, and what factors slow sedimentation? Why is an understanding of vertical migration important for studies of climate change?
8. Why is human exploitation of fish aggregations a contributing factor to overfishing? How do fish aggregations make it difficult for oceanographers to obtain an accurate census of fish populations?

Explore Online

 Visit www.mhhe.com/chamberlin1e for access to chapter quizzing, key term flash cards, video clips, interactive activities, and more. Further enhance your knowledge with web links to chapter-related material!

Exploration Activity 12-1

Exploring the Evolution of Whales (Cetaceans) Through Critical Thinking

Question for Inquiry

How did climate change drive the evolution of whales?

Summary

The study of evolution involves many aspects beyond the scope of this text. However, the cetaceans offer an ideal example in the study of how the vertebrate body plan was modified from a terrestrial form to an aquatic one. The exploitation of a shallow marine habitat by a land mammal and its transition to a fully aquatic animal that rapidly diversified throughout the world ocean describes the macroevolution of cetaceans. In this activity, you will explore the kinds of evidence that scientists are using to piece together the history of cetacean evolution and the role of climate change as a selective force.

Learning Objectives

By the end of this activity, you should be able to:

- Describe the paleoclimate of the Eocene-Oligocen transition and the scientific evidence on which it is based
- Explain the various types of evidence used to determine the evolutionary history of whales
- Use your evidence to defend the idea that climate change was an important factor in the evolution of whales

Materials and Preparation

Access to scientifically credible web sources and peer-reviewed scientific journals is required for this activity. Your instructor may provide appropriate articles or you may research them on your own. If you are uncertain how to find peer-reviewed scientific journals on the internet or in the library, please ask your instructor or consult a librarian. There are numerous websites that claim to be scientific in an attempt to discount the theory of evolution. So use caution and common sense when exploring this subject. If the website appears to be advocating a position, it is likely not a scientific source.

Part I: Analyzing Information with a Critical Eye

Read the statements below and explain in your words what you think each of them mean. What questions arise in your mind in reading each statement? Create a list of facts and information that you would need to verify each claim. Which statements use the language of science and which ones do not? How can you know if a statement is scientific or not?

1. "In one sense, evolution didn't invent anything new with whales. It was just tinkering with land mammals. It's using the old to make the new."[1]

2. "The reason evolutionists are confident that mesonychids gave rise to archaeocetes, despite the inability to identify any species in the actual lineage, is that known mesonychids and archaeocetes have some similarities. These similarities, however, are not sufficient to make the case for ancestry, especially in light of the vast differences. The subjective nature of such comparisons is evident from the fact so many groups of mammals and even reptiles have been suggested as ancestral to whales."[2]

3. "There simply are no transitional forms in the fossil record between the marine mammals and their supposed land mammal ancestors . . . It is quite entertaining, starting with cows, pigs, or buffaloes, to attempt to visualize what the intermediates may have looked like. Starting with a cow, one could even imagine one line of descent which prematurely became extinct, due to what might be called an 'udder failure'."[3]

4. "We have seen that there are nine independent areas of study that provide evidence that whales share a common ancestor with hoofed mammals. The power of evidence from independent areas of study that support the same conclusion makes refutation by special creation scenarios, personal incredulity, the argument from ignorance, or "intelligent design" scenarios entirely unreasonable. The only plausible scientific conclusion is that whales did evolve from terrestrial mammals."[4]

Sources:

[1] Available at: www.pbs.org/wgbh/evolution/library/03/4/l_034_05.html. (Scientific viewpoint) © 2001 WGBH Educational Foundation and Blue Sky Productions, Inc.

[2] Available at: www.trueorigin.org/whales.asp. (Creationist viewpoint) © 1998 Creation Research Society

[3] Available at: www.talkorigins.org/features/whales/. (Creationist viewpoint) © 2001 Raymond Sutera

[4] Available at: www.talkorigins.org/features/whales/. (Scientific viewpoint) © 2001 Raymond Sutera

For the layperson and non-specialist (even scientists), the arguments can be confusing and overwhelming. Science is a works in progress based on logical, self-consistent interpretations of multiple independent lines of available evidence. Even so, scientists get things wrong. But unlike faith-based proponents, scientists are willing to change their ideas in light of new evidence. A useful rule of thumb is to ask yourself whether the author is describing evidence and offering an interpretation (a scientific view) or whether the author is criticizing an interpretation and offering no alternative explanation (a non-scientific view). Scientists, even when they disagree, use data and evidence to support their ideas. Creationists attempt to deny the existence of the evidence but offer little in its place. It's a difficult and emotionally charged subject. Approach it with an open mind and decide for yourself which approach works best for understanding how the natural world works. Here are two websites that can help you explore the creationist versus evolutionist viewpoints controversy:

Edward Babinksi, www.edwardtbabinski.us/

Talk Origins, www.talkorigins.org/

Part II: Exploring the Eocene-Oligocene Transition

The span of geologic time during which modern whales evolved, the so-called the Eocene-Oligocene (E-O) transition (~33.7 mya), was a time of great change in Earth's climate.

During this time—between 45 and 24 million years before present—Earth's climate shifted from a relatively warm "hothouse" climate to its current relatively cool "icehouse" climate. It was also during this time that the archaeocetes made their transition from land to sea and the modern cetaceans diversified throughout the world ocean. As Thewissen (1998) puts it: "Cetaceans originated when a Paleogene land mammal underwent a dramatic shift in biological attributes in order to accommodate an enormous shift in habitat." In this part, you will explore the tectonic, oceanographic, and climate history of the Eocene-Oligocene transition.

1. Go to the Paleomap project and spend a few moments reading the home page to become familiar with the types of information this site offers. As you proceed with your research into the climate and conditions of the E-O transition, you may wish to explore some of the additional tools provided here.

2. Start with Climate History. Take notes on the different types of evidence (geological and biological) and the methods (review section on mapping the past positions of continents, www.scotese.com/method1.htm) that scientists use to determine paleoclimates. Review the chart of climate through geologic time. How many "hothouses" and "icehouses" have occurred in Earth's past? Were they of similar duration?

3. Use the list of links at the left side of the page to learn about conditions from the Paleocene to the Miocene. Pay particular attention to the changes in oceanographic conditions during these epochs. Note the types of evidence used to support the type of climate that existed. Summarize this information and print or save the maps. You will use this information in the final part of this activity.

4. Visit these websites to gain additional insights and information on climate during the Eocene and Oligocene:
   a. The Paleontology Portal, www.paleoportal.org/time_space/period.php?period_id=8
   b. The Museum of Paleontology at University of California, Berkeley, www.ucmp.berkeley.edu/cenozoic/cenozoic.html

c.   PBS Evolution website, www.pbs.org/wgbh/evolution/change/deeptime/cenozo.html

5. Several books also provide excellent information on these epochs. We recommend the following (but there are most certainly others):

   a.   *Evolution of the Earth* by Donald Prothero and Robert Dott, 7th edition, 2004, McGraw-Hill Higher Education, Dubuque, IA.

   b.   *The Book of Life* by Stephen Jay Gould, 2001, W.W. Norton and Company, New York

Part III: Exploring the Paleontological Evidence

Several lines of evidence document the evolution of cetaceans. Like you did in Part II, you will use scientific sources to build a body of evidence, in this case, the paleontological evidence for the evolution of cetaceans.

1. Although Wikipedia entries may vary in the quality of their presentation of scientific evidence (meaning you should never solely rely on Wikipedia as a source of scientific information), their entry on cetacean evolution provides a good overview. Explore this page and its links to gain a basic understanding of the terminology and current hypotheses for the evolution of cetaceans. Take notes!

2. Go to the Thewissen Lab home page (www.neoucom.edu/DEPTS/ANAT/Thewissen/). Dr. Thewissen is one of the foremost authorities on cetacean evolution and his work is recognized worldwide. Explore Whale Origins section. Take notes on the evidence used to support current hypotheses on cetacean evolution. There are extensive links on the left side of the page. Use them to gain a better understanding of whale evolution and a sense of the evidence on which current scientific understanding is based.

3. From the Thewissen Lab home page, explore the Digital Library of Dolphin Development. Study carefully the images of a developing dolphin embryo in the Organ Development section. How does developmental evidence support our understanding of the evolution of whales? Take notes!

4. Visit Talk.Origins and build upon your understanding of the various kinds of evidence and how they support the modern view of cetacean evolution. See especially *The Origin of Whales and the Power of Independent Evidence*, www.talkorigins.org/features/whales/

Part IV: Argue Your Case!

If you have done a thorough job, you now have multiple lines of evidence on which you can make inferences about the evolutionary history of cetaceans. Use this evidence to prepare a 7-10 page paper or a 45-minute presentation that answers this question: How did climate change drive the evolution of whales? Discuss our paper/presentation with your classmates and instructor and refer to the Rubric for Analyzing and Presenting Scientific Data and Information to improve your work. (See Exploration Activity 1-1.)

13 Ocean Productivity

Primary producers can be found in the water column and attached to the seafloor in the lighted regions of the world ocean.

Questions to Consider

1. How have methods for measuring primary productivity changed over historical times, and what challenges remain in determining global rates of productivity?

2. What temporal and spatial factors control the availability of light for primary productivity in the world ocean?

3. What temporal and spatial factors control the availability of nutrients for primary productivity in the world ocean?

4. What role do zooplankton play in the biological carbon pump?

5. How does vertical mixing affect light and nutrient availability for primary productivity?

6. What is the role of the world ocean in the global carbon cycle and climate change?

The Microbial Revolution

In the mid–1970s, biological oceanographers reported an astounding discovery: tiny, photosynthetic microbes, smaller than any previously seen, abounded in the ocean. What's more, these microbes contributed at least half of the total productivity of phytoplankton in the ocean. The discovery of new and diverse forms of microbial life in the ocean has fueled what has been called the **microbial revolution.** There is wide consensus that the microbial revolution represents one of the most significant achievements of recent decades. In many ways, it is a revolution in progress: seemingly every year, new observations of microbial life alter our traditional ways of thinking about the ocean.

The rate of production of organic matter by phytoplankton and other photoautotrophs is called **primary productivity,** a term borrowed from agriculture. Although phytoplankton represent the major primary producer in the world ocean, other primary producers may be dominant in certain locations, especially coastal zones. The seaweeds or **macroalgae** can be found in the rocky intertidal and the shallow waters of the coastal ocean and estuaries. The **zooxanthellae** of corals sustain coral reef communities. **Marine plants,** true plants with roots and leaves, include the seagrasses, surfgrasses, and salt marsh plants, among others. These primary producers also provide food and shelter to a wide variety of marine organisms, although their contributions to carbon flux on a global scale are less than that of phytoplankton. Within hydrothermal vents, cold seeps, and subterranean seafloor communities, both symbiotic and pelagic chemosynthetic organisms supply energy and materials to hosts and consumers. Chemosynthetic primary producers represent an important, nonphotosynthetic source of production in ocean food webs. In this chapter, we focus on the dominant primary producer in the world ocean, the phytoplankton.

From Cell Counts to Satellites

Studies of phytoplankton productivity of the world ocean have been a central topic of oceanographic research for more than 150 years. During that time, oceanographers have learned a great deal about the physical, chemical, and biological factors that control phytoplankton productivity. A critical goal of these studies is to understand how changes in phytoplankton productivity affect Earth's climate, ocean food webs, biogeochemical cycles, and marine sediments. Yet, as we shall see, progress in this field is slow owing to the enormous scales of time and space over which measurements must be taken (see inside front cover). Technical difficulties and interpretation of results from different methodologies also challenge oceanographers attempting to provide accurate measurements of global primary productivity.

The first efforts focused on comparisons of the distribution and abundances of major species that could be identified and enumerated using a microscope. German anatomist and physiologist Johannes Muller (1801–1858) was one of the first to examine seawater samples in a microscope. He also invented the first tow net—now known as the **plankton net** (**Figure 13-1**). Victor Hensen (1835–1924), the "father" of modern biological oceanography, coined the term **plankton** in his classic 1887 publication. He also posed the question that continues to drive much of biological oceanography today: how much fish production can be supported by a given level of primary production?

During these early times, the relationships between the cycles of abundance of phytoplankton, zooplankton, and fishes became firmly established. William McIntosh (1838–1931), a Scottish zoologist who participated in the *Challenger* Expedition, was among the first to publish a "calendar of the plankton" noting the seasonal changes in their abundance. British oceanographer James Johnstone (1870–1932) published a calendar in 1908 for the western coast of England. While admitting that meteorological or other factors might delay or accelerate the calendar, Johnstone noted that the "fauna and flora of the plankton do not vary fortuitously, but there is a very definite order of succession in the nature and abundance of the organisms found throughout the year."

An increasing emphasis on measuring the properties of the biological systems as opposed to quantifying individual species led to further developments in our understanding of primary productivity. Borrowing techniques from other fields of science (a common practice even today), ocean scientists found new ways to measure photosynthesis in the ocean. In 1927, Norwegian oceanographers Thorbjørn Gaarder (1885–1970) and Haaken Hasberg Gran (1879–1955) developed the **light-and-dark bottle method** for determining rates of photosynthesis (**Figure 13-2**). This method relies on changes in the concentration of dissolved oxygen in clear and opaque bottles incubated for a period of time at different depths within the water column. The work of these scientists in the fjords of Oslo set the stage for modern determinations of primary productivity in the ocean.

Early oceanographers also focused on techniques for estimating the concentration of specific elements, such as carbon, to more easily measured molecules, such as pigments. In 1938, Dalhousie University's Gordon Riley (1911–1985) presented conversion factors for relating the chlorophyll content to organic carbon, reporting that 0.88 micrograms of chlorophyll represented the equivalent of 0.0033 milligrams of carbon. In 1942, oceanographers Martin Johnson (1893–1984), Harald Sverdrup (1888–1957), and Richard Fleming presented similar data in *The Oceans*, considered by some to be the most influential oceanography textbook of all time. In this volume, they summarized a number of **elemental equivalents,** including carbon, nitrogen, phosphorus, pigments, wet and dry plankton, settling volume, and oxygen. The most well-known elemental equivalent is the **Redfield ratio,** C:N:P, 106:16:1 (**Figure 13-3** and see Chapter 6).

Some scientists abandoned phytoplankton altogether to estimate production, turning instead to the removal of dissolved elements in seawater as an indirect means for determining growth rates. As early as 1899, scientists recognized that the growth of diatoms was regulated by light and biologically important nutrients. Their work built upon Liebig's law of the minimum, the principle that states that growth is limited by the factor that is available in the least amount (see Chapter 6). Scientists reasoned that phytoplankton must decrease the concentration of particular elements and that by measuring changes in the concentration of elements over some period of time (and accounting for other processes), the growth of phytoplankton could be determined. As one early example, Leslie Cooper (1930–1972), working in the Irish Sea in 1938, estimated production by comparing the winter maximum and summer minimum of phosphate, the difference scaling approximately to the growth of phytoplankton for that year.

The real boost to determinations of oceanic primary productivity came with Steeman-Nielsen's 1951 *Nature* paper, "Measurement of Production of Organic Matter in the Sea by Means of Carbon-14." Carbon-14 (^{14}C), a radioactive isotope of carbon, acts as a tracer for carbon fixation in photosynthesis. By adding trace amounts of **^{14}C-bicarbonate** to a seawater sample containing phytoplankton and allowing the bottle to incubate in sunlight (either on deck, in

■ **FIGURE 13-1** A bongo net, a type of plankton net that collects two samples simultaneously.

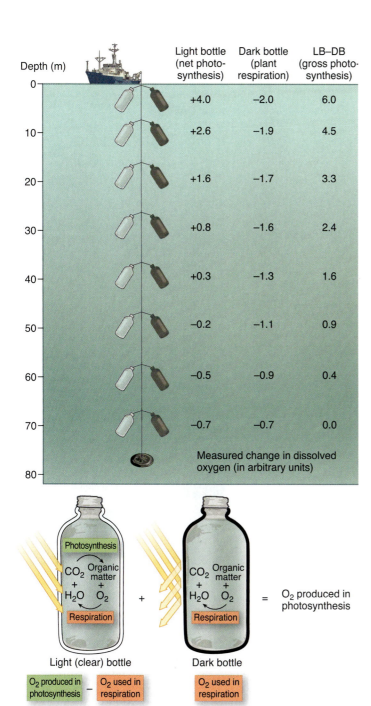

Depth (m)	Light bottle (net photo-synthesis)	Dark bottle (plant respiration)	LB–DB (gross photo-synthesis)
0	+4.0	−2.0	6.0
10	+2.6	−1.9	4.5
20	+1.6	−1.7	3.3
30	+0.8	−1.6	2.4
40	+0.3	−1.3	1.6
50	−0.2	−1.1	0.9
60	−0.5	−0.9	0.4
70	−0.7	−0.7	0.0

Measured change in dissolved oxygen (in arbitrary units)

Light (clear) bottle — Photosynthesis — CO_2 + H_2O ⇌ Organic matter + O_2 — Respiration

Dark bottle — CO_2 + H_2O ⇌ Organic matter + O_2 — Respiration

= O_2 produced in photosynthesis

| O_2 produced in photosynthesis | − | O_2 used in respiration | | O_2 used in respiration |

■ FIGURE 13-2 The light-dark bottle method of determination of primary productivity using oxygen. The decrease in oxygen in the dark bottle (DB) equates to respiration of the phytoplankton, bacteria, and other organisms. The change in oxygen in the light bottle (LB) represents the net difference between rates of photosynthesis and respiration. Note that this method is best suited for conditions where productivity is high.

the water column, or in a laboratory), ^{14}C becomes fixed in the phytoplankton cells at a rate that is proportional to their rate of photosynthesis. When the cells are separated from the seawater on a filter and placed in a scintillation cocktail that flashes when ^{14}C emits radiation, the amount of ^{14}C can be determined and rates of primary productivity can be calculated (**Figure 13-4**). More than 50

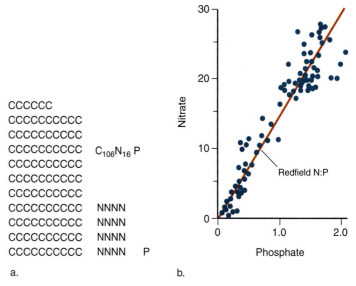

```
CCCCCC
CCCCCCCCCC
CCCCCCCCCC
CCCCCCCCCC    C₁₀₆N₁₆ P
CCCCCCCCCC
CCCCCCCCCC
CCCCCCCCCC
CCCCCCCCCC    NNNN
CCCCCCCCCC    NNNN
CCCCCCCCCC    NNNN
CCCCCCCCCC    NNNN    P
```
a.   b.

■ FIGURE 13-3 The Redfield ratios. When plotted against each other, carbon and nitrogen can be described mathematically by a line with a slope of 106 carbon atoms for every 16 nitrogen atoms. Knowledge of elemental equivalents provides a kind of "principle of constant proportions" for biological oceanographers: measurement of the concentration of one element enables estimation of the concentration of other elements. Of course, oceanographers seek to understand why particulate organic matter in the world ocean conforms to this ratio and under what conditions the Redfield ratio varies. Simply put, elemental equivalents are just another tool for understanding the origins and fate of organic matter in the ocean (data from Sterner and Elser, 2002).

years of using ^{14}C have provided a good understanding of the physical, chemical, and biological processes that affect rates of primary productivity in the ocean.

Improvements in the determination of rates of primary productivity necessitated better methods for estimating the **standing crop** or biomass of phytoplankton. In 1965, oceanographer Ozzy Holm-Hansen developed the fluorometric determination of chlorophyll a. Acetone or ethanol extracts of the photosynthetic pigments are placed in intense blue light, and the resultant red emission, called **stimulated fluorescence**, relates to the concentration of chlorophyll a (**Figure 13-5**). A couple of years later, oceanographer Carl Lorenzen demonstrated that the stimulated fluorescence of natural populations of phytoplankton (called *in vivo* stimulated fluorescence) also provided a rough estimate of chlorophyll biomass. Lorenzen's technique paved the way for the instantaneous measurement of phytoplankton biomass using *in situ* **fluorometers**, submersible instruments that stimulate chlorophyll fluorescence and register the intensity of fluorescence emission. Fluorometers are routinely used on CTDs and have been deployed on a number of different platforms, including moorings, ROVs, AUVs, and gliders.

Since the 1980s, the sun-stimulated or **natural fluorescence** of phytoplankton has been studied for determining rates of primary productivity. Oceanographers first recognized the natural fluorescence of phytoplankton in the mid–1970s with the development of radiometers that were sensitive to chlorophyll fluorescence (i.e., about 670–680 nm). Initial work focused on using natural fluorescence to estimate chlorophyll concentration, but research by oceanographer Dale Kiefer at the University of Southern California suggested that

1. Take predawn water samples at different depths.

2. Fill clear, polycarbonate bottles with water from each depth.

3. Innoculate with ^{14}C.

4. Immediately filter one bottle for nonbiological uptake, the blank.

5. Deploy bottles at depth where sampled originally. Incubate from dawn to dusk.

6. Retrieve bottles and filter contents.

7. Transfer filter to vial with scintillation cocktail.

8. Read radioactivity from ^{14}C-containing phytoplankton in scintillation counter.

9. Calculate productivity.

■ FIGURE 13-4 The carbon-14 method for determination of primary productivity. Unlike oxygen methods, this highly sensitive method enables measurements of primary productivity under low-productivity conditions. Typically, oceanographers take a "blank" at the start of the measurement to account for any non-biological uptake of carbon-14 (i.e., adsorption on particles). The incubations are also kept sufficiently short to prevent respiration of newly incorporated carbon-14. Dark bottles are not commonly used in this method.

$$\text{Carbon uptake} = \frac{^{14}C \text{ in photoplankton on filter} \times \text{available carbon} \times 1.05 \text{ (fractionation factor)}}{\text{Total } ^{14}C \text{ added}}$$

■ FIGURE 13-5 A flask of chlorophyll stimulated by light (flask in foreground) will glow red due to fluorescence of the chlorophyll molecules. In the absence of sufficient stimulation (flask in background), the extracted chlorophyll appears green. The intensity of fluorescence is proportional to the concentration of chlorophyll under constant light. Fluorometric determination of chlorophyll is standard procedure on research vessels conducting studies in biological oceanography.

natural fluorescence might be used to estimate rates of photosynthesis. Working aboard Jacques Cousteau's world-renowned *Calypso* off the coasts of Tahiti and Moorea in 1987, oceanographer Sean Chamberlin compared the emission of natural fluorescence with rates of primary productivity based on ^{14}C. Chamberlin demonstrated for the first time a quantitative relationship between natural fluorescence and primary production. In the twenty-first century, natural fluorescence is being measured from space aboard the Terra and Aqua satellites (**Figure 13-6**). The instruments aboard these satellites will enable oceanographers to obtain the first global-scale estimates of near-surface primary production.

The history of methods described here represents a small sampling of more than a century and a half of research on phytoplankton and primary productivity. By no means is this history comprehensive. Rather, it illustrates the historical development of a few major methods still in use today. In modern times, at least six major approaches and dozens of techniques are being applied to estimate rates of prima-

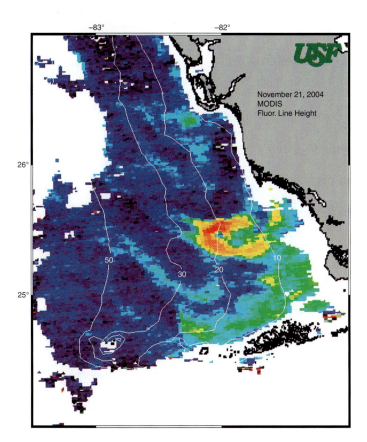

November 21, 2004
MODIS
Fluor. Line Height

■ FIGURE 13-6 Measuring natural fluorescence from space. Both the Aqua and Terra satellites house an instrument called MODIS (Moderate Resolution Imaging Spectroradiometer), which measures, among other properties, the sun-stimulated fluorescence of chlorophyll between 673 and 683 nm. Measurements of natural fluorescence continue to show promise and still remain the only method that measures the fluorescence of phytoplankton under natural illumination.

ry productivity in the world ocean (**Table 13.1**). For a more detailed summary of these methods, please consult our website.

Major Types of Phytoplankton

The phytoplankton (literally *plant wanderer*, though they are not classified as plants) include both prokaryotic bacteria and single-celled eukaryotes. Combined, they may be divided into at least nine distinct groupings of **algae,** defined as photosynthetic autotrophs lacking roots and leaves (a functional definition that distinguishes them from higher plants). Because of their size, they are often referred to as microalgae to distinguish them from seaweeds, or macroalgae. All phytoplankton have the ability to convert inorganic carbon into organic compounds using dissolved

TABLE 13.1 Various Methods Used to Measure Primary Productivity in the World Ocean

1. Observations on individual species (microscopy), their properties (flow cytometry), or their genomes (molecular methods)

2. Observations on chemical compounds, including compounds dissolved in seawater (nutrients) and within the phytoplankton themselves (pigments, elemental composition)

3. Observations using photosynthetic tracers in incubations, including natural tracers (oxygen, carbon dioxide), radioisotope tracers (^{14}C, ^{32}P), and stable isotopes (^{13}C, ^{15}N)

4. Observations of phytoplankton fluorescence, including active (stimulated fluorescence, pump-and-probe, FRRF, Lidar) and passive techniques (natural fluorescence)

5. Observations of ocean color, including in-water techniques (Secchi disk, spectroradiometers) and above-water techniques (airborne or satellite measurements of water-leaving radiance)

6. Observations of particles, including suspended particles (beam transmissometers) and sinking particles (sediment traps)

TABLE 13.2 Size Classification of Phytoplankton

Name	Size*	Representatives
Picophytoplankton	0.2–2 μ m	Cyanobacteria (e.g., *Prochlorococcus, Synechococcus*)
Nanophytoplankton	2–20 μ m	Small diatoms (e.g., *Thalassiosira*), coccolihophorids (e.g., *Emiliana huxleyi*)
Microphytoplankton	20–200 μ m	Large diatoms (e.g., *Cosincodiscus*), dinoflagellates (e.g., *Gonyaulax, Noctiluca*)

*Note that some researchers find it useful to distinguish ultraplankton in the 2–5 μ m size range. This separation enables a closer look at microorganisms involved in the microbial loop, a subsystem of planktonic ecosystems that is dominated by protozoans, bacteria, and small phytoplankton, including cyanobacteria.

substances and radiant energy from the Sun. Beyond that, however, their similarities quickly vanish. From their size and their shape to the composition and arrangement of their photosynthetic pigments, the phytoplankton encompass a diverse group of organisms that have evolved a variety of strategies for surviving in the oceanic environment.

Because phytoplankton readily lend themselves to separation using filters, scientists often classify phytoplankton (and other plankton) according to the major size classes of the organisms (**Table 13.2**). Of these three groups, the picophytoplankton have gained the most recent attention. Now they are believed to be the most abundant phytoplankton in the world ocean. Before the late 1970s, their existence was virtually unknown. The discovery of two distinct types of picophytoplankton, the prokaryotic **cyanobacteria** *Synechococcus* (discovered by John Waterbury in 1979) and *Prochlorococcus* (discovered by Shirley Chisolm in 1988), fundamentally changed the way scientists conceived of ocean ecosystems. Scientists now recognize that cyanobacteria contribute to a new type of food web known as the microbial food web (see Chapter 14).

Synechococcus, about 1.5–2.5 μm in size, seems to prefer tropical and subtropical waters (such as the Sargasso Sea) at high light intensities, although it may be found throughout the world ocean, including polar regions. *Prochlorococcus*, at less than 0.7 μm in diameter (about one-hundredth the width of a human hair), is the smallest known photoautotroph on Earth. It has been found to inhabit waters between 40° N and 40° S (cold temperatures may be lethal) in at least two ecotypes: one that prefers high light and one that prefers low light. Both species appear to have very small genomes yet exhibit considerable plasticity in their ability to adapt to varying oceanic conditions. Both may be capable of nitrogen fixation and may even be capable of heterotrophy.

Taxonomic classification of marine phytoplankton continues to evolve as more species are successfully cultured and their genomes are sequenced. In addition, debates over the naming of subgroups and even the major groups—are they phyla or divisions?—keep the field in a state of flux. Nevertheless, a fairly robust listing of the nine major groups that one might encounter at sea is provided in **Table 13.3**. While it is not intended that you memorize these groups, they are provided here as a starting point for further exploration by those students and instructors who have a greater interest in this topic.

Photosynthetic pigments may also be used to identify and classify different types of phytoplankton. Pigments are the chemicals that give color to our world. In the case of marine phytoplankton, these pigments capture radiant energy from the Sun. The primary light-absorbing pigment in all oxygen-producing autotrophs, including the cyanobacteria, is **chlorophyll a.** This molecule absorbs light in the waveband that penetrates the deepest into the ocean about 440 nm (**Figure 13-7**). Its concentration in seawater is easily determined using instruments that detect fluorescence. Determinations of the concentration of chlorophyll *a* provide one of the best and most widely adopted measures of phytoplankton biomass (i.e., the quantity of phytoplankton) in the world ocean.

Because it is present in all phytoplankton, chlorophyll *a* provides little information on the species present in a given location. To distinguish major groups, scientists measure other kinds of chlorophylls and **accessory pigments,** light-harvesting pigments other than chlorophylls that assist photosynthetic reactions and enable phytoplankton to adapt to different light environments. Chlorophytes can

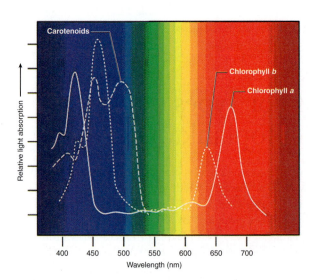

■ **FIGURE 13-7** Absorption spectra for chlorophylls *a* and *b* and carotenoids. Carotenoids function in photoprotection in many species of phytoplankton.

TABLE 13.3   Taxonomic Classification of Phytoplankton

Domain	Taxonomic group	Common name	Occurrence	Marine phytoplankton representatives
Bacteria	Cyanobacteria	Cyanobacteria	Abundant, oceanic and coastal	*Synechococcus, Nostoc, Trichodesmium, Prochlorococcus*
Eukarya	Discomitochondria	Euglenids	Rare, coastal?	*Eutreptia, Eutreptiella*
	Dinomastigota	Dinoflagellates	Abundant, oceanic and coastal, may form unialgal blooms	*Gonyaulax, Pfeisteria, Ceratium, Noctiluca, Gymnodinium, Dinophysis*
	Rhodophyta	Red algae	Rare, coastal	*Rhodosaurs marinus*
	Diatoms	Diatoms or golden-brown algae	Abundant, oceanic and coastal, may form unialgal blooms	*Nitzschia, Asterionella, Thalassiosira, Chaetoceros, Coscindodiscus*
	Haptomonada	Coccolithophorids	Abundant, oceanic mostly, may form unialgal blooms	*Emiliana huxleyi, Gephryocapsa, Prymnesium*
	Chlorophyta	Green algae	Common, oceanic and coastal, abundant in freshwater	*Halosphaera, Pyramimonas, Scenedesmus*
	Cryptomonada	Cryptomonads	Common, mainly coastal	*Cryptomonas, Chroomonas*
	Chrysomonada	Silicoflagellates	Occasionally abundant, probably cosmopolitan	*Dictyocha, Distephanus*

Source: Taxonomic groups based on Margulis and Schwartz (1998, 3rd ed.) following the listing of Falkowski and Raven, (1997), based on John (1996), Mabberly (1987), and Parsons, Takahasi, and Hargraves (1984, 3rd ed.).

be identified by the presence of chlorophyll *b*. A form of chlorophyll *b* occurs in the cyanobacteria, but advanced techniques for separating and identifying pigments in the field, namely, **high-pressure liquid chromatography** (HPLC), now allow scientists to tell these pigments and many others apart. Chlorophyll *b* enables these algae to absorb a greater percentage of blue-green light.

Another type of chlorophyll, chlorophyll *c*, may be found in diatoms and coccolithophorids. Chlorophyll *c* absorbs green light and further extends the ability of these organisms to exploit different light environments. Diatoms contain another important accessory pigment, **fucoxanthin,** by which they may be easily distinguished from other phytoplankton. By increasing their cellular concentrations of fucoxanthin, diatoms increase their ability to capture light at low-light intensities. It is these pigments combined that give diatoms a golden-yellow or brown appearance and how brown tides, blooms of diatoms, got their name.

Some dinoflagellates apparently manufacture fucoxanthin, but they may be differentiated from the diatoms by the presence of **peridinins,** the reddish pigment responsible for red tides. Fucoxanthin and peridinin belong to a major class of photosyn-

thetic pigments called **carotenoids.** Carotenoids serve dual functions in many phytoplankton: some carotenoids, such as fucoxanthin, peridinin, and beta-carotene, deliver absorbed solar energy to the main photosynthetic machinery. Other carotenoids, such as the xanthophylls, protect the photosynthetic machinery from burning up; they serve a photoprotective function.

Primary Productivity

All organisms on Earth depend directly or indirectly on primary producers for energy and matter. **Photoautotrophs** convert inorganic carbon into organic compounds using radiant energy from the Sun and dissolved substances in water. The phytoplankton are one example. **Chemoautotrophs,** organisms that use chemical energy to synthesize organic carbon, are also considered primary producers. The symbiotic chemoautotrophic bacteria found in hydrothermal vents are an example.

A Generalized Scheme of Photosynthesis

Photosynthesis produces the chemical energy that drives the cycling of matter in nearly all of Earth's ecosystems. The description by Kirk (1994) emphasizes the energetic aspects of photosynthesis in two fundamental sets of reactions: 1) **light reactions,** the chemical reactions that are driven by sunlight; and 2) **carbon-fixation reactions,** the chemical reactions that use the energy produced by the light reactions to manufacture sugars and other energy-yielding products (**Figure 13-8**).

Overall, we can combine the light reactions and carbon fixation reactions into a general equation for photosynthesis

$$CO_2 + H_2O + light \longrightarrow (CH_2O) + H_2O + O_2$$

If we include the major limiting nutrients and apply the Redfield ratios for phytoplankton (106 C:16 N:1 P), we get:

$$106 CO_2 + 122 H_2O + 16 HNO_3 + H_3PO_4$$
$$\longrightarrow (CH_2O)_{106} + [NH_3]_{16} + H_3PO_4 + 138 O_2$$

Even when written this way, this equation leaves out a lot of steps. Nonetheless, it symbolizes the important aspects of photosynthesis, namely, that inorganic carbon dioxide and water are transformed using light energy into organic carbon compounds, yielding oxygen as a by-product. It also emphasizes the role of biologically important nutrients, those dissolved substances required for the growth of phytoplankton. In the oceanic environment, the availability of both light and nutrients governs rates of photosynthesis.

Note that if we reverse the first equation, we end up with an equation that describes **respiration,** the breakdown of organic matter in the presence of oxygen to yield carbon dioxide and water. Photosynthesis and respiration are interdependent processes that balance the exchange of materials and energy both within phytoplankton and their consumers. Phytoplankton, like all aerobic organisms on Earth, must respire and consume oxygen to synthesize the cellular products that allow them to survive, grow, and reproduce.

Types of Productivity

Ecologists define **gross primary productivity** (GPP) as the rate of conversion of inorganic carbon into organic carbon (i.e., photosynthesis) and **net primary productivity** (NPP) as the organic carbon remaining after the metabolic needs of the autotrophs have been met. Net primary productivity, therefore, represents the organic carbon available to other trophic levels, that is, the heterotrophs. Some ecologists define **net community production** (NCP) to describe the amount of organic carbon available for export after the needs of all trophic levels have been met. This distinction becomes important in aquatic ecosystems because it is virtually impossible to measure separate rates of autotrophic and heterotrophic respiration.

In the most general sense, GPP can be related to NPP and R by the following equation:

$$NPP = GPP - R$$

where R represents the combined respiratory losses from all members of the ecosystem under study.

Oceanographers also distinguish between new production and regenerated production, following from the classical studies of oceanographers Richard Dugdale and J. J. Goering in 1967. **New production** refers to phytoplankton growth based on a limiting nutrient, such as nitrate, the nitrogen compound that generally limits rates of photosynthesis in the ocean. Upwelling of nitrate or other physical processes that bring nitrate to the upper ocean cause "new" growth of phytoplankton. By contrast, **regenerated production** refers to phytoplankton growth on ammonia, a nitrogen waste product produced by heterotrophic organisms that feed on phytoplankton. Ammonia-based primary production exists on "recycled" nitrogen and does not represent new phytoplankton growth. The distinction between new production and regenerated production stems from a desire to track the fate of carbon in ocean food webs and to identify carbon available for export to the deep sea. **Export production,** the percentage of new carbon that is exported to another part of the ocean, occurs via sinking and the biological pump. While some of this carbon is available to deep-sea organisms, a portion of export production may be permanently buried in sediments on the seafloor (see Chapter 5).

Controls on Primary Productivity

The concept of a limiting factor is an extremely important one as it defines the factor or factors that control the growth of phytoplankton at any given moment. In many cases, the limiting factor

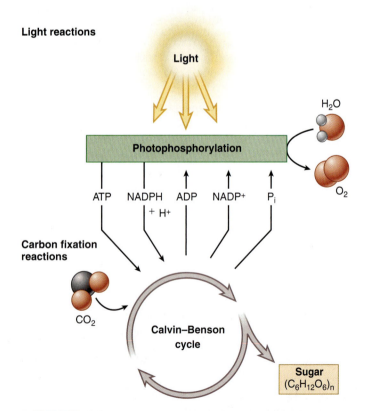

Light reactions

Carbon fixation reactions

■ **FIGURE 13-8** The overall scheme of photosynthesis.

may switch. For example, during winter, light may limit growth despite an abundance of nutrients (**Figure 13-9**). During summer, when light is plentiful, nutrients may limit growth (**Figure 13-10**). In ecosystems, the definition of a limiting factor may be compounded by the observation that limiting factors for one species may not be the same for another. For example, diatoms may be limited by silica while cyanobacteria are not. Thus, a limiting factor may alter the **species composition**, the diversity of species and their abundances in an ecosystem.

Photosynthetic Light Limitation

As discussed in Chapter 7, the intensity of light diminishes with depth in the water column approximately in accordance with Beer's law. At the same time, water, suspended particles, and dissolved substances selectively absorb and scatter different wavelengths of light so that the spectrum of light changes with depth. The wavelengths of light that are available for photosynthesis at a particular depth are referred to as **photosynthetically available radiation** (PAR). The decrease in PAR with increasing depth has important ramifications for the growth of phytoplankton. Where PAR diminishes rapidly, phytoplankton growth may be confined to the near-surface waters. Where PAR penetrates more deeply, phytoplankton growth may occur at greater depths. The intensity of light at a particular depth determines whether photosynthesis can keep up with the metabolic demands of the phytoplankton. The point at which the rate of photosynthesis equals the rate of respiration is called the **compensation point.** Typically, the 1% light level corresponds to the compensation point for most phytoplankton. For that reason, the 1% light level defines the lower limit of the **euphotic zone,** the region in which net photosynthesis occurs.

Dissolved Inorganic Nutrients

The major elements required for the growth of phytoplankton include those elements that are essential for all living organisms, primarily the ones that meet structural, catalytic, and other func-

tions (**Table 13.4**). The macronutrients considered essential for phytoplankton growth include carbon, hydrogen, oxygen, nitrogen, phosphorus, silica, magnesium, potassium, and calcium. Note, however, that some elements, such as silica and calcium, will be in greater demand by certain species, such as diatoms and coccolithophorids, respectively. Cyanobacteria and chlorophytes may not be limited at all by these elements. Trace elements include iron, manganese, zinc, boron, sodium, molybdenum, chlorine, vanadium, and cobalt. Other trace elements may be required by individual species. Any nutrient may limit the growth of phytoplankton if it is not supplied in a quantity that meets their metabolic demands.

Temporal patterns in the concentration of nutrients in the surface waters of the world ocean may be attributed directly to the seasonal growth of phytoplankton. Because rates of photosynthesis tend to be higher at the surface than at depth, nutrients tend to be removed more quickly at the surface than they are at depth. In fact, surface waters may become completely depleted of nutrients, giving rise to a water column feature known as the **nutricline.** The nutricline is that region in the water column where the concentration of nutrients changes rapidly. Above the nutricline, the concentration of a limiting nutrient will be undetectable. Phytoplankton growth and nutrient removal from the water column also vary with the seasons. During winter, when the water column is well-mixed, nitrate reaches its maximum concentration in the water column owing to the deep convective mixing of surface waters. On the other hand, summer stratification of the water column may prevent nutrient replenishment, causing depletion of nutrients in surface waters.

Zooplankton and Nutrients

A number of biological processes may replenish nutrients in the water column. The feeding mechanism of larger zooplankton produces broken and uneaten bits and pieces of phytoplankton, a phenomenon known as **sloppy feeding.** The food zooplankton eat eventually is egested in the form of membranous and occasionally ornate **fecal pellets,** essentially, packages of partially digested phytoplankton. These fecal pellets may sink rapidly out of the euphotic

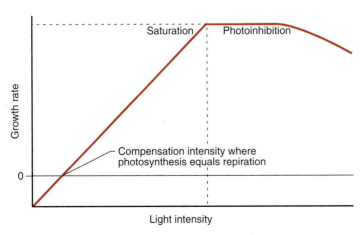

■ FIGURE 13-9 Light-limited growth of phytoplankton. As light intensity increases, the rate of photosynthesis increases. At some point, the intensity of light exceeds the photosynthetic capacity. Additional increases in light intensity inhibit growth, a process called photoinhibition.

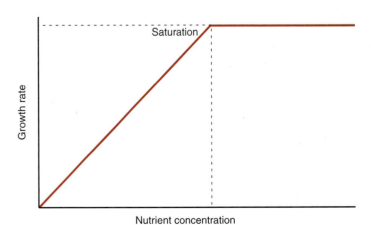

■ FIGURE 13-10 Nutrient-limited growth of phytoplankton. Where a particular nutrient, like nitrate, controls the growth of phytoplankton, increases in the concentration of the nutrient increase the growth rate. At the saturation point, additional increases in nutrient concentration have no effect on growth.

TABLE 13.4 A Few Biologically Important Nutrients for Phytoplankton Growth

Nutrient	Formula	Availability	Used by
Phosphate	PO_4^{3-}	May be the ultimate limiting nutrient over geologic time; seasonally depleted in some waters	All phytoplankton
Nitrogen	N_2	Abundant but unusable for most phytoplankton	Nitrogen-fixing cyanobacteria
Nitrate	NO_3	Often depleted in surface waters (seasonally); abundant in deep waters	Many phytoplankton but requires energy for its uptake
Ammonium	NH_4+	Dependent on grazers; higher availability during summer	Most phytoplankton
Iron	Fe	Limiting in ocean regions where atmospheric deposition is low	All phytoplankton
Silicic acid	SiO_4	Seasonally depleted in surface waters at high latitudes, in upwelling zones, and elsewhere	Diatoms, radiolarians, and silicoflagellates
Calcium Carbonate	$CaCO_3$	Typically abundant; may be depleted under bloom conditions	Coccolithophorids, foraminifera, corals, most mollusks
Trace metals	Various	Depends on region and conditions	Likely, all phytoplankton require different trace metals

zone, or they may be eaten by other zooplankton, a form of feeding called **coprophagy.** Zooplankton also excrete dissolved compounds, such as ammonium, that may be used by phytoplankton when other forms of nitrogen are not available.

Marine Bacteria and Nutrients

The activities of marine bacteria in biogeochemical cycles play an important role in the supply of nutrients to phytoplankton. One way that marine bacteria make nutrients available to phytoplankton is through the breakdown of dissolved and particulate organic matter. This process, known as **remineralization,** is a major source of dissolved nutrients in the ocean. Breakdown of organic matter below the euphotic zone releases nutrients into deeper waters and makes them available to phytoplankton when these waters are upwelled or mixed with surface waters. Remineralization also completes the short-term organic carbon cycle whereby organic carbon is decomposed to inorganic carbon, that is, carbon dioxide (see Chapter 6). Bacteria also participate in other important biogeochemical cycles, such as the **nitrogen cycle** (**Figure 13-11**). Oceanographers have long known that *Trichodesmium,* a species of cyanobacteria, carries out 25 to 50% of the world ocean's nitrogen fixation, the process by which nitrogen gas is converted to organic nitrogen. Blooms of *Trichodesmium* can even be viewed from space. Yet, scientists were surprised to find that other coccoid species of cyanobacteria, such as *Synechococcus*

and *Prochlorococcus,* may also be involved in nitrogen fixation. It is now believed that many types of marine bacteria, including heterotrophic bacteria and bacteria living in the guts of zooplankton, may carry out nitrogen fixation in the ocean. Researchers are also discovering diverse forms of bacteria involved in the conversion of ammonium to nitrate (via nitrification) and the conversion of nitrogen back to nitrogen gas (via denitrification). Research on the metabolic activities of marine microbes is an exciting new area in biological oceanography.

Zooplankton Grazing

Grazing on phytoplankton by zooplankton represents a direct loss of biomass rather than a limit on the rate at which phytoplankton photosynthesize. This distinction is important because phytoplankton can maintain high rates of growth and divide rapidly and still not increase in numbers or biomass. If zooplankton feeding keeps pace with the rate at which phytoplankton divide, then zooplankton biomass may increase while phytoplankton biomass remains the same. The degree to which zooplankton feeding controls phytoplankton growth rates and biomass depends on a great number of factors, including the size and type of the zooplankton, their life cycle, and temperature, among others (**Figure 13-12**). Predation on zooplankton by other organisms may exert its own set of controls. Thus, while zooplankton grazing has some effect on primary productivity, the magnitude of these effects under a given set of circumstances may be uncertain.

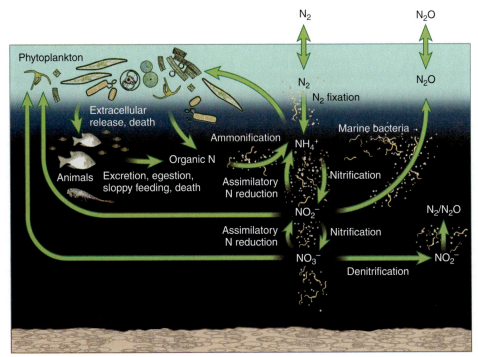

■ FIGURE 13-11 The oceanic nitrogen cycle is one example of the way in which elements are stored and exchanged between various living and non-living reservoirs. Rarely are such biogeochemical cycles simple! Many of the "steps" of the nitrogen cycle—its conversion from one form to another—produce a form of nitrogen that can be used by phytoplankton.

Sedimentation

In Chapter 5, we introduced Stoke's law, a mathematical formulation that describes the sinking rate of particles. While phytoplankton rarely conform to Stoke's law, it provides a rough estimate of the rate at which phytoplankton may sink in the ocean. Sinking or sedimentation of phytoplankton becomes quite important during large blooms of phytoplankton when zooplankton grazing cannot keep pace with the production of phytoplankton biomass. Dead, dying, decaying, and partially grazed phytoplankton along with zooplankton fecal pellets create what is collectively known as **phytodetritus.** This sinking "rain" of particles may be further processed at mid-water depths before sinking to the seafloor, where it provides an important food source for benthic organisms (see Chapter 14).

Vertical Mixing

Physical processes provide mechanisms for the resupply of nutrients to the euphotic zone by mixing nutrient-rich deeper waters with nutrient-depleted surface waters. Physical processes also moderate primary productivity by changing the light field experienced by the phytoplankton. Vertical mixing—the upward and downward movements of seawater—acts like a conveyor belt that carries phytoplankton deeper or shallower in the water column. As a result, the intensity of light and even the availability of nutrients may change. Vertical mixing results from a number of processes, many of which were discussed in previous chapters. Phytoplankton living in the surface mixed layer will be subject to its motions. As water within the mixed layer is set in motion, phytoplankton suspended in that water will experience the full range of light intensities from the surface of the ocean to the depth of the mixed layer. In a shallow mixed layer, the overall irradiance experienced by phytoplankton will be higher than in a deep mixed layer. Thus, rates of photosynthesis within a shallow mixed layer will be greater than rates of photosynthesis within a deep mixed layer.

The relationship between mixed layer depth and productivity was first noted by Gran and Trygve Braarud (1903–1985) in 1935. They noted that phytoplankton populations would not increase if rates of photosynthesis did not meet the metabolic demands of the phytoplankton. They defined what is known as the **compensation depth,** the depth at which the rate of photosynthesis equals the rate

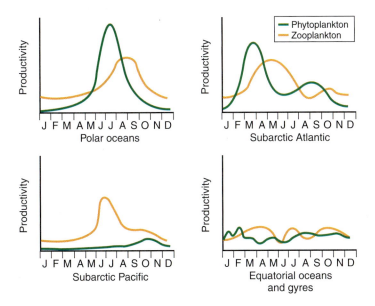

■ FIGURE 13-12 Comparison of the effects of zooplankton grazing on phytoplankton biomass in the North Pacific and North Atlantic. In the North Pacific, the presence of microzooplankton and copepod grazers at the start of the spring bloom limits increases in phytoplankton biomass. In the North Atlantic, delayed hatching of nauplii permit phytoplankton biomass to increase without significant grazer control.

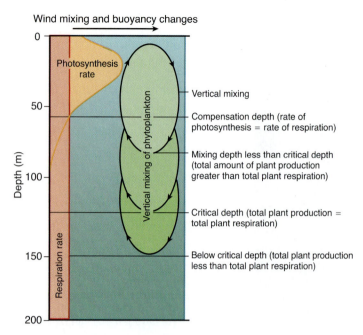

■ FIGURE 13-13 Vertical mixing, the compensation depth, and the critical depth.

of respiration (**Figure 13-13**). Phytoplankton could still sink below the compensation depth as long as the time they spent there (experiencing a net loss of organic carbon) did not exceed the time they spent above the compensation depth (experiencing positive gains in photosynthesis). In 1953, Harald Sverdrup formalized the concept of the **critical depth.** The critical depth is the deepest that phytoplankton can be mixed to where their total production equals their total respiration. Note the subtle difference between critical depth and compensation depth. The compensation depth is the depth at which the instantaneous rates of photosynthesis and respiration are equal. The critical depth is the depth above which the total productivity that phytoplankton experience in the water column is equal to their total respiration. These concepts established the foundations on which biological oceanography rests today.

The Seasonal Cycle of Primary Productivity

As the phytoplankton go, so goes ocean food webs. The structure and functioning of ocean food webs are tightly coupled to the activities of the primary producers and the factors that affect them. Much of what we know about ocean food webs comes from studying seasonal changes in populations of primary producers and correlating those changes with variations in environmental factors. In Chapter 8, we explored stratification of the water column and formation of the seasonal thermocline in response to seasonal changes in solar radiation. Here, we revisit the seasonal thermocline and explore the effects of this physical process on the chemistry and biology of the water column (**Figure 13-14**).

During winter, temperate regions typically lack a seasonal

thermocline as negative buoyancy caused by atmospheric cooling of surface waters keeps the water column well mixed. This deep mixed layer combined with short day length and low sun angles means that the light available for photosynthesis is minimal. Although dissolved nutrients may be abundant (as a result of mixing of surface and deep waters), primary productivity is light-limited. In this situation, the critical depth is shallower than the depth of the mixed layer. In spring, the physical conditions for phytoplankton growth become quite favorable. Heating of the sea surface stratifies the water column and forms a seasonal thermocline. In turn, the mixed layer becomes more shallow. Increased day length and higher sun angles make more light available, and the critical depth deepens. When the depth of the mixed layer becomes shallower than the critical depth, conditions favor the growth of phytoplankton and they bloom. Just like the spring blossoming of trees and shrubs in terrestrial environments, the **spring bloom** of phytoplankton sets off a profusion of biological activity. Grazing on phytoplankton by zooplankton and their larvae increases. The growth and reproduction of zooplankton stimulate increases in the growth and reproduction of carnivores such as squids, fishes, and marine mammals. Microbial processes intensify, as heterotrophic bacteria absorb new pools of dissolved organic matter and break down the new bits of organic debris. Production of marine snow, fecal pellets, phytoplankton aggregates, and detrital particles also becomes more common. These sinking particles of organic matter may sustain pelagic and benthic ecosystems for months. The spring bloom may be observed from shore or most dramatically by satellites, as the abundant phytoplankton turn the ocean a recognizable shade of green.

The spring bloom does not last forever, however. The demand for dissolved nutrients as a result of increased rates of primary productivity outstrips the rate at which those nutrients can be resupplied. Where rates of primary productivity are highest—typically, at or near the surface—one or more dissolved nutrients may be completely removed by phytoplankton. If nitrate is the macronutrient used up most quickly, then the unavailability of nitrate limits primary productivity for those phytoplankton that require nitrate (e.g., diatoms). Other nutrients, such as silicate or iron, may also limit primary productivity. When this occurs, primary productivity is said to be nutrient-limited. Despite the abundance of light, rates of primary productivity and the biomass of phytoplankton plummet in the absence of nutrients. The depth of the nutricline deepens after the spring bloom to a point where light becomes limiting. Photoadaptation may sustain phytoplankton biomass and further deepening of the nutricline, most often observed as the depth of the subsurface chlorophyll maximum. Primary productivity during summer may be minimally sustained by those nutrients that diffuse across the nutricline or that reach the euphotic zone through physical processes (eddies, breaking internal waves, etc.). In the Sargasso Sea, cyanobacteria and other forms of picophytoplankton may reach their peak abundance during summer, owing to their ability to fix nitrogen or sustain photosynthesis on other forms of nitrogen other than nitrate. Jellyplankton and marine protists that are able to filter picophytoplankton may also reach peak abundance. Populations of crustacean zooplankton, namely copepods, decrease as they cannot be sustained by picophytoplankton that are too small for their filtering appendages.

Fall brings yet another shift in physical and chemical conditions that stimulate primary productivity. Cooling of surface waters in fall

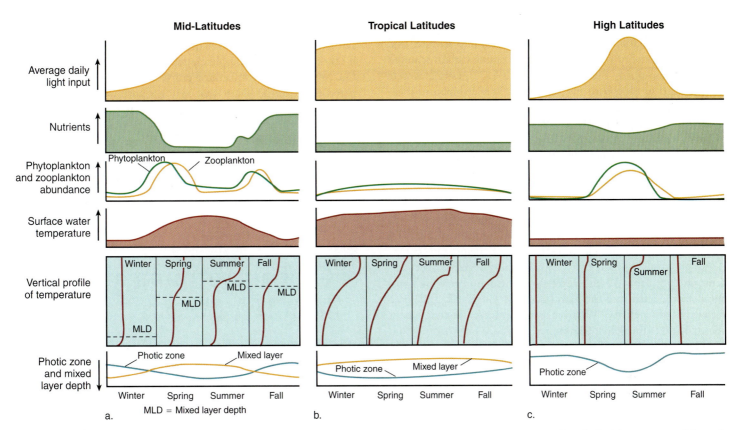

■ FIGURE 13-14 Vertical distribution of physical, chemical, and biological properties during the seasonal cycle in (a) temperate (b) tropical and (c) polar oceans. In many ways, this figure illustrates the fundamental functioning of ocean food webs and the close interactions between physical, chemical, and biological processes in the world ocean (after Segar).

may deepen the mixed layer and erode the seasonal thermocline. Mixing of nutrient-depleted surface waters with nutrient-replete deeper waters may raise concentrations of limiting nutrients in surface waters. If phytoplankton remain above their critical depth, a **fall bloom** may occur. Depending on conditions, the fall bloom may rival the spring bloom in intensity, but higher populations of grazers in fall (due to their growth and reproduction in spring and summer) generally limit increases in phytoplankton biomass. Fall blooms tend to disappear more quickly as the water column destabilizes under continued cooling and light intensity diminishes during the bloom.

Global Primary Productivity

Global estimates of oceanic primary productivity are essential for addressing the impacts of human-induced climate change. Human activities release about 5.5 billion metric tons (plus or minus 0.5 tons) of carbon dioxide into the atmosphere annually through combustion of fossil fuels. Another 1.6 metric tons (plus or minus 1 ton) of carbon dioxide are released through deforestation (e.g., burning of tropical rain forests) and other land-use practices. Of the 7.1 billion metric tons released, approximately 3.3 billion tons accumulate in the atmosphere. The remaining 3.8 billion metric tons enter either terrestrial or oceanic systems, where the carbon is removed from the atmosphere. To understand what happens to that carbon dioxide and where it goes requires knowing how much is converted to organic carbon by phytoplankton in the ocean and whether it is buried "permanently" in sediments or simply re-released through respiration back to the atmosphere.

Satellite-based estimates of primary productivity based on chlorophyll coupled with mathematical models have greatly improved our ability to determine oceanic net primary productivity (**Figure 13-15**). Net primary productivity is the preferred quantity because

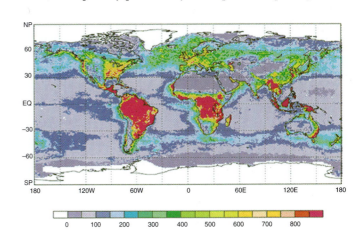

■ FIGURE 13-15 Annual global net primary productivity in grams carbon per square meter.

this is the amount available to herbivores and higher trophic levels (after phytoplankton respiration has been taken into account) and the relevant quantity when discussing the fate of carbon removed from the atmosphere. Before satellites (pre–1978), estimates were largely based on shipboard measurements. While shipboard measurements continue to provide fundamental information on oceanic photosynthesis, especially the effects of light, nutrients, and temperature, satellite measurements have provided the most complete estimates of global productivity. By the mid–1990s, oceanographers concluded that oceanic productivity was twice what they had previously estimated, opening the door for new insights into the role of the oceans in the carbon cycle.

Oceanographers now estimate that approximately 103 billion metric tons of carbon dioxide are fixed in the ocean. Of that amount, some 45 to 50 billion metric tons of organic carbon represent net primary productivity. By comparison, terrestrial ecosystems produce about 52 billion tons of net carbon per year. Thus, the net productivity of the world ocean is roughly equivalent to terrestrial productivity. What is perhaps most startling about this statistic is that phytoplankton account for only 0.2% (about 1 billion metric tons) of the total autotrophic biomass on Earth. How does such a small standing crop of phytoplankton match the productivity of all the grasses and forests on land?

Part of the answer lies in the turnover time of phytoplankton. **Turnover time** is the time it takes to completely replace the standing crop of phytoplankton or carbon or any other measurable quantity. A simple way to understand this is to consider a dozen people in line at a checkout counter. If one person is served every five minutes and a new person gets in line at the same rate, then the turnover time is 60 minutes. In other words, it takes one hour to completely replace the dozen people in line. On average, the phytoplankton crop turns over in one week, compared to terrestrial plants that may not turn over for months to decades. Although the turnover of phytoplankton is high, tropical rain forests remain the most productive regions on Earth when considered in terms of productivity per acre of real estate. High productivity is spatially limited in the oceans to regions of upwelling and estuaries. Thus, the combination of turnover rate and the spatial distribution of autotrophs is important when comparing ocean and land productivity.

Global Warming and the Iron Hypothesis

While the limiting nutrient for most of the world ocean appears to be nitrate, other nutrients, including iron and silicate, are clearly limiting in at least some parts of the ocean. The prospect of **iron limitation,** championed by John Martin (1935-1993) in the 1980s, stimulated considerable efforts to determine the role of iron in the ocean, especially in **high-nutrient, low-chlorophyll** (HNLC) regions. The primary source of iron in the ocean is through wind dispersal of iron-containing particles from the

TABLE 13.5   Comparison of Rates of Primary Production in Various Parts of the World Ocean Versus Major Ecosystems Worldwide

World ocean NPP (units $\times\ 10^{15}$ grams carbon year^{-1})		Terrestrial NPP (units $\times\ 10^{15}$ grams carbon year^{-1})	
Seasonal			
January to March	11.3		11.2
April to June	10.9		15.7
July to September	13.0		18.0
October to December	12.3		11.5
Biogeographic			
Oligotrophic	11.0	Tropical rain forests	17.8
Mesotrophic	27.4	Broadleaf deciduous forests	1.5
Eutrophic	9.1	Broadleaf and needleleaf forests	3.1
Kelps and seaweeds	1.0	Needleleaf evergreen forests	3.1
		Savannas	16.8
		Perennial grasslands	2.4
		Broadleaf shrubs with bare soil	1.0
		Tundra	0.8
		Desert	0.5
		Cultivated	8.0
TOTAL	48.5		56.4

Source: Based on Field et al. (1998).

continents. A number of early oceanographers had speculated that regions where aerial deposition of iron is low may limit productivity. As Martin demonstrated, iron concentrations in certain regions of the world ocean were extremely low (two orders of magnitude lower than predicted by these early researchers). Until recently, studying iron limitation in the world ocean has been nearly impossible because most oceanographic equipment (wires, ship hulls, bottle parts, etc.) contains iron! Collection of uncontaminated waters to test this hypothesis was a major impediment. As clean techniques were developed in the 1970s—techniques that allowed the collection and storage of water uncontaminated by shipboard iron—oceanographers demonstrated that iron concentrations in natural waters were indeed very low in HNLC regions. Several dedicated experiments were then conducted in the equatorial Pacific (IronEx I), the subarctic Pacific (IronEx II), and the Southern Ocean (SoFex, Southern Ocean Iron Experiment) to determine the extent to which iron limitation governs productivity in these waters (**Figure 13-16**). All of these studies demonstrated that iron-enhancement of samples stimulated the growth rates of phytoplankton, especially diatoms. The export of carbon out of the mixed layer, however, was highly variable. Experiments to date have not demonstrated that large carbon exports accompany iron enrichment. Dugdale and colleagues from other institutions have also shown that silicate regulates phytoplankton growth in upwelling regions of the equatorial Pacific. They propose designating such regions as low-silicate HNLC regions "to avoid the impression that all major nutrient concentrations are sufficiently high to support new production."

The demonstration of iron limitation in these waters has provided important insights into global oceanic productivity over geologic time and forces that produce climate change. For example, the Southern Ocean is presently a HNLC region, but during the last glacial maximum (about 18,000 years ago), there was more dust and an estimated 50 times more iron. At that time, phytoplankton flourished and the biological pump was active. Ice core records indicate that carbon dioxide was low, consistent with the idea that high availability of iron stimulates high oceanic productivity, which in turn reduces atmospheric CO_2. Iron and carbon dioxide determinations from ice core data for the past 140,000 years reveal opposite trends: when iron increases, carbon dioxide decreases. These results suggest that iron plays a major role in the abundance of atmospheric carbon dioxide. However, their exact relationship remains the subject of ongoing research.

The proposal to fertilize the ocean with vast amounts of iron to reduce atmospheric carbon dioxide has gained its share of hopefuls and naysayers. Several commercial enterprises are being developed to outfit ships with equipment to dump large amounts of iron in HNLC regions. The hope is that phytoplankton growth would be greatly enhanced and atmospheric carbon dioxide sequestered to the deep sea. The oceanographic community has expressed considerable interest, but oceanographers remain concerned about the feasibility of these efforts and potentially negative side effects on ocean foods webs. It will be interesting to see how these important issues will be addressed in the coming years.

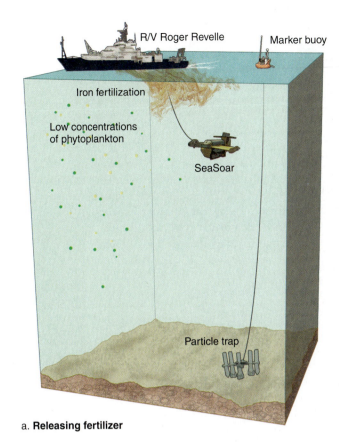

a. **Releasing fertilizer**

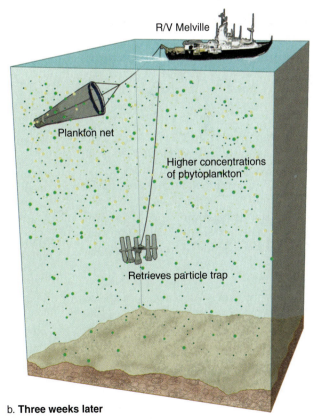

b. **Three weeks later**

■ **FIGURE 13-16** Iron enrichment experiments.

Ocean Color Gets Hyper!

One of the key observational tools of sighted humans is vision. Visualized shapes and sizes are used to detect objects that can be useful, harmful, or merely interesting. Add the sensing of spectral light, known as color, and even more information is logged into our brains. Few oceanographic measurement methods within the past few decades have advanced as rapidly as ocean color sensing. As is often the case, optical technologies have enabled the ability to solve existing problems and stimulated new questions and research topics. Today, ocean optics is being used to provide data sets for far-ranging fundamental and applied problems, such as marine ecology, biogeochemical cycling, pollution, and climate change. For example, ocean optics can be used to detect occurrences of harmful algal blooms (**Figure 13a**).

Perhaps you have flown in an airplane over the ocean and noticed the beautiful range of colors of the sea surface. Fascinated by such observations, oceanographers sought to use the spectral light emerging from the sea surface to estimate concentrations of phytoplankton. The light that emerges from beneath the sea surface depends on the ratio of light that is backscattered versus absorbed by waters lying beneath the sea surface, a quantity called reflectance. In oceanic waters unaffected by terrestrial runoff, the emerging light corresponds with chlorophyll *a*, an indicator of phytoplankton biomass. Emergent light at the blue end of the spectrum indicates near-surface waters with relatively low concentrations of chlorophyll *a*, while green emergent light indicates higher concentrations of phytoplankton.

In coastal waters, the presence of dissolved and particulate materials from rivers, land runoff, and resuspension of seafloor sediments interferes with the reflectance measurement of phytoplankton. In these waters, the determination of chlorophyll *a* is not so simple. A new technique, hyperspectral remote sensing, spans the entire visible light spectrum (400–700 nm) to "fingerprint" surface waters. Hyperspectral fingerprinting allows oceanographers to distinguish different phytoplank-

ton groups or even species. Characterization of the types of the ocean bottom is also important, especially where coral reefs, kelp beds, and seagrasses may be affected by climate change. Tracking coastal pollutants is another potential application. Hyperspectral instruments are being deployed from moorings and other *in situ* platforms in conjunction with aircraft and satellites. As it has done in the past, ocean optics continues to provide new tools for observing and understanding the world ocean.

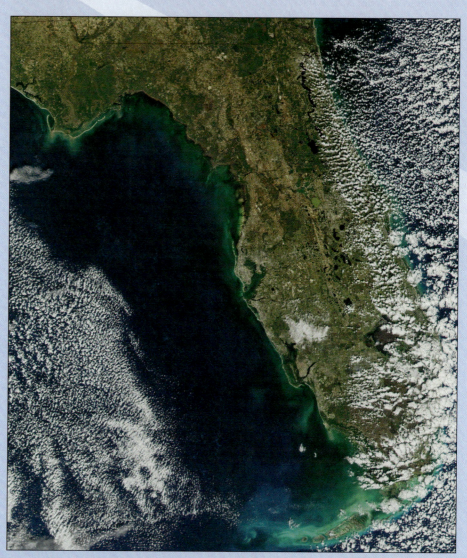

FIGURE 13a Natural color image of a "black" tide (also known as a "red" tide), off the west coast of Florida in September 2001 (discernible as a dark "edge" along the west Florida shelf). A bloom of the toxic dinoflagellate *Karinia brevis* caused the discoloration of the water. These blooms often lead to massive kills of fish and other organisms.

YOU Might Wonder

1. **Sometimes at night during a red tide, the waves give off a bluish glow. How come?**

   Red tides are caused by blooms of dinoflagellates, many of which are bioluminescent (see Chapter 12). For reasons that are not well understood, turbulence stimulates their bioluminescence, that is, the blue glow. Some evidence exists that bioluminescence may deter predators. Bioluminescence in the waves can also be caused by blooms of dinoflagellates that do not cause red tides, most notably, *Noctilua*, a naked dinoflagellate. Many microscopic marine organisms bioluminesce and cause a blue glow in the crashing waves or in the wake of a boat or ship.

2. **Why are polar oceans so productive compared to tropical oceans? It seems just the opposite of terrestrial environments.**

   The key to understanding the difference between polar and tropical oceans lies in nutrients. Deep mixing of polar waters replenishes nutrients to surface waters annually. In tropical oceans, persistent stratification limits the mixing of nutrients into surface waters. Thus, when the ice melts and daylength and sun angle increase, conditions are ripe for the growth of phytoplankton. Such conditions rarely occur in tropical oceans.

3. **In addition to iron fertilization of HNLC regions for reducing atmospheric carbon dioxide, what other schemes have been proposed to combat global warming?**

   Some people have proposed a solid-CO_2 penetrator, which involves freezing carbon dioxide from the atmosphere and storing it beneath the soft sediments of the seafloor. This design calls for shaping blocks of frozen carbon dioxide into "torpedoes" and allowing them to free-fall to the seafloor. Their streamlined shape would allow them to penetrate the soft sediments and form clathrates, essentially blobs of frozen gas. Preliminary work on various delivery systems appears encouraging, yet some studies suggest that the carbon dioxide would not remain frozen. Other studies have suggested that frozen carbon dioxide would negatively affect benthic marine organisms. Thus, like iron fertilization, it is unlikely that this scheme will prove useful for remediating greenhouse house emissions.

Key Concepts

- Phytoplankton are the major producers of organic matter in the world ocean, otherwise known as primary producers.
- Phytoplankton include a diverse range of species from photosynthetic bacteria to mixotrophic protists.
- The discovery of picophytoplankton changed the way oceanographers view energy and material transfers in the world ocean.
- A number of different techniques are used to measure primary production in the ocean to span the large temporal and spatial scales of change.
- A number of physical and chemical factors affect rates of primary production in the world ocean.
- The seasonal cycle of primary productivity drives energy and material transfers in ocean food webs.

Terms to Remember

Critical Thinking

1. Compare and contrast the pico- and nano-sized phytoplankton with the micro-sized phytoplankton in terms of their productivity over temporal and spatial scales?

2. Trace the development of methods for studying primary productivity from cell counts to satellite observations. Compare and contrast three different methods for determining primary productivity. Discuss the advantages and disadvantages of each.

3. Describe the physical and chemical conditions that may lead to light-limited versus nutrient-limited growth of phytoplankton. How do phytoplankton populations adapt to these two different "states"?

4. What is the critical depth hypothesis and how does it explain the spring bloom of phytoplankton in temperate waters?

5. Discuss the differences between new and regenerated production. Under what physical and chemical conditions does each occur? What is the role of zooplankton in providing nutrients to phytoplankton?

6. Discuss the role of vertical migration in the transfer of carbon from surface waters to the deep ocean.

7. Why is an understanding of temporal and spatial scales critical to determining the rate processes that affect the organic and inorganic carbon cycle?

8. Debate the positive and negative consequences of iron fertilization of the world ocean for reducing atmospheric carbon dioxide. Discuss the gaps in our knowledge and how we might go about better understanding if iron fertilization is feasible.

Explore Online

 Visit www.mhhe.com/chamberlin1e for access to chapter quizzing, key term flash cards, video clips, interactive activities, and more. Further enhance your knowledge with web links to chapter-related material!

Exploration Activity 13-1

Exploring Phytoplankton Through Investigation and Modeling

Question for Inquiry

What factors cause phytoplankton to vary over space and time in the world ocean?

Summary

Of all the technological advances in oceanographic research in recent decades, perhaps none has brought the world ocean into sharper focus than satellite measurements of ocean color. With the ability to observe changes in the concentration of chlorophyll simultaneously with changes in sea surface temperature, oceanographers have developed a much-improved understanding of the interactions of biological and physical processes in the world ocean. Such studies have direct bearing on the food we eat (ocean productivity and fisheries), the air we breathe (ocean productivity and atmospheric gases), and the temperature of our habitat (ocean productivity and climate change), to name a few. Here we explore satellite measurements of chlorophyll concentrations and corresponding sea surface temperature data in different biomes in the world ocean.

Learning Objectives

By the end of this activity, you should be able to:

- Acquire and interpret maps and time series of satellite-derived sea surface chlorophyll concentration
- Acquire and interpret maps and time series of sea surface temperature
- Make inferences about the relationship between chlorophyll concentration and sea surface temperature
- Use maps of satellite-derived sea surface chlorophyll concentrations to explore hypotheses concerning the factors that cause variations in time and space

Materials and Preparation

For this activity, you will need access to the Internet and a spreadsheet program. Alternatively, your instructor may provide a printout of figures and a worksheet sufficient for completing all or part of this activity.

Part I: Concept Mapping Phytoplankton Growth

Because concept maps help you to make connections between properties and processes, they are a good place to start exploring the spatial and temporal factors that control the abundance of phytoplankton in the world ocean. Refer to Exploration Activity 4-1 if you are unfamiliar with concept mapping.

1. On a single sheet of paper (or using concept mapping software), draw a concept map that answers the question: What are the characteristics of phytoplankton? Start with phytoplankton as your central box and create separate boxes for each major type of phytoplankton. For each type, create more boxes that further describe and define them. For example, diatoms require nitrate and silica for growth and contain fucoxanthin as an accessory pigment (draw as separate boxes, along with other descriptive characteristics). Use the materials in this chapter to expand and refine your map. When you are finished, draw a single box that includes all of your boxes. Label this box "chlorophyll."

2. On another sheet of paper (or as a separate file), draw a concept map that answers the question: what factors affect the growth of phytoplankton? Start with "phytoplankton growth" as your central box and create separate boxes for physical, chemical, and biological factors. Use the materials in this chapter and previous chapters (for example, see Chapter 6 on biologically important nutrients; Chapter 7 on buoyancy and light; Chapter 8 on wind stress, sea ice, and the physical structure of the water column; Chapter 9 on upwelling; Chapter 12 on plankton and nekton, the chlorophyll maximum, aggregation, etc.) to map factors and processes that regulate phytoplankton growth. You should notice that many of the concepts presented in this book relate to phytoplankton in some way.

3. Compare your two concept maps. Write a 1-page summary describing how the "processes" (your second

concept map) affect the "properties" (your first concept map) over space and time. Your description may be weak and incomplete at this time. The idea here is to start making connections between various physical, chemical, and biological processes and their interactions in the world ocean.

Part II: Exploring Global Maps of Chlorophyll and Sea Surface Temperature

In this part, you will explore spatial and temporal scales of variability in sea surface chlorophyll and temperature in the world ocean as measured the SeaWiFS (Sea-Viewing Wide Field-of-View Sensor) and MODIS (Moderate Resolution Imaging Spectroradiometer) sensors aboard Earth-orbiting satellites.

1. First, it will be helpful to be familiar with the SeaWiFS and MODIS instruments and their mission. Go to http://oceancolor.gsfc.nasa.gov/SeaWiFS/ and click on Spacecraft Information. You may also want to check out the current location of the satellite that is carrying the SeaWiFS instrument. Next, view the tutorial Ocean Research from MODIS on the MODIS website, http://modis-ocean.gsfc.nasa.gov/edu/researchintro/. You will be performing an analysis quite similar to the one presented in the tutorial. Take notes!

2. Data products for SeaWiFS and MODIS are provided by NASA at Ocean Color Web, http://oceancolor.gsfc.nasa.gov/. Many of the resources here support scientific research and can get quite technical. However, there are some excellent resources for students and non-technical users as well. From the home page, click on Image Gallery. You may want to spend some time exploring the imagery of hurricanes and storms, sediments, dust, coastlines, and other phenomena. Many of these images you will find here complement materials discussed in this textbook. Otherwise, click on the SeaWiFS Biosphere Animation, a time series of 32-day composites (combined satellite images) presented as an image file (called an animated GIF). We highly recommend that you download the large version of the animation (centered over the Americas or eastern Asia and Australia). Depending on your connection, it may take several minutes or longer to download it. It's well worth it! While you are waiting, study the frames of the global distribution of chlorophyll as they download (the images that display on the page as your computer is downloading the animation). Use the color key to pick out areas of high chlorophyll and areas of low chlorophyll. Do the same for terrestrial vegetation. Make notes.

3. When the animation downloads, study it for several minutes (or longer), then answer the following questions:
   a. Describe the spatial distribution of chlorophyll in general. What regions exhibit the highest concentrations of chlorophyll? the lowest?
   b. How do the regions of high and low chlorophyll change over the months and years?
   c. What global trends in chlorophyll concentration are apparent over the seasonal cycle? Over multiannual cycles?
   d. Compare and contrast terrestrial vegetation with ocean chlorophyll. Do they vary in step with each other or are they different?
   e. Return to the main animation page and view the small animations of El Nino, Hawaii, and Madagascar. Take notes on what you observe.

4. Return to the Ocean Color Web home page. Click on the Level-3 browser linked on the left side of the page. You should be on a page labeled Level-3 Standard Mapped Images. There are several things to note about navigating this page. Images are displayed starting with the current month (by default). A menu at the top of the page provides links to a number of different variables (which you are encouraged to explore). Below the menu is a series of small rectangles, arranged sequentially according to month and year. The current month (the one displayed on the page you are viewing) is highlighted in a different color. This feature is useful for browsing different months. Starting at the right-hand side of the page, you will see images composed of daily and 3-day composites, and weekly, monthly, seasonal, and yearly composites (if available). You may need to scroll down and across to see all of the images. Go to the month-year menu and click on the previous month (which will give you a monthly composite). View any one of the daily images (click on PNG) and make notes on what you see in the image, noting high and low chlorophyll regions and black areas (where no data exist). Do the same for the 3-day, weekly, and monthly composites. Why do the images "improve" when the daily images are averaged into 3-day, weekly, and monthly composites. What infor-

mation is lost by combining and averaging images? Test your idea by comparing the images again.

5. Your "mission" is to track seasonal variability in three different biomes using the SeaWiFS and MODIS Filled Level-3 Rolling 32-day Composites. You can find these data at the top of the Level-3 Standard mapped Images you viewed above. Note that in the "filled" data, the black regions (where no data exist) have been removed through data processing that uses data from other images to "fill in" missing values (it's a bit more complicated than that but the point to stress here is that NASA is not making up data!). When you get to the page with the filled composites, you will also see a link for Aqua-MODIS SST or sea surface temperature data from the MODIS sensor aboard the Aqua satellite. You will use these data to interpret the changes you observe in chlorophyll concentrations in the three biomes. Refer to Figure 12-10 and choose three locations in the world ocean with each of the following biomes: 1) upwelling system (equatorial or coastal); 2) stratified system (seasonal or permanently); and 3) subpolar or marginal ice zone. For your locations in each of these biomes, choose a filled level-3 rolling 32-day composite image over a one-year period (data are archived back to September 1997) during winter, spring, summer, and fall (choose any month within the season but be ready to defend in scientific terms why you chose one month over another), and answer the following questions:

a. How do chlorophyll concentrations change over the four seasons in each biome? (Provide actual numbers. You may wish to create a table for each biome.)

b. How do chlorophyll concentrations compare between biomes in any one season?

c. How do sea surface temperatures change over the four seasons in each biome? (Provide actual numbers.)

d. How do sea surface temperatures vary between biomes in any one season?

e. Observe chlorophyll concentrations and sea surface temperatures in each biome during the other months of the year (the months you did not choose). Look for evidence of a bloom or sudden decrease in sea surface temperature that might not have occurred in the months you chose. Make notes on these "episodic" events, including when and where they occurred and the values for chlorophyll and temperature you observe.

f. By the end of this part, you should have compiled data from a minimum of twelve (12) images, 4 for each biome, and a maximum of 36 images (if you record observations on each month). You may want to print or save some or all of these images for future reference.

Part III: Linking Physical and Biological Variability

Your observations of chlorophyll concentrations and sea surface temperatures over the seasonal cycle in three different biomes should have started you thinking about factors responsible for variations in phytoplankton over space and time. If not, review your data and repeat your observations with this "thought" in mind. Review the materials in this chapter (or even look ahead to Chapter 14). Look over your concept maps and try to see where your data fit in. Answer the following questions:

1. Do increases in chlorophyll always accompany episodes of cold water?

2. Do stratified systems always exhibit low chlorophyll concentrations?

3. Do polar systems always exhibit a single bloom?

4. What factors might be responsible for the differences between each of your three biomes?

5. How do your observations of the three biomes "fit" the classic model of phytoplankton productivity? (See Figure 13-14.)

6. Based on your prior knowledge of factors that change the physical structure of the water column (buoyancy changes, wind stress, upwelling, etc.) and factors that alter the chemistry of the upper ocean (mixing, upwelling, nutrient utilization, nutrient regeneration, etc.), speculate on the causes of variability in chlorophyll concentrations in each of the three biomes.

7. Write a 3-5 page paper that hypothesizes on the factors that might have caused the spatial and temporal variability observed in your study. Use your chlorophyll and temperature data to support your hypotheses. In your paper, include a brief proposal of the types of data you would need to confirm or refute your hypotheses. How might you go about conducting such a study? Discuss your ideas and proposal with your classmates and instructor. Use the Rubric for Analyzing and Presenting Scientific Data and Information to improve your work. (See Exploration Activity 1-1).

14

Ocean Food Webs

Large predators may hunt cooperatively to "herd" their prey, which often create spectacular bait balls. This is just one example of the many kinds of trophic interactions that occur in the world ocean.

Questions to Consider

1. How do geological, physical, and chemical processes alter the structure and functioning of food webs and the flows of energy and materials through them?

2. What is trophic efficiency and how does it apply to food pyramids based on numbers, biomass, or energy?

3. What are the major characteristics of the pelagic ecosystem, and how does it respond to seasonal changes in physical and chemical factors?

4. Why is the transitional food web important for studies of climate change?

5. What are trophic cascades and how do human activities contribute to them?

6. What challenges face fisheries oceanographers, and how are they attempting to meet them?

Ocean Ecology

The world ocean abounds with life. From the tiniest microbe to the largest whale, life manages to eke out an existence in the ever-changing ocean environment. Survival depends on an ability to respond to the time- and space-varying geological, physical, chemical, and biological processes in a given habitat (see inside front cover). For oceanographers, knowledge of the interactions of organisms with their environment and with each other is critical for understanding the flows of energy and cycles of materials in the world ocean. The study of the interactions between organisms and their environment (including each other) is called **ocean ecology.** The term *ecology* literally means "the study of households," so ocean ecology includes every aspect of the ocean that affects the lives of organisms.

One tool used by ocean ecologists is an illustration of the trophic (i.e., feeding) relationships of organisms, a conceptual model called a **food web.** The arrangement of organisms into food webs provides a means for comparing different ocean ecosystems (spatial comparisons) or the same ocean ecosystem at different times (temporal comparisons). Ideally, our understanding of food webs allows oceanographers to predict what happens when ocean conditions change or when species are eliminated through overfishing or habitat destruction. Unfortunately, even the simplest of goals, discovering what organisms are present and what role they perform, has not been an easy task for oceanographers.

In this chapter, we explore the structure and dynamics of the ocean food webs from the point of view of an ocean ecologist. Borrowing the tools of terrestrial and aquatic ecology, we attempt to identify the role of different organisms in ocean food webs. We also explore how changes in ocean conditions may increase or decrease the growth and abundance of various organisms in a particular ocean food web. An understanding of ocean food webs is fundamentally important to understanding the role of the ocean in the global biogeochemical cycles and climate change. At an applied level, an understanding of the dynamics of ocean food webs is vital for fisheries management and for identifying the effects of anthropogenic disturbances, including overfishing and habitat destruction. Our study of ocean food webs incorporates nearly everything we have covered in this textbook so far. It synthesizes geological, physical, chemical, and biological principles, and asks you to apply them to an understanding of how ocean food webs work.

The Foundations of Ecology

One of the first ecologists to describe "patterns" in the abundances of different groups of organisms was Charles Elton (1900-1991). In 1927, he proposed and discussed two important concepts that set the foundation for modern ecological research: the concept of **food chains,** the linear transfer of energy and material from plants to animals, and the **pyramid of numbers,** a graphical description of the numbers of organisms at a particular level in a food chain. Food chains express the idea inherent in "big fish eat little fish," a proverb that dates at least to the 1500s (if not to antiquity). Elton defined food chains as "chains of animals linked together by food, and all dependent in the long run upon plants." He also pointed out that because larger animals tend to eat smaller ones, there would be an upper limit to the number of links in the food chain, a number he estimated as around five. Each link in the food chain, called a **trophic level,** represents the principal mode of nutrition of an organism (i.e., autotroph, herbivore, primary carnivore, secondary carnivore, top predator). Elton also defined food cycles, now known as food webs. *Food web* refers to the full range of feeding relation-

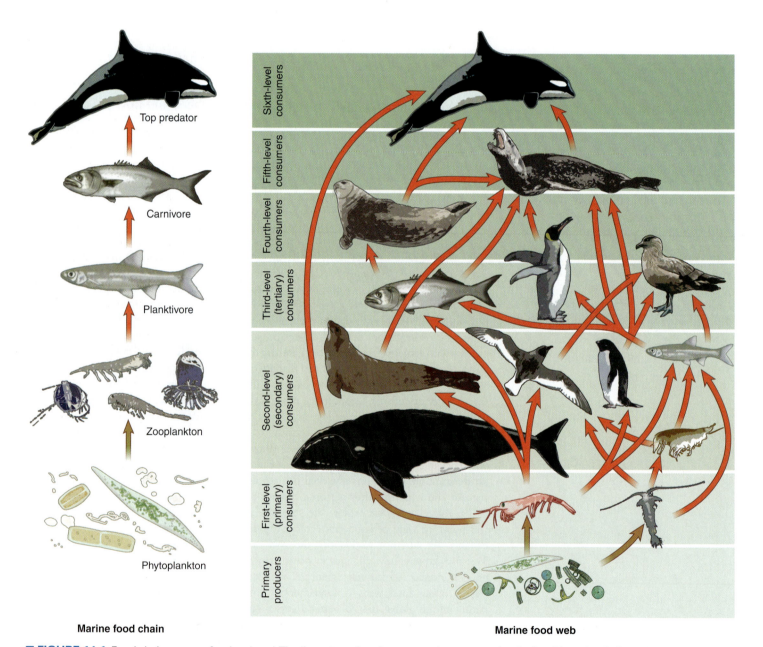

Marine food chain

Marine food web

■ **FIGURE 14-1** Food chains versus food webs. a) The linear transfer of energy and matter can be depicted in a simple food chain. Each organism or group of organisms in the chain represents a trophic level. In the open ocean, these chains may become quite long. b) Food webs include all of the possible pathways of exchange of energy and materials among organisms. Food webs are especially useful where organisms feed at multiple trophic levels during different stages of their life.

ships that may occur between organisms. In food webs, energy and material flows take on complex, weblike patterns, often obscuring clear-cut definitions of trophic levels (**Figure 14-1**).

Elton supported his idea of the pyramid of numbers with an observation ascribed to Alfred Wallace in 1858 that if lower trophic levels support higher trophic levels, they must be more abundant (i.e., herbivores are more numerous than carnivores; small animals are more numerous than larger animals). The **trophic pyramid** rapidly gained acceptance as a tool for understanding ecosystem structure (**Figure 14-2**). Soon after Elton's publication, ecologists demonstrated that pyramids could also be built on the basis of biomass (the weight of the organisms), called **pyramids of biomass,** or energy flow (solar radiation or chemical energy), called **pyramids of energy.**

Ecologists almost just as quickly recognized that not all pyramids were pyramidlike: some were like a tiered column, some bulged in the middle, and yet others were inverted. One case was the exception: pyramids of energy. Raymond Lindeman in 1941 was the first to suggest that a pyramid of energy provided the most appropriate description of the dynamics of food webs. Pyramids of energy take into account the efficiency with which energy and material are transferred from one level to the next. Because chemical transformations always involve a loss of energy (in accordance with the second law of thermodynamics), pyramids of energy retain their pyramid shape.

Trophic efficiency is defined as the ratio of energy in one trophic level divided by the energy in the next lowest trophic level, or:

$$\frac{E_t}{E_{t-1}}$$

where E is the energy in trophic level t or $(t-1)$, the trophic level below it. The trophic efficiency represents the efficiency of transfer of energy between trophic levels and provides a theoretical basis for calculating the productivity that a given food web will support (**Figure 14-3**).

Early ecologists also introduced the view of life as a system. In 1935, Oxford University Professor Sir Arthur Tansley described the concept of an ecosystem in his paper entitled "The Use and Abuse of Vegetational Concepts and Terms":

> It is the systems so formed which, from the point of view of the ecologist, are the basic units of nature on the face of the earth. Our natural human prejudices force us to consider the organisms . . . as the most important parts of these systems, but certainly the inorganic "factors" are also parts—there could be no systems without them, and there is constant interchange of the most curious kinds within each system, not only between the organisms but between the organic and the inorganic. These ecosystems, as we may call them, are of the most various kinds and sizes.

Lindeman in 1941 expanded Tansley's concept and defined an ecosystem as "the system composed of physical-chemical-biological processes active within a space-time unit of any magnitude." Lindeman's definition serves as the basis for the modern-day concept of an **ecosystem,** a community of organisms and the abiotic environment with which they interact. This definition proves difficult, however, within the three dimensions of a fluid environment such as the world ocean. For this reason, oceanographers have adopted a view of ecosystems based on the flow of energy and the cycling of materials within characteristic groups of organisms. In this view, an ocean ecosystem may be defined as a characteristic biota that develops in response to a particular set of physical, chemical, and biological factors and which exhibits identifiable flows of energy and matter. The study of ocean ecology rests to a large degree on defining the types and abundances of organisms present in a particular ocean habitat or under a particular suite of conditions (see Chapter 15).

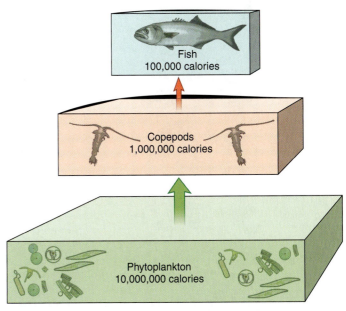

■ FIGURE 14-3 Trophic efficiency. In a three-level pyramid, a trophic efficiency of 10% means that 10,000,000 calories of phytoplankton will only support 1,000,000 calories of copepods which in turn will support 100,000 calories of fish.

■ FIGURE 14-2 (a) Trophic pyramids may be based on numbers, biomass or energy.

Bottom-Up or Top-Down Control?

A key question in ocean ecology is when, where, and how physical, chemical, and biological factors control the structure and dynamics of food webs. Traditionally, it has been assumed that ocean food webs respond to a kind of **bottom-up control,** what some call "the very small controlling the very large" (**Figure 14-4**). Energy is captured from the environment by plants and transferred to higher trophic levels in a one-way flow. The seasonal growth of zooplankton on the spring phytoplankton bloom is a good example. However, **top-down control,** "the very large controlling the very small," may play a role in structuring some food webs (**Figure 14-5**). Top-down control occurs when predators control the abundance of herbivores, which permits plants to flourish. A good example is the predation of sea urchins (an herbivore) by sea otters (a predator), which reduces grazing of urchins on kelp (an autotroph) and allows it to flourish. A third type of control, **wasp-waist control** (referring to the narrow waistline of the wasp relative to its head and abdomen), has also been hypothesized in which organisms in the "middle" of the trophic pyramid control the abundance of organisms above and below them (**Figure 14-6**). This structure emphasizes the role of intermediate trophic levels. Anchovies, for example, may control the abundance of trophic levels above and below them. Thus, an ecosystem under wasp-waist control exhibits dual effects: changes in intermediate-level populations cause changes in top predators and primary producers.

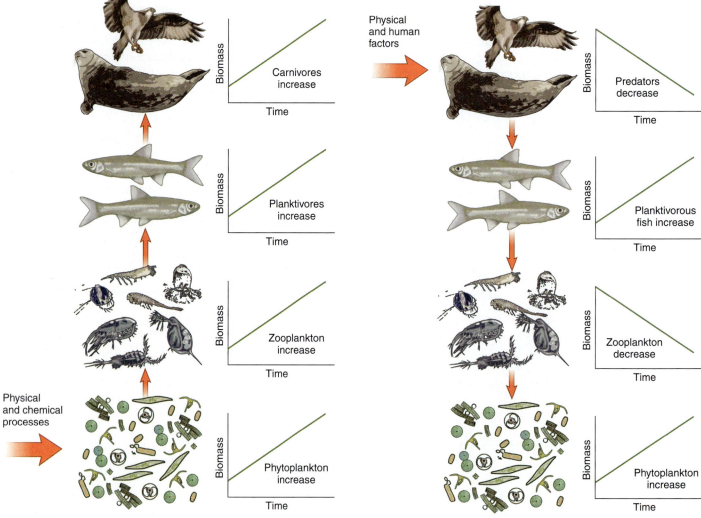

■ **FIGURE 14-4** Bottom-up control of a food web.

■ **FIGURE 14-5** Top-down control of a food web.

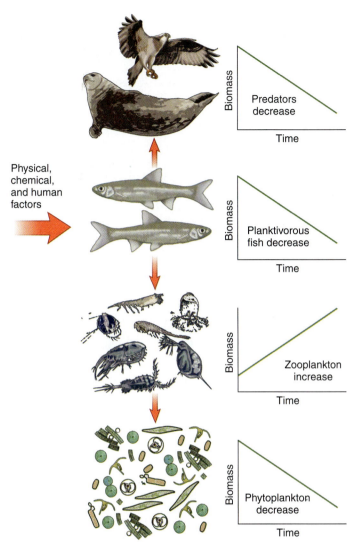

■ **FIGURE 14-6** Wasp-waist control of a food web.

ocean food webs. Rather, small zooplankton, the nanoplankton, were important both as consumers of small phytoplankton, the picophytoplankton, and as contributors to a detrital food web involving heterotrophic bacteria and gelatinous plankton (e.g., jellyfish and salps). A number of important "microbial" papers in the mid–1970s established a new type of food web known as the **microbial food web.** Oceanographers now accept that two fundamentally different but closely connected food webs operate in the world ocean.

Our early twenty-first-century view of ocean food webs is that they consist of two fundamental components: 1) the classical food web, a metazoan, grazing-based, "macroscopic" food web; and 2) the microbial food web, a bacteria and protistan-based, "microscopic" food web. Despite the historical prejudice, the microbial food web is now considered to be the dominant food web in the ocean. At least 60 to 75% of the primary production in the ocean is consumed by protistan zooplankton (e.g., ciliates and flagellates) less than 200 μ m in size. Although these two components of ocean food webs have largely been studied independently, they exist, for the most part, as a single, interdependent food web. Dave Karl at the University of Hawaii asserts that "the classic marine food chain . . . may be considered as a variable phenomenon in a sea of microbes." Attempts to identify the continuum of types from classic food webs at one extreme and microbial food webs at the other represent the cutting edge of ocean ecology research.

The Classical Food Web

In the classical food web (**Figure 14-7**), the large phytoplankton occupy a position as **autotrophs** at the base of the trophic pyramid. Herbivorous grazers feed upon the phytoplankton and occupy the second trophic level as **primary consumers.** Subsequent trophic levels represent predators (or carnivores), with **top predators** (sometimes known as apex predators) occupying the peak of the trophic pyramid. In this scheme, energy from the Sun and matter in the form of dissolved inorganic nutrients are fixed into organic carbon by phytoplankton and subsequently transferred up the food chain by consumers to the top predator, with losses at each level. Lost energy dissipates as heat, while lost matter (in the form of excretion, egestion, shedding, or death) is recycled by a group of organisms known as **decomposers.** The maximum number of trophic levels in any ecosystem may be restricted by the losses of energy at each level. In general, from three to six trophic levels may be found in any given ocean ecosystem. Upwelling systems where phytoplankton are consumed by krill or planktivorous fishes that in turn are fed upon by baleen whales represent three-level systems. Open-ocean ecosystems with phytoplankton, protozoans, zooplankton, squid, fishes, and sharks (or toothed whales) represent six-level systems.

Our modern-day understanding of the classical food web represents, in many ways, the culmination of more than a hundred years of research aimed at understanding the relationship between plankton and fishes. As pointed out by Michael Landry, most textbooks continue to depict ocean food webs in the classical "style," similar to

A Tale of Two Food Webs

From the time of the *Challenger* Expedition, ocean food webs were thought to consist of large phytoplankton (e.g., diatoms and dinoflagellates), crustacean zooplankton (e.g., copepods and krill), small and large fishes, and other predators, such as marine mammals and birds. This view, eloquently encapsulated in John Steele's 1974 book, *The Structure of Marine Ecosystems,* dominated oceanographic thinking for more than a century. The **classical food web,** as it is now known, also provided the centerpiece of fisheries management from the early 1900s. However, as often occurs in science, new evidence transforms old ideas into new ideas. In the same year Steele's book was published, Larry Pomeroy published his landmark paper, "The Ocean's Food Web: A Changing Paradigm." In this paper, Pomeroy argued that the classic concept of the food web did not adequately explain the structure of

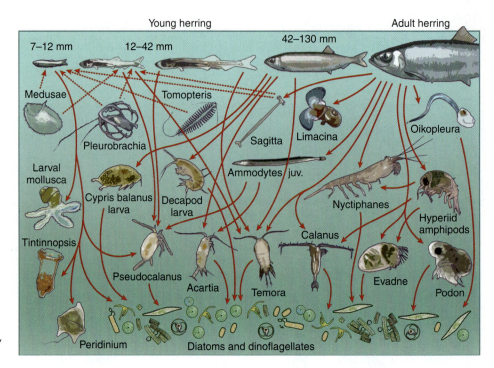

■ **FIGURE 14-7** The classical food web, based on Hardy (1924).

the herring food web published by Sir Alistair Hardy (1896–1985) in 1924. This view of ocean food webs has changed dramatically, however. Despite resistance among some scientists, a complete understanding of ocean food webs and even the classical food web cannot be achieved without taking into account the microbial food web.

The Microbial Food Web

The microbial food web (**Figure 14-8**) consists largely of free-living heterotrophic marine bacteria and microzooplankton grazers; small protists, such as flagellates and ciliates; and mixotrophic dinoflagellates. The bacterioplankton provide a source of food for these animals, which may, in turn, be eaten by larger zooplankton, such as copepods. Bacterioplankton may also supply food to gelatinous grazers whose filters and mucous nets are sufficient to capture them or the particles to which they may be attached. The key component of the microbial loop is the availability of dissolved organic material (DOM) that supplies nutrition for the growth of bacterioplankton. The dissolved pool of organic materials takes on a number of different forms (e.g., dissolved organic carbon, or DOC, and dissolved organic nitrogen, or DON) and comes from a number of different sources. The primary source of DOM is the phytoplankton: as much as 50% of the carbon fixed by phytoplankton ends up in the dissolved pool. Terrestrial sources from rivers and aeolian inputs may also be important. Two important transformations occur as a result of the microbial loop: 1) uptake of DOM/DOC by bacterioplankton, bringing "lost" carbon back to the living; and 2) regeneration of inorganic nutrients that are important to the growth of phytoplankton. Oceanographers estimate that up to half of the primary production of the ocean may be "channeled" through the microbial loop. Thus, this source of carbon and its effect on ocean food webs is at least as significant as the carbon fixed by phytoplankton. Regeneration of dissolved nutrients

involves biological processes that convert organic materials into inorganic compounds that can be utilized by phytoplankton. Excretion of ammonium (and probably phosphate) by protozoan micrograzers and

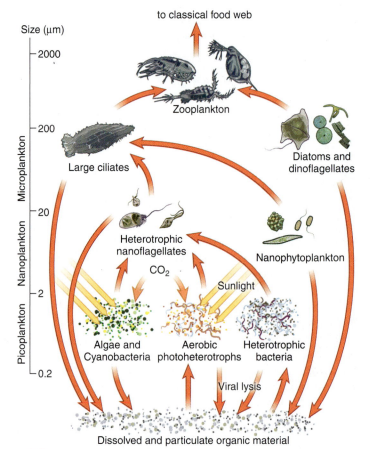

■ **FIGURE 14-8** The microbial food web.

Shaping the Future of Large Marine Ecosystems

SPOTLIGHT

14.1

In an effort to understand the effects of multidecadal climate and ocean variability on marine fisheries, the National Oceanic and Atmospheric Administration (NOAA) adopted in 1986 the concept of a large marine ecosystem, or (LME). Ten LMEs exist within US waters and at least 64 have been defined worldwide (**Figure 14a**). The waters encompassed by the LMEs worldwide account for some 90% of the global fisheries catch and represent coastal regions that are significantly impacted by water quality deterioration, overexploitation, and habitat destruction. Thus, LMEs have become a focus for assessment, monitoring, and management of coastal ecosystems in addition to their importance for fisheries resources.

A major emphasis in the establishment of LMEs is how they respond to shifts in oceanographic conditions over multidecadal scales, what have come to be known as regime shifts. For reasons that are not completely understood, ocean conditions oscillate between various "steady-states" that favor certain species of fishes. For example, off the western United States, periods of below average water temperatures (i.e., the cold regime) that span multiple years favor anchovy populations, whereas above average temperatures (i.e., the warm regime) favor sardines. Understanding the response of LMEs to regime shifts is essential for establishment of sustainable fishing quotas and proper management of fisheries resources. Overfishing of sardines during a regime shift from warm to cold waters (which favor anchovies) may have been responsible for the "crash" of the sardine fishery in Monterey, California, in the 1950s. A failure to understand the response of LMEs to regime shifts may have led to the collapse of other fisheries as well, like the North Sea and Grand Banks cod fisheries.

In a compilation of data from 29 LMEs by NOAA-Fisheries, 2005, 14 were found to suffer primarily from the impacts of overfishing, while 13 were principally affected by climate change. Secondary impacts resulted from either overfishing or climate change, respectively, and eutrophication. The implication is that management decisions in nearly half of the studied LMEs will need to focus on fishing practices, while those in the other half must be guided by an understanding of regime shifts. Note that fisheries biomass yields are often the yardstick by which "success" in an LME is measured. Fisheries management remains the principal driver of LME efforts. In cases where catch has been reduced, fish stocks appear to be rebounding. However, a lack of long-term data on fisheries yields in managed LMEs and a scarcity of data on fishing effort invite caution when interpreting the success of managed LMEs. Many challenges face LME management, including fishing down the food web (targeting lower trophic levels) and genetic alteration of stocks (human-induced selection for smaller-sized fish). At the same time, allowance for nonfishing goals, such as recreational diving, whale watching and other nature viewing, and preservation of species for scientific study, often pits fishers against other interests. Nonetheless, LME management tools provide a framework by which all interested parties may participate in the decision-making process.

FIGURE 14a The California Current Large Marine Ecosystem spans a region from Vancouver Island to Baja Mexico. Natural oceanographic variability dominates this LME, although overfishing, urban runoff, habitat destruction, and other human impacts are increasingly responsible for altering biomass, food web structure, and biodiversity of marine organisms.

metazoan zooplankton makes a significant contribution to regenerated production. The breakdown of organic matter by bacterioplankton is also important for returning nutrients to the water column. The dissolved nitrate present in deeper waters comes largely from the activities of bacterioplankton.

Steady-State and Transitional Food Webs

Blooms of phytoplankton and other organisms may occur any time that physical and chemical conditions support their growth. For example, episodic mixing events (winds, storms, buoyancy forcing) may halt or stimulate blooms, depending on whether the water column is nutrient-replete or nutrient-depleted, respectively. Similarly, delayed or early-onset events—prolongation of winter conditions or an early winter—may delay the spring bloom or eliminate the fall bloom. Recognizing this, oceanographers have defined two contrasting pelagic food webs: **steady-state food webs,** typical of stable physical and chemical conditions; and **transitional food webs,** characteristic of unstable or time-varying physical and chemical conditions. Under steady-state conditions, an equilibrium exists between the biological components of the food web and their environment. Typically, a mix of classical and microbial components is present with environmental conditions only slightly favoring one or the other. These food webs export little organic material to other ecosystems. In contrast, transitional food webs represent an uncoupling between bottom-up and top-down control that occurs when physical or chemical conditions in the water column change. Either the classical component or the microbial component may be favored and become dominant. The spring bloom is a good example of a transitional food web that favors the classical component. Here, phytoplankton proliferate rapidly (bottom-up control) in the near-absence of grazers (top-down control). Once the grazers increase their populations, they regain control over phytoplankton populations, and a steady-state food web is reestablished. Exports of organic matter under transitional conditions may be significant when the classical component is favored.

Representative Ocean Food Webs

In Chapter 13, we explored the seasonal cycle of primary productivity and its impact on food webs in a temperate pelagic ocean ecosystem. Here, we explore four additional ocean ecosystems whose dynamics illustrate the response of food webs to episodic physical, chemical, and biological factors.

Upwelling Ecosystems

Upwelling systems are among the most productive ecosystems in the world ocean. Combined, they contribute more than 10% of the global ocean primary productivity and produce at least 20% of the world's fisheries catch. Five major coastal upwelling systems can be found in the world ocean (**Figure 14-9**). Not surprisingly, these systems correspond to the sites of major fisheries. Upwelling systems provide a good example of transitional food webs. Episodes of upwelling bring nitrate from deeper waters to the surface (see Chapter 9), stimulating a bloom of phytoplankton, typically diatoms. The bloom of phytoplankton, in turn, stimulates other components of the food web. Copepods and euphausiid shrimp increase in biomass and, if the upwelling is sustained, may produce a new generation. Otherwise, the energy gained from the bloom may be stored in eggs that eventually develop into juveniles awaiting the next bloom event. Small planktivorous fishes, especially anchovies and sardines, quickly follow the bloom and increase their biomass. These organisms may serve as food for higher-level predators, such as seabirds and marine mammals.

A unique feature of upwelling systems is the way in which they vary over both time *and* space. From nearshore to offshore, upwelling systems may exhibit gradients in physical, chemical, and biological properties, particularly where upwelling is prolonged or offshore transport is extensive. Like a conveyor belt, upwelled water

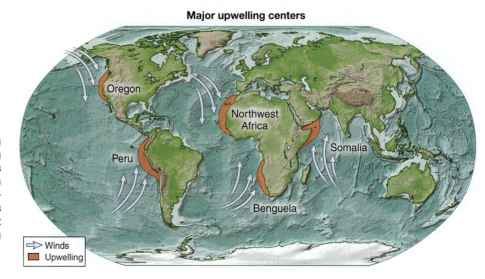

■ FIGURE 14-9 Location of major upwelling systems in the world ocean and prevailing winds that cause upwelling. These systems are among the most productive fisheries in the world. Nonetheless, they are highly sensitive to climate change. Interdecadal events like El Nino can prevent upwelling and shut down the fisheries, often with devastating economic and human consequences.

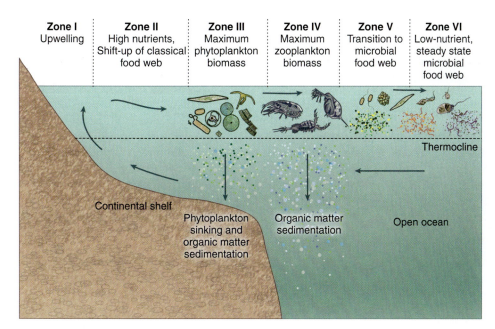

Zone I
Upwelling

Zone II
High nutrients,
Shift-up of classical
food web

Zone III
Maximum
phytoplankton
biomass

Zone IV
Maximum
zooplankton
biomass

Zone V
Transition to
microbial
food web

Zone VI
Low-nutrient,
steady state
microbial
food web

Thermocline

Continental shelf

Phytoplankton
sinking and
organic matter
sedimentation

Organic matter
sedimentation

Open ocean

■ **FIGURE 14-10** Conceptual model of temporal and spatial variability in food web dynamics of an upwelling event. (After MacIsaac et al., 1985).

exhibits distinct "stages" of succession from nearshore to offshore. Oceanographers define four zones in an upwelling system (**Figure 14-10**). In the initial stage (Zone I), upwelling brings nitrate-rich water toward the surface. The newly upwelled water may contain components of the classical and microbial food webs and mix with components already present in surface waters. After a period of photoadaptation and chemical acclimation, called **spin-up,** large phytoplankton, namely diatoms, begin to divide and their biomass increases (Zone II). Continued phytoplankton growth and increases in biomass reach a maximum in the third stage (Zone III). By this time, the upwelled water may be some distance offshore. At the same time, bacterioplankton, fueled by increases in dissolved and particulate carbon, begin to multiply rapidly as well. Zooplankton and higher organisms aggregate in response to the bloom and begin to graze. During the latter stages of the bloom (Zone III and Zone IV), sinking of phytoplankton and production of "detrital" components by the classical food web (marine snow, detritus, zooplankton fecal pellets, etc.) contribute to a rapid fallout of organic materials below the euphotic zone. At some point, nitrate is depleted by the phytoplankton. Stratification of the upwelled water through surface warming prevents replenishment of nitrate, and the bloom begins to decline. Subsequently, components of the classical food web decline. However, bacterioplankton and other components of the microbial food web reach their peak abundance in this zone, the productivity of nano- and pico-sized phytoplankton being sustained on regenerated nutrients, namely ammonium (Zones V and VI). Within a week or two, the entire process has peaked and declined. Nonetheless, the transfer of energy and materials produced during the bloom may continue for weeks as higher predators feed on the now-increased biomass of lower trophic levels.

The productivity of a particular upwelling event depends on the intensity and duration of the upwelling episode. Relaxation of upwelling soon after it begins may shorten the bloom and reduce its impact on higher trophic levels. Sustained winds may create jets or filaments that extend hundreds of miles offshore, prolong-

ing the period of productivity and the dynamics of the food web within it. Regional differences in physical, chemical, and biological processes further determine the magnitude of productivity that a particular upwelling system will support. Climatic factors, such as El Niño and multidecadal shifts in ocean temperature, may also alter these systems.

Ice-Edge Ecosystems

Another highly dynamic component of pelagic ecosystems can be found in the **seasonal ice zone** of polar oceans. This zone of ice, defined as the region between the maximum and minimum extent of sea ice in an ocean, exhibits some of the highest biomass and highest short-term rates of productivity observed in the world ocean. This productivity is driven by physical and chemical processes associated with the melting ice edge in the ice zone (limited to the region of local influence of sea ice). Like upwelling ecosystems, ice-edge ecosystems favor classical components of the pelagic food web whose abundance varies in time and space, that is, a transitional food web. Ice-edge ecosystems also provide fascinating insights into the interaction of physical, chemical, and biological processes in the ocean.

During winter, sea ice forms across polar oceans to an extent governed primarily by the distribution of surface water masses but also affected by winds, currents, and climate. In the Southern Ocean, sea ice is confined to waters south of the Polar Front and averages 19 million km² (7 million square miles) at its maximum extent, more than twice the size of Australia. The sea ice zone of the Arctic is smaller, about 7 million km² (2.7 million square miles). Sea ice may be inhabited by a variety of organisms, despite the extremely low levels of light. In fact, an entire community of microorganisms, called **sea ice microbes** (SIMCO, for short) often can be found within the ice. Sea ice algae embedded in the ice support krill during winter. Under-ice winter-feeding krill, in turn, provide food for epipelagic fishes and even mesopelagic fishes, which ascend to the surface to

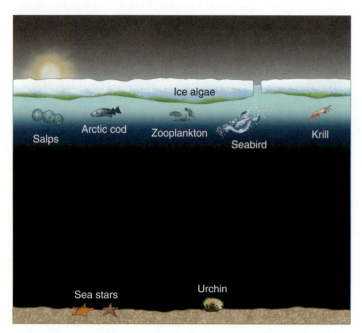

■ FIGURE 14-11 The marginal ice zone ecosystem in fall/winter.

feed in winter (**Figure 14-11**). These fishes may even supply food to seabirds feeding along the ice edge, a surprising result discovered by examining the gut contents of birds collected during winter in the Weddell Sea. Thus, the winter-time ice ecosystem may be spatially compressed with many of its components feeding in a rather narrow vertical band along the restricted horizontal zone of the ice.

As spring nears and light intensities and daylength increase, the surface mixed layer becomes more shallow either by warming or by inputs of fresh water from melting ice. At the same time, the penetration of light through the water column increases dramatically as the ice cover is removed. These conditions combined with near-continuous daylight often generate spectacular **ice-edge blooms** of phytoplankton. Norwegian ocean ecologist Egil Sakshaug points out that the depth of the mixed layer has a strong influence on the timing of the bloom. A shallow mixed layer may initiate a bloom four to six weeks earlier than a deep one. Sea ice melting also exerts considerable influence over the bloom. The timing and pace of sea ice melting determines the speed at which conditions develop that promotes a bloom. Oddly, cold winters may produce more rapid and intense blooms because the ice edge extends well beyond its normal limits and begins to melt earlier (due to the warmer water on which it floats). During warm winters, when sea ice is confined to colder water, the ice-edge bloom may develop later and more slowly. The breakup of sea ice, which may also be facilitated by winds, waves, and currents, also allows light to penetrate within the cracks and leads of the ice zone. These conditions favor the rapid growth of phytoplankton at or near the ice edge. It also stimulates the growth of ice algae, which, after being melted out of the ice, may serve as an additional source of biomass to the ice-edge ecosystem. As the ice edge retreats, the bloom follows, creating what has been called a "bloom on the move" (**Figure 14-12**) Typically, this pattern results in a spatially distributed ice-edge ecosystem, where the biological community near the ice edge represents "young" stages of development and those farther away from the ice edge are more "mature."

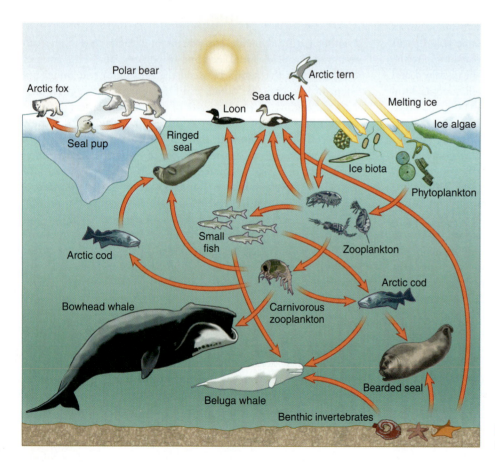

■ FIGURE 14-12 The ice edge ecosystem in spring/summer.

The biomass of phytoplankton in ice-edge ecosystems may increase ten-fold in the space of a week. Chlorophyll concentrations from 10 to 30 mg m^{-3} are not uncommon, and values as high as 100 mg m^{-3} have been observed. In Arctic waters, nearly half the annual primary production may be associated with ice-edge blooms. In the Antarctic, ice-edge blooms may account for a third of the annual primary productivity. These high rates of productivity and their very rapid development mean that ice-edge ecosystems may export large amounts of organic material to benthic ecosystems. Quantifying the "biological pump" in polar oceans, especially the Southern Ocean, is a major focus of polar oceanographers. The speed at which the bloom develops determines the degree of coupling between primary producers and grazers, and the rate at which organic matter is exported. Where the bloom develops rapidly, grazers may lag behind and a significant portion of the production may sink. A more slowly developing bloom may allow grazers time to ramp up growth rates and production. Export production may be reduced or confined to the upper ocean under these circumstances.

Termination of the bloom usually results from nutrient limitation in the mixed layer. At such time, a prominent subsurface chlorophyll maximum may develop at the nutricline (see Chapter 12). In Arctic waters, nitrate, silicate, and phosphate may be reduced to undetectable concentrations during a bloom. However, in Arctic basins fed by North Pacific waters, the oldest and most nutrient-rich water mass in the world ocean, nutrient concentrations before the bloom (i.e., the winter maximum) may be quite high. Thus, termination of a bloom depends not only on rates of photosynthesis but also on the availability of nutrients in the first place. Overall, nutrient concentrations in the Southern Ocean tend to be three to five times higher than Arctic Ocean concentrations. Yet, Southern Ocean productivity may be limited by the availability of iron, which is required by the enzyme for nitrate uptake. The Southern Ocean has been identified as an HNLC (high-nitrate, low-chlorophyll) region whose productivity may be controlled by wind patterns and continental processes that affect the airborne transport of iron.

Midwater Food Webs

The productivity of surface waters also sustains organisms that live in the abyss, especially in the mesopelagic and bathypelagic regions of the world ocean. Whether these organisms migrate upward to the surface to feed (see Chapter 12) or wait for food to "rain down" from above, their activities correspond to episodic and seasonal changes in productivity in the upper ocean. Here, the classical and microbial components of the ocean food web work in concert, quickly processing and breaking down large bits of organic detritus into finer and finer particles that eventually sink to the seafloor.

Although primary producers are absent, their by-products can be found in many forms (**Figure 14-13**). Egestion of solid waste or **fecal pellets** by zooplankton accompanies their feeding activities. These particles of semiprocessed organic matter sink rapidly from the euphotic zone. They are quickly consumed, however, by coprophagous organisms that can utilize the organic matter present in fecal pellets. Particles may be processed several times over as one group of organisms finds nutrition in the waste products of another. Zooplankton in surface waters also create fragments of phytoplankton through **sloppy feeding.** This incomplete ingestion of food

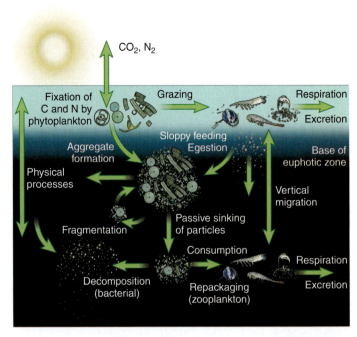

■ **FIGURE 14-13** The mid-water food web and its role in the biological pump.

releases the dissolved contents of phytoplankton cells (providing food to bacteria) and produces a rain of organic particles. Small protists, such as flagellates, may consume these particles, metabolize them, and release them as **picopellets,** colloidal-sized particles that become part of the dissolved pool. Oceanographers estimate that from 10 to 30% of ingested phytoplankton carbon may be released as DOC during feeding by protists. This amount is equivalent to or exceeds the amount of DOC produced through phytoplankton leakage. These compounds represent important nutrients for bacterioplankton. Physical and biological processes may also contribute to a form of oceanic detritus known as **marine snow** (for its resemblance to snow), which may include fragments of phytoplankton, fecal pellets, allochthonous (land-derived) material, and other particulates. The discarded mucous feeding traps of larvaceans, a type of invertebrate chordate, contribute significantly to marine snow.

Of particular interest to oceanographers is the amount of organic matter that eventually makes it to the seafloor and the rate at which it sinks. Production of fecal pellets (and their by-products) and formation of marine snow may accelerate or slow the sinking of particles. The vertical migration of marine organisms may accelerate or slow the sinking rate of particles. Vertical migrators act as a kind of biological pump transporting sediments between surface waters and midwater depths in a single night. Twenty years of sediment trap observations at the Bermuda Atlantic Time Series site have revealed extensive reworking of particles in the waters beneath the euphotic zone. While strong seasonal signals are evident in the mesopelagic, biological activities generate a fine rain of particles with a low carbon content and relatively homogeneous composition by 3200 m (10,499 feet). The degree to which particles are packaged and processed, and their rates of sinking in the water column are an active area of research, especially as these processes affect the transport of carbon to the deep sea.

The activities of larger organisms, such as deep-sea fishes and abyssal jellies, play a largely unknown role in the dynamics of mid-water food webs. Here dwell what ichthyologist C. P. Idyll calls "the most grotesque and improbably monstrosities." Although these depths were once thought to be scarcely inhabited, observations by ROVs have revealed an abundance of strange jellyfish and other unusual invertebrates. Some gelatinuous species have the ability to filter fine particles, an adaptation well-suited to midwater zones. Other species maintain low metabolic rates and may even shrink if food supplies run low. Deep-sea fishes employ a sit-and-wait strategy to finding food. Many are armed with fangs that extend beyond their mouth cavities and are equipped with lighted appendages for luring prey. Fortunately, most of them are quite small, a few inches to a few feet, owing to the scarcity of food at these depths and the high premium on energy efficiency. Other than observations and collections from net tows, little is known about most midwater species. Given their dependence on surface waters, it seems likely that these food webs respond in a bottom-up fashion to seasonal and episodic variations in export production.

Deep-Sea Benthic Food Webs

At least 60% of the Earth's surface lies beneath 2 km (1.2 miles) of water in darkness. Despite its designation as the one of the most common environments on our planet, we know less about the abyssal seafloor than the surface of Mars. Nonetheless, research by a small number of determined benthic oceanographers over the past 100-plus years has shed some light on this habitat. This is the domain of the deep-sea benthos, the organisms that dwell on the seafloor in the abyss. Broadly speaking, the deep-sea benthos can be divided into soft-bottom habitats (e.g., muds, oozes, other sedimentary environments), "vent" habitats (hydrothermal vents, cold seeps, and similar habitats), and "other" habitats with unique geological or physical features (e.g., islands, seamounts, trench environments, etc.). Of these, the soft-bottom benthos comprises the greatest proportion, although this environment is more heterogeneous than its broad categorization implies.

The soft-bottom benthos has been the subject of study since the mid–1800s when the father-son team, Michael (1805–1869) and G. O. Sars (1837-1927) collected and described nearly 100 deep-water benthic species in Norwegian fjords. The proposal by Edward Forbes that the deep sea was lifeless (see Chapter 12) stimulated efforts to prove him wrong. These scientists were not disappointed as several studies in the late 1800s, including the *Challenger* Expedition, revealed the ubiquity of life on the seafloor. Early studies also established the strong theoretical foundations of benthic oceanography. As one example, it was hypothesized that species diversity of the benthos decreased markedly with depth. In other words, there was a gradient in the number of species from shallow water (high diversity) to the deep sea (low diversity). That hypothesis was rejected in the mid–1960s in studies by Howard Sanders (1921–2001) and Robert Hessler. Their work in the North Atlantic (and work since that time) has revealed an almost unimaginable diversity in the deep sea. Some researchers have even suggested that species diversity on the abyssal seafloor rivals that of tropical rain forests! Species diversity of the deep-sea soft-bottom benthos and its causes continue to be a subject of intense debate among oceanographers.

Another example of hypothesis testing in benthic oceanography pertains to food web dynamics in the deep-sea benthos. In the first half of the 1900s, research by British marine biologists suggested that deep-sea populations reproduced continuously, brooded their young, and were largely unaffected by seasonal processes owing to the uniformity of conditions in the deep sea. Subsequent studies appeared to support their hypotheses. However, research by British marine biologist John Gage (1939–2005) and others in the Rockall Trough (off the coast of Scotland and Ireland) revealed a marked seasonality in the reproductive cycles of several invertebrates. Why would organisms in a uniform environment exhibit seasonal cycles? The answer came when time-lapse photography studies in the nearby Porcupine Seabight revealed a blanket of phytodetritus covering the seafloor several days after blooms of phytoplankton. Much like in pelagic and midwater food webs, the soft-bottom benthos receives matter and energy from photosynthetically derived organic matter. The activity of benthic species, especially sea cucumbers and sea urchins, increases markedly when phytodeteritus reaches the seafloor. Modern studies support the conclusion that the reproductive cycles of many deep-sea benthic species are synchronized with episodic pulses of phytodetritus. This seasonality has been observed in temperate and polar benthic environments but does not appear to occur in tropical oceans.

Phytodetritus is not the only food and energy source to the deep sea. Episodic **food falls** in the form of dead whales and other animal carcasses deliver a sudden bounty of food to the benthos (**Figure 14-14**). Mobile benthic species such as hagfishes, rattails, and amphipods, quickly aggregate on food falls and scavenge the carcass. A succession of organisms can be observed as the carcass is reduced over the months and years. The surrounding sediments become enriched with the "detritus" of these feeders, stimulating the activities of benthic polychaetes and other invertebrate species. The bones are colonized by bacteria that feed on the oily marrow and produce hydrogen sulfide as a by-product. This sets the stage for the appearance of chemosynthetic bacteria capable of utilizing hydrogen sulfide as an energy

■ **FIGURE 14-14** The carcasses of whales provide an abundant albeit episodic food source to the deep-sea benthos. Their skeletons give rise to a community of unique organism whose energy is supplied through chemosynthesis.

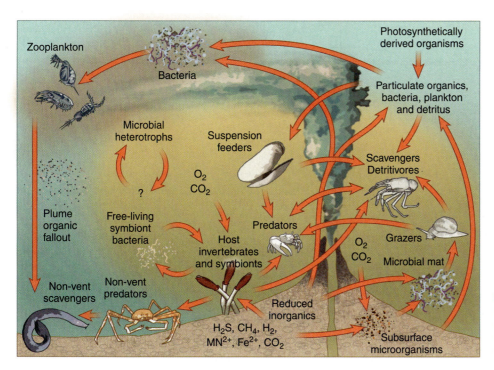

Zooplankton

Bacteria

Photosynthetically derived organisms

Particulate organics, bacteria, plankton and detritus

Microbial heterotrophs

Suspension feeders

O_2
CO_2

Scavengers Detritivores

?

Plume organic fallout

Free-living symbiont bacteria

Predators

Host invertebrates and symbionts

Grazers

O_2
CO_2

Microbial mat

Non-vent scavengers

Non-vent predators

Reduced inorganics

H_2S, CH_4, H_2,
MN^{2+}, Fe^{2+}, CO_2

Subsurface microorganisms

■ **FIGURE 14-15** The food web of a hydrothermal vent. Contributions of energy and matter from a variety of sources has been recognized as oceanographers continue to study these ecosystems.

source. Other organisms may feed on these bacteria, creating what has been called a **sulfophilic community.** In July 2004, oceanographers reported the discovery of an unusual worm that houses bacteria in its tissues to aid in the degradation of whale bone.

Since their discovery in 1977, hydrothermal vents have continued to surprise researchers (see Chapter 4). While chemosynthetic symbionts support most vent organisms, including the familiar tubeworms, clams, mussels, and scallops, evidence exists that free-living chemosynthetic bacteria and particulate matter from the surface make some contribution to vent food webs. Oceanographers now recognize a complex web of trophic interactions involving several sources of primary production (**Figure 14-15**). Suspension feeding on free-living heterotrophic bacteria, free-living chemosynthetic bacteria, and even phytodeteritus provides nutrition to "feather-duster" polychaete worms, anemones, and other species. Surface deposit feeding on microbial mats, microbes growing on rocky surfaces, and microbes that settle out of the water column is common among many mollusk, crustacean, and worm species. Crabs, shrimps, and lobsters consume a range of food types from bacteria to other invertebrates. Opportunistic species, such as fishes and cephalopods, participate as higher predators in vent food webs. In addition to making direct observations using ROVs, vent scientists collect animals and measure the presence and abundance of stable isotopes and chemical "biomarkers," molecules that indicate the food source of an animal. Biomarkers allow researchers to infer the relative importance of different food types in the nutrition of an animal.

The dynamics of vent food webs are intimately connected to the dynamics and life cycles of the vents. Fast-spreading ridges typically produce hydrothermal vents that last on the order of years, while vents on slow-spreading ridges may last decades. The rate of fluid flow and the fluid chemistry influence the types of species that colonize vents and their abundances. On some vents, ecological succession has been observed. Mobile species, such as

crabs, may be the first inhabitants of a newly formed vent. Soon thereafter, vent tubeworms settle on the vent and rapidly grow and proliferate. Mussels and clams follow and eventually displace the worms as the dominant species, forming a "climax" community. This community survives as long as the vent continues to produce fluids. When the vent wanes or stops producing fluids, the tubeworms, mussels, and clams perish. Scavenging species, such as crabs and gastropods, may increase in abundance for a time, but these, too, disappear in the absence of a source of primary production. Eruptions of lava on the seafloor may also terminate a vent community. The vent fields Clam Bake and Tubeworm Barbecue owe their name to sudden death by lava eruption.

Fisheries Oceanography

Fisheries oceanography concerns the study of the abundance and distribution of fish stocks in relation to the physical and chemical environment of the ocean. The abundance of fishes in a particular habitat will be affected by those factors that affect ocean food webs, that is, the food supply of fishes. They may also be affected by physical, chemical, and biological factors that affect the recruitment and survival of fish larvae. Humans, too, have a direct effect on fish stocks by removing fish from the ocean, often at unsustainable rates. **Fisheries management** attempts to use scientific information supplied by fisheries oceanographers to regulate the catch of fishers to achieve a sustainable yield where removed populations are replaced by growth and reproduction. Unfortunately, fisheries management is where science, economics, and politics clash. Despite global efforts to manage fish stocks, at least 70% of the world's fishing grounds are overexploited, heavily exploited, depleted, or recovering from depletion (**Figure 14-16**).

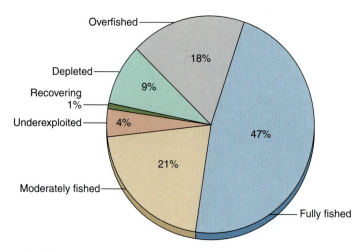

■ **FIGURE 14-16** Status of the world's fisheries in 1999.

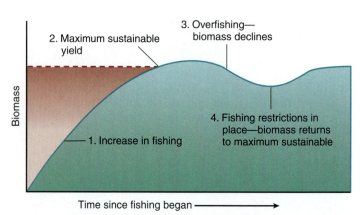

■ **FIGURE 14-17** The maximum sustainable yield, in theory, sets a limit on number of fishes that can be caught according to the "replacement rate" (taking into account growth, recruitment, death, etc) under existing conditions. The problem is that determination of the "replacement" rate for a species depends on a number of interacting factors, including natural climate variability. Attempts have been made to manage the catch from year to year but ultimately this approach has led to overfishing of species.

Oceanography at the turn of the twentieth century was solidly focused on the links between the plankton and the fisheries. By the mid–1950s, a focus on individual fish species by fisheries biologists and an expanded role for oceanography as a whole in military and geophysical interests diminished the attention given to fisheries oceanography. Establishment of the United Nations Food and Agriculture Organization (FAO) fisheries statistics gathering efforts in the 1950s further promoted a single-species focus. By the 1960s, fisheries was viewed as the answer to the world's growing food problem with the wildly optimistic goal of 350 million tons annually, the "generous sea," as Daniel Behrman (1924–1990) put it in 1969, "pouring protein into the mouths of my 10 billion descendants." Yet, as efforts to reduce catch by limiting and closing fishing grounds were met with tremendous resistance for the economic hardships they brought to fishers, a more comprehensive and inclusive approach to fisheries has gained popularity. Surprisingly, fisheries oceanography has reemerged as a highly active discipline in the twenty-first century, despite having been practiced for more than 100 years.

Fisheries management changed from the early years of trying to understand the factors responsible for variations in fish stocks to one of "how much fish can humans take from the sea?" This change in perspective led to the concept of the single-species **maximum sustainable yield** (MSY), defined as "the largest catch that can continuously be taken from a stock under current environmental conditions" (**Figure 14-17**). The MSY accounts for natural variations in **recruitment,** the new adult-sized or reproductively mature fish entering the fisheries. The debate over MSY is a long and protracted one. Determination of the MSY invites overfishing a fish stock to see if it can recover. Fisheries regulations now state that the single-species MSY is a "limit to be avoided rather than a target to be exploited."

The effects of fishing go far beyond changes in the abundance of a particular fish stock. Fishing alters the age and size distribution of the stock (by mainly taking large adults), their sex ratios, and their genetic composition. In one of the most startling examples, researchers have shown that fishing on northern cod led to the evolution of smaller individuals that reproduce at a younger age. Since a moratorium on cod fishing in 1992, cod stocks have remained small-sized. **Fishing-induced evolution** has been observed in other species as well. The implication of these results is that without preservation of the natural genetic variability of species, large-sized fish of a heavily exploited species may not return.

Trophic Cascades

In the simple food chain model, the big fish eats the smaller fish that eats the even smaller fish and so on. In this model, each organism or group of organisms occupies a specific trophic level, an integer from 1 for primary producers up to 5 (or more) for predators. A key aspect of this model is the efficiency of transfer of energy and matter from one tropic level to the next, that is trophic efficiency. Estimation of the trophic efficiency for a given pelagic system provides the means to estimate the abundance (biomass) of fishes that can be supported by a given biomass of primary producers. Though measurements are extremely difficult to come by, most commonly, a value of 10 to 20% is used. Rough calculations of this sort, though seemingly simplistic, actually have provided much of the rationale for sustaining high catches of fishes. Even today, some scientists estimate that the potential catch is near 227 metric tons compared to the 80 metric tons currently being harvested.

An "improvement" to this model takes into account the complete diet of a species (or group) as it may feed at multiple trophic levels during its lifetime. These **fractional trophic levels,** which vary from 1 to 5 or more in tenths (i.e., 3.1, 3.2, 3.3, etc.), represent a step toward the incorporation of a more realistic food web structure in pelagic ecosystems. Using this approach, scientists have shown that since 1950, the trophic level of global fisheries catch has declined by about 0.1 per decade. This decline, according to these scientists, represents a transition from catching predatory, piscivorous fishes (e.g., tuna, swordfish, sharks) to an increasing take of lower trophic level, planktivorous fishes and invertebrates (e.g., anchovies, krill, etc.). This phenomenon, nicknamed "**fishing down the food web,**" could eventually lead to a collapse of the fisheries as the "foundations" of the food web are whittled away (**Figure 14-18**).

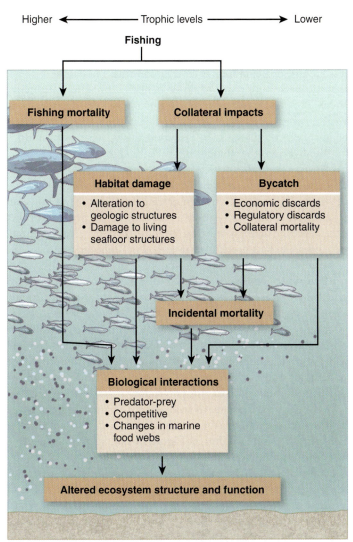

Higher ← Trophic levels → Lower

Fishing

Fishing mortality

Collateral impacts

Habitat damage
- Alteration to geologic structures
- Damage to living seafloor structures

Bycatch
- Economic discards
- Regulatory discards
- Collateral mortality

Incidental mortality

Biological interactions
- Predator-prey
- Competitive
- Changes in marine food webs

Altered ecosystem structure and function

■ **FIGURE 14-18** Fishing and its impacts have severely depleted higher trophic levels. Fishermen, to sustain a living, must fish on less desirable species at lower trophic levels. The trend towards fishing on lower trophic levels has been termed fishing down the food web (after Morgan and Chuenpagdee, 2003).

At first glance, fishing directly on lower trophic levels should increase the potential catch. However, this type of fishing has unintended and unexpected consequences, ones that have not been appreciated until recent times. Removal of a species from an ecosystem often relieves or enhances competitive interactions among remaining species. Species that may once have been kept in check by a predator may now flourish in its absence and compete more effectively for food and other resources. Such a consequence may result in **trophic cascades,** the "domino effect" whereby removal of a key species, usually a predator at a higher trophic level, alters the diversity of the food web and the abundances of individual species at lower trophic levels. Among the most well-known examples are the keystone predators, the sea otter-urchin-algae cascade, whereby the presence of the sea otter keeps sea urchins in check, allowing for a greater abundance and diversity of algae. Trophic cascades have been suggested to account for a rise in squid and jellyplankton populations worldwide as removal of carnivorous and planktivorous fishes has reduced predation on these organisms (especially squid) and increased the abundance of their prey (e.g., zooplankton).

Ecosystem Approach to Fisheries Management

Since the late 1980s, global fisheries have been in decline and at least half of the world's fishery resources are overexploited. At the start of the twenty-first century, the notion of the ocean's inexhaustible bounty has been proven false. In serial fashion, fishers have overexploited resource after resource. From the collapse of the Peruvian fishing grounds in the 1970s to the closure of the Grand Banks in the 1990s, fisheries management has been unable to conduct fishing in a sustainable fashion. The unfortunate response has been to fish out known resources and move on to exploit new resources, such as is now occurring in the Southern Ocean and along the coasts of West Africa and Southeast Asia. Like the whalers of the nineteenth century, the fishers of the twenty-first century have moved farther and deeper in an attempt to maximize their take of an ever-dwindling resource. The reasons for global overfishing are many-fold and involve policies and practices well beyond the influence of the individual fishers, a population that may number some 36 million people worldwide. Governments, agencies, commercial stakeholders, the public, and "pirates" may all share part of the blame for the global overfishing crisis.

The limitations of the MSY as a tool for single species management has led to adoption of an **ecosystem approach** to fisheries. This approach acknowledges "uncertainties in the biotic, abiotic, and human components of ecosystems" and allows for the diverse societal uses of fish stocks (i.e., fishing versus recreation or tourism). Stimulus for this approach in the 1990s followed from the adoption and enforcement of several international agreements, including the 1982 Convention of the Law of the Sea, the UN Fish Stocks Agreement, and the FAO Code of Conduct. The ecosystem approach takes into account the oceanographic and ecological factors that sustain fishes and, expanding the definition from its typical scientific usage, emphasizes the multiple uses to which fishery resources may be put, including recreation and tourism.

Bycatch reduction has been a major focus of fisheries over the past few decades. Some 20 million tons of unwanted fish may be discarded annually, either because regulations prevent capture of these fish or because it is not economically feasible for fishers to process them. Seabirds, marine mammals, and turtles are also susceptible to bycatch and measures are being taken to reduce the numbers of these animals killed as a result of fishing.

Many factors affect the recruitment of fishes. Increasingly, fisheries biologists are recognizing the importance of fish migration for successful mating and reproduction. Hundreds of fish species migrate annually, including economically important species (anchovy, mackerel, cod) and sharks. Mating aggregations have been known for some time, but their relationship with oceanographic conditions has not been well studied. Subpopulations of North Sea plaice, which aggregate during spawning season, appear to use tidal currents to guide their larvae to suitable nursery grounds. Larval dispersal also helps explain "why mothers matter," as one scientist puts it: older mothers produce larvae with larger internal oil droplets that help sustain them during dispersal. Removal of older, larger females reduces survivorship among larval fishes.

Habitat destruction often ranks high on the list of negative impacts on fisheries. Some fishers claim that habitat destruction is *the* cause of reductions in fish stocks (as opposed to overfishing), but

a.

b.

■ FIGURE 14-19 A deep coral garden, before (above) and after (below) trawling. As nearshore and pelagic fishing grounds become depleted, fisherman pursue fishing grounds farther and deeper from their home ports. Bottom trawling in many regions of the world ocean is destroying deep coral communities.

historical and modern evidence tells us otherwise. Nonetheless, habitat destruction is a contributor to reductions in fish stocks and a factor affecting their survival. Bottom trawling on shelf regions has repeatedly plowed the sea bottom and destroyed nursery habitat for larval fishes (**Figure 14-19**). Even where such practices have been stopped, it may take years to decades for the habitat to recover. Bottom trawling also threatens newly discovered deep coral gardens, an ecosystem about which we know little but which undoubtedly serves to provide habitat for commercially important fishes, such as rock grouper. Dead zones created through eutrophication kill fish locally, and marine toxins, such as polychlorinated compounds and birth control hormones, that enter the ocean through land runoff and sewage discharge may lower the reproductive output of fishes or make them sterile. Equally destructive are gear losses that result in **ghost fishing,** unattended nets floating through the sea, capturing and killing any fishes, birds, and mammals in their path. Ghost fishing remains a serious problem in many parts of the world ocean, although efforts are being taken to reduce incidences in local waters, such as Puget Sound in Washington.

Ecological interactions among species may be one of the most difficult factors to consider in the ecosystem approach to fisheries. Ecological theory is not as well developed for marine environments as it is for terrestrial and freshwater environments, and indications exist that marine systems may function differently than other ecosystems. While bottom-up control of phytoplankton is well-supported, it is less clear whether zooplankton populations are regulated by the availability of phytoplankton or by predation, that is, top-down control.

Biodiversity, the number of species within a given ecosystem, also may prove difficult to factor into operational management of fisheries. Scientific understanding of the role of biodiversity in marine ecosystems is limited. Even the number of species in the world ocean is uncertain as vast regions remain unexplored. In general, a belief exists that high diversity reflects a "healthy" ecosystem, and low diversity appears to be a property of disturbed or "unhealthy" ecosystems. But ecosystem health may be a matter of perspective.

Ecosystem resiliency provides a measure of the ability of populations to recover from natural disturbances. Natural climatic events, such as El Niño, decadal-scale oscillations (North Atlantic Oscillation, Pacific Decadal Oscillation), and multidecadal shifts in oceanographic conditions, called **regime shifts,** may alter physical and chemical conditions to an extent that severely reduces some populations and favors others. The degree to which overfishing contributes to the crash of fish populations following regime shifts is hotly debated. Long-term evidence exists from sediments for regime-shift-induced fluctuations in sardine and anchovy populations off the coast of California (**Figure 14-20**). Surprisingly, anchovies are the most abundant pelagic fish over at least two millennia. Most fisheries researchers and history books point to overfishing as the cause of the collapse of the sardine industry in Monterey Bay, yet a regime shift during this period may have contributed to the large-scale replacement of sardines by anchovies. Researchers differentiate

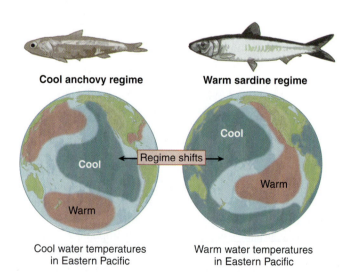

■ FIGURE 14-20 Natural climate variability brings about "stable" conditions which may persist over decades. Changes from one "steady state" to another have been termed regime shifts. Multidecadal climate variability may be responsible for increases and decreases in certain fish species, i.e., cold water anchovies or warm-water sardines. One challenge of fisheries management is to allow fish populations to "weather" regime shifts by reducing their exploitation when oceanographic conditions limit their ability to recover.

Can We Responsibly Farm the Ocean?

In the view of many scientists, the world ocean can no longer sustain the extreme fishing pressures to which it is subjected. At least 70% of global fish stocks are fully or near-fully exploited. While the rate of overfishing slowed in the 1990s, little evidence exists that fish stocks will start to return to historical levels (pre–1970s and earlier) any time soon. Faced with a fishing crisis, the world has increasingly turned to aquaculture, the controlled cultivation of marine organisms, to meet the world demand for fish and fish products (**Figure 14b**). More than a billion people already depend on fish as their primary source of protein. Indeed, in 2002, aquaculture provided nearly 40 million tons of the 130 million tons of fish consumed worldwide. Most of the increase in food production since the 1990s has come from aquaculture, which grew at a rate of 9.1% annually (driven largely by China), the fastest-growing food sector in the world. Demand for fish products worldwide is expected to triple by the year 2025.

In view of the growing demand for seafood and the current dependency of the United States on seafood imports (about 70% of all seafood in the United States is imported), the US Senate introduced in June 2005 the National Offshore Aquaculture Act. Announcement of the bill set off a firestorm of debate among environmentalists, fishers, policy makers, scientists, and the general public concerning the benefits and consequences of offshore aquaculture. How will aquaculture and its attendant waste products impact adjacent marine ecosystems? How will aquaculture impact existing wild fisheries, especially those that have strived to maintain sustainable yields? How will state and federal governments share jurisdiction over aquaculture beyond the 3-mile state boundaries in the Exclusive Economic Zone?

Experiences with shore-based aquaculture have brought mixed results. For example, construction of shrimp and fish farms in the Philippines, Thailand, Ecuador, and other countries comes at the expense of ecologically important mangrove forests, which are destroyed in the process of building the farms. In a British Columbia fjord, salmon farms were found to harbor populations of parasitic sea lice that infect migrating wild juvenile salmon. Release of feces, fish-food waste, and antibiotics used to prevent illness in farmed fishes negatively impacts the water quality of surrounding pelagic and benthic ecosystems. Escapement of nonnative farmed species into wild stocks has resulted in harm to wild stocks, which may not be able to compete for food and habitat. Many farmed species, especially shrimp and carnivorous fishes, require wild fish as food or seed stock.

At the same time, fishers view promotion of aquaculture with great skepticism. Fisheries that have strived to produce markets for wild-caught fishes, such as the Alaskan salmon, worry that increased competition and reduction of prices will undercut their highly managed and relatively costly practices. Impacts of offshore fish farms on native pelagic and bottom fishes are unknown and raise concerns among fishers who depend on these stocks to make a living. State and federal officials have squared off over the "rights" of federal agencies to permit construction of fish farms beyond the 3-mile boundary of state waters, especially where those fish farms may harm or limit other uses. Concerns have also been raised over the lack of a regulatory framework for developing and managing offshore fish farms. Clearly, a number of challenges remain for offshore fish farming to produce net positive results as a food source.

Nonetheless, other countries are rapidly moving forward with offshore aquaculture. More than 100 tuna pens have been built offshore in southern Australia. In the Sea of Cortez, Mexican tuna ranchers have built a multimillion-dollar industry. In 2005, the United States imported approximately $11 billion in seafood and exported approximately $3.7 billion worth, an annual deficit of almost $7.5 billion. The high demand for seafood in the United States will continue to place economic and political pressure on the development of aquaculture. Whether the National Offshore Aquaculture Act of 2005 jump-starts US offshore aquaculture production or whether environmental, economic, and political concerns delay its implementation remains to be seen. With advanced oceanographic monitoring techniques and careful attention to the location of farms and the species raised, some of the issues surrounding offshore aquaculture, especially water quality issues, can be addressed. Perhaps the cooperative efforts of all concerned parties can make responsible and productive fish farming a reality in the twenty-first century.

FIGURE 14b Inside an offshore fish pen.

these periods as the cool anchovy regime, or La Vieja, and the warm sardine regime, or El Viejo, to emphasize the role of water temperatures (at least as a proxy) in the recruitment of these populations. From the mid–1970s to the mid–1990s, California experienced a warm sardine regime. Since the mid– to late–1990s, a cooler anchovy regime has prevailed. Regime shifts may also have different time scales in different regions. Between 1947 and 1992, sardines were completely absent off the coast of British Columbia. Their return

after 45 years marks a regime shift in oceanographic conditions that possibly favor the growth of diatoms, an important food source for larval sardine fishes. Similar linkages between pelagic food web components and climate variables have been observed throughout the world ocean. For scientists and policy makers, regime shifts present a difficult problem for the management of fisheries. Unraveling the degree to which regime shifts versus overfishing alter the abundance of fish stocks may be one of our greatest challenges.

YOU Might Wonder

1. How can I find out which fish in my supermarket are not threatened by overfishing?

Figuring out how to make smart choices at the supermarket is a great way to put pressure on suppliers to market only fishes whose populations can sustain fishing. Unfortunately, not many healthy and productive fisheries are left. A good source of information for smart seafood choices is the Seafood Watch website, maintained by the Monterey Bay Aquarium (www.mbayaq.org/cr/seafood-watch.asp). It lists dozens of popular fish and seafood species and the threats posed by fishing and other practices. It also lists a number of ways you can get involved to educate yourself and others about the perils of overfishing.

2. How can scientists be certain that the reductions in fish populations are not natural? Why blame the fishers?

That's a great question and one that is currently the subject of considerable debate. Worldwide, global fisheries production is at an all-time high. While fishers and fisheries managers acknowledge that overfishing can be locally important, little agreement exists that overfishing is a global problem. Many studies such as point to long-term variability in fish populations associated with climate change. So called "herring years" and similar fishers' catch-phrases have been known since the sixteenth century. Nonetheless, there are an alarming number of warning signs that the global fisheries is overfished. Shifts in the size structure of commercial fishes and their inability to return even

when the fisheries are closed are one sign. Another is the shift in species fished. Species once considered "trash" fish are now fished regularly. At the same time, comparisons of the dynamics of fish populations in marine reserves with fished populations provide a kind of baseline against which fishing production can be better evaluated. Many, if not most, oceanographers and fisheries scientists consider overfishing to be a major problem. Yet, as with climate change, the scale and scope of evidence required to support that conclusion require more research. Hopefully, these questions can be answered before it's too late.

3. How hard is it to become a fisheries oceanographer?

Taking an oceanography class is a great first step! Most persons who work in this field pursue a degree in biology, fisheries, or oceanography. Some take a more technical route, learning the tools of fisheries managers, from fish collection to aquaculture. The field is ripe with opportunity, especially since so much of the world depends on fisheries for protein. The US Fish and Wildlife Service has a lot of information for prospective fisheries personnel (www.fws.gov/hr/hr/Careers_FWS.htm). NOAA's SeaGrant offers graduate fellowships to study population dynamics and marine resource economics (www.nsgo.seagrant.org/funding/fisheriesgradfellowship.html). Check out other resources on the web or e-mail us. We'd be happy to help you find your way to this most important and satisfying career.

Key Concepts

- Knowledge of the interactions of organisms with their environment and with each other is critical for understanding the flows of energy and cycles of materials in the world ocean.
- Trophic pyramids may be based on numbers, biomass, or energy.
- The early twenty-first-century view of ocean food webs is that they consist of the classical food web, a metazoan, grazing-based, "macroscopic" food web; and the microbial food web, a bacteria and protistan-based, "microscopic" food web.
- Controls of the structure and dynamics of food webs may be top-down, bottom-up, or wasp-waist.
- Oceanographers recognize two types of pelagic food webs:

  steady-state food webs, typical of stable physical and chemical conditions; and transitional food webs, characteristic of unstable or time-varying physical and chemical conditions.
- Temporal scales of variability dominate many types of ecosystems, leading to variations in the types and abundances of species present at any given time.
- Overfishing has resulted in genetic changes in fish stocks, fishing down the food web, trophic cascades, and a number of other harmful effects on ocean food webs.
- Ecosystems management of fisheries and the establishment of marine reserves promise to restore and protect depleted species.

Terms to Remember

Critical Thinking

1. Compare and contrast food chains and food webs in terms of the flows of energy and materials in ocean ecosystems. Why are food webs more difficult to study?
2. How is the field of terrestrial ecology now being applied to ocean ecology? What are the limitations of a "terrestrial" point of view when studying ocean ecosystems?
3. Describe the relationship between the classical food web and the microbial food web. How do they each respond to changes in physical factors, such as stratification or destratification of the water column and seasonal changes in light intensity and penetration?
4. Compare and contrast an ice-edge ecosystem with an upwelling ecosystem. Why are transitional food webs necessary for the transfer of energy and materials from lower trophic levels, such as phytoplankton and zooplankton, to higher trophic levels, such as sharks, whales, and humans?
5. Discuss the role of physical factors in a bottom-up food web. What is the role of physical factors in a top-down food web? A wasp-waist food web?
6. Identify and discuss the impacts of humans on ocean ecosystems. What role do fisheries oceanographers play in assessing these impacts? What recommendations would you make to prevent overexploitation of marine resources?

Explore Online

 Visit www.mhhe.com/chamberlin1e for access to chapter quizzing, key term flash cards, video clips, interactive activities, and more. Further enhance your knowledge with web links to chapter-related material!

Exploration Activity 14-1

Exploring Food Webs Through Conceptual and Mathematical Models

Question for Inquiry

How do oceanographers use models to better understand and predict the dynamics of food webs?

Summary

Although conceptual models of food webs such as the ones illustrated in this chapter help oceanographers to identify the ecological roles of organisms, a quantitative understanding is necessary to understand and predict changes in population sizes as a result of natural or human-caused factors. Though we have avoided throughout this textbook much of the math that accompanies oceanographic research, rest assured that mathematics is essential for understanding the world ocean. Mathematical modeling of ocean food webs takes many forms, from simple growth models to complex coupled physical-biological grid models. In this activity, we explore a few simple models to gain an appreciation for the predictive capability that even the simplest equations provide.

Learning Objectives

By the end of this activity, you should be able to:

- Identify the trophic roles of organisms and illustrate the pathways of flows of energy and matter in food webs
- Compare and contrast the structure of food webs
- Make inferences about the effects of natural and human factors on productivity at different trophic levels in a food web
- Manipulate simple equations and graph their results
- Test various hypotheses using simple equations
- Express an understanding of the importance of mathematical modeling for predicting the impacts of humans on ocean food webs

Materials and Preparation

This activity requires the use of a spreadsheet program, although some parts may be completed using hand calculations and a calculator. To facilitate your work, we invite you to download from our website the Excel spreadsheets that accompany this activity. Alternatively, your instructor may provide data tables and figures to allow you to complete the conceptual questions in this activity. Some of the materials in Chapter 15 apply to this activity. Be sure to use them even if your instructor has not yet assigned that chapter!

Part I: Concept Mapping of Food Webs

Concept maps are ideal for exploring food web structure and trophic relationships. You may wish to review Exploration Activity 4-1 if you are unfamiliar with concept mapping. In this part, you will create concept maps of major ocean ecosystems.

1. Review Figures 14-1, 14-2, and 14-3 and make sure you are familiar with the classification of organisms according to their trophic level and the concept of trophic efficiency. You will create a multi-level concept map starting with primary producers at the bottom of the page. Use boxes to represent individual organisms and arrows to indicate who eats whom. Draw your arrow from prey to predator to indicate the direction of flow of energy and matter. Arrange your boxes so that organisms belonging to the same trophic level appear on the same general horizontal line (resembling the marine food web shown in Figure 14-1). If you are unfamiliar with some of the organisms and their trophic role, consult a marine biology textbook, Internet sources, or your instructor. Use the relevant pages in the text and other sources to construct trophic maps for three of the following ecosystems:
   a. coastal ocean pelagic ecosystem
   b. open ocean oligotrophic pelagic ecosystem
   c. midwater ocean ecosystem
   d. coastal soft-bottom benthos ecosystem
   e. abyssal soft-bottom benthos ecosystem
   f. coastal rocky bottom ecosystem (intertidal, near-shore, or offshore)
   g. hydrothermal vent ecosystem
   h. whale fall ecosystem
   i. Antarctic ice-edge ecosystem
   j. Arctic ice-edge ecosystem
   k. coastal upwelling ecosystem
   l. equatorial upwelling HNLC ecosystem
   m. coastal estuarine ecosystem
   n. salt marsh ecosystem
   o. mangrove ecosystem

2. Next to each ecosystem, list the geological, physical, and chemical factors that govern them. Rank the importance of each of these factors (1 = most important) in terms of the productivity of different trophic levels within the food web. In other words, which factors do you think have the greatest effect on the abundance of phytoplankton, fishes, or top predators?

3. Number each trophic level (1 = primary producer). Refer to Figures 14-4 through 14-6 and, on a separate sheet of paper, list plus/minus or 0 (no effect) to indicate the effects of the following processes on the abundance of each trophic level in each of your three ecosystems:
   a. increases in solar radiation
   b. deepening of the mixed layer
   c. vertical stratification of a previously unstratified water column
   d. upwelling of nutrient-rich water
   e. the passage of a hurricane
   f. iron fertilization
   g. melting of sea ice
   h. harvesting of small fishes, such as anchovies or sardines
   i. whaling
   j. ghost nets
   k. runoff of nutrient-rich sewage
   l. increased delivery of fine sediments to the surface waters

4. For three of the factors listed above, defend your reasoning. Write a short paper that summarizes why you think a particular factor (for example, iron fertilization) will increase or decrease the abundance of organisms in a particular trophic level. Include in your summary a statement of how you might test your "hypotheses."

Part III: Exploring Mathematical Models

Observations and data provide the fundamental information on which our understanding of the world ocean is based. Yet this information is most useful when it can be used in a quantitative manner using models to test hypotheses and predict future states of a property or system. Just as an equation for the speed of your car (distance × time) can help you predict your arrival time at your favorite destination (distance × time/total distance), equations of ocean processes can help oceanographers predict the behavior of a system at a future time. Most importantly, models test how well oceanographers understand how the ocean works and give them insights into the types of new observations needed to improve their understanding. In this activity, we explore some of the equations that oceanographers use to model different aspects of ocean food webs. While a complete set of equations for food webs is beyond the scope of this text, an introduction to the simplest of equations can provide a much deeper appreciation for the application of mathematics to oceanography. In fact, you might even find it fun! The equations listed below serve as a starting point for stu-

dents and instructors who wish to explore quantitative aspects of oceanography by constructing simple models in spreadsheets or a programming language. For those that wish to explore these models by manipulating the different variables, we have spreadsheet versions on our website.

1. *Beer's Law* (see Chapter 7). This equation (and its variants) allows oceanographers to predict the intensity (and spectrum, ideally) of the underwater light field from surface irradiance and the attenuation coefficient. It is widely used in models of primary productivity and carbon flow in food webs. For a given value of surface irradiance, compare the equation output for three values of attenuation corresponding to the open ocean, a coastal ocean, and an estuary. How does the depth of the euphotic zone change in each location? What are the implications for primary productivity? Compare and contrast the factors that alter the attenuation of light in each of these three locations.

Light Intensity at depth (Iz) = Light Intensity at the surface (Io) × the exponent (e) raised to negative power of the attenuation coefficient (Kd) × the depth

$$Iz = Io\, e^{-kz} \qquad \text{Beer's Law}$$

Typical surface irradiance = 1500 microEinsteins per square meter per second

Typical Kd values (440 nm): oceanic = 0.04 m^{-1}; coastal = 0.1 m^{-1}; estuarine = 8 m^{-1}

2. *Photosynthesis versus Irradiance (P vs I) equation* (see also Figure 13-9). This equation (which take different forms) provide estimates of gross productivity (P) at a given light intensity. Two key variables determine the "shape" of the P vs I curves: the maximum rate of photosynthesis at saturating irradiance (Pmax); and the initial slope of the curve, $\Delta P/\Delta I$, symbolized as alpha, α. The values of Pmax and α determine the response of phytoplankton under a given set of light conditions. High values of Pmax enable phytoplankton to utilize bright light, such as might occur near the ocean surface. On the other hand, low values of α allow phytoplankton to take advantage of low-light environments, such as occurs near the bottom of the euphotic zone. In fact, values of Pmax and α change with species, temperature, and even with the photoadaptive state of a species (see section on photoadaptation in Chapter 12). Several forms of the P vs I curve may be found in the scientific literature. The Platt and Jassby (1976) equation can be used if photoinhibition is not observed (or ignored). This equation fits a hyperbolic tangent function (introduced in Chapter 9) to the P vs I

relationship. Observe and describe the shape of the curve for different values of Pmax and α. Couple this equation with Beer's Law and calculate rates of productivity for light intensities typical of oceanic, coastal, and estuarine waters. How do overall rates of productivity compare (calculated as the integral of P over the euphotic zone or the sum of P as calculated at discrete depths)? What knowledge is important when predicting rates of productivity at different locations in the world ocean using P vs I curves?

Rate of gross productivity (P) = rate of maximum productivity (Pmax) × hyperbolic tangent of (alpha (α) × light intensity (I) ÷ rate of maximum productivity)

$$P = Pmax \tanh(\alpha I / Pmax) \qquad \text{P vs I curve}$$

Representative Values for Pmax and α.

Pmax (mg carbon mg per mg chlorophyll a per hour)	α (mg carbon per mg chlorophyll a per hour/microEinsteins per square meter per second)	Biome
2-14	0.01 – 2	World ocean
0.75 – 4	0.5 - 2	Polar
6-10	0.005 – 0.01	Tropical, subtropical, subpolar
0.2 – 1.0	0.02 – 0.06	Oceanic
9-17	0.006 – 0.13	Coastal

Source: Modified from *Biological Oceanography*, Second Edition, Carol Lalli and Timothy Parsons, The Open University, 1997. Note that these values vary widely with oceanic conditions and species. They are provided here to be illustrative rather than as a source of real-world values. Consult scientific references to obtain the most appropriate values for these constants.

3. *Michaelis–Menten Nutrient-Limited Growth* (see also Figure 13-10). The Michaelis-Menten formulation (1913), derived and published in English by Briggs and Haldane (1925), has been widely applied to enzyme kinetics and biochemistry. Oceanographers adopted the formulation to describe the nutrient- and light-limited growth of unicellular organisms, including phytoplankton and other microbes. In essence, it describes the growth rate of phytoplankton (μ) as it varies with the concentration of the biologically important nutrient that controls growth, or the limiting factor. The equation can be applied to studies of nutrient utilization by phytoplankton, resource competition among different phytoplankton, and biological models of productivity and food webs in the ocean, among other applications. The "shape" of the nutrient-limited growth equation depends on two main factors: the nutrient concentration at which growth is maximal, or μmax, and the concentration at which the growth rate equals μmax/2, or the half-saturation constant (KN). Like the P vs I curve, the growth-nutrient curve varies with species, their nutrient "history" (short- versus long-term nutrient deprivation), the element that is limiting, temperature, and other factors. By combining equations for two species, the competitive outcome within a particular nutrient environment (meaning, who survives and who perishes) can be examined. An excellent description of Michaelis-Menten kinetics as applied to the growth of phytoplankton in a laboratory vessel can be found here: http://web.pdx.edu/~rueterj/courses/esr473/notes/competition.html. This website also provides a spreadsheet for exploring resource competition. Here you may wish to explore the shape of the growth curve for different values of μmax and KN and hypothesize on the oceanographic conditions that may favor one set of values over another (for example, an upwelling ecosystem versus open ocean ecosystem).

nutrient-limited growth rate (μ) = (max growth rate (μmax) × nutrient concentration [N]) ÷ (half saturation constant (K_N) + nutrient concentration)

$$\mu = \mu max\ [N]/K_N + [N] \qquad \text{Michaelis-Menten formulation}$$

Representative Values for μmax and K$_N$ Under Nitrate Limitation

μmax (generations per day)	K$_N$ (micromoles)	Biome
0.1 – 3	0.01 – 2	World ocean
0.3 – 0.7	2.5 - 5	Polar
0.4 - 1	2 - 10	Upwelling coastal
1 - 3	0.02 – 0.06	Upwelling oceanic
0.1 – 0.2	0.5 - 2	Oligotrophic oceanic

Source: Modified from *Biological Oceanography*, Second Edition, Carol Lalli and Timothy Parsons, The Open University, 1997. Note that these values vary widely with oceanic conditions and species. They are provided here to be illustrative rather than as a source of real-world values. Consult scientific references to obtain the most appropriate values for these constants.

4. *Exponential and Logistic Growth Equations.* The growth of unicellular and multicellular organisms may be described in terms of the geometric progression of individuals (2,4,8,16,32, etc.), or doublings, that may be represented by an exponentially increasing curve, or exponential equation (exactly like the one we used for Beer's Law). If there were no restrictions placed on reproduction, exponentially growing populations would multiply infinitely. However (and fortunately), there are limitations on the rate at which organisms multiply. At some point, populations reach the maximum number that a given environment can sustain, a quantity known as the **carrying capacity.** The growth of populations limited by their carrying capacity is described by an S-shaped curve, or logistic growth. The logistic growth equation forms the basis for models of the human population and the carrying capacity of Earth, ecological models of the growth of populations of organisms, predator-prey interactions, and the maximum sustainable yield of fisheries, and much more. An Internet search on exponential and logistic growth will yield many useful resources. It is worthwhile to explore the behavior of these equations graphically in a spreadsheet program using values for constants found in the literature or on the Internet.

exponential growth = change in population size over time ($\Delta P/\Delta t$) = growth factor (r) × initial population size (Po) which may be represented by

$$P = Po\, e^{-rt} \qquad \text{exponential equation}$$

Here the "growth factor" relates to the intrinsic rate of increase of a population (often referred to as the intrinsic growth rate), usually determined experimentally or by observation and reported in units per day. Try values of 0.2 to 1.2 and simulate the exponential growth of phytoplankton. Speculate on the advantages and disadvantages of faster or slower growth rates for any species.

logistic growth = change in population size over time ($\Delta P/\Delta t$) = intrinsic growth rate (r) × population size (P) × (1− population size ÷ the carrying capacity) which may be represented by

$$P(t) = K\, Po\, e^{-rt} / K + (Po\,[e^{-rt}-1]) \qquad \text{logistic equation}$$

5. *Predator-Prey and NPZ Models.* For modeling interactions among predators and their prey, the most frequently used set of equations are those developed by A.J. Lotka and Vito Volterra, otherwise known as the Lotka-Volterra predator-prey model. This equations has been applied to a wide range of ecological and oceanographic problems, including phytoplankton-zooplankton interactions and fisheries. This model is often incorporated into more complex models of ecosystem and food web interactions, including nitrogen-phytoplankton-zooplankton (NPZ) models. A excellent resource for learning more about the application of predator-prey equations to NPZ models is *Biological Oceanography* by Charles Miller (Blackwell Publishing, 2004). We have included a simple NPZ spreadsheet model on our website.

15 Humans and the Coastal Ocean

Millions of people visit the coastal ocean annually where they enjoy a wide variety of recreational activities. Unfortunately, these uses strongly impact the structure and functioning of coastal ecosystems, oftentimes with negative consequences for marine organisms and humans.

Questions to Consider

1. What defines the coastal ocean, and what geological, physical, chemical, and biological factors and processes differentiate it from the open ocean?

2. Why is an understanding of temporal and spatial scales of variability important when studying coastal environments and the effects of humans on them?

3. What defines a particular coastal ecosystem, and what properties do coastal ecosystems share in common?

4. What challenges do different coastal ecosystems present for their management and preservation?

The Crowded Edge of the Sea

The **coastal ocean** includes all waters where the simultaneous presence of terrestrial and ocean properties can be detected. More than half of the US population lives within 80 km (50 miles) of the coastal ocean. This population and its activities exert tremendous pressure on coastal habitats and resources (**Table 15.1**). Coastal economies in the United States employed more than 2 million people in 2000 and generated more than a trillion dollars, nearly one-tenth of the total US gross domestic product. Human development and alteration of river flows have severely changed the natural processes that replenish beaches. Some 70% of the world's beaches suffer erosion. Without the protective barrier of beaches, coastlines and coastal communities are more susceptible to flooding, storm surge, and changes in sea level. At the same time, runoff of human-made debris, chemicals, and waste into the ocean has led to widespread increases in waterborne diseases, eutrophication, and alteration of marine food webs. Overharvesting of marine resources, both commercial and recreational, has severely depleted fish stocks. Dredging, deforestation, and other forms of habitat alteration have contributed to additional declines. Though efforts are being made to mitigate these impacts, many scientists believe that human impacts on coastal environments have reached crisis proportions.

The need to properly manage coastal resources has invigorated the field of **coastal oceanography,** the study of coastal waters. So complex are the interactions in coastal environments and the temporal and spatial scales over which they occur (see inside front cover) that oceanographers have only recently begun to understand how they function. In many ways, the open ocean has proven simpler to study and understand than the coastal ocean, despite its close proximity to oceanographic research institutions. New coastal observing systems and coastal research initiatives are bringing a renewed scientific focus on coastal environments.

TABLE 15.1
Human Uses of the Coastal Ocean

Coastal region	Type of habitat	Human uses
Offshore	Continental shelf	Fishing Oil and gas production Mineral extraction Transportation Sand and gravel dredging Sewage dumping Recreation Tourism–whale watching Coral reefs Tourism–snorkeling, diving Fishing Reef animal husbandry Shell collecting Quarrying
	Marine reserves	Tourism–kayaking, diving Marine research Coastal regeneration
Nearshore	Sandy shorelines	Recreation Fishing Sand and gravel mining Multiuse pier construction Coastal engineering Tourism Sporting events
	Rocky shorelines	Fishing Kelp harvesting Recreation Tidepooling Rock collecting Tourism
Estuaries	Aquatic	Fishing Recreation Tourism Transportation Multiuse construction Coastal engineering Aquaculture
	Mud flats	Shellfishing Bird-watching

A Brief History of the Environmental Movement

A brief summary of the evolution of scientific and public concerns for coastal environments frames modern-day efforts to address human impacts. Before the early 1960s, public concern for environmental issues was largely confined to creation and protection of parks and wilderness regions for outdoor recreation. Then, in 1962, the publication of Rachel Carson's *Silent Spring* struck a public nerve. It marked, in many ways, the beginning of the modern environmental movement. Carson's "silent spring" brought attention to the harmful effects of pesticides and other toxic human-made compounds that were being released into the nation's waters.

Television played an important role in raising awareness and promoting environmental activism. On April 22, 1970, Earth Day was born, an event that served to highlight the negative effects of human activities on the environment. Only three months earlier, President Richard Nixon had signed the National Environmental Policy Act, which established the Environmental Protection Agency (EPA). The EPA was charged with oversight of impacts of federal projects on the environment. Passage of the Clean Air Act (1963) and revised Clean Air Act (1970), the Clean Water Act and Ocean Dumping Act (1972), the Marine Mammal Protection Act (1972), the Endangered Species Act (1973), the Safe Water Drinking Act (1974), and similar legislation in subsequent years quickly and firmly established the environmental movement in the United States.

Almost as quickly, a movement against government regulation of environmental protection was born. Since the late 1970s, pro- and antienvironmental regulation groups have been locked in a struggle to balance environmental and economic interests. By the twenty-first century, the environmental movement has become an amalgam of global, national, and local organizations with considerable legal competency and strong public support (**Figure 15-1**).

■ **FIGURE 15-1** Coastal Cleanup Day began in Oregon in 1984 when Judie Neilson, a concerned citizen, organized more than 2,800 volunteers to collect beach debris. A year later, the state of California adopted California Coastal Cleanup Day. In 1989, the Ocean Conservancy, an environmental advocacy group, brought together participants worldwide for International Coastal Cleanup Day. Since that time, increasing public awareness of beach debris has led to more stringent measures to prevent it.

TABLE 15.2 — Summary of Critical Action Recommendations of the U.S. Commission on Ocean Policy

Improve governance	❏ Establish National Ocean Council
	❏ Create President's Council of Advisors on Ocean Policy
	❏ Strengthen NOAA
	❏ Develop voluntary, regional ocean councils
	❏ Coordinate management of offshore waters
Support science	❏ Double the federal budget for ocean science
	❏ Implement an Integrated Ocean Observing system
Promote education	❏ Support and coordinate formal and informal ocean science education
	❏ Create a national ocean education office, Ocean.ED
Coordinate management	❏ Strengthen coastal and watershed management
	❏ Set measurable goals for reducing human impacts on coastal and oceanic waters
	❏ Reform fisheries management
	❏ Engage in ocean policy issues at the international level
Implement recommendations	❏ Establish Ocean Policy Trust Fund

So, too, the political, legal, and financial resources of the economic-interests movement have grown. Perhaps nowhere is the debate between these two groups greater than over issues surrounding the growth, development, and protection of coastal ecosystems and their resources. Coastal zone issues, including coastal resource utilization, habitat protection, coastal development, beach nourishment, pollution, and many others, have spurred decades-long, bitterly fought battles between the environmental, economic, political, and regulatory interests. Publication in 2004 of the US Commission on Ocean Policy's final report, *An Ocean Blueprint for the 21st Century*, which contained 212 recommendations on ocean and coastal policy, underscores the importance of this debate for the future of coastal waters (**Table 15.2**).

What Is the Coastal Ocean?

The strip of land of varying width from the edge of the sea inland to the first terrain unaffected by the sea is called the **coast.** A coast may be marked by rocky cliffs, dunes, sandy beaches, estuaries, lagoons, marshes, and river deltas. In some regions, the coast may extend considerable distances inland where meandering inlets or narrow fjords penetrate the continents. The seaward edge of the coast, the **coastline** (or shoreline), represents the high-tide boundary of the land. A cliff, a berm, or a sand dune may mark this boundary. In some places, the coastline may be difficult to follow, especially where estuaries and river deltas occur. As well, the coastline may be quite dynamic over temporal scales, changing over the course of days to millennia.

Because of differences in their properties and patterns of circulation, it is often convenient to divide the coastal ocean into two components: coastal waters and estuarine waters. **Coastal waters** represent undiluted salty waters on the seaward side of tidal fronts, river plume fronts, and other fronts formed where freshwater discharges meet oceanic waters. Coastal waters typically extend to the edge of the continental shelf, although this distinction depends on the degree to which the adjacent landmass and seafloor influence the properties of the nearby water masses. Arid coastal regions with very little freshwater input may exhibit very little difference from oceanic waters. At the other extreme, regions where rivers discharge enormous volumes of water, such as the Ganges and Amazon Rivers, may extend their influence beyond the continental shelf. **Estuarine waters** occur on the landward side of tidal fronts and may vary in salinity from nearly oceanic to nearly fresh. Estuarine waters include waters mixed within or formed by tidal lagoons and estuaries where streams, rivers, ice melt, or other freshwater discharges mix with saline coastal waters.

Although these definitions serve oceanographers quite well, societal interests require a term that defines the region where human impacts are greatest. The term **coastal zone** has come to define the lands and waters along the coast that share a common ecology, that is, a region where terrestrial and oceanic ecosystems interact. In addition to the lands adjoining the coast, the coastal zone includes tidal rivers, estuaries, the shore, and the inner continental shelf.

Anatomy of a Coastal Ecosystem

As we learned in Chapter 14, an ecosystem is commonly defined as a community of organisms and the nonliving environment with which they interact. However, this definition applies better to terrestrial ecosystems than oceanic ones. In a fluid, three-dimensional environment such as the ocean, the distinction as to what constitutes a community is not always apparent among organisms that sink or float, swim and migrate, activate or hibernate, and live attached or free. Oceanographers have suggested an alternative definition for ocean and coastal ecosystems. According to Duke Marine Lab's Richard Barber, an ecosystem is "the unit of biological organization interacting with the physical environment such that the flow of energy and mass leads to a characteristic trophic structure and material cycles." The key words here are *trophic structure,* the arrangement of organisms according to who eats what or whom, and *material cycles,* the biogeochemical cycles that govern the movement of elements within an ecosystem. The oceanographic study of ecosystems in the world ocean largely concerns identification of their trophic structure and quantification of the cycling of materials and energy within and between them (**Figure 15-2**).

The geology of a particular region often defines the **habitat,** the place where an organism lives. Marine organisms modify habitats and, in some cases, create them. Mangrove trees feature an external root system that traps sediments and builds coastlines. Corals extract dissolved calcium carbonate and create reefs. Plate tectonics, waves, tides, currents, and sedimentary processes also shape coastal habitats. The physical characteristics of a coastal ecosystem, the **physical factors,** will be defined by the range and variability of oceanographic and meteorological conditions. Solar radiation, winds, tides, currents, evaporation, precipitation, river flow, and factors that affect the buoyancy of water will play a role in defining the physical characteristics of a coastal ecosystem. **Chemical factors** define the chemical characteristics of an ecosystem. Salinity, pH, availability of nutrients, dissolved gases, and other chemical factors define where organisms may live. **Biological factors** include physiological tolerances and the interactions among organisms. Competition for space, food, resources, and mates alters the distribution and abundance of organisms. Benthic and pelagic organisms are integral to the structure and functioning of coastal

ecosystems. In fact, **benthic-pelagic coupling,** the exchange of energy and materials between the benthos and pelagos, differentiates coastal ecosystems from open-ocean ecosystems. Two-way, tightly coupled interactions between benthic and pelagic organisms are common in coastal ecosystems.

Temporal and Spatial Scales of Variability

Throughout this text, we have emphasized the importance of time-varying and space-varying components of various geological, physical, chemical, and biological processes. These temporal and spatial scales of variability have a great impact on the "snapshot" of a coastal ecosystem that we observe at a given location and time. Coastal ecosystems are highly dynamic in time and space, presenting an even greater challenge for oceanographers. Keeping these scales of variability in mind enables scientists and resource managers to interpret observed changes in an ecosystem.

Spatial scales in coastal ecosystems range from centimeters to kilometers. Rills and ripples over a sand bar, indentations and cavities in a hard substrate, or differences in coral types represent a form of microtopographic relief that may be important for biological processes. A given stretch of a rocky coast may include a small embayment with a beach and exhibit characteristics of both beach and rocky substrate ecosystems. The morphology of river deltas and sandy shorelines may be altered by a single storm, thus disturbing and altering the local ecology. Tectonic processes, volcanic eruptions, and the emergence of islands produce large-scale spatial changes in landforms resulting in a sequence of physical, chemical, and biological processes.

Temporal scales also vary over a large range from minutes to millennia. The passage of a series of large waves may permanently

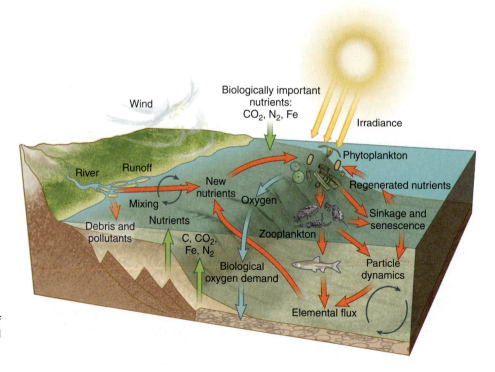

■ **FIGURE 15-2** Idealized representation of important geological, physical, chemical, and biological exchanges in coastal ecosystems.

reshape a shoreline. Tidal variations combined with waves produce near-daily alterations in the shape of many sandy beaches. Blooms of algae or seaweeds and settling of mangrove seeds or invertebrate larvae can transform a coastal ecosystem in days and weeks. Successional changes in organisms as a result of hurricane-induced disturbance (such as in the Florida Everglades) or maturation and growth of mangrove islands may occur over months and years. Longer time scales bring variations in the accretion or erosion of sediments, changes in the shape and location of river deltas, the growth of corals, isostatic adjustments, and the rising and falling of sea level in response to climate.

Studies of coastal ecosystems must take into account the relevant scales of variability for the particular scientific question being addressed. Scales relevant for coastal engineers (preparing for the ten-year storm) may be different from those of an oyster fisher, who may be concerned with year-to-year fluctuations in catch. Geologists may be mostly interested in millennial-scale changes over broad expanses of a coastline, while an ecologist may require daily information over very small patches of a reef. Modern-day studies can provide insights into how different processes interact (e.g., waves and longshore transport) and inform studies of past coastal environments by providing plausible mechanisms that gave rise to a particular ancient feature (such as an ancient shoreline). At the same time, studies of past environments provide the context for interpreting modern landforms. The rise of sea level some 6000 years ago submerged coastlines and river valleys, creating new estuaries and commencing a new round of marine geomorphic processes on land once high above the sea. Understanding the origins of coastal ecosystems can provide insights into their future and help differentiate natural variations from anthropogenic ones.

Classification of Coastal Ecosystems

Historically, the study and classification of coastal environments have been approached on a discipline-by-discipline basis. The most widely used are **geomorphic classifications,** which are based on the shape (morphology) of coasts and the processes that formed them (e.g., Shepard's classification). **Hydrographic classifications** emphasize physical characteristics of the water body, whether they are well-mixed, highly stratified, or in between (e.g., Pritchard's classification); whether their spring tide range is macrotidal (greater than 4 m or 13 feet), mesotidal (2–4 m or 6.6–13 feet) or microtidal (less than 2 m or 6.6 feet); whether they are exposed to waves or protected; or combinations of all these factors (e.g., Dronker's classification). **Habitat and community-based classifications** generally recognize the predominant substratum (e.g., sandy beach, mud flat, rocky intertidal) and dominant organisms (e.g., seagrass bed, mangrove forest, coral reef).

New coastal classification systems have emerged in the last decade in response to limitations inherent in historical classification schemes. Coastal oceanographers and estuarine scientists and resource managers have begun to work together to adopt multidisciplinary approaches that recognize the interactions among geological, physical, chemical, and biological processes in coastal ecosystems. Protection and management of coastal ecosystems require a common language for coastal resource managers, scientists, and the public. Considerable discussion is required for the development of terms, their definitions, and their relative importance. Researchers in Australia, Europe, and the United States have proposed systems that attempt to link coastal geomorphology, hydrodynamics, and ecology. The US system, the Marine and Estuarine Ecosystem and Habitat Classification (MEEHC) system developed by NOAA, employs a 13-level hierarchical scheme for differentiating marine ecosystems. For example, San Francisco Bay and Chesapeake Bay, both large estuaries, occur in temperate zones (Level 1) and represent continental, aquatic, estuarine systems with benthic and pelagic components (Levels 2–7). Subsequent levels (8–13) define increasingly detailed information, ideally leading to a unique set of descriptions. Although San Francisco Bay and Chesapeake Bay share similar properties, many characteristics of these estuaries are quite different, including their tectonics (active versus passive), growth (shrinking versus growing), physical oceanography (bounded by eastern versus western boundary current), climate (Mediterranean versus northern temperate), and more. Ultimately, the aim of the MEEHC system is to provide a tool for assessing the susceptibility of coastal and estuarine ecosystems to human impacts.

Exploring Coastal Ecosystems

Coastal ecosystems are arguably among the most complex ecosystems on Earth. Our brief and generalized descriptions here are only intended to guide interested readers toward a deeper exploration of their own examples. One means for understanding these ecosystems is by comparing the similarities and differences in their geologic setting, physical oceanography, chemical transformations, and biological interactions, subjects covered in previous chapters. You may also consider the temporal and spatial scales of change observed within them, especially as they pertain to human impacts, such as habitat destruction and climate change. Consider, too, how humans have altered or disrupted the structure and functioning of these ecosystems. Educating yourself about the issues surrounding a coastal ecosystem is one of the most important steps you can take as a science-literate person.

Beach Ecosystems

Beaches make up nearly 40% of the world's coastlines, though beaches may also be associated with other types of shorelines, such as rocky cliffs, or coastal ecosystems, such as estuaries. Most of us are familiar with beach-dwelling organisms such as clams, sand dollars, beach hoppers, and mole crabs. But pelagic organisms are important as well. Beach ecosystems where adjacent rocky shores are minimal may depend solely on plankton as a food supply. Reproduction of benthic organisms often includes a planktonic stage where larvae may feed in the plankton. Thus, consideration of beach ecosystems must include factors that affect both their benthic and pelagic components (**Figure 15-3**).

As we learned in Chapter 10, beaches represent a dynamic balance between waves and sediments, both affecting the shape and character of a beach. In consideration of beach ecosystems, it may be useful to review the nearshore processes affecting beaches. Beach ecosystems have been historically considered as an example of an **open ecosystem,** one that is dependent on imports of organic materials to sustain its populations. Researchers suggest, however, that some beach ecosystems may, in fact, be characterized as **semiclosed or closed ecosystems,** sustained by productivity within their local confines. Whereas open-beach ecosystems rely on imports of **beach wrack,** the broken bits and rafts of seaweeds or seagrasses, to supply food to animal populations, semiclosed-beach ecosystems may reply on primary production from **surf zone diatoms,** phytoplankton species that prefer high-energy surf conditions, or other local sources of phytoplankton. In all likelihood, beach ecosystems benefit from a variety of sources of organic material, including terrestrial sources, delivered by rivers, runoff, or wind transport. The degree to which a beach ecosystem is open or closed may vary with location and season.

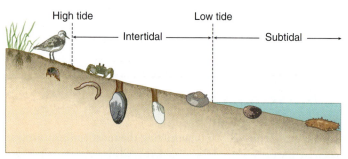

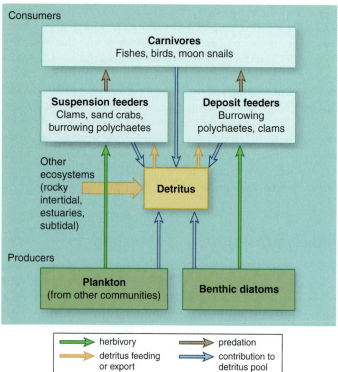

| herbivory | predation |
| detritus feeding or export | contribution to detritus pool |

■ **FIGURE 15-3** Generalized concept of a beach ecosystem.

Another important aspect of beach ecosystems concerns their **stability,** the degree of variability associated with sediment movements. Sandy beaches tend to be less stable than rocky beaches or mud flats. Movements of sand affect the ability of organisms to build structures in which to live. Rocks and cobble in relatively moderate- to low-energy wave environments may be stable and support populations that more resemble rocky intertidal communities than sandy ones. Where wave action frequently moves them, however, larger sediments will be unstable and devoid of life. At the other extreme, fine-sediment sandy and muddy beaches may be highly stable, leading to very fine-scale interactions among organisms and high diversity. Where waves or tidal currents frequently suspend fine materials, however, conditions may be less suitable for certain types of organisms, especially filter feeders sensitive to turbid conditions.

Rocky Shores

Rocky and cliffed shorelines make up nearly three-quarters of the world's coastlines, although many of those coastlines simultaneously feature some type of beach. Associated with rocky cliffs are rocky or hard substrate shorelines that provide habitat for marine organisms. This substratum also typically extends into nearshore and offshore regions. The geologic setting for the rocky shore depends on the composition of the substrate and its slope relative to the seafloor. Rocky shores may be composed of hard granitic cliffs that resist erosion and may be little changed since the rise of sea level. Most rocky shores, however, consist of rocks that erode to some degree and feature prominent coastal landforms, such as **sea stacks** (residual small islands along a coast), **sea caves** (excavations in a sea cliff), **sea arches** (rocks through which an opening occurs), and **blowholes** (geysers of water created by the trapping of air by incoming waves). Rocky shores also exhibit gradations in slope from very steep to nearly flat. Near-vertical cliffs, called **plunging cliffs,** may create impressive underwater "walls" that appear to plunge into the abyss. In other regions, erosion of rocky cliffs may give rise to **shore platforms,** horizontal or sloping layers of rock or hard substrate (such as limestone or chert) at the base of a rocky cliff. Shore platforms may be formed by waves, shore weathering, dissolution of rock, or **bioerosion,** the breakdown of rocks by organisms. Most shore platforms exhibit some type of **tide pool,** a seawater-filled depression in rocks visible when the tide retreats.

One of the distinctive features of rocky shore ecosystems is **zonation,** the clustering of organisms into distinct bands along the shoreline, usually according to tide height (**Figure 15-4**). So distinctive are these bands that marine ecologists often classify intertidal organisms based on where they live: supratidal (the splash zone occasionally wetted by waves), high intertidal, middle intertidal, lower intertidal, and subtidal zone. The exact tide heights corresponding to each zone depend on the tidal range and wave exposure. Intertidal zonation has been studied intensely by ecologists interested in the interactions between organisms and their environment. These studies have led to the broad conclusion that the upper limit of an organism's distribution is controlled by physical factors, while the lower limit of an organism's distribution in a particular zone is controlled by biological factors (predator-prey interactions, competition for space and food, etc.).

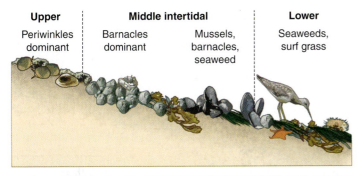

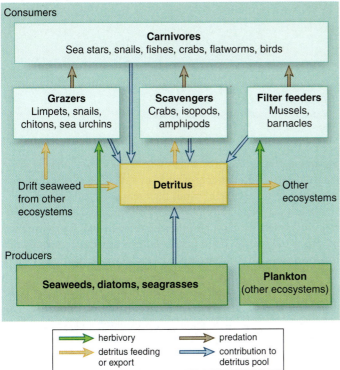

■ **FIGURE 15-4** Generalized features of a rocky shore ecosystem.

Studies of rocky intertidal ecosystems have provided a great deal of the theoretical foundation for understanding ocean ecology. Working off the coast of Washington in 1969, Robert Paine demonstrated by exclusion experiments that by feeding on mussels and opening space for other species to inhabit, the sea star *Pisaster ochraceous* increased the diversity of intertidal communities. This study and others led him to the concept of the **keystone predator,** a species whose presence alone alters the structure of the community in which it occurs, including prey size, abundance, diversity, distribution, and composition. When the species is removed, the community shifts ("collapses") to a less diverse assemblage (hence the analogy to a keystone in an arch; when the keystone is removed, the arch collapses). Yet marine ecologists increasingly recognize that "bottom-up" factors, such as oceanographic processes affecting nutrient and food availability, supply, and distribution, may be important in structuring nearshore marine communities. In reality, the pattern of organisms on rocky shores may be better described as a shifting mosaic of top-down and bottom-up factors. Interannual variations in oceanographic processes that alter the settlement rates of larvae may be, in some instances, more important than local pro-

cesses in controlling the distribution and abundance of some intertidal organisms. Human impacts may exert an even greater influence in some regions. As is often the case, the keystone predator concept is just one of many factors that must be taken into account when developing models of coastal ecosystems.

Estuaries and Tidal Lagoons

Part land and part sea, estuaries and their accompanying wetlands suffer from a limited public and scientific understanding. Historically, estuaries were viewed as coastal swamps that were valuable only insofar as they could be developed into housing, airports, shopping malls, power plants, marinas, and the like. In the latter decades of the twentieth century, estuaries came to be viewed as "nature's filtration system," capturing sediments and removing pollutants before their discharge into the ocean. Their role as a breeding ground and nursery for fishes and other organisms gained appreciation. As their extent decreased, especially along the western coast of the United States, their importance as a flyway for migratory birds was realized. Billions of dollars have now been committed to restore estuaries and wetlands in the United States alone.

The most commonly cited definition of an estuary belongs to Donald Pritchard (1918–1999), who defined an **estuary** as a "semienclosed coastal body of water with a free connection to the open sea and within which seawater is measurably diluted with fresh water from land drainage." This definition excludes estuaries with periodic or infrequent freshwater inputs, such as those found in parts of Mexico, Australia, and Africa, and "wet" tropical estuaries with high volumes of freshwater that mix outside of a "semienclosed" basin. Nevertheless, estuaries represent regions where fresh and salt water mutually influence each other and may include **coastal lagoons,** which periodically have openings to the sea. Note that not all estuaries are associated with rivers. Semienclosed basins and lagoons with significant freshwater runoff from land are also considered estuaries. Glacially carved fjords, such as those found in the Pacific Northwest and along the New England coast, may receive freshwater inputs from seasonal snow and ice melt.

Hydrographic classifications of estuaries are common for describing the most general patterns of circulation (**Figure 15-5**). Pritchard defined a **salt-wedge estuary** as one in which freshwater flows seaward over the top of seawater that maintains a distinct pycnocline (and halocline). Mixing of salt water upward at the pycnocline results in a deficit of seawater and creates a net landward flow of seawater up the estuary. Tides may moderate the intensity of freshwater and saltwater flows, but the flows retain their net seaward and landward directions, respectively. Where tidal volumes of seawater, called the **tidal prism,** exceed freshwater flows, such as in macrotidal regions, and where vigorous mixing occurs, the estuary is said to be **well-mixed.** Well-mixed estuaries exhibit a horizontal and general smooth gradient of salinities from their seaward to landward sides, but vertical gradients in water column properties (temperature and salinity) are small or not apparent. Isohalines in well-mixed estuaries may be nearly vertical. Waters move into the estuary from surface to bottom at flood tide and out of the estuary during ebb tide. In mesotidal regions or regions where some mixing due to other processes occurs, an estuary may be **partially mixed.** Partially mixed estuaries represent the intermediate state between the two extremes of the highly stratified salt-wedge estuary

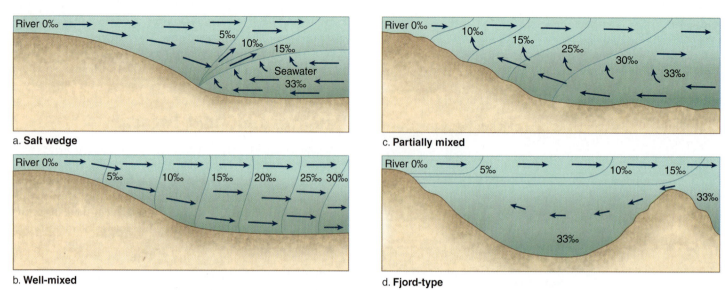

a. **Salt wedge**

b. **Well-mixed**

c. **Partially mixed**

d. **Fjord-type**

■ **FIGURE 15-5** Pritchard's hydrographic classification of estuaries. a) salt wedge; 2) well-mixed; 3) partially mixed; 4) fjord type.

and the well-mixed estuary. In some cases, high rates of evaporation of surface waters create a deficit of water such that seawater flows nearly continuously into an estuary. These estuaries are called **inverse or hypersaline estuaries.**

In truth, estuaries represent a dynamic balance between oceanic, atmospheric, and terrestrial processes, especially those that govern the circulation of water within them (**Figure 15-6**). Freshwater inputs will nearly always vary seasonally and depend on the extent of the watershed that drains into an estuary. In arid regions, freshwater inputs may only be occasional, while regions with substantial rainfall or where large rivers empty will experience a continuous supply of freshwater. The volume of freshwater input will affect the buoyancy properties of estuarine waters and influence their pattern of circulation. Tides exert considerable influence in most estuaries, especially in macrotidal estuaries where tidal ranges are large. The action of tides and tidal currents may enhance mixing of estuarine waters and influence buoyancy-driven circulation. Waves, too, may play a role in mixing, especially for large estuaries where wind waves can develop or for

estuaries exposed to large coastal waves. Winds also act to mix estuaries and change their patterns of circulation. Larger, exposed estuaries may be more susceptible to winds than semiprotected estuaries.

A major aspect of the biological character of estuarine ecosystems concerns the relative importance of their pelagic versus benthic primary producers (**Figure 15-7**). In many estuaries, seagrasses and seaweeds may be the dominant primary producers by which the estuary's food webs obtain their nutrients. In other estuaries, phytoplankton may be dominant. In some cases, phytoplankton and benthic algae/plants may compete with each other for light and nutrients. Eutrophication may stimulate the growth of both phytoplankton and benthic plants. Increases in turbidity due to human perturbations that increase sediment loads may limit light penetration and reduce the growth of bottom-dwelling seaweeds and seagrasses. Here again, the dynamics of geological, physical, chemical, and biological processes may alter the temporal and spatial patterns observed in the abundance of estuarine organisms. Human processes that alter or interfere with any of these processes may have unexpected and unintended impacts.

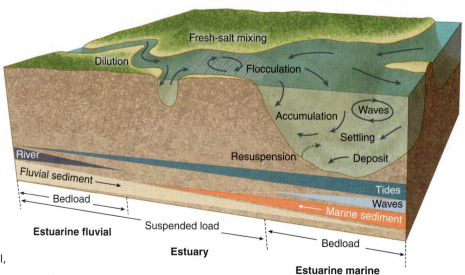

■ **FIGURE 15-6** Generalized geological, physical, and chemical features of estuaries.

Open water in channels
and muddy bottom

Consumers

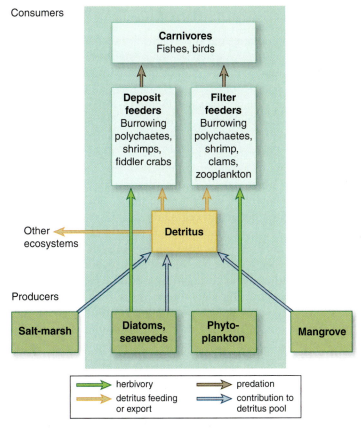

Carnivores
Fishes, birds

Deposit feeders
Burrowing polychaetes, shrimps, fiddler crabs

Filter feeders
Burrowing polychaetes, shrimp, clams, zooplankton

Other ecosystems

Detritus

Producers

Salt-marsh

Diatoms, seaweeds

Phyto-plankton

Mangrove

→ herbivory
→ detritus feeding or export
→ predation
→ contribution to detritus pool

■ **FIGURE 15-7** Generalized estuarine ecosystem.

Mangrove and Salt Marsh Ecosystems

Mangroves are one of the few trees that thrive in salt water (**Figure 15-8**). Though only four species grow in the southeastern continental United States, more than 60 species of mangrove trees can be found on sheltered tropical and subtropical coastlines world-wide. Mangrove trees may reach heights of more than 30 m (100 feet) and may form dense canopies several miles wide. The trunks feature extensive prop roots, while their subsurface root systems project ver-tical "breathing tubes," called **pneumatophores,** which stick above the surface of the sediments and allow the plant to respire. These adaptations allow mangroves to inhabit soft sediment environments and withstand periodic submergence by high tides. The root systems

a.

b.

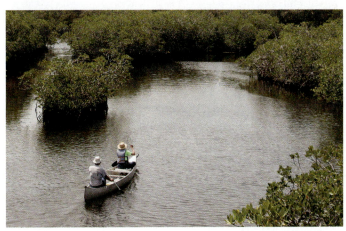

c.

■ **FIGURE 15-8** (a) Mangrove trees. (b) A floating mangrove seed. (c) Mangrove forest.

of mangroves serve an important ecological role, providing shelter for numerous invertebrate and fish species. Mangroves act to stabilize sediments and may even increase land area by moderating water flows and increasing rates of sediment deposition. Mangrove forests are highly productive, and their leaf litter and detritus are a source of food to marine organisms within and beyond their confines. In many regions of the world, mangrove forests provide an important source of wood and other products for humans.

Though simpler and lacking woody parts, salt marshes are similar to mangrove forests in their structure and function. Though many types of marsh plants, marsh grasses, and seagrasses may be found in a given location, they typically form dense stands of a single species in zones or "islands" corresponding to their tolerance of saline conditions and submergence. Like mangroves, salt marsh plants may trap sediments and facilitate the formation of mud flats and salt marsh terraces. Similarly, salt marshes provide habitat and a food source for numerous marine organisms (**Figure 15-9**). Both mangrove and salt marsh ecosystems are remarkably stable and may occupy the same stretch of coast for thousands of years.

Mangrove and salt marsh ecosystems exhibit features of both terrestrial and marine ecosystems. They have been recognized as serving three roles in coastal environments: 1) ecosystem-food web functions; 2) biogeochemical functions; and 3) anthropogenic functions. As a source of protection, they may act as a nursery for juvenile fishes and other immature marine organisms. Their high productivity attracts numerous herbivores, which in turn act as a food source for invertebrate and vertebrate predators. They also create a stable habitat for benthic communities living on or below the sediment surface. Organic materials deposited on the muddy bottom serve as a source of food for a rich community of decomposers that mediate the cycling of nutrients and important elements. They also serve as natural catchments for pollutants and sediments and thereby improve water quality. Mangrove and salt marsh ecosystems also may serve as a habitat for commercially important species. Perhaps most important, they provide a refuge for numerous species of terrestrial and aquatic wildlife for the enjoyment and education of people of all ages.

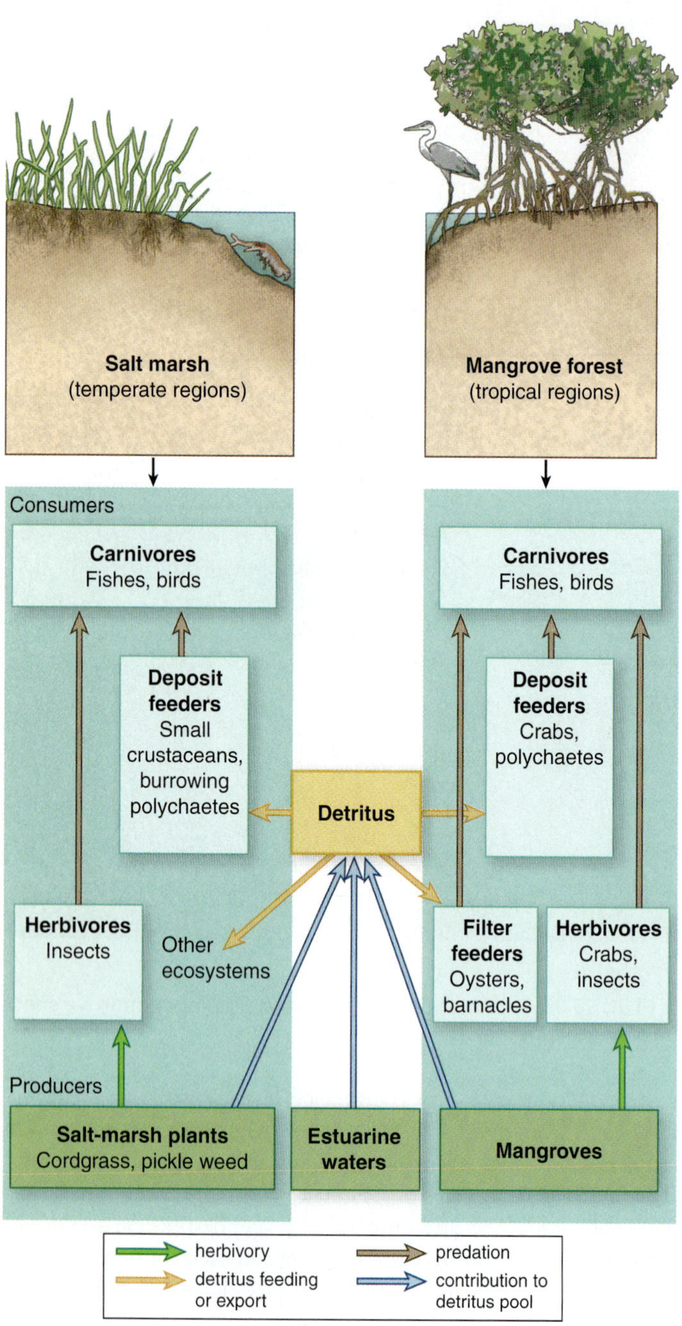

Delta Ecosystems and Coastal Wetlands

Coastal deltas are formed at the mouths of rivers where sediments brought from upstream accumulate at a rate faster than they are removed by oceanic processes. The size and shape of a delta depend on the relative importance of waves, tides, and river flow in transporting, depositing, and eroding sediments (**Figure 15-10**). Typically, a delta consists of a series of bifurcating channels that spread outward through or around a wedge or lobe of subaerial sediments. The bird's-foot pattern of the Mississippi River delta is a classic **digitate delta,** where freshwater flows dominate fluvial processes. Tide-dominated deltas may produce **lobate deltas** (Ganges-Brahmaputra-Meghna delta, the largest in the world), highly tapered, almost capillarylike in their appearance. Wave-dominated deltas may give rise to **cuspate deltas** where longshore currents move sediments along the coastline (Danube River delta in the Black Sea or the Ebro delta in Spain). Alternatively, intense wave energy may create **blunt deltas** (São Francisco River, Brazil) or **blocked deltas** (Senegal River, West Africa) where wave action produces barrier islands at the mouths of rivers.

Although coastal deltas are not commonly treated as ecosystems *per se*, an ecosystem approach proves useful for understanding the critical role they play in coastal processes. In that sense, these regions are often called **coastal wetlands,** fine-sediment environments along the coast whose emergent or occasionally emergent landforms are partially or wholly vegetated or inhabited by benthic organisms. Coastal wetlands include mangroves, salt marshes, and tidal flats, and the term has gained popularity in discussions of human impacts on these environments. Coastal delta ecosystems share similarities to mangrove and salt marsh ecosystems, but they also exhibit characteristics of beach ecosystems, especially in the patterns of movements of their sediments. They may also function like estuaries as a natural filtration system for land-derived pollutants.

The Mississippi River delta serves as a prime example of an ecosystem that has undergone major changes in the past century. Formed where the Mississippi River empties into the Gulf of Mexico along the lower southeastern portion of Louisiana, this delta covers approximately 521,000 acres, of which 420,000 are open water. In recent decades, this region has experienced a rapid rise in sea level, almost 1m (about 3 feet) per century. This is largely due to geologic subsidence but also is a result of natural processes (e.g., hurricanes and climate change) and human activities, including levee building, canal excavation, and removal of subsurface petroleum products and groundwater. As a result, some 113,000 acres of coastal wetlands have

■ **FIGURE 15-9** The generalized mangrove and salt marsh ecosystem.

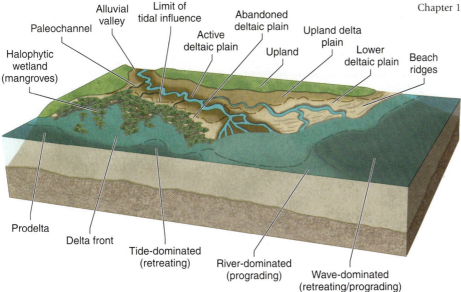

■ FIGURE 15-10 Anatomy of a delta.

been lost to the sea in the last 60 years. At least three major problems arise as a result of this loss: 1) increased susceptibility of coastal regions to marine processes, especially wave-, tidal-, and storm-induced flooding (e.g., Hurricane Katrina); 2) reduced trapment of sediments and pollutants discharged to the Gulf of Mexico; and 3) loss of coastal wetlands habitat for fisheries and waterfowl. Impacts on human recreation, tourism, and natural aesthetics are also important. Alarmed by these potential impacts, Louisiana has requested $14 billion in federal aid to commence restoration of coastal wetlands, similar to projects underway in Chesapeake Bay and the Florida Everglades.

Spits, Barriers, and Barrier Islands

Beaches that extend outward from the edge of a bay, river mouth, or island are called **spits.** A variety of spits can be found along the shoreline, such as simple spits, hooked spits, island spits, and even double spits (**Figure 15-11**). Spits typically represent the extension of a beach as a result of the longshore current and wave action (see Chapter 10) that deposits sediments where an abrupt change in shoreline direction occurs. Spits may even extend across a bay or bedrock embayment to create a **barrier,** a ridge of sand or gravel that fronts the shoreline.

Barriers may create a coastal lagoon. They may also close off an embayment (such as a cove) and form pocket beaches, dune barriers, or other types of coastal barriers. When a spit joins an offshore island and connects it to the mainland, a **tombolo** is formed. Along shorelines where the predominant swell approaches from two different directions, a cape may be formed, formally known as a **cuspate foreland.** Cape Hatteras and Cape Lookout along the coast of North Carolina are thought to have formed as a result of waves approaching at high angles relative to the shoreline. Wind data and modeling efforts support this idea.

One of the major coastal landforms that results, in part, from longshore transport and wave action are **barrier islands.** Barrier islands represent elongated, shore-parallel islands of sand or gravel with a bay, lagoon, or marsh separating them from the mainland. Barrier islands may be found all along the East and Gulf coasts of the United States, as well as along the Arctic and Gulf coasts of Alaska. The Outer Banks of North Carolina are certainly the most well-known barrier islands, although the Padre Island National Seashore and adjacent barrier island in Texas merit recognition as the longest stretch of undeveloped barrier island in the world at more than 161 km (100 miles) long. Barrier islands and their management remain the focus of a hotly contested battle between scientists, environmentalists, citizens, policy makers,

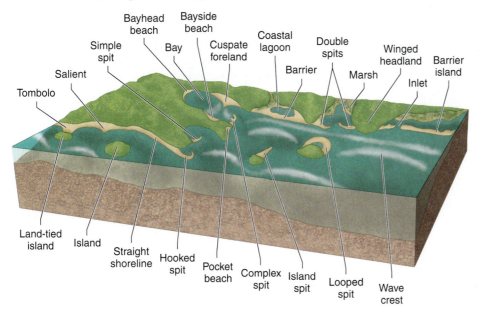

■ FIGURE 15-11 Features of spits.

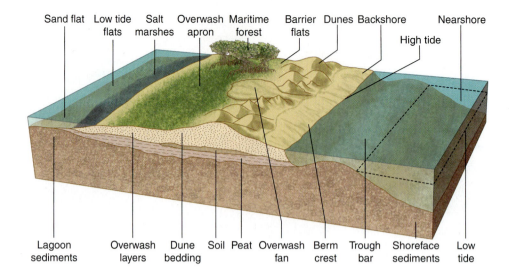

Sand flat Low tide Salt Overwash Maritime Barrier Dunes Backshore Nearshore
flats marshes apron forest flats
High tide

■ FIGURE 15-12 Anatomy of a barrier island.

Lagoon Overwash Dune Soil Peat Overwash Berm Trough Shoreface Low
sediments layers bedding fan crest bar sediments tide

and developers (among others), all with different interests. Fueling the debate are uncertainties regarding the dynamic nature of barrier islands and the processes by which they grow, shrink, and migrate.

The general morphology of a barrier island system helps us appreciate its dynamics (**Figure 15-12**). The barrier island represents the main component. Formed from sand or gravel that once resided on a beach, the island may extend below sea level down to about 9m (30 feet) and above sea level to several tens of feet, depending on the height of its sand dunes. The island exhibits a beach on its shoreward side where foreshore (lower beach) and backshore (upper beach) processes both play important roles. The landward-facing side of the barrier island, called the **backbarrier,** may be bordered by a marsh, lagoon, or bay. The dynamics of the backbarrier side are typically less energetic than the seaward side but depend on the type of tides (macrotidal, mesotidal, or microtidal) and the fetch provided by the water body. Extensive tidal flats may be present here. The barrier crest, the emergent part of the barrier island, typically shows a fore-to-back gradation from dunes to vegetation to a fanlike deposit of sediments formed from storm overwash, called an **overwash apron.** The breaking of waves and storm surge over the barrier crest is an important process in barrier island formation and dynamics. Storm waves and surge also create breaks between barrier islands, resulting in multiple barrier islands along a coastline. Once formed, the ebb and flood of tides further carve these channels, resulting in tidal inlets. Deltas may form along at both ends of the inlet. Sand pushed into the inlet by the flood tide results in a large flood-tide delta on the lagoon side of the inlet. A smaller ebb-tide delta will form on the seaward side. Wave action and the magnitude of the tidal currents influence the size of the ebb-tide delta, which may extend outward from hundreds of feet to several miles.

Sea-level rise is a major factor in the formation of barrier islands (**Figure 15-13**). Repeated episodes of sea-level fall (during glaciations) and rise (during interglacials) have alternately exposed and inundated landforms. Rising sea levels flood river valleys, creating headlands and embayments. Headlands, subject to erosion by wave action, produce sediments that may be transported by longshore currents to create spits. Extension of the spits across embayments creates barriers and coastal lagoons. A continued rise in sea level may overwash the barrier and expand the backbarrier lagoon, sepa-

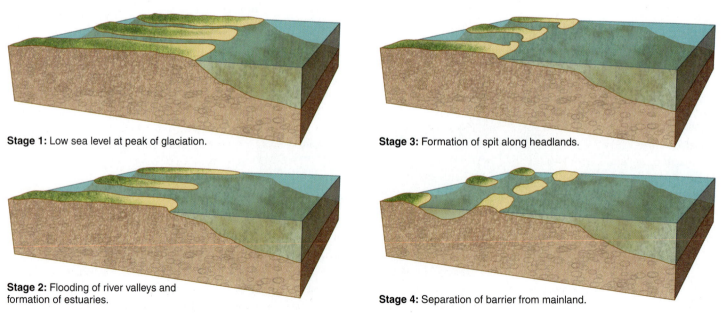

Stage 1: Low sea level at peak of glaciation.

Stage 2: Flooding of river valleys and formation of estuaries.

Stage 3: Formation of spit along headlands.

Stage 4: Separation of barrier from mainland.

■ FIGURE 15-13 Sea level rise and the formation of barrier islands.

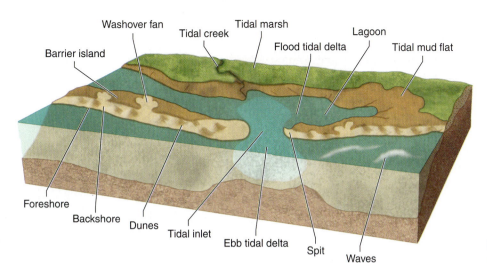

■ **FIGURE 15-14** Wave action and the dynamics of barrier islands.

rating the barrier from the mainland. Episodic phenomena, notably storm surge and hurricanes, may contribute to wave erosion of sediments on the barrier beach face and deposition of sediments on the backbarrier as an overwash apron. The net result of overwash is to elevate the barrier crest and move the barrier island toward the mainland. This sequence of events is thought to have produced the barrier islands along the Currituck Peninsula on the Virginia-North Carolina border. Barrier islands whose shorelines are moving landward are called **transgressive barriers,** and they may be recognized by a sedimentary sequence in which marine deposits overlie lagoonal and marsh deposits. Some barrier islands, such as Galveston Island in Texas, maintain a rate of supply of sediments that exceeds the rate of sea-level rise. In this location, the barrier island has moved seaward, depositing lagoon sediments on top of marine ones. These types of barrier islands are called **regressive barriers.**

Wave energy is the ongoing, dynamic driver of barrier island evolution (**Figure 15-14**). Longshore and cross-shore transport of sediments shapes the beach profile and the net movement of sediments up and down and along a shoreline. Wave energy and direction, coupled with tides (which alter the placement of wave energy on a beach face), determine when and where sediments move. Longshore currents may alter the shape of tidal deltas and even cause tidal inlets to migrate or close.

Tidal inlets alter the supply and distribution of sediments by creating deltas. Sediments from ebb-tide deltas may migrate to the beach of a barrier island. Flood-tide deltas typically accrete to the backbarrier side of a delta. If the inlet closes, the ebb-tide delta may disperse, whereas the flood-tide delta may remain as a feature of the marsh or lagoon. The tidal range for a given location appears to determine the number and spacing of tidal inlets. Where tidal range is greater, a greater number of inlets (or larger inlets) accommodate the greater tidal volume, resulting in shorter barrier islands. Nonetheless, too great of a tidal range (e.g., macrotidal) appears to limit the formation of barrier islands.

Winds may assist in the buildup of barrier islands by blowing sand from the beach onto dunes and increasing their height. Where winds blow onshore, greater dune development has been observed. Barrier islands oriented parallel to prevailing winds may experience sand movements along the length of the island but little buildup of

dunes. Like overwash, winds can be an important factor in the landward migration of barrier islands. Vegetation acts to stabilize barrier islands and limit the movement of sediments. Dune plants tend to trap sediments and build dunes. Forests may even form on the backbarrier side of barrier islands. Planting of dune grasses and building of dune fences may also stabilize sediments and cause a buildup of a dune. These mitigations, however, tend to redistribute sediments on a barrier island, rather than contribute new sediments. By preventing storm overwash, the natural cycle of barrier island replenishment is interrupted, resulting in loss of sand offshore from the beach.

The combined effect of these processes is **barrier island migration,** the natural movement of barrier islands. Rising sea level generally means landward migration of barrier islands through erosion of sediments on the beach side, deposition of sediments on the lagoon side, and buildup of sediments on the barrier crest. Until the 1960s, barrier island migration was not recognized as a natural process, resulting in attempts to limit the erosion. However, since that time, efforts have been made to allow natural processes to dominate. On a global basis, barrier islands appear to be shrinking in size over the past 100 years. Erosion on both the beach and lagoon sides has resulted in thinning of barrier islands on the Outer Banks. The causes are uncertain, although reductions in sand supply caused by damming of rivers, building of inlets and jetties, and dredging likely are responsible for part of the problem. Increases in the rate of sea-level rise may be another, whether natural or human-induced. As Orrin Pilkey points out, the thinning of barrier islands actually makes them more susceptible to overwash, which may replenish their sediments. Thus, barrier islands may maintain a kind of long-term natural equilibrium with sea-level rise.

Coral and Algal Reef Ecosystems

Coral reef ecosystems flourish in what otherwise might be considered the ocean deserts. Nutrient concentrations in the tropical oceans where corals occur are vanishingly small, yet these organisms—part animal, part algae—build the largest living structures on Earth. Because phytoplankton abundance is limited in these oligotrophic waters, corals and coralline algae can take advantage of the abundant light that penetrates the transparent waters. Marine biologist

Richard Murphy calls coral reefs "cities under the sea." Indeed, coral reefs exhibit a structure and function not unlike cities. Coral reefs represent a type of **biogenic coast** (i.e., a secondary coast or "organic reef" in Shepard's classification), constructed by the activities of marine organisms. The location and extent of these coastal structures depend intimately on the physical, chemical, and biological processes that moderate the growth and decline of the organisms that build them. Thus, unlike primary or "tectonic" coastlines, whose structure depends to a large degree on geological processes, these coastlines are created and shaped by marine organisms. Nevertheless, geologic processes determine the initial location of coral reefs, which require some type of hard structure in shallow water on which to form. Plate tectonics governs the birth and death of coral reef ecosystems as plates move reefs into more or less desirable environments, respectively.

Coral reef ecosystems are among the most diverse on our planet with perhaps as many as a million species of organisms or more inhabiting their domain. This diversity seems surprising, given the relative rarity of their habitat: coral reefs occupy some 284,900 km^2 (110,000 square miles) of Earth's surface, less than 1.2% of the total continental shelf area in the world ocean. Most coral reefs are found in the Indo-Pacific, although a small number (less than 8% of all coral reefs) can be found in the Caribbean and Atlantic. Patterns of biodiversity in coral reef ecosystems largely reflect diversity in the **stony corals** (*Scleractinia*), a relative of sea anemones and jellyfish and one of the major reef builders in the world ocean. The other major reef builder, the **coralline algae,** though often overlooked, may dominate some habitats and create "coral-like" carbonate reefs (e.g., Bermuda).

Stony corals depend on **zooxanthellae,** single-celled photosynthetic microbes, for the energy by which the coral skeleton is built. This cooperation among organisms is termed **symbiosis,** the association of organisms for their own or their mutual benefit. Numerous examples of symbiosis exist on coral reefs. As much as 85% of the living biomass on a reef may belong to zooxanthellae and coralline algae. As primary producers and integral components of reef-building corals, algae indirectly or directly control the construction of coral reefs. Both zooxanthellae and coralline algae carry out photosynthesis, using light energy to fix carbon dioxide into chemical energy. Zooxanthellae, however, provide energy and other compounds to their animal host, the coral, which carries out calcification (**Figure 15-15**). Thus, zooxanthellae indirectly build reefs through their association with corals. In return, the coral animal supplies to the symbiont a source of protection and important dissolved nutrients—some of which are obtained by filter feeding on plankton. Coralline algae may occupy the surface, crevices, cavities, and sides of coral reefs and directly deposit calcium carbonate in their cell walls. Coralline algae also produce sticky surfaces that attract sediments when they die, further contributing to reef construction. When reef-building animals die, they become part of the coral reef and provide a surface on which other reef-building organisms may grow. In this way, coral reefs become quite massive and exist in the same location for hundreds, if not thousands, of years.

A classification scheme for coral reefs still in use today was first proposed by Charles Darwin in 1842 (**Figure 15-16**). Darwin defined **fringing reefs** as shore-attached or nearshore reefs formed when corals colonize the rocky coastline of a continental shelf or an island. Fringing reefs grow most rapidly in shallow waters, typically up to the depth of the mean low tide. Another type of reef formation, the **barrier reef,** occurs when a **lagoon,** a narrow body of shallow water, separates the reef from the shoreline. Barrier reefs often protect a coastline and

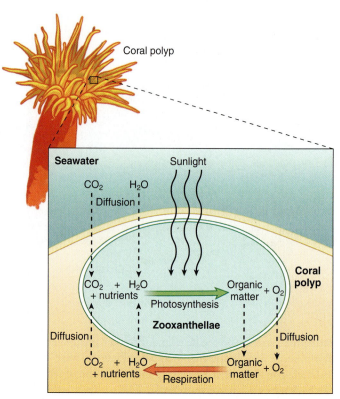

■ **FIGURE 15-15** The symbiosis of zooxanthellae and coral animals.

exhibit their own morphology, typically a **fore reef,** facing the ocean, a **back reef,** facing the land, and a **reef crest** in between. Because physical and biological processes differ on the fore and back parts of the barrier reef, these parts often exhibit a number of secondary structures that contribute to different reef morphologies. Sedimentary processes on the fore and back parts of the reef will differ as well. Coral knolls, or **patch reefs,** often form as small reefs inside barrier reefs or atolls. **Ribbon reefs,** narrow, broken reefs that form along the edge of the continental shelf as part of a barrier reef system, feature channels between them that may be used for navigation. Some reefs appear independent of a landmass. **Bank or platform reefs,** common in the Caribbean, may form in this way. Low-profile **coral cays** form from the accumulation of sedimentary debris from a coral reef and exhibit visible differences between their windward and leeward sides. Coral cays are a young feature, having formed over the past 6000 years, but they represent an important part of reef ecosystems.

On larger temporal and spatial scales, the distribution, extent, growth, and decay of coral reefs may be governed by plate tectonics, ocean processes, and climate. **Atolls,** annular-shaped, open or closed reef systems with a central lagoon rising from deep water, have been at the center of a debate about the role of various processes in structur-

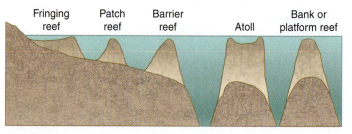

■ **FIGURE 15-16** Principal types of coral reefs.

ing coral reefs (**Figure 15-17**). Even their definition is under dispute as some scientists regard many atolls as **oceanic faros,** reef structures with a central enclosed lagoon or depression whose origins differ from "true" atolls. Charles Darwin was among the first to postulate that atolls represented the final stage as volcanic islands with fringing reefs and barrier reefs subsided. However, one puzzle of Darwin's model was the shallow nature of atoll lagoons. Erosion of reef materials and sedimentation into the lagoon were insufficient to account for this observation. Another hypothesis, called the solution hypothesis, was first advanced in the 1950s but has gained favor in recent years. In the solution model of atoll formation, subaerial dissolution of carbonate materials by rainfall, called **karst topography,** generates the "saucer-shaped" morphology of atolls during periods of sea-level lowering (i.e., glaciations). One intriguing outcome of this model is the young age it proposes for atolls, perhaps no more than 2000 years. According to one scientist, the late formation of atolls also explains a 1500-year gap in the migration of people throughout Oceania. Simply put, there were no islands to inhabit before about 2000 years ago.

Though we most often think of coral reefs in terms of processes that build them, it is equally important to consider the processes that erode or destroy them. Coral reefs are largely unconsolidated with only about 30% of the materials in a reef deposit representing "framework" materials. The remainder of the reef structure consists of pore spaces and sands. As the dead reef materials are buried, interactions between pore water and reef materials may cause precipitation or dissolution of carbonates. **Coral reef diagenesis,** chemical alterations of the corals and associated sediments, is of great interest to scientists trying to determine the role of coral reefs as a source or sink of global carbon. Biodeposition and bioerosion play a role as well. Five types of organisms may be responsible for the growth and decay of coral reefs:

1. Constructors
2. Binders
3. Bafflers
4. Destroyers
5. Dwellers

Constructors and binders, organisms that cause sediments to stick together, include the corals and coralline algae, along with the contributions of reef sand builders, such as benthic foraminifera and a green seaweed called *Halimeda*. An early observation from reef drilling operations was that reef cores were rarely solid. Yet as much as 50

to 80% of reef cores may be composed of carbonate sands. *Halimeda*, which precipitates calcium crystals (aragonite) within its blades, may be a significant source of these sands, especially along deep fore reefs and back reefs. Bafflers include organisms that alter the flow patterns over a reef and thereby increase rates of sedimentation. Most structural organisms fall into this category as well, although fleshy algae, bacteria, invertebrates, and fishes may directly or indirectly moderate sedimentary processes. Organisms that graze on or bore into corals may be considered destroyers. A number of invertebrate and vertebrate species fall into this category, including humans. Coral reef dwellers, organisms that live within, upon, or above coral reefs, may participate in an indirect fashion. Overfishing is now recognized as one of the major causes of the global decline of coral reefs. Removal of herbivorous fishes has allowed fleshy algae, a competitor for light and nutrients, to flourish over the surface of coral reefs. Reductions in light and increases in sedimentation as a result of these algae essentially "smother" the corals and they die. Thus, the ecology of coral reefs is very important to an understanding of the growth and decay of coral reef ecosystems (**Figure 15-18**).

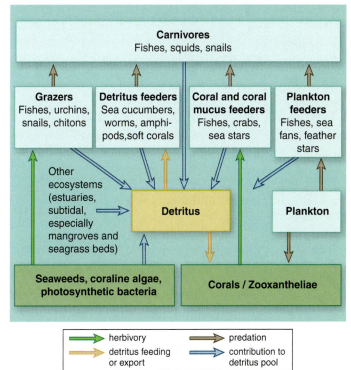

FIGURE 15-18 The generalized coral reef ecosystem.

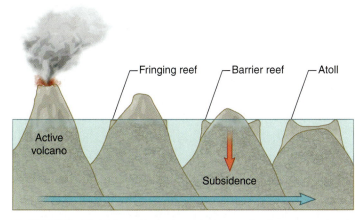

■ FIGURE 15-17 Darwin proposed that atolls form as volcanic islands subside. New data indicate that atolls may have multiple origins, including dissolution of carbonate platforms during sea level fall.

Coastal Shelf Ecosystems

Like the open ocean, coastal shelf environments display complex, three-dimensional structure and dynamic interactions between geological, physical, chemical, and biological processes (**Figure 15-19**). Much of what we learned about pelagic and benthic ocean food webs in Chapter 14 can be applied to an understanding of coastal shelf ecosystems. Broadly shaped by their geological structure, coastal shelf ecosystems maintain recognizable and persistent patterns in their physical, chemical, and biological properties. Coastal pelagic habitats may extend from the outer edge of the surf zone to the continental

shelf and beyond at depths from 2 to 200 meters (6.6–656 feet). Pelagic habitats may be mostly photic, aphotic, or somewhere in between, depending on the conditions that affect light penetration. Coastal benthic habitats refer to any habitat provided by the seafloor on the continental shelf. Coastal benthic habitats are extremely diverse, ranging from sandy and muddy seafloors, to cobble and gravel bottoms, to rocky platforms and vertical walls. They may be formed by any of a number of geological processes, including volcanism, seismo-tectonism, hydrothermal processes (including cold seeps), mass wasting, and sedimentation. Physical processes, especially currents along the seafloor, may influence and shape these habitats. Even

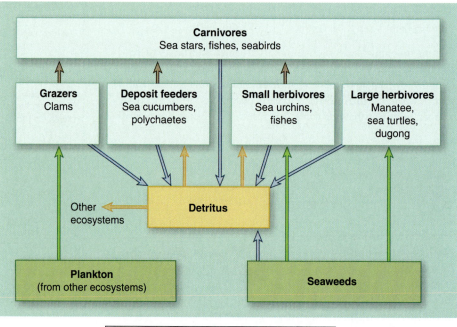

■ **FIGURE 15-19** The generalized coastal shelf ecosystem.

biological processes, such as reef construction (by mollusks, deep-water corals, and other invertebrates) or small-scale soft-sediment processing, may be important. Recent efforts to describe and define these habitats have benefited from the use of ROVs, AUVs, manned submersibles, and scuba.

Exploring Coastal Issues

Many coastal regions experience **eutrophication,** an overabundance of nutrients, especially nitrate, that stimulates the excessive growth of phytoplankton, seaweeds, and marine plants. If the overproduction of organic matter exceeds the ability of a food web to consume it, organic matter builds up on the seafloor. Aerobic decomposition of these materials combined with limited water circulation can deplete dissolved oxygen and cause hypoxia or **anoxia,** the absence of oxygen. Eutrophication is responsible for the occurrence of large zones of anoxia, called **dead zones** (although anaerobic organisms continue to persist). Eutrophication of the Gulf of Mexico, largely as a result of agricultural activities in the Mississippi River basin, has contributed to a seasonally extensive dead zone covering thousands of square kilometers. Eutrophication may also stimulate excessive blooms of phytoplankton, the so-called red tides and brown tides, some of which may be toxic. **Harmful algal blooms** (HABs) are on the rise worldwide, probably as a result of eutrophication and other human-induced changes in marine ecosystems.

Eutrophication and Coastal Dead Zones

Every year during summer in North America, tens of thousands of square kilometers of the coastal ocean lose their life-sustaining oxygen. In a matter of hours, marine organisms inhabiting these waters must migrate to oxygenated waters or perish. For attached plants and animals that depend on oxygen, there is little choice. Even those animals that are capable of swimming often find themselves trapped. Fish kills occur frequently in these waters, making the situation worse as the decomposition of the organic matter increases oxygen demand and releases additional nutrients, the culprit of eutrophication.

Since the 1960s, the occurrence of hypoxia in coastal waters has doubled nearly every decade. More than 146 coastal regions around the world suffer from seasonal or persistent low- or no-oxygen conditions (**Figure 15-20**). At least 11 major dead zones occur in the United States, with numerous smaller occurrences in estuaries, harbors, embayments, and other waterways. Hypoxia affects more than just a few unfortunate marine organisms: economic consequences result from the loss or relocation of fishing grounds, changes in species fished, loss of habitat for larval and juvenile fishes, and other factors affecting fisheries production. Recreational losses occur when the sight and smell of hypoxic waters discourage water activities and tourism. One major stumbling block in the search for a solution to hypoxia is the lack of quantitative data on economic losses.

As we have learned, hypoxia results from eutrophication, which occurs primarily from increased loads of nitrogen and other nutrients delivered to coastal waters from urbanization, agriculture, and industry. Nearly all hypoxic zones occur near population centers and regions where watersheds have been altered and developed. While agriculture and industry share a portion of the responsibility for increased nutrient loads—56% of the nutrient loads in the Mississippi River come from agriculture—urban horticultural practices (fertilization of greenways, golf courses, lawns, and shrubs), incomplete sewage treatment (nonremoval of nitrogen before discharge), and the destruction of wetlands (which can trap nutrients) further add to the problem. Fundamentally, any practice that accelerates the release of nutrients from the land (e.g., deforestation) or hinders its removal by natural processes (e.g., pavements) can increase nutrient loading in a watershed. The second-largest dead zone in the world ocean occurs seasonally in the northeastern Gulf of Mexico over an area of some 7770 to 22,015 km^2 (3000–8500 square miles). In the summer of 2005, the Gulf of Mexico dead zone encompassed an area of more than 11,840 km^2 (4554 square miles), almost the size of Connecticut (**Figure 15-21**). The region affected extends from near the mouth of the Mississippi River to the Louisiana-Texas border at depths between 6.1 to 26 m (20–85 feet). The size of the 2005 dead zone places it near the annual mean value of 12,700 km^2 (4800 square miles), based on studies since 1985. The size of the dead zone was larger than predicted by NOAA models that take into account measured nutrient loads in May and June in two rivers of the Mississippi watershed. NOAA researchers predicted a dead zone size of 3626 km^3 (1400 square miles) for 2005. In contrast, a model developed by Eugene Turner of Louisiana State University that is based on the May nutrient load and takes into account the previous year's nutrient loads predicted a size of 16,058 km^2 (6200 square miles), an overestimate of the observed 2005 dead zone.

Differences between modeled predictions and field observations illustrate an incomplete understanding of all of the factors that determine the size and longevity of coastal dead zones. Episodic events in June–July 2005, such as Hurricane Dennis and tropical storms Arlene and Cindy, alter the seasonal patterns of water column mixing and watershed runoff, factors that models do not presently take into account. Other factors, such as the intensity of

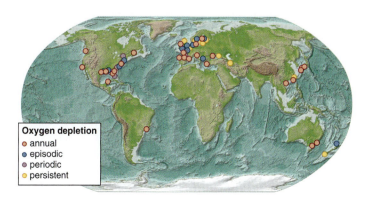

■ FIGURE 15-20 Hypoxic zones in various regions of the world ocean.

■ **FIGURE 15-21** Areal extent of hypoxia (< 2.0 mg/L) in bottom waters of the Gulf of Mexico in July 2005.

surface warming, changes in coastal circulation, and the temperature of bottom waters, may play a role for which present observational and predictive capability is insufficient. Nonetheless, these models help oceanographers to narrow down the list of possible factors that may affect hypoxic zones. Most likely, improvements in these models will come from better characterization of water column processes through increased observational capability and incorporation of additional physical, chemical, and biological factors in model runs.

Improved understanding of the temporal and spatial variations in dead zones will also assist efforts to mitigate their occurrence and extent. Proposals to limit the release of nitrogen from upstream agriculture in the Mississippi River basin have met with fierce resistance. At the same time, some scientists hypothesize that reductions of 40% would be required to contain the problem. A lack of quantitative data on fisheries losses associated with the Gulf of Mexico dead zone limits a cost-benefit analysis of upstream controls. New research is aimed at understanding how fishes and shrimp react behaviorally and physiologically to hypoxic waters. In addition, alternative solutions, such as construction of wetlands or vegetation buffer zones, are being explored. As one researcher puts it, "If we are going to ask farmers in the Mississippi drainage basin to reduce nutrient loading, we need to show that it will solve the problem."

HABs: Coming to a Shore Near You

In Alfred Hitchcock's classic 1963 movie *The Birds*, people in a small coastal California town are terrorized for no apparent reason by flocks of deranged birds. Hitchcock's idea for the film came from a news item reported on August 18, 1961, by the Santa Cruz *Sentinel* newspaper:

> A massive flight of sooty shearwaters, fresh from a feast of anchovies, collided with shoreside structures from Pleasure Point to Rio del Mar during the night. Residents were awakened about 3 a.m. today by the rain of birds, slamming against their homes. Dead and stunned seabirds littered the streets and roads in the foggy, early dawn. Startled by the invasion, residents rushed out on their lawns with flashlights, then rushed back inside, as the birds flew toward their light. When the light of day made the area visible, residents found the streets covered with birds. The birds disgorged bits of fish and fish skeletons over the streets and lawns and housetops, leaving an overpowering fishy stench.

What caused the strange behavior of these birds? At the time, no one knew for sure. Subsequent "mysterious" deaths of seabirds and marine mammals in Monterey in 1991, however, led scientists to suspect blooms of the diatom *Pseudo-nitzschia* (pronounced sue-doe-NITZ-chee-uh). This organism is one of more than 300 species of phytoplankton that proliferate as harmful algal blooms (HABs). HABs include all blooms of microscopic algae and seaweeds that cause harm to ecosystem function, human health, or coastal economies. The most significant impacts come from the 60 to 80 species of phytoplankton (primarily, dinoflagellates) that produce toxins and the fewer species with spines and other cellular structures that irritate or damage fish gills. When inhaled or ingested by humans, the toxins produced by HABs may cause a range of symptoms from mild discomfort to death. Although hundreds of cases of illness from HABs have been reported in the United States over the past several decades, only a few human deaths have occurred. Wildlife losses may be especially severe, with single events leading to the deaths of hundreds of marine birds and mammals or millions of fishes. By far, the greatest impacts of HABS occur as loss of revenue during closures of fishing grounds or beaches, loss of aquaculture stocks (directly through fish kills or indirectly by contaminating shellfish), and restricted recreational activities and tourism. In the United States, HABs cost an estimated $49 million per year, including costs associated with human health care, lost revenues to commercial fishing, tourism, and recreational interests, and costs for monitoring of blooms.

The frequency of HABs has risen significantly since the 1970s (**Figure 15-22**). Particularly severe blooms in the 1980s and 1990s stimulated greater funding and coordination of scientific efforts to identify the species responsible for HABs and their causes. These efforts have advanced considerably our understanding of the physical, chemical, and biological factors that contribute to HABs. Proliferation of a particular algal species requires a particular suite of physical and chemical conditions. Unfortunately, no single factor provides a reliable indicator of when, where, and how long a bloom will occur, or even if it will be harmful. Not all blooms produce toxins, even among toxin-producing species. Apparently, genetic diversity within the same species may determine whether a bloom is toxic or not. It is the unique confluence of temporal and spatial scales of variability in physical, chemical, and biological factors and the response of specific genotypes to those conditions that determine the occurrence of HABs. Through global cooperative efforts, new capabilities can be developed for observing, modeling, and predicting HABs. At the same time, public education about the dangers of HABs and dissemination of real-time information on HAB conditions will better ensure public safety and reduce economic losses along our shores.

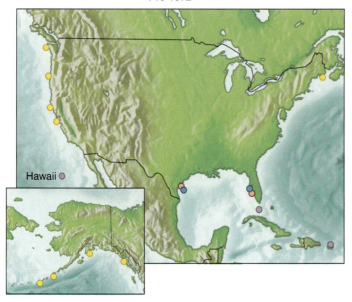

Pre-1972

Hawaii ●

Post-1972

Hawaii ●

HAB occurrence in the United States
● neurotoxic shellfish poisoning
● paralytic shellfish poisoning
● amnesic shellfish poisoning
● fish kills
● ciguatera
● brown tide

■ **FIGURE 15-22** The occurrence of HABs is on the rise in the United States and elsewhere around the globe.

Marine Protected Areas

In 1901, geologist, naturalist, conservationist, and Sierra Club founder John Muir wrote in *Our National Parks:*

> The tendency nowadays to wander in wilderness is delightful to see. Thousands of tired, nerve-shaken, overcivilized people are beginning to find out that going to the mountains is going home; that wilderness is a necessity; and that mountain parks and reservations are

useful not only as foundations of timber and irrigating rivers, but as foundations of life.

In 2005, Muir's sentiments about terrestrial wilderness areas might equally apply to the marine wilderness. A growing movement to establish marine protected areas (MPAs) has begun to demonstrate positive results for the conservation of marine life and remediation of severe overfishing. MPAs have enhanced public awareness of and interest in the inherent beauty and complexity of marine ecosystems beyond their role as a provider of food and other resources. While less than 1% of US waters are protected, marine protected areas such as California's Channel Islands National Marine Sanctuary, the Florida Keys National Marine Sanctuary, and the Northwestern Hawaiian Islands Coral Reef Ecosystem Reserve are quickly gaining status as marine versions of Yosemite and Yellowstone National Parks.

Paramount to the argument for creating marine reserves is scientific data that demonstrate the benefits to fishing and marine ecosystem management. A number of studies have shown that marine reserves increase the density, biomass, size, and diversity of many species within the reserve (**Figure 15-23**). Larger sizes of female fishes have been shown to increase the output and viability of fish larvae, further enhancing the benefits of marine reserves. In a few studies, clear evidence exists that marine reserves can export commercially important species to adjacent waters. Marine reserves work best for species with limited dispersal ranges, such as invertebrates and reef species, but may not be as successful for species with a wide dispersal range. On the other hand, too large of a marine reserve may not be ideal for spillover of commercially important species to adjacent waters. Many small reserves may prove more beneficial than a few large reserves because they export larvae to surrounding waters and enhance recruitment. Successful design of marine reserves requires establishment of observational and monitoring tools within the reserve and adjacent waters. A better understanding of marine reserves, their structure and function, and the factors that affect them will better enable scientists and resource managers to design and create them for the improvement of fisheries and benefit of all who enjoy them.

■ **FIGURE 15-23** Large predatory fishes, like this grouper, have enjoyed a resurgence in marine no take reserves, like Molasses Reef in the Florida Keys.

Coastal Ocean Observing Systems: Sentinels for Science and Public Safety

The importance of weather to human affairs has long been recognized. One of the primary objectives of US National Weather Service and the World Meteorological Organization is to provide information and forecasts about atmospheric conditions that may endanger human lives and cause damage to property. Weather forecasting places many demands on scientists and weather observers, who may be responsible for tracking and predicting any number of episodic and extreme phenomena, including tornadoes, hurricanes, typhoons, and winter storms. Though less appreciated, forecasting of ocean weather is equally important and even more daunting. The number of oceanographic processes and variables of interest far outnumber those of the atmosphere. The diversity of oceanic hazards and public safety concerns are considerable. Because more than 60% of the world population now resides within the coastal zone, the necessity for timely delivery of accurate oceanographic data to provide nowcasts and forecasts for industry, tourism, recreation, and public safety has become more widely recognized. To this end, a number of local, state, federal, and international public and private agencies have begun to commission and develop coastal ocean observing systems (COOS).

These systems will have many components, and virtually all portions of the world ocean will be systematically observed using a broad range of sampling platforms. Clearly, implementation of COOS is a formidable proposition, but on the positive side, it is bringing together oceanographers with a multiplicity of disciplines and interests from all over the globe. In the not-too-distant-future, the entire coastline of the United States will be equipped with land-based and *in situ* platforms supported by satellite observations to provide comprehensive temporal and spatial information on ocean temperature, salinity, currents, waves, sea level, bottom topography, phytoplankton (including harmful algal blooms), and sediments (**Figure 15a**). Coastal oceans interact with the open ocean, so efforts are under way to incorporate data from offshore ocean observatories to maximize weather and climate forecasting capability and to integrate these observations with other Earth-observing systems.

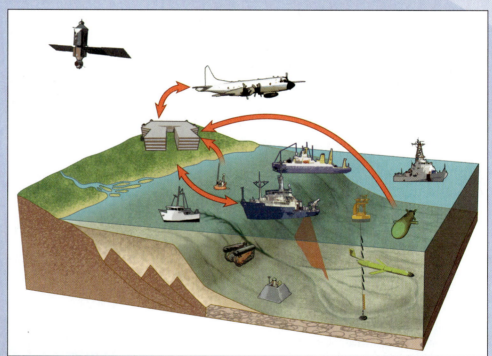

FIGURE 15a Concept of an integrated coastal ocean observing system.

YOU Might Wonder

1. **I spend a lot of time in the water and, on occasion, I develop rashes and feel sick. Is it from sewage in the water?**

   You likely are experiencing surfer's itch or swimmer's eruption, which has a long history. At first, public officials denied that there was anything unhealthy about coastal waters. Yet new methods of identifying particular strains of bacteria and efforts to develop epidemiological data on the occurrence of these conditions suggest that runoff from land and possibly discharges from sewage may be making some waters unhealthy. Heal the Bay and Surfrider Foundation have been especially active in identifying health hazards on beaches and educating the public about their occurrence. Unfortunately, there is no real way to know ahead of time if the water is unhealthy. The best remedy is to avoid ocean waters two to three days following heavy rains and to always rinse thoroughly after swimming or surfing.

2. **How can coral reefs be so productive if they exist in such nutrient-poor waters?**

   The key to coral reef productivity lies in a highly efficient recycling system. The coral-zooxanthellae symbiosis ensures that waste products from the coral are made directly available to the zooxanthellae. In pelagic systems, excretion by zooplankton permits diffusion and dilution of nutrients. The high biodiversity of coral reef ecosystems also ensures that nearly every bit of energy and matter is utilized. Coral reef ecosystems are a good example of highly efficient coupling of the flows of energy and matter in an ecosystem.

3. **I want to do something to help preserve coastal ecosystems. What can I do?**

   That's an admirable goal! There are a number of things that you and your friends, relatives, and classmates can do to help. The first and most important step is to educate yourself and others about the issues facing your local coastlines. Consulting your oceanography instructor would be a good place to start. Websites for public agencies (coastal commissions, parks and recreation, fisheries, environmental health and protection, etc.) often provide a list of issues and post news related to local environmental issues. Environmental organizations such as the Ocean Conservancy, the Surfrider Foundation, Heal the Bay, National Resource Defense Council, Sierra Club, and many others have local chapters that focus on specific issues. As you learn more, think about a specific issue or two that especially concern you. Become active with organizations that address those issues. Participate in Earth Day and other coastal awareness activities. Attend meetings of local government agencies. Start your own campaign. Don't be shy about going to a college library and researching the science behind your issues. It may take a few readings, but knowledge of the scientific data can help you fully understand the complex issues that face our coastal environments. And feel free to contact one of us if you desire additional information on a particular issue or the science behind it. It takes a concerted effort by everyone to provide the responsible stewardship that our coastlines deserve.

Key Concepts

- New coastal observing systems and coastal research initiatives are bringing a renewed scientific focus on coastal environments.
- Two-way, tightly coupled interactions between benthic and pelagic organisms are common in coastal ecosystems.
- New coastal classification systems have emerged in the last decade in response to limitations inherent in historical classification schemes.
- Educating yourself about the issues surrounding a coastal ecosystem is one of the most important steps you can take as a science-literate person.

Terms to Remember

anoxia, 329

atolls, 326

back reef, 326

backbarrier, 324

bank or platform reefs, 326

barrier, 323

barrier island migration, 325

barrier islands, 323

barrier reef, 326

beach wrack, 318

benthic-pelagic coupling, 316

bioerosion, 318

biogenic coast, 326

biological factors, 316

blocked deltas, 322

blowholes, 318

blunt deltas, 322

chemical factors, 316

coast, 315

coastal lagoons, 319

coastline, 315

coastal ocean, 313

coastal oceanography, 313

coastal waters, 315

Critical Thinking

1. Compare and contrast the coastal ocean and the open ocean. What special challenges does the coastal ocean present for coastal oceanographers?

2. Compare and contrast two coastal classification systems. Identify the benefits and drawbacks of each system for managing coastal resources.

3. Why is beach erosion such a concern to coastal communities? What measures can be taken to reduce or eliminate beach erosion? Use the Internet or other printed resources to discover what measures are being taken along a nearby coastline.

4. Pick one coastal ecosystem described in this chapter. Describe its trophic structure and identify the pathways of energy and material flows. Discuss the temporal and spatial factors that alter the flows of energy and materials in this ecosystem.

5. Pick one coastal ecosystem described in this chapter. Describe the ways in which geological processes, physical processes, chemical processes, and biological processes shape this ecosystem.

6. Pick one coastal ecosystem described in this chapter. Describe the ways in which physical processes shape this ecosystem.

7. Pick one coastal ecosystem described in this chapter. Describe the ways in which chemical processes shape this ecosystem.

8. Pick one coastal ecosystem described in this chapter. Describe the ways in which biological processes shape this ecosystem.

9. Pick one coastal ecosystem described in this chapter. Describe the ways in which human processes shape this ecosystem.

10. Pick one coastal ecosystem described in the chapter. Identify the existing and potential impacts of human activities on this ecosystem. Prioritize the impacts from worst to least damaging. For your top-three impacts, discuss the possible consequences if no action is taken.

11. Pick one coastal ecosystem described in the chapter. Formulate a scientific research plan to identify human impacts on this ecosystem. Include in your plan a means for establishing an environmental baseline for your ecosystem (hint: think of historical records or studies). Defend the importance of your study for public health and enjoyment of coastal waters.

Explore Online

 Visit www.mhhe.com/chamberlin1e for access to chapter quizzing, key term flash cards, video clips, interactive activities, and more. Further enhance your knowledge with web links to chapter-related material!

Exploration Activity 15-1

Exploring Human Impacts on the World Ocean Through Critical Thinking

Question for Inquiry

How are humans impacting the coastal ocean?

Summary

Throughout this textbook, we have highlighted scientific evidence that human activities are having significant and harmful impacts on the world ocean. Two of these impacts, global warming and overfishing, have received greater attention (commensurate with the threat they pose) but other impacts are also important, including eutrophication, habitat destruction, release of pharmaceuticals, metals, and other toxins, marine debris, runoff of human and pet waste into coastal waters, noise pollution, ship impacts, and others. In view of your newly acquired knowledge and understanding of the world ocean, and your experience in the analysis and application of oceanographic data, we invite you to demonstrate your abilities as non-scientist "expert" in a particular human impact that concerns you. This activity complements Exploration Activity 1-1, so if you have not completed that one, you may wish to substitute it here.

Learning Objectives

By the end of this activity, you should be able to:

- Acquire, analyze, and present multiple sources of scientific information
- Use scientific evidence to support a particular policy view or directive
- Propose a scientifically valid approach to solving a particular problem that has resulted from human impacts on the world ocean

Part II: Research and Analysis

Use this textbook, the Internet, or news sources to find an issue related to human impacts on the world ocean about which you can feel passionate. Research your topic using the materials in this textbook (the index is a great place to start), scientific references in the library (ask your librarian for help), or on our website (see the Resources section). It bears repeating: be sure to use verifiable sources of scientific information. You should especially note the scientific evidence that the impact is harmful. Look for before- and after studies or comparisons of impacted versus non-impacted sites. Try to find data where a particular site can be compared to a previous baseline (i.e., baseline studies). Note any uncertainties in the data and areas where additional research is being planned. In addition, look for ways in which the issue is being addressed. Has the "threat" reached the general public? What are the scientific and political limitations for addressing the problem? Who are the groups involved on either side of the issue and what are their agendas? What possible solutions are being proposed or implemented? In short, become an expert on the issue in every possible manner

Part III: Present Your Issue

Write a 10-page position paper and prepare a 15-minute oral presentation that addresses every aspect of the issue from the scientific evidence of a problem to the possible impacts on natural and human systems to proposed or existing policy solutions. Your paper should be thorough, thoughtful, and substantive. You should make use of and present a variety of scientific, popular, and political resources. Use the Rubric for Analyzing and Presenting Scientific Data and Information (see Exploration Activity 1-1) as a guide to writing and improving your paper and presentation.

16

Future Explorations

The future of the world ocean lies in the hearts and minds of ocean explorers, like you!

If we are wise we will keep our eyes on the depths of the ocean as well as on the heavens, and seek answers in each to the mysteries of the other. One of the momentous developments of twentieth-century science is the realization that there is a fundamental unity to the whole universe. We cannot understand the Moon without understanding the Earth, for land and sea, Earth and star, planet and Sun are all pieces of one vast and unified whole.

—C. P. Idyll, *Abyss*

To Know What We Know and
   What We Do Not Know

Seeing the Deep

Listening to the Deep

Sensing the Deep

Life as an Oceanographer:
   Then, Now, and Into
   the Future

To Know What We Know and What We Do Not Know

Chinese philosopher and scholar Confucius wrote: "To know that we know what we know, and [to know] that we do not know what we do not know, that is true knowledge." Knowing what we know and what we do not know is one of the most challenging aspects of science, especially ocean science. In fact, so critical is this question to ocean scientists that Jesse Ausbel of the Alfred Sloan Foundation organized a conference of ocean scientists from around the world to get their opinion on three major problems: 1) What is known about the ocean? 2) What is unknown about the ocean? 3) What is unknowable about the ocean? The third question emerges from chaos theory, discussed in Chapter 8. Despite our best efforts, the future state of complex systems may be unknowable. Ocean scientists, however, can tackle the other two questions, and these two questions impact the future exploration of the world ocean.

In Chapter 2, we learned about James Hutton and his idea that the present is the key to the past. In that same sense, the present is the key to the future. Knowledge of the past and present states of the world ocean helps frame the question of what is known and what is not yet known. Our present understanding of the world ocean allows us to build conceptual and mathematical models for predicting the future state of the world ocean. At the same time, knowledge of the limits of our knowledge makes us painfully aware of the limits of our technological capabilities: the ongoing problem of undersampling, an inability to reach the deepest depths, the sensor limitations of satellites and buoys, and the challenges of sampling and identifying ocean microbes are just a few examples.

Throughout this textbook, we have emphasized the interconnectivity of ocean processes and the challenges of observing the ocean over the full range of temporal and spatial scales. This emphasis and challenge will continue to drive ocean science research in the twenty-first century. The challenge is more than a technological one, although technology will continue to play a major role. Oceanographers are also challenged by what George Philander calls "knowing too much." Despite their best efforts, oceanographers, like other scientists, favor particular ways of thinking and individual perspectives. Traditional approaches and models can hamper progress in science, what science philosopher Thomas Kuhn might call the "problem of the paradigm." Science often progresses when established ways of thinking are shown to no longer fit the facts. Modern oceanography has undergone a few such "revolutions" since the time of the US *Exploring* and *Challenger* expeditions. Who knows what revolutions the future of ocean science holds? While this brief chapter makes no predictions about upcoming revolutions in ocean science, we do highlight a few areas where the promises of technology and the development of new conceptual understandings may impact the near-term future of ocean science.

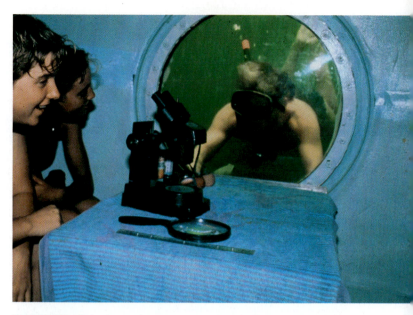

■ **FIGURE 16-1** :An underwater experience seems like a visit to another world.

(**Table 16.1**): including the discovery of hydrothermal vents in 1977 (using a camera sled and a submersible), the finding of the *Titanic* in 1985 (using an ROV), and the first observation of a living giant squid in 2005 (using a camera and bait). At the

Seeing the Deep

When Jacques Cousteau entered the "silent world" with his crude aqualung in 1943, the world of undersea exploration entered a new era. Now humans, or "menfish," as Cousteau called them, could spend extended periods of time underwater roaming as freely as time, pressure, and air capacity permitted. Though arguably as important as Neil Armstrong's first steps on the Moon, no television crews or news reporters were present on that momentous day. Instead, Cousteau enjoyed his first aqualung "flight" collecting lobsters. To celebrate this historic moment, Cousteau's wife, Simone, and his partners and co-inventors, Phillipe Taillez and Frederic Dumas, feasted while begging Cousteau for every little detail of his experience.

Those of us fortunate enough to have personally visited the undersea world—either as skin or scuba divers—know that it offers a special sense of serenity and wonder that cannot be experienced anywhere else (**Figure 16-1**). In recent decades, oceanographers have increasingly relied on crafts and instruments to obtain a close-up and personal view of the ocean. High-resolution imaging systems deployed on spacecraft, underwater cameras mounted on ROVs and AUVs, and time-lapse cameras affixed to benthic sleds have provided stunning new views of the world ocean. Cousteau brought the undersea world into the public's eye in the decades of the sixties, seventies, and eighties. A similar window to the deep has begun to stimulate greater public awareness of the issues facing the world ocean. Consider the recent *visual* ocean discoveries that have had an impact on the public

Year	Discovery
	TABLE 16.1 Select History of Ocean Discoveries Using Visual Technology
1934	William Beebe and Otis Barton directly observe deep-sea organisms from a bathysphere.
1943	Jacques Cousteau makes his first dive using an aqualung and soon after begins recording his dives with an underwater video camera.
1960	Jacques Picard and John Walsh visit the Mariana Trench in the bathyscaphe *Treiste* and report their observations to awaiting television crews on their return.
1974	The US submersible *Alvin* and the French submersible *Archimede* obtain visual evidence of basalt on the Mid-Atlantic Ridge, confirming a key prediction of plate tectonics theory.
1985	Bob Ballard finds the *Titanic* using the robotic vehicle *Jason*.
1995	The Japanese ROV *Kaiko* visits the Mariana Trench and transmits images of life on the seafloor.
2005	Japanese scientists take the first pictures of a live giant squid.

same time, we should not forget that visual technology includes imaging systems that capture a view of the very small. Towed camera systems, such as the video plankton recorder, have been critical for collecting long-term information on the distribution of plankton. Increasingly, high-speed, high-resolution camera systems are being used in the laboratory and the field to observe and understand microbes and larger organisms. Linked to image processing systems and high-speed computers, these visual systems may one day be able to identify the organisms and provide information on their size distribution and biomass. There is little doubt that visual exploration of the world ocean will continue to play a major role in oceanographic research and in the public's perception of ocean science.

Listening to the Deep

In Chapter 4, we introduced echosounders and similar technology for mapping the seafloor. In Chapter 7, we learned that this technology is also being used for mapping the thermohaline structure of the water column. In fact, the application of active acoustic technologies for observing the world ocean has progressed rapidly in recent decades. The acoustic doppler current profiler (ADCP) was developed in the 1980s for measuring current speed and direction. ADCPs are now routinely deployed aboard moorings, research vessels, and other platforms for quantifying the movements of seawater in great detail. Confirmation of the Ekman spiral (Chapter 9) has been possible because of this technology. Active acoustic technologies also provide highly detailed views of the temporal and spatial distributions of zooplankton, enabling oceanographers to study patchiness and vertical migration. Future systems may even allow for real-time identification of species. Of course, active acoustic technologies remain an important component of the communication systems of deployed scientific instrumentation. Acoustic transmission of data will be particularly important in the deployment of cabled observatories that depend on autonomous sensors and robotic vehicles.

Oceanographers are also finding new ways to deploy and use passive acoustic technologies. NOAA's Pacific Marine Environmental Laboratory has developed hydrophones encased in ceramic that can be deployed from ships to collect continuous sound data for periods up to a year. Hydrophone systems are also being deployed on cabled observatories. These technologies have already proven successful for detecting undersea earthquakes and undersea landslides. They are also being used routinely to monitor marine mammal behavior (**Figure 16-2**). No doubt some day you will be able to listen to different parts of the world ocean at home over the Internet or perhaps even over your cell phone or portable music device. Because sound travels over great distances in the world ocean, acoustic technologies will continue to serve the needs of oceanographers. Future applications will only be limited by the

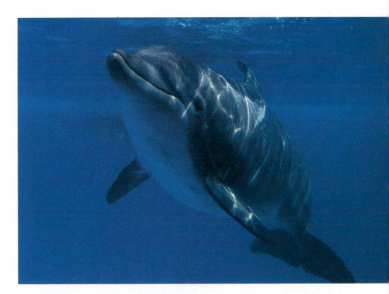

■ **FIGURE 16-2** Recordings of the sounds and activities of marine mammals and other organisms may one day allow us to understand and perhaps even speak to these animals.

creativity of engineers and scientists for whom acoustic solutions for observing the world ocean provide the best choice.

Sensing the Deep

Perhaps the biggest improvement in oceanographic sampling has been in the development and deployment of optical and chemical sensors (**Figure 16-3**). The application of radiometers and other optical devices for probing, measuring, and observing the ocean has vastly improved our understanding of ocean processes. This new generation of sensors has provided unprecedented information on physical and biological processes from the very small (e.g., small-scale turbulence, microbial patchiness) to the very large (e.g., coastal jets and mesoscale eddies). At the same time, these instruments have opened a new window on temporal scales of variability in the ocean. Continuous seasonal records of the physical and biological structure of the water column, once unimaginable, are now common at the few sites where moored instruments are deployed. These records have revealed the frequency and importance of episodic and extreme events in the ocean (see Chapter 8). Bio-optics has also emerged in recent decades as a vibrant and vital oceanographic subdiscipline grappling with ocean productivity and climate change. Flow-through systems that use lasers to detect and sort various types of phytoplankton are proving useful for characterizing their abundance and distribution. Sensors for the routine and continuous measurement of many important chemical factors, including salinity, dissolved oxygen, and

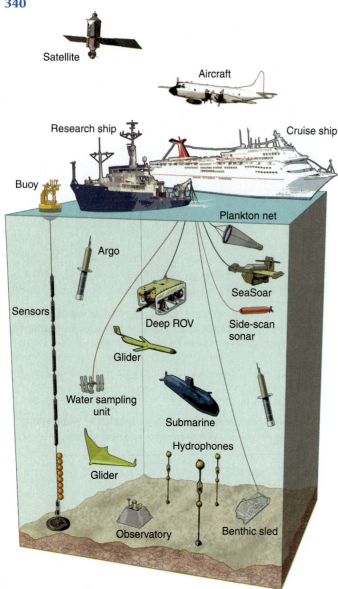

Satellite

Aircraft

Research ship

Cruise ship

Buoy

Plankton net

Argo

SeaSoar

Sensors

Deep ROV

Side-scan sonar

Glider

Water sampling unit

Submarine

Hydrophones

Glider

Observatory

Benthic sled

■ FIGURE 16-3 New generations of sensors are being routinely deployed from a wide variety of platforms in the world ocean. They provide an unprecedented view of ocean processes. A comparison of 20th century oceanography with 21st century oceanography is like the difference between three-channel, black-and-white television and hundreds-of-channel, high-definition, color television. Our view of the world ocean has been expanded enormously.

biologically important nutrients, are spurring development of a wider range of sensors that can be deployed on autonomous robotic platforms. Coupled with artificial neural networks, these chemical systems may one day alert the public to the presence of harmful algal blooms (HABs; see Chapter 15), the onset of hypoxia, and the accidental release of pollutants. Perhaps some day molecular "sensors" will be able to detect the presence and activity of specific RNA sequences, providing a better understanding of microbial food webs.

Life as an Oceanographer: Then, Now, and Into the Future

To see how far oceanography has come in a short time and to get some idea of where oceanography might go in the near future, consider the typical life of a US oceanography student in the early 1970s. In addition to taking the requisite classes in geological, physical, chemical, and biological oceanography, this student was required to go to sea aboard a mostly uncomfortable ship where he or she provided logistic or sampling support to a rather serious crowd of ocean scientists. Geological oceanographers were still grappling with the ramifications of plate tectonics. Physical oceanographers obtained vertical profiles using STDs or CTDs and plotted T-S diagrams by hand or by cumbersome programming using punch cards on computers that took up half a room. Chemical oceanographers still mostly relied on discrete sampling using messenger-activated sampling bottles for water samples and performed laborious bench-top analyses to determine dissolved oxygen or major nutrient concentrations. Biological oceanographers collected plankton with nets (squishing the jellyplankton in the process) and were unaware of hydrothermal vents and microbes smaller than a diatom. Satellite oceanography was still on the drawing board and viewed by many oceanographers of the day as a huge waste of money. (Oceanography from *space?*) In the twenty-first century, an oceanography student still has to take classes, but he or she might never go to sea. Instead, opportunities to pursue a career in ocean science can be found in a vast numbers of subdisciplines that support operational and research platforms that report data to ground stations. A student especially good in mathematics might build ocean circulation models, coupled air-sea models, or ecosystem models. Molecular oceanography has revived the need for highly trained laboratory technicians and specialists, and offers a wide open frontier for the discovery of new forms of life. Of course, a student can still go to sea, but new generations of oceanographic vessels feature a few more creature comforts and much improved food! The command-and-control and mechanical systems of these ships allow precise satellite navigation, stationary positioning, and deployment of moorings, AUVs, ROVs, gliders, and profiling floats, along with the usual array of CTDs, nets, and sampling bottles. Most ships now even offer e-mail services. Sampling grids are routinely coordinated with the latest satellite images of sea surface temperature or chlorophyll so that oceanographers now exactly which features of the ocean they are sampling. A lot has changed in a few short decades.

Future oceanographic explorers will be able to coordinate and utilize the entire sampling toolbox of ocean instruments. Henry Stommel once envisioned an ocean filled with autonomous sampling platforms (a near reality by some measures), but ships, moorings, robotic vehicles, and satellites will be equally indispensable to future oceanographers. Ships of opportunity and volunteer observing ships, such as cargo vessels and cruise ships, are already being outfitted with autonomous sampling systems to collect oceanographic data. A new generation of ships, such as SWATH vessels, submersibles, and even submarines, will deliver oceanographers and their instruments to hard-to-reach locations in a timely manner (**Figure 16-4**). Greater flexibility of scheduling and the availability of "on-demand" vessels will allow oceanographers to go to sea when unusual and episodic events happen, such as undersea earthquakes, HABs, coastal upwelling, and mesoscale eddies. The vast array of sensing platforms and communications systems will deliver their data in near real time via the Internet. These data may be downloaded and viewed by scientists, emergency response personnel, and the interested public. Most important, these data will be incorporated into reports on current conditions and forecasts of ocean weather. Perhaps one day soon, the "Ocean Weather Channel" will provide a new career path for young oceanographers.

While the challenges of deploying, maintaining, and coordinating ocean technologies and their data streams remain considerable (**Table 16.2**), oceanographers look to the future with great hope and enthusiasm. Greater public interest and increased governmental funding for ocean sciences will help oceanography make the transition from a largely research-based discipline to one with ongoing operational support that provides continuous and synoptic oceanographic data over broad temporal and spatial

■ **FIGURE 16-4** New platforms for sampling the ocean will be needed in the 21st century. Cruise ships and other vessels of opportunity are already being outfitted with autonomous ocean sampling systems.

scales. At the same time, better visualization of oceanographic data (how about Google Ocean?), greater dissemination of information to coastal managers and emergency personnel, and better education about issues facing ocean science will serve to heighten public awareness of the importance of the world ocean in their daily lives. The future is bright for the next generation of oceanographers.

TABLE 16.2   Current and Future Challenges of Ocean Science

- Processes of interest involve variability over temporal and spatial scales spanning several magnitudes—in some cases up to ten orders of magnitude.
- Many of the key variables and rates remain inaccessible via present technologies.
- Contemporaneous (synoptic) sampling over three-dimensions is needed but is rarely possible.
- Cumbersome water sampling remains necessary for many, if not most, biological and chemical measurements.
- Multiplatform, nested sampling schemes coupled with models are essential to fill in information in the requisite time-space continuum.
- Many chemical elemental (or compound) analyses and biological measurements at the species or group levels are not possible at present using autonomous sampling platforms.
- Real-time data acquisition is typically deemed very important, if not essential, because of societal and management needs and is critical for expendable or vulnerable platforms; data telemetry, however, is a limiting factor for many broadband sensors (e.g., seismic, hyperspectral optics and acoustics, imaging). Fiber-optic cables on the seafloor can be effectively used in some situations.
- Power requirements for some advanced sensors and systems remain restrictive for some of the *in situ* autonomous sampling platforms. Power cables can be utilized in combination with data telemetry fiber-optic cables.

Source: Dickey and Bidigare.

Explore Online

Visit www.mhhe.com/chamberlin1e for access to chapter quizzing, key term flash cards, video clips, interactive activities, and more. Further enhance your knowledge with web links to chapter-related material!

Appendix 1: Commonly Used Oceanographic Symbols

c = phase speed of a wave

C_{10} = empirical drag coefficient

$°C$ = degrees in Celsius units

f = Coriolis parameter = $2\Omega \sin \theta$

e = exp = 2.72

$°F$ = degrees in Fahrenheit units

g = acceleration of gravity on Earth = 9.8 m sec^{-2}

h = water depth

I_z = light intensity at depth z

I_0 = light intensity at sea surface

K = degrees in Kelvin units

K_d = diffuse attenuation coefficient of light

pH = scale used for determining acidity of a liquid

S = salinity in units of psu, ppt, or ‰

T = temperature

U_{10} = wind speed measured at 10 m above the sea surface

Z_{SD} = Secchi depth

z = vertical coordinate for distance beneath sea surface

α = thermal expansion coefficient

β = salinity expansion coefficient

Δ = symbol for difference in following variable

ρ = density of sea water

ρ_a = density of air

σ_t = sigma-t = (ρ–1000 kg/m^3) using *in situ* temperature for ρ

σ_θ = sigma-t = (ρ–1000 kg/m^3) using potential temperature, θ, for ρ

θ = potential temperature

τ = wind stress

ϕ = latitude in degrees

Appendix 2: Metric Prefixes

Prefix	Symbol	Name	Magnitude	Multiply by
yotta-	Y	1 septillion	10^{24}	1,000,000,000,000,000,000,000,000
zetta-	Z	1 sextillion	10^{21}	1,000,000,000,000,000,000,000
exa-	E	1 quintillion	10^{18}	1,000,000,000,000,000,000
peta-	P	1 quadrillion	10^{15}	1,000,000,000,000,000
tera-	T	1 trillion	10^{12}	1,000,000,000,000
giga-	G	1 billion	10^{9}	1,000,000,000
mega-	M	1 million	10^{6}	1,000,000
kilo-	k	1 thousand	10^{3}	1,000
hecto-	h	1 hundred	10^{2}	100
deka-	da	1 ten	10	10
deci-	d	1 tenth	10^{-1}	0.1
centi-	c	1 hundredth	10^{-2}	0.01
milli-	m	1 thousandth	10^{-3}	0.001
micro-	μ	1 millionth	10^{-6}	0.000,001
nano-	n	1 billionth	10^{-9}	0.000,000,001
pico-	p	1 trillionth	10^{-12}	0.000,000,000,001
femto-	f	1 quadrillionth	10^{-15}	0.000,000,000,000,001
atto-	a	1 quintillionth	10^{-18}	0.000,000,000,000,000,001
zepto-	z	1 sextillionth	10^{-21}	0.000,000,000,000,000,000,001
yocto-	y	1 septillionth	10^{-24}	0.000,000,000,000,000,000,000,001

Examples: Write the approximate depth of the ocean, 4000 m, in decimeters, centimeters, millimeters, micrometers, and nanometers. Note that powers of ten add when multiplied and subtract when divided. For example, $10^3 \times 10^7 = 10^{10}$ and $10^5/10^3 = 10^2$.

4000 m (10 dm/m) = $4 \times 10^3 \times 10$ dm = 4×10^4 dm

4000 m (10^2 cm/m) = $4 \times 10^3 \times 10^2$ cm = 4×10^5 cm

4000 m (10^3 mm/m) = $4 \times 10^3 \times 10^3$ mm = 4×10^6 mm

4000 m (10^6 μm/m) = $4 \times 10^3 \times 10^6$ μm = 4×10^9 μm

4000 m (10^9 nm/m) = $4 \times 10^3 \times 10^9$ nm = 4×10^{12} nm

Appendix 3:
Useful Conversions

TABLE A.1 Common Units

Quantity	SI or mks name	mks symbol	cgs name	cgs symbol
length	meter	m	centimeter	cm
mass	kilogram	kg	gram	gm
time	second	sec	second	sec
temperature	Kelvin	K	Celsius	C or °C

Note: SI = International System of Units
mks = meter-kilogram-second
cgs = centimeter-gram-second

Physical Conversions

TABLE A.2 Length

Length	cm	m	km	in	ft	mi	nt mi
1 centimeter =	1	10^{-2}	10^{-5}	0.394	3.28×10^{-2}	6.21×10^{-6}	5.40×10^{-6}
1 meter =	100	1	10^{-3}	39.4	3.28	6.21×10^{-4}	5.40×10^{-4}
1 kilometer =	10^5	1000	1	3.94×10^4	3.28×10^3	0.621	0.540
1 inch =	2.54	2.54×10^{-2}	2.54×10^{-5}	1	8.33×10^{-2}	1.58×10^{-5}	1.37×10^{-5}
1 foot =	30.5	0.305	3.05×10^{-4}	12	1	1.89×10^{-4}	1.64×10^{-4}
1 mile = 1 statute mile =	1.61×10^5	1.61×10^3	1.61	6.34×10^3	5280	1	0.870
1 nautical mile =	1.85×10^5	1.85×10^3	1.85	7.29×10^3	6.08×10^3	1.15	1

Others: 1 nm = 1×10^{-9} m, 1 μ m = 1×10^{-6} m, 1 mm = 1×10^{-3} m
  1 fathom = 6 ft = 1.83 m, 1 yard = 3 ft = 0.915 m, 1 light-year = 9.460×10^{12} km
Example: Convert 5 meters to inches.
  5 m = (5 m) (39.4 in / 1 m) = 197 in
  (Note how m in numerator and denominator divide out so that final units turn out to be the desired inches.)

TABLE **A.3** Area

Area	meter²	cm²	ft²	in²
1 square meter =	1	10^4	10.8	1550
1 square centimeter =	10^{-4}	1	1.08×10^{-3}	0.155
1 square foot =	9.29×10^{-2}	929	1	144
1 square inch =	6.45×10^{-4}	6.45	6.94×10^{-3}	1

Others: 1 square mile = 2.79 × 10^7 ft² = 640 acres, 1 acre = 43,560 ft², 1 hectare = 10^4 m² = 2.47 acres
Example: Convert 10 square meters to square inches.
10 m² = (10 m²) (1550 in²/1 m²) = 15,500 in²
(Note how m² in numerator and denominator divide out so that final units turn out to be the desired in².)

TABLE **A.4** Volume

Volume	m³	cm³	L	ft³	in³
1 cubic meter =	1	10^6	1000	35.3	6.10×10^4
1 cubic centimeter =	10^{-6}	1	1×10^{-3}	3.53×10^{-5}	6.10×10^{-2}
1 liter (L) =	1×10^{-3}	1000	1	3.53×10^{-2}	61.0
1 cubic foot =	2.83×10^{-2}	2.83×10^4	28.3	1	1.73×10^3
1 cubic inch =	1.64×10^{-5}	16.4	1.64×10^{-2}	5.79×10^{-4}	1

Others: 1 ml = 1 cm³, 1 US gallon = 4 US quarts = 8 US pints = 128 fluid ounces = 231 in³
Example: Convert 3 cubic meters to liters.
3 m = (3 m³) (1000 L/ 1 m³) = 3000 L
(Note how m³ in numerator and denominator divide out so that final units turn out to be the desired liters.)

TABLE **A.5** Mass

Note that we include pounds in units of mass even though a pound, which is commonly treated as a mass unit, is actually a force unit. To be correct, the conversion "1 kg = 2.21 lb" really means that a mass of 1 kg weighs 2.21 pounds where the acceleration of gravity, g, is 9.81 m sec⁻². The English unit system uses a true mass unit of a "slug."

Mass	g	kg	lb
1 gram =	1	10^{-3}	2.21×10^{-3}
1 kilogram =	10^3	1	2.21
1 pound =	454	0.454	1

Others: 1 US ton = 2000 pounds, 1 metric ton = 1000 kg
Example: Convert 5 pounds to kilograms.
5 lb = (5 lb) (0.454 kg/ 1 lb) = 2.27 kg
(Note how pounds in numerator and denominator divide out so that final units turn out to be the desired kilograms.)

TABLE **A.6** Density

Density	kg m³	gm cm³
1 kilogram per meter³ =	1	10^{-3}
1 gram per centimeter³ =	10^3	1

Others: 1 pound per foot³ = 16.0 kg/m³, 1 pound/in³ = 2.77×10^4 kg/m³; note definition of pound in Table A.5.
Example: Convert 4 kilograms per cubic meter to grams per cubic centimeter.
4 kg/m³ = (4 kg/m³) [(10^{-3}gm/cm³)/(1 kg/m³)] = 4×10^{-3} gm/cm³
(Note how kg/m³ in numerator and denominator divide out so that final units turn out to be the desired grams per cubic centimeter.)

TABLE A.7 Time

Time	y	d	h	min	sec
1 year =	1	365.25	8.77×10^3	5.26×10^5	3.16×10^7
1 day =	2.74×10^{-3}	1	24	1440	8.64×10^4
1 hour =	1.14×10^{-4}	4.17×10^{-2}	1	60	3600
1 minute =	1.90×10^{-6}	6.94×10^{-4}	1.67×10^{-2}	1	60
1 second =	3.17×10^{-8}	1.16×10^{-5}	2.78×10^{-4}	1.67×10^{-2}	1

Example: Convert 4 years to days.
4 years = (4 years) (365.25 days/year) = 1461 days
(Note how years in numerator and denominator divide out so that
final units turn out to be the desired days.)

TABLE A.8 Speed (Velocity)

Speed (Velocity)	ft/sec	km/h	m/sec	mi/h	cm/sec	knot = kt
1 foot per second =	1	1.10	0.305	0.682	30.5	0.59
1 kilometer per hour =	0.911	1	0.278	0.621	27.8	0.54
1 meter per sec =	3.28	3.6	1	2.24	100	1.92
1 mile per hour =	1.47	1.61	0.45	1	44.7	0.87
1 centimeter per second =	3.28×10^{-2}	3.6×10^{-2}	10^{-2}	2.24×10^{-2}	1	0.0194
1 knot = 1 kt =						
1 nautical mile per hour =	1.69	1.85	0.52	1.15	51.4	1

Example: Convert 2 knots to centimeters per second.
2 knots = (2 knots) [(51.4 cm/sec)/1 knot)] = 103 cm/sec
(Note how knots in numerator and denominator divide out so that
final units turn out to be the desired centimeters per second.)

TABLE A.9 Force

Force	dyne	newton	lb
1 dyne =	1	10^{-5}	2.25×10^{-6}
1 newton =	10^5	1	0.225
1 pound =	4.45×10^5	0.448	1

Other: 1 ton = 2000 lb
Example: Convert 100 pounds to newtons.
100 lb = (100 lb) (0.448 newton/1 lb) = 448 newton
(Note how pounds in numerator and denominator divide out so
that final units turn out to be the desired newton.)

TABLE A.10 Pressure and Stress

(Note: pressure = force or weight per unit area)

Pressure and Stress	atm	dyne cm^{-2}	pascal = 1 newton/m^{-2}	lb/in^{-2}	bar	millibar
1 atmosphere =	1	1.01×10^6	1.01×10^5	14.7	1.01	1013
1 dyne per square centimeter =	9.87×10^{-7}	1	0.1	1.41×10^{-5}	10^{-6}	10^{-3}
1 pascal = 1 newton/m^2 =	9.87×10^{-6}	10	1	1.45×10^{-4}	10^{-5}	10^{-2}
1 pound per square inch =	6.81×10^{-2}	6.90×10^4	6.90×10^3	1	7.09×10^{-2}	70.9
1 bar =	9.87×10^{-1}	10^6	10^5	14.1	1	1000
1 millibar =	9.87×10^{-4}	10^3	10^2	1.41×10^{-2}	10^{-3}	1

A rough rule of thumb for the upper ocean is that 1 atm translates to about 10 m of water depth; so at 200 m, pressure would be about 20 atm.

Example: Convert 1000 pounds in^{-2} to newtons m^{-2} or pascals

1000 lb/in^2 = (1000 lb/in^2) [(6.90×10^3 pascals)/(1 lb/in^2)] = 6.90×10^6 pascals

(Note how lb/in^2 in numerator and denominator divide out so that final units turn out to be the desired pascals.)

TABLE A.11 Energy, Work, and Heat

Energy, Work, Heat	erg	joule	cal
1 erg =	1	10^{-7}	2.39×10^{-8}
1 joule =	10^7	1	0.239
1 calorie =	4.19×10^7	4.19	1

Example: Convert 5 joules to calories.

5 joules = (5 joules)(0.239 cal/joule) = 1.20 cal

(Note how joules in numerator and denominator divide out so that final units turn out to be the desired calories.)

TABLE A.12 Power

(Note: power = force times velocity or energy per unit time)

Power	cal/sec	kW	watt
1 calorie per sec =	1	4.19×10^{-2}	4.19
1 kilowatt =	239	1	1000
1 watt =	0.239	10^{-3}	1

Example: Convert 1000 calories per second to kilowatts.

1000 cal/sec = (1000 cal/sec) [(4.19×10^{-2} kW)/(cal/sec)] = 41.9 kW

(Note how cal/sec in numerator and denominator divide out so that final units turn out to be the desired kilowatts.)

Reference
Halliday, D., R. Resnick, and J. Walker. 2001. *Fundamentals of Physics.* 6th ed. New York: John Wiley and Sons, Inc., New York.

Appendix 4: Useful Oceanographic Formulas

pH Scale

The pH scale is used to determine the acidity of a liquid and is formally defined as the negative logarithm of the concentration of the hydrogen ion, H^+.

$$pH = -\log [H^+]$$

pH values vary between 0 and 14.

Salinity Determinations and Conversions

Salinity may be defined as the number of grams of dissolved material in a kilogram of water ($gm\ kg^{-1}$). A chemical titration may be performed to measure chlorinity or the concentration of the chloride ion, Cl^-. The law of constant proportions or Dittmar's principle may be applied to calculate salinity, S, according to

$$S = 1.80655 \times Cl^-$$

with units of $gm\ kg^{-1}$ or parts per thousand (ppt = gm per 1000gm) or ‰ – .

More recently, electrical conductivity has become a more common measure of salinity. In this case, the practical salinity unit (psu) is used as the salinity unit. For practical purposes, oceanographers typically use psu, ppt, and ‰ interchangeably.

Example: Determine salinity for a measured chlorinity of 20 ppt.

$$S = 1.8 \times Cl^- = 1.8 \times (20ppt) = 36\ ppt = 36\ ‰ = 36\ psu$$

Computation of Residence Time

Residence time calculations are useful for many different oceanographic problems. The idea is to obtain an estimate of the time a water parcel or mass of a given element resides in a particular reservoir (such as the world ocean).

$$\text{Residence time (years)} = \text{mass of element (kg)}/ \text{rate of input (kg/year)}$$

Temperature Conversions

Converting between temperature units:

Celsius to Fahrenheit

$$°F = 1.8 \times °C + 32$$

Example: Convert 10°C to °F

$$°F = (1.8) \times (10) + 32 = 50°F$$

Fahrenheit to Celsius

$$°C = (0.56) \times (°F - 32)$$

Example: Convert 92°F to °C

$$°C = 0.56 \times (92-32) = 34°C$$

Celsius to Kelvin

$$K = °C + 273.2$$

Example: Convert 20°C to K

$$K = °C + 273.2$$

$$K = 20 + 273.2 = 293.2 \text{ K}$$

Light Extinction with Depth (Beer's law)

The diffuse attenuation coefficient of light, K_d in m^{-1}, can be estimated using a Secchi disk according to

$$K_d = 1.44/Z_{SD}$$

where Z_{SD} is the Secchi depth in meters.

Beer's law describes the exponential decrease in light intensity, I_z, with depth, z (m), or

$$I_z = I_0 \times exp^{(-K_d z)}$$

Where I_0 is light intensity at the surface, *exp* is equal to e = 2.72, and K_d is the diffuse attenuation coefficient of light.

Potential Temperature

Potential temperature, θ, is defined as the temperature of a parcel of water brought from depth to the sea surface without exchanging heat with the surrounding water (adiabatically). The *in situ* and potential temperatures of a parcel of water at depth are different because of compressibility effects. Thus, it is important to use potential temperature for studies involving comparisons of water masses from different depths. Standard temperature and potential temperature differences are greatest for the deepest ocean waters but do not exceed about 1.5°C.

Density

Density (ρ) is defined as the mass of water per unit volume (kg m^{-3} or gm cm^{-3}).

The density of seawater is a function of temperature, salinity, and pressure:

$$\rho = \rho (T, S, P)$$

Exact formulations are quite complex. However, for the upper ocean, a reasonably accurate density is given by the approximated equation of state, or:

$$\rho = \rho_0 [1 + \alpha (T = 27) + \beta (S = 36)]$$

where ρ is density at depth and ρ_0 is density at the surface (kg m^{-3}), T is temperature (°C), and S is salinity (ppt or ‰ or psu) and the constants, $\alpha = -3.3 \times 10^{-4}$ and $\beta = 7.6 \times 10^{-4}$. A typical value for ρ_0 is 1.0235×10^3.

Example: Determine density for the upper ocean for a sample of seawater with a salinity of 37 ppt and a temperature of 21°C.

$$\rho = \rho_0 [1 + \alpha (T–27) + \beta (S–36)]$$

$$\rho = 1.0235 \times 10^3 [1 + (-3.3 \times 10^{-4})$$
$$(21–27) + (7.6 \times 10^{-4}) (37–36)]$$

$$\rho = 1.0235 \times 10^3 [1.0002704]$$

$$\rho = 1.0238 \times 10^3$$

Answer: density = $\rho = 1.0238 \times 10^3$ kg m^{-3}

Sigma-t

Oceanographers have created a notation to simplify representations of density using the following definitions:

Sigma-t $=\sigma_t = (\rho – 1000$ kg $m^{-3})$ in mks units and using *in situ* temperature for density calculation

Sigma-t $=\sigma_t = 1000 (\rho – 1.000$ gm $cm^{-3})$ in cgs units and using *in situ* temperature for density calculation

Sigma-θ $=\sigma_\theta = (\rho – 1000$ kg $m^{-3})$ in mks units and using *in situ* potential temperature for density calculation

Sigma-θ $=\sigma_\theta = 1000 (\rho – 1.000$ gm $cm^{-3})$ in cgs units and using potential temperature for density calculation

Example: Given a density of 1029 kg m^{-3}, what is σ_θ (sigma-theta)?

$$\sigma_\theta = (\rho – 1000 \text{ kg m}^{-3})$$

$$\sigma_\theta = (1029 \text{ kg m}^{-3} – 1000 \text{ kg m}^{-3})$$

Answer: $\sigma_\theta = 29$ kg m^{-3}

Hydrostatic Pressure

Hydrostatic pressure, P, is the weight of water per unit area and may be computed using the following formula:

$$P = \rho g h$$

where ρ is density of sea water in kg m^{-3}, g is acceleration of gravity = 9.8 m sec^{-2}, and h is depth.

Example: Given a water density of 1020 kg m^{-3}, what is the pressure at a depth of 100 m?

$$P = \rho g h$$

$$P = (1020 \text{ kg m}^{-3}) (9.8 \text{ m sec}^{-2}) (100 \text{ m})$$

Answer: $P = 1.098 \times 10^6$ N m^{-2}

Example: How many atmospheres of pressure would be felt at 100 m? Recall that each 10 m interval of water depth produces about 1 atm of pressure. Therefore:

$$p = (1 \text{ atm}/10 \text{ m}) \times (100 \text{ m}) = 10 \text{ atm}$$

What would the approximate pressure be at the bottom of the Mariana's Trench? Use a depth of 11,000 m for rough approximation, or:

$$p = (1 \text{ atm}/10 \text{ m}) \, (11000 \text{ m}) = 1100 \text{ atm}$$

Wind Stress

Wind stress is like pressure but is directed tangentially to the ocean surface, whereas air pressure is directed perpendicular to the surface. Wind stress has units of a force per unit area. It is important for determining momentum transfer across the air-sea interface and propagation of waves and currents. A simple formula for estimating wind stress, τ, in N m^{-2}, from wind speed measurements at 10 m above the sea surface, U_{10} (m sec^{-1}), is given by:

$$\tau = \rho_a \, (C_{10}) \, (U_{10})^2$$

where ρ_a is air density (approximately 1.26 kg m^{-3}) and C_{10} is an empirical drag coefficient (unitless) with a value taken to be 1.4 $\times$ 10^{-3}. The exact formulation for wind stress and the drag coefficient is the subject of intense study, especially for high wind conditions such as those accompanying hurricanes and typhoons.

　Example: Compute wind stress for hurricane force winds of 40 m sec^{-1} or about 90 mi h^{-1}.

$$\tau = \rho_a \, (C_{10}) \, (U_{10})^2$$

$$\tau = (1.26 \text{ kg m}^{-3}) \, (1.4 \times 10^{-3}) \, (40 \text{ m sec}^{-1})^2$$

$$\text{Answer: } \tau = 2.82 \text{ N m}^{-2}$$

Surface Gravity Waves

Useful theories for idealized surface gravity waves have been developed (see discussion in Chapter Nine). One of the useful general formulas for phase velocity, c, of surface gravity waves is given by

$$c = \{[(gL)/(2\pi)] \, \tanh(2\pi \, h/L)\}^{1/2}$$

where g is the acceleration of gravity = 9.8 m sec^{-2}, L = wavelength, π = 3.14, h = water depth, and L = wavelength.

The hyperbolic tangent function can be written in a general form as

$$\tanh(x) = [[e^x - e^{-x}/e^x + e^{-x}]]$$

so in our formula, we simply insert x = 2π h/L.

　With given values of wavelength, one can use a hand calculator or write a simple computer program to compute wave phase speed for surface gravity waves.

　The exact equation for wave speed can be simplified under limiting conditions:

1. If the computed value of tanh(x) or tanh(2π h/L) is large, that is, h is greater than L/2, then tanh(2π h/L) is about equal to 1. In this case, the phase speed for the so-called "deep-water wave" is simply

$$c_{deep} = [(gL)/(2\pi)]^{1/2}$$

Using the relation between wave speed, wavelength, and wave period, T, or c = L/T, we can derive the "deep-water" expressions

$$c_{deep} = 1.25 \, L^{1/2}$$

and

$$c_{deep} = 1.56 \, T$$

2. If the computed value of tanh(x) or tanh(2π h/L) is small, that is, L is greater than 20h or L/20 is greater than h, then tanh(2π h/L) is about equal to 2π h/L. In this case, the phase speed for the so-called "shallow-water wave" is simply

$$c_{shallow} = (gh)^{1/2}$$

This is an interesting result as the shallow water wave depends only on water depth. Note that long ocean waves such as tsunami and tides satisfy the shallow water approximation. Their phase speeds depend only on water depth.

Appendix 5: Reading a Map

Maps are representations of the surface of our planet, including the sea floor. Map reading is extremely important to oceanography. Maps allow us to navigate across the world ocean, locate important features, calculate distances from one point to the next, and represent properties of the world ocean, among other uses. There are many different ways in which a spherical globe can be projected on a flat surface. Different projections serve different purposes. In this textbook, many of our maps are depicted as the pseudocylindrical **Mollweide projection**, a projection which accurately represents the areas of geographic regions. You will also find the rectangular **Mercator projection**, which tends to distort higher latitudes.

Most maps employ a grid system of horizontal and vertical lines for precisely locating a particular point on the map. Lines of **latitude** encircle the globe in the same direction as the equator, i.e. horizontally. Lines of latitude are parallel to each other; they never cross, so they are sometimes called parallels. The equator is a parallel of latitude. Lines of **longitude** are vertical on a map projection. They converge at the poles.

When we divide the globe up into lines of latitudes and longitudes, it is convenient to assign each line of latitude or longitude the geometrical unit of measure called a **degree**(°). Degrees are nothing more than a way to divide up a circle (or a sphere) into equal sections. For example, as we travel north of the equator, each parallel of latitude (represented as a circle) is assigned a degree between zero (0) and ninety (90). The latitude of the north pole is 90° north, represented as 90° N. Scientists have also adopted a system where northern hemisphere latitudes are given positive values and southern hemisphere latitudes are given negative values. For example, 30° North latitude is +30° latitude and 30° South latitude is -30° latitude. Degrees of longitude are measured from the **prime meridian** or the adopted zero meridian. The meridian of 0 degrees longitude runs through the original site of the Royal Observatory at Greenwich, England. Other prime meridians have been designated in the past, including Oslo, Norway, Paris, France, Rome, Italy, and Washington D.C. By international agreement in 1884, the Greenwich prime meridian was chosen. East of the Observatory, lines of longitude are given positive values. West of the Observatory, lines of longitude are given negative values. The point where they meet halfway around the globe is called the **International Date Line** (180°).

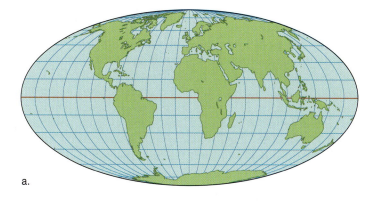

a.

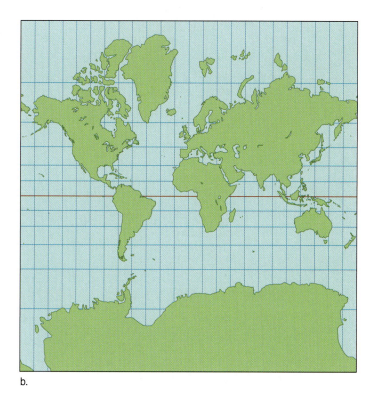

b.

■ **Figure A-1** (a) A Mollweide projection. This projection is most useful for depicting the global distribution of properties. (b) A Mercator Projection. Note the exaggeration of the size of Antarctica. This projection is most useful for representing the shapes of continents.

Latitude is a convenient way to divide up our globe in terms of climate. These **climate zones** are important for understanding physical and biological processes in the ocean. The **tropical zone** occupies the region between 23.5° S and 23.5° N. The **temperate zone** exists as the region 23.5° and 66.5° S and 23.5° and 66.5° N. The **polar zones** can be found between 66.5° and 90° S and 66.5° and 90° N. The tropics corresponds to the region of the most direct solar radiation. The northern limit of the tropics, the **Tropic of Cancer**, marks the location of the sun overhead on June 20, the parallel of latitude at 23.5° N. The **Tropic of Capricorn** marks the position of the sun overhead on December 21, the parallel of latitude at 23.5° S. Similarly, the **Arctic Circle** marks the parallel of latitude at 66.5° N, the land of the midnight sun during northern hemisphere winter. The **Antarctic Circle**, the parallel of latitude at 66.5° S, is the southern hemisphere equivalent. Note that the exact location of these boundaries varies with the Milankovitch cycle.

On a historical note, when people started putting together this grid system, latitude was pretty easy to figure out. Astronomers had given us an excellent representation of the movement of our planet around the sun and knowing the time of year, one could calculate the angle of the sun at sunrise and noon and figure out position in terms of latitude. However, figuring out longitude wasn't so easy. The "search" for longitude is a fascinating tale of discovery as told in Dava Sobel's *Longitude*, published in 1995.

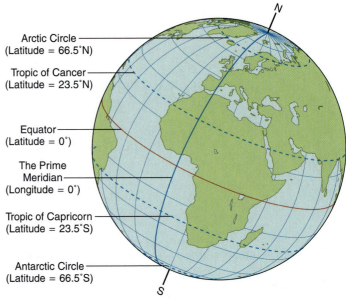

Arctic Circle
(Latitude = 66.5°N)

Tropic of Cancer
(Latitude = 23.5°N)

Equator
(Latitude = 0°)

The Prime
Meridian
(Longitude = 0°)

Tropic of Capricorn
(Latitude = 23.5°S)

Antarctic Circle
(Latitude = 66.5°S)

■ **Figure A-2** The climate zones.

Glossary

A

abrupt climate change an extreme change in climate

absorption the conversion of light (radiant) energy into other energy forms such as chemical or thermal energy; the fraction of power (energy per unit time) of a beam of light that is absorbed per unit length is defined as the absorption coefficient (units of m^{-1})

abyssal fan a fan-shaped, sedimentary feature often found at the base of submarine canyons

abyssal hills hummocky, hill-like features of the abyssal seafloor

abyssal plains flat, featureless regions of the seafloor

abyssopelagic the vertical life zone beneath the bathypelagic, approximately 4000 to 7000 m in depth

accessory pigments photosynthetic pigments other than chlorophyll that absorb different wavelengths of visible light in a photoprotective or photoadaptive capacity

acid a substance that donates protons in solution, lowering the pH; common examples include vinegar and battery acid

acoustic thermometry the use of sound to determine the mean temperature of seawater between two points, a source and a receiver, at different locations in the world ocean; variations in the speed of sound as a function of temperature provide the basis for this method

active acoustics a technique that relies on the production of sound to determine seafloor depth and the characteristics of the seafloor or other objects

active margins continental margins characterized by tectonic activity

activity model of the scientific method a model of the scientific method characterized by multiple scientific activities and interactions among scientists

actualism the "modern" philosophy of geology incorporating uniformitarianism while acknowledging a role for episodic and catastrophic events

adiabatic change a change in heat due to pressure; occurs without the exchange of heat with the surroundings

aerosols minute airborne particles that form the nucleus of cloud droplets

aggregation an accumulation or schooling of organisms by physical or biological processes; for example, schooling of fishes

Agulhas Retroflection an energetic region off the tip of South Africa that produces frequent mesoscale eddies

air-sea interface the region where the atmosphere and the ocean meet

albedo the reflectivity of portions of Earth

algae photosynthetic, eukaryotic organisms, single-celled or multicellular, that do not generally produce differentiated tissues

allometric relationship the relationship between size and growth in aquatic systems, most often measured as respiratory rate

amphidromic point the point of rotation for Kelvin waves, that is, tides

anemometer an instrument that measures wind speed

angular velocity the rate at which an object revolves around a fixed axis

anions negatively charged ions

anoxia a complete absence of dissolved oxygen

Antarctic Bottom Water one of the most dense water masses in the ocean; typically formed in winter in the Weddell Sea

Antarctic Circumpolar Current the current that flows around Antarctica

Antarctic Convergence the boundary between Antarctic surface waters and surface waters from the three major basins

anthropogenic CO_2 carbon dioxide released as a result of human activities; for example, the burning of fossil fuels

anticyclonic air moving around a high pressure system

Archaea a domain of microorganisms found throughout the world ocean; thought to have evolved early in Earth's history

Archean cratons pieces of the continents that formed during the Archaean period

Archimedes principle the principle stating that the buoyant forces on a floating object are equal to the weight of the displaced fluid

Argo system a global system of nearly 3000 autonomous floats that alternately rise and descend to collect oceanographic information

ascent rising from depth to the surface

aseismic ridges oceanic ridges on which no observable tectonic activity takes place

asteroid belt a region between Mars and Jupiter filled with small planetary bodies

asthenosphere the weak and plastic upper layer of Earth's mantle; the region of Earth's interior below the lithosphere and the "lubricant" over which the plates move

Atlantic-type margins passive margins

atmosphere gases held close to a planet's surface by the force of gravity

atmospheric circulation cells large-scale movements of air within Earth's atmosphere

atolls ring-shaped coral islands thought to originate during the sinking of volcanoes or the erosion of carbonate platforms

atomic mass the total number of protons and neutrons in an atom

atomic number the total number of protons in an atom

attenuation a decrease in light or sound that may occur from absorption or scattering

attribution problem the discrepancy between measurements of sea level based on tide gauge stations versus satellites

authigenic forming in place; for example, chemical sediments

autotrophs organisms that manufacture their own nutrition from carbon dioxide and water

autumnal equinox the date when the Sun appears to cross from the Northern Hemisphere to the Southern Hemisphere

axial tilt the tilt of Earth relative to its plane of orbit around the Sun

axial valley see *rift valley*

B

backbarrier   the landward side of a barrier island

back reef   the part of a reef farthest away from the ocean

backwash   the return flow from a wave on the beach face

Bacteria   one of the three domains of organisms on Earth; microscopic, single-celled, prokaryotic organisms inhabiting all regions of the world, including the subterranean seafloor

bacterioplankton   planktonic bacteria important in the mineralization of organic compounds

bank or platform reefs   typically, independent, flat-topped reefs in shallow seas far offshore, found on banks or shallow regions of a sea

baroclinic   a state of the water column that generates a horizontal pressure gradient because isobars and isopycnals are not parallel; in this state, seawater density causes variations in hydrostatic pressure with depth at different locations

barotropic   a state of the water column when isobars and isopycnals are parallel; pressure changes with depth are primarily due to the increasing mass of the water column rather than changes in seawater density

barrier   see *barrier island*

barrier island   an offshore island or emergent sand bar along a coastline

barrier island migration   the typically landward movement of barrier islands as a result of wave action, tides, sediment supply, and other processes

barrier reef   an offshore reef that parallels a coastline

base   a substance that accepts protons in a chemical reaction, raising the pH; common examples include baking soda and ammonia

bathymetric chart   a map depicting the depth of the seafloor below sea level

bathypelagic zone   an ocean life zone in the open ocean at depths between 1000 and 4000 m; typically, home to permanent, bioluminescent residents of the abyss

beach   a shoreline with unconsolidated material

beach nourishment   a preventative or remedial measure to add sediments to beaches starved of their natural supply

beach profile   the slope of a beach from the beach face to the backshore

beach wrack   rafts of seaweed or other organic debris stranded on a beach

Beaufort Wind Force Scale   an estimate and ranking of wind speeds according to their effects on the ocean or land objects, such as trees

Beer's law   a mathematical formulation that describes the exponential decrease in light intensity with depth in the ocean

Benguela Current   the cold eastern boundary current of the subtropical South Atlantic gyre

Benguela Niño   the Atlantic equivalent of the Pacific El Niño, similarly disrupting productivity and fisheries

Benguela upwelling system   one of five major upwelling centers in the world ocean; located off the coast of central West Africa

benthic-pelagic coupling   the exchange of energy and materials between the benthos and pelagic zones in coastal and estuarine environments

benthos   the group of organisms that spend most of their life attached to, living upon, or living within the rocks or sediments of the seafloor

berms   sediment barriers built on a beach through wave action, typically observed as a small cliff or sudden change in the beach slope

bicarbonate   one of the forms of dissolved carbon in seawater; a salt containing HCO_3^-

bimodal distribution   a graph that exhibits two peaks in its distribution

biodiversity   the number of species inhabiting a given region

bioerosion   the breakdown of rocks by organisms

biogenic coast   a coastline formed primarily from the activities of organisms, such as corals

biogenous sediments   biological in origin

biogeochemical cycles   the transfer of elements between various earth, atmosphere, and ocean reservoirs as a result of biological, chemical, geological, and physical processes

biogeography   the study of the distribution of organisms in relation to climate or oceanographic conditions

biological elevator   the two-way transfer of carbon between surface and deep waters in the ocean

biological factors   biological characteristics of an ecosystem, such as physiological tolerances and competition for space, food, resources, and mating, that influence the distribution and abundance of organisms

biologically important nutrients   substances necessary for the growth of phytoplankton

biological pump   the downward transfer of carbon from surface waters to the deep sea

biological weathering   the dissolution, fragmentation, and breakdown of rocks into sediments caused by the activities of organisms

bioluminescence   the natural light emitted by organisms

biomes   regions of the world ocean with definable physical, chemical, and biological characteristics

bio-optical oceanography   the study of the relation of the biota to the optical properties of seawater

biosphere   the sum of all life on Earth and the processes governed by it

bioturbation   the physical or chemical disturbance or alteration of sediments as a result of the activities of organisms

blackbody radiation   the radiation emitted by a perfect emitter, an object that emits 100% of the energy it absorbs

black smoker   a chimneylike formation on a hydrothermal vent that emits plumes of dark minerals

blocked delta   a delta eroded to the coastline by high wave energy and deflected by the presence of a longshore spit

blowhole   a geyser of water created by trapping of air by incoming waves; the air hole through which a marine mammal breathes

blunt delta   a prograded delta that results from the rapid dispersion and sorting of sediments by waves and longshore transport

bottom-up control   a flow of energy and materials in food webs where the primary producers control the growth and abundance of higher tropic levels

bottom waters   dense water masses formed in polar regions and distributed along the seafloor in the major ocean basins

Brazil-Malvinas Confluence   one of the most energetic regions of the world ocean; where the Brazil current meets the Malivans (Falkland) current

Brazil Current   the southerly flowing western boundary current along the eastern coast of South America

buffer   a substance that inhibits a change in pH (or other properties)

buoyancy   the rising or sinking of water masses (or objects) according to their density

burglar alarm defense   a behavior of some bioluminescent organisms that embeds their predators in a sticky bioluminescent mucus to attract their predators

bycatch reduction   technologies or fishing practices that reduce the practice of harvesting unwanted or unintended species of fish

C

¹⁴C-bicarbonate   a radioactive compound used to trace carbon assimilation in aquatic photosynthetic microorganisms

calcium carbonate compensation depth   the depth above which calcium carbonate accumulates in sediments

California Current   the eastern boundary current that flows along the western coast of the United States

calories   a unit of specific heat; the heat energy needed to raise the temperature of one gram of water by one degree Celsius

capillary waves   very small wavelength waves (less than 1.7 cm)

carbonaceous meteorite   a carbon-containing meteorite that contains small amounts of water

carbonate   a solid form of carbon, such as calcium carbonate

carbonate platforms   shelf-attached or shallow water structures formed from the shells of organisms or by reef-building organisms such as corals

carbon cycle   the storage and transfer of carbon among various living and nonliving earth reservoirs; both short- and long-term carbon cycles are recognized

carbon-fixation reactions   the dark cycle or Calvin-Benson cycle in photosynthesis whereby the energy-yielding molecules of the light reactions are used to construct carbon backbones for the synthesis of organic compounds

carbonic acid   one of the forms of dissolved carbon dioxide in seawater

carotenoids   a group of photosynthetic pigments mostly involved in protecting the reaction centers of photosynthesis

Cascadia subduction zone   one of the most dangerous tsunami-producing regions of the Pacific; a subduction zone off the coast of Oregon

cations   dissolved ions with a positive charge

Celsius scale   a scale of measurement for temperature in which the freezing point of water is set at 0 degrees and the boiling point is set at 100 degrees

central water masses   generally less-dense water masses that occupy the central gyres of the ocean basins

Challenger Deep   the deepest location in the world ocean, found in the Mariana Trench

***Challenger* Expedition**   one of the first global expeditions dedicated wholly to oceanography, undertaken by the British from 1872 to 1876

change of state   a transformation in the physical state (gas, liquid, solid) of a substance

chaos theory   a theory for describing the large effects of small perturbations on the future state of a system

chemical   of or relating to a change or transformation in the molecular properties of substances

chemical factors   the chemical characteristics of an ecosystem, such as salinity, pH, availability of nutrients, and dissolved gases, that influence where organisms may live

chemical oceanography   the study of the chemistry of the world ocean

chemical properties   characteristics of a substance on an atomic (microscopic) level that involve changes in the composition of a substance

chemical sediments   sediments forming directly from dissolved compounds in seawater

chemical weathering   the dissolution, fragmentation, and breakdown of rocks into sediments caused by chemical processes, such as acid rain

chemistry   the science of the composition, structure, properties, and behavior of matter

chemoautotrophs   organisms that use chemical energy to synthesize organic matter, as opposed to photoautotrophs, which use light

chirp subbottom profiler   a wide-frequency, small-footprint echosounder used for detecting features beneath the seafloor

chlorophyll *a*   the primary pigment involved in oxygenic photosynthesis

choke point   any region of the world ocean that exerts significant control over world ocean circulation

cladistics   the most widely accepted classification method for testing the evolutionary relationships among organisms

classical food web   a pelagic food web that includes mostly visible organisms from diatoms to herring; the most widely known and studied food web before the mid-1970s

classification   a method for grouping organisms into similar categories on the basis of size, lifestyle, chemical composition, anatomy, or other characteristics

climate   the average weather from short to long time periods at specific locations or over the entire globe

climatological variability   the range of a particular climate factor (such as temperature) at a particular location and over a specific time interval

climatology   the study of climate

coast   the strip of land of varying width from the edge of the sea inland to the first terrain unaffected by the sea

coastal jets   narrow filaments of upwelled water that meander from a coastline seaward

coastal lagoon   a type of estuary consisting of seawater trapped behind a beach or bar that receives small or occasional inputs of water from land sources or precipitation

coastal ocean   all of waters where the simultaneous presence of terrestrial and ocean properties can be detected

coastal oceanography   the study of the coastal ocean

coastal waters   undiluted salty waters on the seaward side of tidal fronts, river plume fronts, and other fronts formed where fresh water discharges meet oceanic waters

coastal wetlands   a tidally influenced, water-inundated region along or near the coast whose soil and vegetation differ from upland regions

coastal zone   the lands and waters along the coast that share a common ecology, that is a region where terrestrial and oceanic ecosystems interact

coastline   the seaward edge of the coast

cold-core rings   an eddy shed from the Gulf Stream whose center is colder than the surrounding water

cold regimes   for a given oceanic region, a multiannual or multidecadal period of colder-than-average seawater temperatures

compensation depth   the depth at which the rate of photosynthetic carbon fixation equals the rate of carbon respiration

compensation point   the light intensity at which net photosynthesis equals zero

compounds   substances composed of more than one element bonded together

community-based classification   an ecosystem classification based on the dominant species; for example, mangrove forest

condensation   the formation of a liquid from a vapor

condensation nuclei   tiny aerosols or dust particles required for the formation of clouds

condensation sequence   the order in which elements condense from gases in a newly formed cooling solar system

conduction   the movement of heat through a solid object

conductivity-temperature-depth instrument   see *CTD*

conservative elements   elements whose relative ratios do not vary regardless of the salinity

constructive interference   a superposition of waves that results in a wave whose amplitude is greater than that of the original waves

continental-continental boundaries   plate boundaries where two continents converge

continental crust   a type of lithosphere of varying composition that makes up the continents

continental drift   the movement of the continents over geologic time

continental margins   the submerged edges of the continents

continental rise   the sediment-laden boundary between continental and oceanic crust

continental shelf   the submerged, gently sloping region of a continental margin from the shore to the shelf break

continental slope   the more steeply sloping region of a continental margin from the shelf break to the continental rise

continental weathering   the physical, chemical, and biological breakdown of continental rocks

continents   the elevated portions of Earth's lithosphere composed of continental crust

convection   the circulatory movements of the atmosphere, ocean, and mantle in response to heat

convection currents   currents generated by heating where the source of heat is lower than the substance being heated

convection model of plate dynamics   the idea that plate motions are driven by convection within the mantle

convergent boundaries   plate boundaries where plates are moving toward each other, such as subduction zones (regions where one plate moves beneath another)

coprophagy   consumption of fecal pellets

coral cays   low-profile coral islands formed from the accumulation of carbonate sediments

coralline algae   a type of red algae that incorporates calcium carbonate in its tissues; may be the dominant reef builder in some locations

coral reef diagenesis   chemical alteration of corals and associated sediments

co-range line   in amphidromic tides, the line that connects equivalent tide heights

core   the innermost layer of Earth's interior

core-mantle boundary   the boundary between Earth's core and mantle

cosmogenous sediment   sediment derived from extraterrestrial processes, such as cosmic dust and meteorites

co-tidal line   in amphiodromic tides, the line that connects equivalent times

counterillumination   a type of bioluminescent camouflage that matches the surrounding light

covalent bond   a type of chemical bond in which electrons in the outer orbitals are shared

crest   the peak of a wave

critical depth   the depth to which a photosynthetic organism can be mixed and still maintain zero or positive carbon fixation when respiratory losses are taken into account

critical layer   the layer of air at the air-sea interface subject to displacement by the motion of waves and important in the transfer of momentum from the wind to the sea surface in the formation and propagation of waves

cross-shore transport   the movement of sediments across the beach face

crust   the surface layer of Earth; may be oceanic or continental

CTD   an electronic instrument for measuring conductivity, temperature, and depth

cumulonimbus convection   a pattern of air movements characterized by rising air and release of latent heat that forms thick, rain-bearing clouds

Curie point   the temperature at which magnetic particles in molten rocks align with Earth's magnetic field

currents   movements of fluids, such as water or air

cuspate delta   a river delta with hornlike extensions

cuspate foreland   a triangle-shaped shoreline cape formed from sediment deposits

cyanobacteria   a type of photosynthetic bacteria found throughout the world ocean in high abundance, previously known as blue-green algae

cyclone   air moving around a low-pressure system

D

daughter isotope   the product of radioactive decay; the element formed from the decay of a radioactive element

dead zones   regions of the ocean devoid of oxygen and aerobic marine life

declination   the angle between the position of the Moon (or Sun) and the equatorial plane; analogous to latitude, declination takes values of 90 degrees at the North Pole and -90 degrees at the South Pole

decomposers   organisms that facilitate the breakdown of organic matter

deep circulation   water movements beneath the permanent thermocline in the world ocean

deep-ocean basin   the region of the seafloor beyond the continental margins

deep scattering layer   any aggregation of organisms that causes high reflectivity of sound pulses

deep waters   waters at depths below intermediate waters

deep-water waves   waves whose propagation is not affected by the seafloor

delta ^{18}O   a unit of measure of the relative abundances of the natural and heavy isotope of oxygen used as a proxy for seawater temperatures

density   mass per unit volume

descent   downward vertical movement in the water column

descriptive classification   a description of ocean sediments based on their visible characteristics

destratification   the breakdown of density layers in the ocean

destructive interference   a superposition of ocean waves that results in a wave with lower amplitude than the original waves

diagenesis   the chemical transformation and alteration of sediments important in the formation of sedimentary rocks

diatomic state   a natural state of certain elements in which two atoms are present

diatoms   a type of photosynthetic microorganism that produces a silicate skeleton; found throughout the world ocean, most often in association with episodic bursts of productivity

diel vertical migration   the daily upward or downward movement of marine organisms in response to light

digitate delta   a fingerlike alluvial deposit found at the mouths of rivers and tidal inlets

dimethyl sulfides   a compound released by certain marine organisms that act as cloud condensation nuclei for the formation of clouds

dispersion   the spreading out of ocean or sound waves

dissolving   the process whereby one substance is dissolved into another to form a solution

disturbing force   a force that generates ocean waves

diurnal tide   a type of daily tidal pattern characterized by one high tide and one low tide

divergent boundaries   plate boundaries where plates are moving away from each other, such as oceanic ridges

doldrums   regions of rising air between the equatorial trade winds characterized by little or no wind

domain   the broadest level of classification of living organisms; one of three major groupings of life, such as Archaea, Bacteria, and Eukarya

downwelling   the downward vertical movement of ocean currents caused by convergence or Ekman transport toward a shoreline

Drake Passage   the strait between the tip of South America and Antarctica that connects the Pacific and Atlantic basins

dynamical physical oceanography   the branch of physical oceanography that quantifies flow patterns and property distributions in the world ocean

dynamical theory of the tides   a quantitative description of tidal motions in terms of their wavelike motions

early faint Sun paradox   the apparent contradiction of the presence of liquid water (instead of ice) on early Earth when the solar output was significantly reduced

E

earth tides   the response of Earth's crust to the daily differential gravitational forces of the Sun and the Moon

Earth's geophysical fluids   the ocean and the atmosphere

ebb tide   the period of time when tidal currents are slack, usually at the peak of the high and low tides

echo sounder   an acoustically active instrument used for mapping the depth of the seafloor, finding submerged objects, or tracking the movements of marine organisms

ecliptic   the apparent path of the sun across the sky over the seasons as viewed by an observer on Earth

ecological stoichiometry   the field of ecology that attempts to quantify and describe ecological systems through a study of the ratios of elements in different components of the system

ecosystem   a unit of organization of biological systems that includes a community of organisms and their nonliving environment; may also be defined as a community of organisms found in association with a particular set of physical, chemical, and biological conditions in the ocean

ecosystem approach   a fisheries management tool that considers the full range of ecosystem processes that support a fisheries and their natural and human-caused variations

ecosystem resiliency   a measure of the ability of an ecosystem to recover from natural or human disturbance

Ekman layer   in Ekman spirals, the region from the ocean surface to the depth where a current vector is pointed in a direction 180 degrees opposite to that of the surface current vector

Ekman spiral   a pattern of wind-induced flow in the ocean where successively deeper layers exhibit a clockwise or counterclockwise rotation as a result of the Coriolis effect

Ekman transport   the net movement of water as a result of an Ekman spiral; in the Northern Hemisphere, 90 degrees to the right of the wind direction; in the Southern Hemisphere, 90 degrees to the left of the wind direction

El Niño   officially, any three-month period during which the sea surface temperature anomaly exceeds 0.5 degrees C

electrical conductivity   the ability of a substance, such as seawater, to transmit an electrical charge

electromagnetic radiation   a form of energy that propagates as a wave and exhibits both electrical and magnetic properties; includes light and heat

electromagnetic spectrum   the full range of wavelengths of electromagnetic radiation from gamma rays (short wavelengths) to radio waves (long wavelengths)

electrons   a subatomic particle with a negative charge surrounding the nucleus of an atom

electron orbitals   the positions occupied by electrons in an atom

electronegative   a molecule or part of a molecule with a negative charge

electropositive   a molecule or part of a molecule with a positive charge

elemental equivalents   elements whose ratios remain in constant proportions, allowing determination of the concentration of one element from measurements of another

elements   substances comprised of atoms of a single type that cannot be divided into other substances

endosymbiosis   the theory that the organelles of eukaryotic cells are derived from primitive bacteria

energy budget   an accounting of incoming and outgoing radiation on Earth

ENSO   an acronym for the El Niño Southern Oscillation, the coupled air-sea phenomenon that generates above-average sea surface temperatures in the equatorial Pacific

epifauna   animals that live upon the seafloor

epipelagic zone   the life zone that exists between the ocean surface and the depth of light penetration

equilibrium theory of the tides   the idea put forth by Newton that the tides could be explained as tidal bulges resulting from the gravitational forces of the Sun and the Moon

equilibrium vapor pressure   the point at which the rate of evaporation and condensation are equal for a given temperature

estuarine waters   waters on the landward side of tidal fronts, including waters mixed within or formed by tidal lagoons and estuaries where streams, rivers, ice melt, or other freshwater discharges mix with saline coastal waters

estuary   a region where fresh and saltwater mutually influence each other

Eukarya   one of the three domains of life, including single-celled organisms with organelles and all multicellular forms of life

euphotic zone   the region of the upper ocean where photosynthesis may occur, typically down to the depth where light intensity is 1% of its surface value

eustatic change   changes in sea level due to increases or decreases in the amount of seawater

eutrophication   the overenrichment of waters with biologically important nutrients that leads to excessive algal growth

evaporation   the conversion of liquid water into water vapor

evaporites   sediments formed from the evaporation of seawater

evolution   the process whereby random changes in the genetic makeup of a species are acted upon by environmental forces over successive generations in a manner that produces new species

expendable bathythermograph   a disposable instrument for obtaining a vertical profile of temperature as a function of depth in the water column

exponential relationship   a mathematical relationship described by the equation $y = a(b)^x$, typically, a curvilinear relationship between two variables such as light intensity and depth

export production   excess carbon produced by a biological system available to other biological systems

extinction coefficient   a number that corresponds to the degree of light removed along a length; typically, in units of per meter

extreme waves   waves with heights in excess of normal wave heights

extreme wave surfing   a type of surfing dedicated to large waves

eye   the central part of a hurricane

eye wall   the clouds that surround the eye of a hurricane

F

Fahrenheit thermometer   the temperature scale most commonly used in the United States in which the freezing point of water is set at 32 degrees and the boiling point is set at 212 degrees

fall bloom   the increase in phytoplankton productivity associated with environmental conditions during autumn in temperate oceans

fecal pellets   zooplankton feces

Ferrel cell   one of the large atmospheric cells in the three-cell model of global atmospheric circulation; typically, appears only intermittently between 30 and 60 degrees latitude

fetch   the distance over which the wind blows in a constant direction and constant speed

fisheries management   a branch of fisheries oceanography dedicated to the study and regulation of techniques for conserving fish stocks

fisheries oceanography   study of the dynamics of fish populations in response to the changing environment of the world ocean

fishing down the food web   the practice of targeting lower trophic levels when higher trophic levels have been overfished

fishing-induced evolution   changes in the gene frequency of a fish population as a result of fishing

flood tide   the period of the tidal cycle when tidal currents are maximal

fluid   a substance that flows

fluorometers   an instrument for measuring the concentration of chlorophyll *in situ* or as an extract in solution

folk forecasting   weather prediction based on folklore

food chain   the linear transfer of energy and materials from one organism to another

food falls   the episodic settling of large food items, such as dead whales, to the deep seafloor

food web   the multiple pathways of energy and material flows between organisms

foraminifera   single-celled, shelled eukaryotes found throughout the world ocean

forced waves   waves that are continuously under the influence of a disturbing force, such as tides

fore reef   the outermost part of a barrier reef facing the ocean

foreshore   the region between the berm crest and the level of the mean low tide

fractional trophic level   a noninteger designation of the position of an organism in a food web that accounts for multiple modes of nutrition throughout its life cycle from juvenile to adult

free waves   waves that travel without additional influence from a disturbing force

fringing reefs   shore-attached or nearshore reefs formed when corals colonize the rocky coastline of a continental shelf or an island

fucoxanthin   a yellow-brown accessory pigment in diatoms and kelps that facilitates photosynthesis

full moon   the period of time when the disk of the Moon is fully illuminated

fully developed sea   a sea state in which the rates of energy addition and dissipation are equal

G

gas hydrates   frozen methane found on or within the seafloor

genetic classification   a sediment classification system aimed at identifying the location and material from which the sediment originated

geobiology   a scientific discipline that focuses on the mutual interactions of geological and biological processes

geoid   the equipotential gravitational surface of Earth

geologic time   the span of time from Earth's origin to the present

geomorphic classification   a classification based on the shape (morphology) of coasts and the processes that formed them, such as Shepard's classification

geophysical fluid dynamics   the study of the motions of fluids

geosphere   the sum of features and processes encompassed by Earth's geology

geostrophic currents   "Earth-turning" currents; the major currents of the ocean gyres that arise as the result of a balance between the Coriolis effect and the horizontal pressure gradient

ghost fishing   the unintended capture and death of marine organisms from abandoned or lost fishing nets that drift unattended in the ocean

Giant Impact model   a model for the formation of Earth's Moon based on the impact of a Mars-sized meteorite

glacial isostatic adjustment   the subsidence or uplift of Earth's crust that results from advancing or retreating glaciers, respectively

global teleconnections   the correlation of oceanic and meteorological events over global scales, such as the higher incidence of Atlantic hurricanes during periods of a Pacific La Niña

global warming   the rise in the mean global temperature of Earth since the 1800s

Gondwana   a supercontinent comprised of Africa, Antarctica, Australia, India, and South America

grain size distribution   the range of sizes of individual sediment particles in a sample

grains   individual particles of sediments

granular   a sediment type composed of particles resulting from the fragmentation of inorganic or organic parent materials

gravity waves   waves for which gravity is the restoring force

Great Ocean Conveyor Belt   a hypothetical and illustrative model of the coupled surface and deep circulation

greenhouse effect   the trapping of heat (long wavelength radiation) by greenhouse gases, such as carbon dioxide, methane, and water vapor

gross primary productivity   the rate of photosynthetic carbon fixation by phytoplankton before respiratory losses have been taken into account

group speed   the speed of a wave train, a group of waves traveling together

Gulf Stream rings   mesoscale eddies shed as a result of the meandering of the Gulf Stream

guyots   flat-topped seamounts

H

habitat   the place where an organism lives

habitat-based classification   an ecosystem classification system based on the predominant substratum; for example, sandy ecosystem

habitat destruction   human-caused alteration of the living conditions of an organism or group of organisms

Hadley cell   an atmospheric circulation cell caused by heating and rising of air at the equator and cooling and sinking of air at subtropical latitudes; gives rise to the trade winds

hadopelagic   the deepest ocean life zone, below 6000 m; the pelagic habitat encompassed by submarine trenches

half-life   the time required for a radioactive isotope to decrease in abundance by one-half

Halimeda   a calcareous green seaweed that plays an important role in the growth of coral reefs

halocline   a region of the water column where a sharp gradient in salinity exists

hard-bottom benthos   bottom-dwelling organisms associated with hard substrates, such as rocks and rocky shelves

harmful algal blooms   high concentrations of toxin-containing or otherwise deleterious phytoplankton

heat   the spontaneous transfer of energy between two systems at different temperatures

Heinrich events   abrupt or sudden cooling of Earth's climate over decades to centuries

heterogeneous accretion model   a model of Earth's formation in which the planet-forming planetesimals are not similar in composition

high-nutrient, low-chlorophyll region   region of the ocean characterized by high concentrations of biological important nutrients, such as nitrate, but low concentrations of phytoplankton chlorophyll, presumably because of nutrient limitation by a trace element, such as iron or silica

high-pressure liquid chromatography   a method used for the separation, identification, and quantification of phytoplankton pigments (and other compounds)

highs   areas of atmospheric high pressure relative to the surrounding pressure

high tide   the highest sea level during a tidal cycle

holoplankton   organisms that spend their entire lives as plankton (suspended drifters)

homeothermy   maintenance of a constant body temperature

homogeneous accretion model   a model of Earth's formation in which the planet-forming planetesimals are nearly identical in composition

horse latitudes   a region of descending air between 30 and 35 degrees latitude where winds are light and variable

hot spots   localized regions of persistent magmatic extrusion through Earth's crust

humidity   a general term for the percentage of water vapor in the atmosphere, most commonly in relation to the saturation vapor pressure, such as relative humidity

Hurricane Alley   a broad region of the tropical North Atlantic in which hurricanes form and propagate westward

hurricane season   the time of year when most tropical cyclones form; officially, from June 1 to November 30 in the Atlantic

hydrated   surrounded by water molecules

hydrologic cycle   the water cycle

hydrogen bond   weak chemical bond formed between the electronegative oxygen atom of one water molecule and the electropositive hydrogen atom of another water molecule

hydrogenous sediment   see *chemical sediments*

hydrographic classification   classification of an estuary based on the physical characteristics of the water body, whether it is well-mixed, highly stratified, or in between (such as Pritchard's classification)

hydrosphere   all of the waters of the Earth

hydrostatic equation   the mathematical formulation that describes water pressure as a function of depth and water density

hydrostatic pressure   the pressure exerted by the weight of fluid in a vertical column

hydrothermal vents   cracks in the seafloor through which seawater circulates as the result of geothermal processes

hypoxia   a condition of low dissolved oxygen concentration in a body of water; typically, less than 2 milligrams per liter

hypsographic curve   the cumulative frequency of the area or percentage of land elevation or depression on a planet

I

ice-edge blooms   blooms of phytoplankton, often intense, that follow the onset of stratification caused by the receding and melting ice edge

ice worm   a type of worm found living in association with frozen gas hydrates

ichthyoplankton   the planktonic larval form of fishes

in situ   in place; in the water under existing conditions

in situ **temperature**   the temperature of seawater in the location where it exists

Indonesian Throughflow   the maze of currents that flow through the Indonesian archipelago thought to represent a choke point of ocean circulation

infauna   animals living within sediments, such as tube-dwelling ghost shrimp

inner core   the solid, innermost part of Earth's core

interdisciplinary science   scientific research that requires knowledge and expertise beyond one's own discipline; see also *multidisciplinary*

interglacial   periods of Earth's geologic history between the ice ages

intermediate water masses   water masses that occur between the surface and deep or bottom waters below the depth of the permanent thermocline

intermediate waves   waves in transition from deep water to shallow water whose depth is less than one-half and greater than one-twentieth of the wavelength of the wave

internal waves   waves that occur at the interface between fluid layers with different densities

International Decade of Ocean Exploration   a major international initiative to advance ocean exploration and responsible use of the ocean from 1971 to 1980

interplanetary dust particles   tiny particles of cosmic dust captured by Earth's gravitational field found in ocean sediments

interstitial organisms   organisms living between sediment grains, such as nematodes

Intertropical Convergence Zone   a region of low pressure and dense clouds near the equator formed where the northeast and southeast trade winds meet

inundation   the maximum landward extent of a tsunami or storm surge measured as the horizontal distance from the shoreline

inverse or hypersaline estuaries   estuaries in which high rates of evaporation cause a nearly constant inflow of saltwater

ionic compounds   substances made up of a cation and anion held together by an ionic bond

iron limitation   the regulation of phytoplankton growth by dissolved iron

iron sink   the migration of molten iron to the core when Earth was in a molten state early in its formation

isobars   contour lines of equal atmospheric pressure

isobaths   contour lines of equal hydrostatic pressure

isopycnals   contour lines of equal seawater density

isostasy   the balance between the mass of the crust and the buoyant forces of the mantle that permit the crust to "float" on the mantle

isostatic adjustment   the rising or subsidence of the crust in response to mass changes, such as glaciers or sea-level change

isotopes   the different atomic weights of a single element

J

jellyplankton   gelatinous, drifting marine organisms, such as medusae or salps

Jovian or gas planets   the four outer planets, composed primarily of gases

karst topography   a land surface formed from the erosion of carbonate platforms; one of the mechanisms thought to have produced atolls

K

Keeling curve   the graph that depicts the rise in atmospheric carbon dioxide since the late 1950s

Kelvin scale   a measurement scale for temperature based on absolute zero, the temperature at which nearly all molecular motion ceases

Kelvin wave   a long-wavelength gravity wave that propagates eastward along the equator or counterclockwise in Northern Hemisphere basins

keystone predator   a species whose presence alone alters the structure of the community in which it occurs

Kullenberg piston corer   a weighted tube with an internal piston that provides sampling of undisturbed, relatively long lengths of sediments

Kuroshio Current   an eastern boundary current along the coast of Japan

Kuroshio Extension   the eastward flow of the Kuroshio Current into the North Pacific

L

lagoon   a shallow body of water trapped behind a reef or sand bar

latent heat   the heat required to change the physical state of a substance

latent heat of fusion   the heat required to change a liquid to a solid

latent heat of vaporization   the heat required to change a liquid to a gas

late veneer   the arrival of water-laden planetesimals or comets in the latter stages of Earth's formation

Laurasia   a supercontinent comprised of parts of Asia, Europe, and North America

law of conservation of energy   the law of thermodynamics that states that energy can neither be created nor destroyed

lead line   a weighted rope or line used for determining depth of the seafloor

leads   cracks in sea ice

light-and-dark bottle method   a commonly used method for determining rates of oxygen production and consumption by microbes in aquatic environments

light reactions   light-capturing photosynthetic processes that supply energy for the formation of organic matter by the dark reactions

limiting factors   any factor that limits the productivity of autotrophs

linear velocity   in the Coriolis effect, the speed at which the Earth is turning

lithification   the formation of rock from sediments or unconsolidated rock

lithogenous   sediments formed from the fragmentation of solid rock

lithosphere   Earth's outermost layer, consisting of the rigid plates and a portion of the upper mantle

lithospheric plates   the rigid blocks of lithosphere that move relative to each other across Earth's surface

littoral cell   a segment of a coastline that includes a sediment source, such as a river, and a sediment sink, such as a submarine canyon

lobate delta   a lobe-shaped (rounded) sediment deposit at the mouth of a river

logging   inspection and recording of observations on a sediment core

longshore current   a wave-induced current that flows parallel to the coastline within the surf zone

longshore transport   the movement of sediments and floating debris by a longshore current

long-term inorganic carbon cycle   processes involved in the formation and dissolution of carbon-containing rocks, such as limestones, over geologic time

lower mantle   the dense and mechanically strong lower part of Earth's mantle known as the mesosphere (also used to describe a layer of the atmosphere)

lows   areas of atmospheric low pressure relative to the surrounding pressure

low tide   the minimum sea level in a tidal cycle

low-tide terrace   the portion of the beach face exposed during low tide

luminous symbionts   bioluminescent bacteria that inhabit the light organs of many marine organisms

lysocline   the depth below which calcium carbonate begins to dissolve due to its undersaturation

M

M2 tide   the semidiurnal tidal component caused by the Moon's gravitational field

macroalgae   multicellular, photosynthetic eukaryotes, including seaweeds

macronutrients   the dissolved chemical substances in greatest demand by photosynthetic organisms

major constituents   those dissolved inorganic compounds in seawater whose concentration exceeds 1 part per million

mantle   Earth's middle layer, occupying the greatest volume of Earth's interior

mantle convection   the circulation of molten materials within Earth's mantle

mantle plumes   conduits of molten material originating from the lower mantle

marine carbonate system   the forms of dissolved carbon dioxide in seawater and the changes in their concentration and equilibrium in response to pH and biological uptake

marine communities   the plants and animals that inhabit the world ocean

marine plants   eukaryotic, multicellular, photosynthetic, marine organisms with roots and leaves

marine snow   a type of flocculent organic material of various origins that drifts and settles in the water column

mass extinction   periodic elimination of a major percentage of Earth's species

mass mortality   the dying of the adult population of a marine animal following reproduction

mattang   a stick chart used for teaching wayfaring, a form of navigation

maximum sustainable yield   the theoretical level of fishing that can be sustained for a particular species of fish

mean lower low water   the average of the lowest tide that occurs in regions with mixed semidiurnal tides

mean sea level   the average height of the ocean relative to a shoreline when the effects of tides, waves, and other ocean disturbances have been removed

mechanical bathythermograph   an automated temperature-measuring device that uses a pen scribe to continuously record depth on a rotating drum

mechanical strength   the susceptibility of rocks to movement and deformation

meddies   Mediterranean eddies

Mediterranean Outflow Water   the water that flows out of the Mediterranean Sea into the North Atlantic

mediterranean sea   a semienclosed basin that has limited exchange of waters with the world ocean

megaplumes   mantle plumes with the greatest volume; thought to have produced large igneous provinces and oceanic plateaus

megathrust earthquakes   highly destructive earthquakes produced by release of energy along subduction zones

megatsunami   highly destructive earthquakes produced by megathrust earthquakes

meridional sections   profiles of ocean properties along lines of longitude

meroplankton   marine organisms that spend only a portion of their life cycle as a planktonic drifter

mesopelagic zone   the "twilight" zone; the ocean life zone at the interface of the photic and aphotic zone, typically from 200 to 1000 m in depth

messenger   a brass weight used for triggering water sampling bottles or other oceanographic instruments at depth

Messinian Salinity Crisis   the drying up of the Mediterranean Sea about 5 to 6 million years ago

meteorology   the study of the weather

methanogens   methane-producing microbes

Mg/Ca thermometry   a technique using the ratios of magnesium and calcium as a proxy for seawater temperature

microbial food web   the largely invisible microbial organisms of pelagic food webs and the transfer of energy and materials that occurs between them

microbial revolution   the scientific realization since the mid-1970s of the importance of microbial forms of life in the world ocean

micronutrients   those dissolved substances required in small amounts by autotrophs

Milankovitch cycle   the millennial-scale variations in the Earth's orbit around the Sun that produce ice ages

minor constituents   dissolved substances in seawater whose concentration is less than 1 part per million

missing carbon problem   the apparent discrepancy between the amount of carbon released by human activities and the smaller amounts found in the atmosphere

mixed layer depth   the depth to which seawater is regularly mixed in the upper ocean

mixed semidiurnal tide   a tidal pattern consisting of two high tides and two low tides of unequal height

model   a conceptual or mathematical framework for exploring and testing hypotheses

modern synthesis   a modern theory of evolution incorporating gene theory and population genetics

mode waters   water masses characterized by weak stratification and a general homogeneity in temperature and density over a broad range of depths

molecular phylogeny   classification of the evolutionary relationships of organisms based on their genetic makeup

molecules   substances composed of two or more atoms

momentum   formally, mass times velocity for an object; informally, the tendency of an object to maintain a straight path

monsoon   the seasonal reversal of winds, often bringing a dramatic change of weather

multibeam echosounder   a device for mapping seafloor features that uses multiple frequencies of sound beamed in a narrow to wide swath

multichannel seismic profiling   a technique using explosive sound to determine the layering of rock strata beneath the seafloor

multichannel seismic system   a system of ship-towed, sound-producing air guns and hydrophones for mapping subseafloor features

multidisciplinary   scientific research that requires the cooperative efforts of scientists from multiple disciplines

Mysticetes   a suborder of cetaceans that includes the baleen whales

N

National Data Buoy Center   an Internet-based reporting system for weather and oceanographic buoys deployed in coastal US waters by the National Oceanic and Atmospheric Administration

National Tidal Datum Epoch   the 16.8-year time period over which measurements of zero tide levels are based

natural fluorescence   the solar-stimulated fluorescence of phytoplankton in natural waters

natural selection   the differential survival and reproduction of a population that results in evolution; the forces that cause evolution

neap tide   the period of time in the monthly tidal cycle when the range of tides is minimal

nekton   marine organisms capable of movement independent of tidal and current flows

net community production   the amount of primary productivity available for export by a community

net primary productivity   the amount of organic carbon available to higher trophic levels after the respiratory utilization by autotrophs has been taken into account

neutrons   elementary uncharged particles found in atomic nuclei

new moon   when viewed from Earth, the period when no Moon is visible at nighttime

new production   gross primary productivity based on new inputs of nitrate to a pelagic ecosystem

nitrogen cycle   the storage and cycling of nitrogen in and between various living and nonliving reservoirs

node   the center point of no tidal displacement around which amphidromic tides rotate

nonconservative elements   substances dissolved in seawater whose concentration varies independently of salinity

North Atlantic Deep Water   an abundant type of deep water formed in the North Atlantic and distributed throughout the world ocean

North Pacific Deep Water   an abundant type of deep water formed in the North Pacific where it is generally confined

nucleus   the central part of an atom composed of protons and neutrons

numerical weather prediction   a type of weather forecasting based on equations that describe the behavior of the atmosphere

nutricline   a region of rapid change in the concentration of a biologically important nutrient

O

ocean-atmosphere system   the coupled physical and chemical interactions between the ocean and the atmosphere

ocean basin   formally, the region of the seafloor other than that occupied by continental margins and oceanic ridges; informally, the depressions in Earth's crust occupied by seawater

ocean ecology   the study of the interrelationships among organisms as influenced by the ocean environment

ocean general circulation model   a gridded mathematical model used for determining present and future heat and mass transports of ocean currents

ocean genomics   the study of the genomes of marine microbes

oceanic crust   a dense, basaltic form of Earth's lithosphere found in ocean basins

oceanic faros   reef structures with a central enclosed lagoon or depression whose origins differ from atolls

oceanic gyres   the central regions of the Northern and Southern Hemisphere oceanic basins around which the broad-scale surface circulation flows

oceanic ridge   an elevation of the seafloor where seafloor spreading occurs along plate boundaries

oceanic trenches   deep and narrow troughs that border the edge of continents and other locations where subduction occurs

oceanic-continental convergent boundary   the plate boundary at which oceanic crust of one plate is subducted beneath continental crust of another plate

oceanic-oceanic convergent boundary   the plate boundary at which oceanic crust of one plate is subducted beneath oceanic crust of another plate

ocean life zones   loosely, depth-based regions of the world ocean characterized by particular conditions of light, temperature, and pressure in which characteristic assemblages of marine organisms may be found

oceanographer   a scientist who studies the ocean

oceanography   the study of the ocean

ocean weather   ocean conditions that currently exist at a given location

Odontocetes   a suborder of cetaceans that includes the toothed whales

open ecosystem   an ecosystem whose boundaries and exchanges of energy and materials are not well defined, such as a beach ecosystem

orbital   the space in which an electron orbits the nucleus of an atom

organelles   intracellular structures involved in the metabolic functioning of a cell, such as mitochondria or chloroplasts

outer core   the liquid outer part of Earth's core

outgassing   the release of gases and water vapor from Earth's interior

oversaturated   a state in which the concentration of a substance or gas exceeds the solubility of seawater

overwash apron   a fanlike deposit of sediments on the landward side of a barrier island

oxygen minimum zone   a region of the world ocean where the concentration of oxygen reaches a minimum at depth, caused by the biological utilization of oxygen at a rate faster than its resupply by physical processes

P

Pacific-type margins   continental margins at a plate boundary dominated by tectonic activity

paleoceanography   the study of ancient oceans and their evolution

paleodeposits   deposits of sediments indicative of past tsunami or other sedimentary processes

paleomagnetism   the study of the orientation of magnetic particles in rocks through geologic time

Pangaea   the supercontinent formed by the coming together of Gondwana and Laurasia approximately 225 million years ago

Panthalassa   the ocean surrounding Pangaea

parent isotope   the radioactive element that decays

partially mixed   an estuary characterized by a weak tidal wedge; intermediate between a salt-wedge estuary and a well-mixed estuary

passive acoustics   the study of underwater sound using hydrophones, that is, by listening to ocean sounds

passive margins   Atlantic-type continental margins not near a plate boundary dominated by sedimentary processes

Patagonian Shelf LME   a highly productive marine ecosystem and fisheries extending from Uruguay to the Straits of Magellan

patchiness   a nonuniform distribution of marine organisms

patch reefs   isolated, small coral reefs

pelagic crenarchaeota   a type of marine Archaea living planktonically in the water column

peridinins   a type of photosynthetic accessory pigment found in dinoflagellates

periodic table of the elements   a grouping of all known chemical elements into rows and tables based on their chemical properties

pH   the negative log of the concentration of hydrogen ions in a solution

photoadaptation   the physiogical or ecological response of phytoplankton to changing light conditions

photoautotrophs   photosynthetic marine organisms

photophores   light-producing organs on squids and fishes

photosynthetically available radiation   the solar radiation between 400 and 700 nm that may be utilized by photoautotrophs

pH scale   a numerical designation between 1 and 14 corresponding to the pH of acids and bases

phylogenetic tree   a treelike grouping of organisms based on their evolutionary relationships

physical factors   physical properties of an oceanic environment, such as temperature and pressure

physical oceanography   the study of physical properties and processes in the world ocean

physical properties   see *physical factors*

physical structure   the layering of water parcels to create stable or unstable physical conditions

physical/mechanical weathering   the fragmentation and breakdown of rocks into sediments caused by physical and mechanical processes, such as ice expansion

phytodetritus   the debris associated with the remains of phytoplankton

phytoplankton   drifting photosynthetic microbes

picopellets   tiny fecal pellets produced by marine protists

planetary differentiation   the separation of a newly formed molten planet into distinct layers in its interior

planetsimals   the embryonic materials of planets

plankton   any organism that is unable to maintain motion independent of the tides and currents

plankton net   a netlike device used to capture plankton

plate boundaries   the narrow regions where plates meet

plate-driven plate tectonics   the idea that the plates organize and drive mantle processes, which results in plate motions

plate dynamics   the study of the processes responsible for plate movements

plate kinematics   the study of the motions of the plates

plate tectonics   the theory that describes the movement of rigid segments of Earth's crust in response to heat flow from Earth's interior

plunging cliffs   near-vertical underwater "walls" that appear to plunge into the abyss

pneumatophores   above-water roots of mangroves used in the exchange of gases

polar cells   in the three-cell model of atmospheric circulation, the pattern of air flow that occurs at the poles

polar easterlies   weak and irregular winds created by descending air near 60 degrees latitude

polar front   the boundary between the cold polar air mass and the warm tropical air mass

polar jet stream   a relatively narrow and persistent, high-velocity westerly wind found along the polar front

polar molecule   a molecule exhibiting positive and negative charges on different ends of the molecule

polar reversal   the reversal of Earth's magnetic poles every few million years

polar wandering   movement of the geographic location of the magnetic pole over time

polynyas   ice-free regions surrounded by sea ice

poorly sorted   sediments whose grain size varies considerably

potential temperature   the temperature of a parcel of water with the effects of pressure removed

practical salinity scale   the standard scale of measurement for salinity

precipitating   the coming out of solution of a dissolved substance

pressure center   the central region of high and low pressure systems

primary consumers   the trophic level that consumes primary producers

primary productivity   the rate of carbon fixation by autotrophs

principle of constant proportions   the constancy of the ratios of the major constituents regardless of the salinity

principle of superposition   the idea that undisturbed sediment layers exhibit a progression of ages from youngest at the top strata to oldest at the bottom strata

principle of uniformitarianism   the idea that processes acting today are the same processes that acted in the past

progressive waves   waves with a forward motion

prokaryotes   organisms lacking organized cellular components

proplyd   short for protoplanetary disk, the precursor to a solar system

proton   an elementary positively charged particle of an atom found in the atomic nucleus

protostar   the precursor to a star

proxy   a measurement of one property that can be used to indirectly infer another property

punctuated equilibrium   a type of evolution characterized by long periods of relative stasis interrupted by short periods of rapid change

pycnocline   a region of the water column characterized by a rapid change in seawater density

pyramid of biomass   a representation of the biomass of organisms in each trophic level from autotrophs at the base to top predators at the apex

pyramid of energy   a representation of the energy content of organisms in each trophic level from autotrophs at the base to top predators at the apex

pyramid of numbers   a representation of the numbers of organisms in each trophic level from autotrophs at the base to top predators at the apex

Q

quarter moon   as viewed from Earth, the period of the lunar cycle when the disk of the Moon appears half full

R

radioactive isotopes   radioactive forms of an element; those that emit elementary particles or energy

radioactivity   the spontaneous emission of radioactive particles from an element

radiolarian   a single-celled eukaryote with a silica skeleton

radiometer   an instrument for measuring the intensity of solar radiation

radiometric dating   the determination of the age of rocks based on the natural decay of radioactive elements within the rock

recruitment   the number of juvenile fishes reaching catchable size in a habitat

Redfield ratio   the near constancy of the ratios of carbon, nitrogen, and phosphorus of particulate and dissolved organic matter in the world ocean

reef crest   the near-surface portion of the reef over which waves break

reference ellipsoid   the ideal shape of the Earth from which measurements of sea level may be based

regenerated production   primary productivity supported by ammonium excreted by zooplankton

regime shifts   multidecadal changes in ocean conditions that cause a geographic displacement of marine organisms

regressive barriers   barrier islands moving seaward and exhibiting accumulation of lagoonal sediments on top of marine sediments

remineralization   the breakdown of organic matter by heterotrophic bacteria to produce dissolved inorganic substances that may be utilized by phytoplankton

residence time   the time that a dissolved substance spends in a particular reservoir

resonance frequency   the natural frequency of oscillation of a body of water, related to the length and depth scales of the basin

respiration   the metabolic breakdown of organic matter by aerobic or anaerobic means

restoring force   the force that dampens and removes ocean waves

reversing thermometer   a type of mercury thermometer in which the flow of mercury is disrupted when turned upside down; useful for making accurate measurements of temperature at depth in the water column

ribbon reef   a relatively narrow, elongate stretch of reef present on a barrier reef

ridge-push, slab-pull model   a model of plate dynamics whereby plate motions result from uplift and gravitational sliding at oceanic ridges and subsidence and gravitational pulling at oceanic trenches

rift or axial valley   a narrow valley that forms along the crest of some oceanic ridges, especially slow-spreading ridges

Rodinia   a supercontinent that formed approximately 1100 million years ago

rogue waves   unusual waves of extreme height that may cause danger or harm to humans and property

rolling   a type of movement exhibited by sediment grains that resembles the movement of a ball down a hill

rosette   the apparatus that holds water sampling bottles in a circular configuration around a CTD

Rossby waves   mostly westward-propagating, long-wavelength planetary waves found in the ocean and atmosphere

run-up   the vertical extent of tsunami or storm surge inundation relative to sea level; also the advance of a wave on a shore

S

S2 tide   the semidiurnal component of tides caused by the gravitational attraction of the Sun

salinity   the concentration of dissolved salts in seawater

salinity-temperature-depth instrument   an instrument for the continuous measurement of salinity, temperature, and depth in the water column

saltation   a type of movement of sediment grains characterized by sudden hops

salt tectonism   the subsurface migration of salt deposits resulting in a hummocky appearance to the seafloor

salt-wedge estuary   an estuary in which freshwater flows seaward over the top of seawater and maintains a distinct pycnocline

Sandström's theorem   the idea that convection in fluids can only occur when the source of heat is lower than the source of cooling

satellite altimeters   satellite-mounted microwave radar instruments used for determining sea surface height

saturation concentration   the maximum concentration of a solute that can be dissolved in a substance

scattering   the change in direction of light rays by molecules or particles in the atmosphere or ocean

schools   aggregations of fishes in which individuals swim synchronously

scientific inquiry   the actions and thought processes that scientists use to formulate and answer questions about nature and how it works

sea arches   sea stacks through which an opening occurs

sea caves   excavations in a sea cliff

sea ice microbes   marine bacteria, diatoms, and other microorganisms found in the interstices and brine channels of sea ice

seafloor spreading   the lateral movement of oceanic crust as it is formed along oceanic ridges

sea level   the level of the ocean relative to Earth's geoid

seasonal ice zone   the region from the Antarctic continent seaward in which sea ice forms and melts

seasonal thermocline   the seasonally varying region of the water column exhibiting a rapid change in temperature

sea stacks   residual small islands along a coast

sea surface height   the distance from the seafloor to the sea surface

sea surface salinity   the concentration of dissolved salts at the surface of the ocean

sea surface temperature   the temperature at the sea surface

sea surface topography   the shape of the sea surface

Secchi disk   a circular white disk used to determine the depth of light penetration in the ocean

sediment   any unconsolidated fragment of inorganic or organic material

sedimentation   the sinking of sediments in the water column and the processes that affect it

sediment budgets   an accounting of the rate of supply of sediment sources and the rate of loss of sediment sinks on a beach

sediment cores   cylindrical samples of sediments

sediment cycle   the reservoirs for sediments and the processes that affect their origination, transport, deposition, and lithification

sediment trap   an instrument used for characterizing and quantifying sinking marine sediments

seiche   a standing wave in an enclosed or semienclosed body of water

seismic reflection profiling   a technique using seismic profiling instruments to identify subseafloor rock layers; may also be used to characterize the density structure of the water column

seismic tomography   a technique for determining the three-dimensional structure of Earth's interior using multiple records of seismic waves

seismic waves   waves of energy produced by earthquakes and other displacements of Earth's crust

semiclosed or closed ecosystems   ecosystems sustained by productivity within their local confines

semidiurnal tide   a daily tidal pattern characterized by two high tides and two low tides of approximate equal height

sensible heat   heat that can be detected by an instrument or human observer

shallow-water waves   waves traveling in water whose depth is less than one-twentieth of the wavelength of the wave

shore platforms   horizontal or sloping layers of rock or hard substrate (such as limestone or chert) at the base of a rocky cliff

short-term organic carbon cycle   the reservoirs and processes involved in the transformation and exchange of newly fixed organic carbon

shortwave solar radiation   solar radiation at short near-visible and visible wavelengths, such as ultraviolet and blue light

sidereal month   the length of time for the Moon to orbit Earth and return to the same position relative to distant stars; about 27.3 days

side-scan sonar   a type of active acoustic device towed behind a ship or deployed on an ROV that provides a high-frequency, detailed view of features and objects on the seafloor

sigma-t   an alternative expression for the density of seawater; equals $(1 - density) \times 1000$

shore platform   a gently sloping hard surface along the shore

significant wave height   the average height of the highest one-third of all waves

sinks   processes that remove a substance or material from a reservoir

size classification   a means for organizing and classifying marine organisms based on their size

slack tide   the period at the height of the high or low tide when the speed of tidal currents is minimal

sloppy feeding   the incomplete grazing of zooplankton on phytoplankton that releases organic particles into the water column

small comet hypothesis   the idea that Earth's water was supplied by numerous small comets

Snowball Earth hypothesis   the idea that the Earth has been completely or nearly completely frozen one or more times in its geologic past

SOFAR channel   sound fixing and ranging channel; a layer of the ocean between 600 and 1200 m (1968–3937 feet) in low and middle latitudes where temperature and pressure conditions confine sound waves and propagate them over very distances of hundreds of kilometers (hundreds of miles)

soft-bottom benthos   marine organisms that live upon, within, or between sediments (soft substrates as opposed to hard substrates)

solar constant   the incident solar radiation per unit area received at the outer atmosphere

solar energy   electromagnetic energy supplied by the Sun, such as light, heat, and other forms of radiation

solar ignition   the onset of nuclear fusion in a newly forming star, accompanied by a tremendous explosion of particles

solar nebula model   the model in which solar systems originate when clouds of interstellar gas and dust were disturbed by the explosion of a nearby star or some other cosmic event

solar spectrum   the types of electromagnetic radiation emitted by the Sun

solar wind   a stream of charged particles emitted by the Sun

solubility   the amount of a solute that can be dissolved by a solvent under a given set of conditions

solute   the substance being dissolved

solvent   the substance into which other substances are dissolved

sources   processes that add a substance or material to a reservoir

Southern Oscillation   periodic variations in the difference in air pressure between Papeete, Tahiti, and Darwin, Australia

spatial scales   dimensions of space from microscopic to global

species composition   the types of species inhabiting a given location

specific heat   the heat required to raise the temperature of 1 gram of water by 1 degree Celsius

spin-up   the time required for phytoplankton to synthesize the metabolic machinery required to respond to a change in environmental conditions that promotes their growth

spiral rain bands   narrow and elongate "walls" of convective rain clouds surrounding a hurricane

spit   a coastal landform composed of a narrow strip of sediments extending into a bay or the ocean; usually formed by longshore transport and shaped by wave activity

spreading loss   the decrease in energy per unit area as a progressive wave propagates and expands along the wavefront

spring bloom   the rapid proliferation of phytoplankton associated with the onset of stratification, typically occurring in spring in temperate latitudes

spring equinox   the first day of spring when the Sun is overhead at the equator in its northerly movement (N. Hemisphere)

spring tide   the period during the tidal cycle when the tidal range is maximal

stability   the degree of motility of sediments in a beach or estuarine habitat

stable isotopes   nonradioactive forms of an element with an unvarying chemical composition

stable water column   a water column whose layers are at neutral buoyancy

standing crop   the abundance of phytoplankton, most often measured as the concentration of chlorophyll

standing wave   a progressive wave that appears to be motionless; formed by two or more progressive waves traveling in opposite directions

steady-state   unvarying in rate; when rates of flows of energy and matter in an ecosystem are constant

steady-state food webs   food webs whose transfer and exchanges of energy and matter are relatively constant

Stefan-Boltzmann formula   a mathematical formula that describes the total radiation emitted by a blackbody at a given temperature

steric change   the portion of sea level variability attributable to changes in seawater density

still water level   the resting state of the sea surface in the absence of a disturbing force

stimulated fluorescence   the red emission of chlorophyll as induced by an artificial light source

Stoke's law   a mathematical formulation to describe the rate of settling of small spheres in a fluid

stony corals   true corals in the Order Scleractinia that produce hard skeletons

storm surge   the rise in sea level associated with storm systems moving onshore

stratification   the layering of water parcels with different densities

strike-slip fault   a horizontal displacement of Earth's crust along a fault

subduction   the underthrusting of one tectonic plate beneath another

submarine canyons   V-shaped cuts in the continental shelf produced by turbidity currents

subolar gyres   oceanic gyres immediately south of the Arctic circle and north of the Antarctic circle

substance   matter that has a definite or constant composition and exhibits distinctive properties

subsurface chlorophyll maximum   a relatively high concentration of chlorophyll associated with the pycnocline or nutricline

subtropical gyres   the major ocean current systems found in subtropical regions, such as the North Atlantic or North Pacific

subtropical jet stream   a fast-moving current of high-altitude air that stretches around the globe on the poleward side of the Hadley cell

sulfophilic community   an association of organisms found on whale falls that depend on chemosynthetic bacteria capable of utilizing hydrogen sulfide as an energy source

Summer Monsoon   the atmospheric pattern that occurs seasonally in some parts of the world as warming of the continental land mass draws wind and moisture from a nearby gulf or ocean

summer solstice   the first day of summer when the Sun appears at 23.5 degrees above the celestial equator (N. Hemisphere)

supercontinent   an aggregation of several continental landmasses; caused by continental drift

superposition   one wave on top of another

supersaturated   when the concentration of a dissolved gas or substance exceeds the solubility for a given set of conditions

surface (upper) waters   waters found at the surface of the ocean above the permanent thermocline

surface circulation   the global-scale water movements that occur at the surface of the ocean

surface wave reflection   the change in path of a progressive wave when it encounters a barrier

surface wind stress   the near horizontal or tangential force exerted by the wind on the surface of the ocean

surf forecasting   a science aimed at predicting surf conditions at beaches

surf zone diatoms   phytoplankton species that inhabit the surf zone

suspension   the temporary floating of sediment particles within a fluid

swash zone   the region along the beach face in which wave run-up and backwash occur

swath beam mapping   a technique for rapidly measuring the shape of the seafloor using a sound source with a wide angle of emission

swells and superswells   elevated regions of the seafloor, sometimes extensive

symbiosis   a cooperative relationship between two or more organisms; may be beneficial to one or both or beneficial to one and harmful to the other

synodic month   the time of one complete cycle of moon phases; the time required for the Moon to orbit Earth and return to the same position relative to the Sun; about 29.5 days

systematics   the study of the evolutionary relationships of life and its changes over time

T

TAO-Triton Buoy Array   a network of weather buoys and oceanographic instrumentation stretched across the equatorial Pacific for monitoring and early detection of El Niño and La Niña events

tektites   melted fragments of Earth's crust produced as the result of a meteorite impact

temperature   a measure of the average kinetic energy of molecules

temperature anomalies   deviations of a specific interval of temperature measurement from the climatic average

temperature-salinity (*T-S*) diagram   a graph of temperature versus salinity used by physical oceanographers for identifying water types and water masses

temporal scales   units of time from fractions of seconds to millennia

terrestrial or rocky planets   the four inner planets in our solar system, including Earth, composed largely of rocky materials

theory of common descent   the theory that explains the common origins of all life on Earth from a universal ancestor

theory of plate tectonics   the idea that Earth's crust can be divided into large rigid segments that move relative to each other

thermal equilibrium   the state when two systems are at the same temperature

thermistors   electronic instruments for measuring temperature

thermocline   a region of rapid change in temperature in the water column

thin layers   regions of the water column where a change in a property is confined to a relatively short depth interval, on the order of a few meters

three-cell model of atmospheric circulation   the classic model of atmospheric circulation based on the theoretical existence of three atmospheric convection cells in each hemisphere

threshold velocity   the fluid velocity above which a sediment particle goes into suspension

tidal bore   a wavelike advance of the tide in a narrow bay or river

tidal bulges   the theoretical accumulations of water that result from the gravitational forces of the Moon and the Sun, as developed by Sir Isaac Newton

tidal constituents   the individual contributions of the gravitational forces of planetary bodies to the tides

tidal currents   flows of water associated with tides

tidal datum   a fixed point on the shore from which tide heights may be measured

tidal day   the time of one complete tidal cycle; about 24 hours and 50 minutes

tidal forces   the horizontal (tangential) gravitational forces of the Moon and Sun that cause the tides

tidal lag   the time of delay between the passage of the Moon overhead and the high tide

tidal period   the time between successive high or low tides; about 12 hours and 25 minutes

tidal prism   the volume of water in an estuary contributed by the ocean, typically estimated as the difference between sea level at the mean high and mean low tides

tidal range   the difference in tide height between high and low tides in a tidal period

tidal wave   the long-wavelength Kelvin wave produced as a result of dynamic forcing of the tides; in essence, the tides

tide gauge benchmark   a fixed marker on a hard substrate on which measurements of tide height can be based

tide height   the vertical height of the tide above the zero tide height, the tidal datum, or a tide gauge benchmark

tide pool   a seawater-filled depression in rocks visible when the tide retreats

tides   the daily periodic rise and fall of the sea surface due to horizontal movements of water

tombolo   a depositional landform that connects an island or rock to a coastline

top-down control   a food web in which the top predators control the abundance and dynamics of lower trophic levels

top predators   the highest trophic level in a food web

towed-camera sled system   a sled-mounted camera system used for photographing the seafloor

trace elements   elements whose concentration in seawater is less than 1 part per million; see *minor constituents*

tracers   substances that can be used to track the movements of water masses and their mixing with other water masses

transform boundaries   a plate boundary where two plates move horizontally relative to each other

transform fault   a type of strike-slip fault that connects offset oceanic ridges and permits motion on a sphere

transgressive barrier   a type of barrier island that moves landward through accumulations of sediments on its landward side or erosion of sediments on its seaward side

transitional food webs   "temporary" food webs that appear in response to changing ocean conditions, such as upwelling food webs and ice-edge food webs

trench suction   a feature of subduction zones whereby the gravitational pull of the subducted plate causes the trench to move towards the subducting plate

trophic cascades   the changes that occur in a food web when one trophic level is disturbed

trophic efficiency   the fractional amount of energy or material transferred from one trophic level to the next

trophic level   broadly, the type of feeding exhibited by an organism or group of organisms in a food web; used to identify the role of organisms in food webs and understand their ecology

trophic pyramid   a graphical representation of the numbers, biomass, or energy in different trophic levels

tropical disturbance   the first stage in the development of a hurricane; a region of unstable atmospheric conditions in the tropics

trough   the low point in sea level during passage of a progressive wave

tsunamigenic   any force that causes a tsunami

tsunami generation   the force that causes a tsunami

tsunami impact   the interaction of a tsunami with a coastline

tsunami propagation   the movement and changes in path of a tsunami from its source

tsunami run-up   the maximum vertical height of a tsunami on an impacted shoreline as measured from sea level

tsunami scattering   changes in the path of a tsunami as a result of refraction and interactions with the seafloor

turbidites   sediment deposits formed from turbidity currents, characterized by graded bedding

turbidity current   a flow of suspended sediments down an underwater slope

turbulence   chaotic fluid flows

turbulent boundary layer   the depth to which wind-induced turbulence penetrates in the water column

turbulent eddies   swirls of water produced as a result of turbulent fluid flows

turnover time   the time it takes the contents of a reservoir to be completely replaced

twilight zone   a term used to describe the mesopelagic zone; the boundary between the photic zone and the aphotic zone

U

Udden-Wentworth scale   a classification system for sediment sizes

ultraslow ridge   an oceanic ridge that exhibits very slow rates of spreading

undersampling   statistically, a lack of sufficient data; oceanographically, an inability to adequately measure the complete range of temporal and spatial scales in the ocean

undersaturated   when the concentration of a substance is less than the saturation concentration

underwater acoustics   the study of sound within the ocean

unstable water column   a water column whose layers are not at neutral buoyancy

upwelling   the movement of subsurface water to the surface caused by Ekman transport or interactions of currents with seafloor features

US Exploring Expedition   one of the first scientific expeditions to explore the world ocean; carried out from 1838 to 1842

V

valence   the number of single bonds that can be formed by an element

vapor pressure   the pressure exerted by a gas above a liquid

Vine-Matthew-Morley hypothesis   an explanation for the pattern of magnetic anomalies on either side of spreading ridges based on sea-floor spreading

virioplankton   planktonic viruses

visible light   solar radiation at wavelengths between 400 and 700 nanometers

vortex ratio   the length-to-width ratio of the tube of a plunging wave, used to characterize and identify different types of plunging waves for surfing

W

Walker circulation   the atmospheric circulation of the equatorial Pacific responsible for El Niño and La Niña

warm-core rings   Gulf Stream eddies composed of warm, Sargasso Sea water surrounded by colder Labrador Current water

warm regimes   for a given oceanic region, a multiannual of multi-decadal period of warmer-than-average seawater temperatures

wasp-waist control   a food web in which organisms at middle trophic levels control the abundance and dynamics of lower and higher trophic levels

water column   an undefined column of water stretching from the surface to depth

water mass   an unspecified volume of water with a narrow range of temperature and salinity

water type   an unspecified volume of water defined by a particular temperature and salinity

wave   a sinusoidal movement of energy along the surface of the ocean or at the interface between water layers with different densities

wave diffraction   the bending or spreading of waves around an object or through a gap

wave energy spectrum   the frequency of wave energies present at a particular time

wave frequency   the number of wave crests passing a given point in a specified period of time

wave focusing   the convergence of wave fronts on an undersea feature or headland

wave height   the vertical distance between the crest and trough of a wave

wavelength   the horizontal distance between wave crests

wave period   the time between successive wave crests

wave refraction   the bending of wave fronts as one part of the wave feels bottom

wave steepness   the ratio of wave height to wavelength

wave tank   a long and narrow container for studying waves

wave trains   groups of waves traveling together

wayfaring   a type of ocean navigation used by the people of Oceania

weather   the instantaneous state of the atmosphere or ocean

weather forecasting   the prediction of weather conditions at a future time

well-mixed   when the water column is homogeneous in all of its properties

well-sorted   a sediment sample whose grain sizes are the same or similar

westerlies   winds blowing at higher latitudes equatorward of the polar front

western boundary currents   surface currents that flow along the western side of ocean basins

western intensification   the increase in velocity of western boundary currents due to variations in the Coriolis effect with latitude

wet magma ocean hypothesis   the idea that primitive Earth's water was temporarily held within the molten interior before its release to the surface

white smokers   hydrothermal vents that emit light-colored particles

whole-genome shotgun sequencing   a technique for identifying gene fragments of natural seawater samples

Wien's law   a mathematical formula that describes the wavelength of peak emission of radiation for a given temperature

Wilson cycle   a description of the life cycle of an ocean basin from opening to closing

winds   movements of air

wind shear   the force exerted by the wind on the sea surface

wind-wave resonance   a dynamic interaction between wind and waves that enhances the rate at which energy is supplied to waves

Winter Monsoon   the seasonal atmospheric condition that causes winds to blow off continental landmasses toward large bodies of water

winter solstice   the first day of winter; when the Sun is located at 23.5 degrees below the celestial equator (N. Hemisphere)

world ocean   all of the waters contained in oceanic basins

world ocean circulation   the coupled surface and deep circulation; the complete circulation of the world ocean

Z

zircon   a colored or colorless neosilicate crystal, often used as a gemstone; also found as inclusions in ancient rocks

zonation   the clustering of organisms into distinct bands along the shoreline, usually according to tide height

zooplankton   animals suspended in water whose movements are generally slower than water motions associated with currents and tides

zooxanthellae   single-celled, symbiotic, photosynthetic protists found in the tissues of stony corals

Credits

Front Matter: Photo by Jonathan Bird/www.jonathanbird.net

Chapter 1

Opener NASA/Goddard Space Flight Center; **Figure 1.2a** W. Sean Chamberlin/Maureen Conte; **Figures 1.2b & c** NOAA; **Figure 1.2d** W. Sean Chamberlin; **Figure 1.4** © CORBIS; **Figure 1.a** © T. O'Keefe/PhotoLink/Getty Images

Chapter 2

Opener NASA/JPL; **Figure 2.1** Dr. Clifford E. Ford; **Figure 2.3** NASA/JPL/Cal Tech 2005; **Figure 2.4** NASA; **Figure 2.7** © Digital Vision/PunchStock; **Figure 2.9** NASA; **Spotlight Figure 2.1**/pg. 18 © PhotoLink/Photodisc/Getty Images; **Figure 2.10** ESA; **Figure 2.15** NASA/CXC/L. Townsly/Harvard; **Figure 2.16** NASA; **Figure 2.b**/pg. 24 NASA-Jet Propulsion Laboratory

Chapter 3

Opener USGS; **Figure 3.a**/pg. 32: © Dynamic Graphics/Jupiter Images; **Figure 3.b**/pg. 44 US Navy Photo

Chapter 4

Opener NOAA/DIL; **Figure 4.2** NOAA; **Figure 4.5** NOAA/NGDC; **Figures 4.10, 4.11** NOAA; **Figure 4.14** Courtesy Sean W. Chamberlin; **Figures 4.15, 4.16** NOAA; **Figure 4.19** Courtesy Dr. Patricia McCrory, USGS; **Figure 4.21** University of Washington; **Figures 4.22, 4.23** NOAA; **Figure 4.24** Map by and courtesy of G. Pavan, CIBRA, University of Pavia, Italy; **Figure 4a** Nancy Penrose, NEPTUNE Program/University of Washington; **Figure 4.25** Bill Haxby; **Figure 4.26** Charles Fisher, Pennsylvania State University

Chapter 5

Opener Courtesy Carey Wein; **Figure 5.2** Dave Caron; **Figures 5.5, 5.7** Ocean Drilling Program; **Figure 5.8a** © Punchstock; **Figure 5.8b** Colleen Chamberlin; **Figure 5.8c** Sean Chamberlin; **Figure 5.12** Richard Goode; **Spotlight Figure 5a** © Brand X Pictures/Punchstock; **Spotlight Figure 5b** Sean Chamberlin; **Figure 5.20** NOAA

Chapter 6

Opener Sean Chamberlin; **Figure 6.1** Sean Chamberlin; **Spotlight Figure 6a** Image Courtesy of Norman Kuring, Sea WiFS Project; **Figure 6.14** Sean Chamberlin; **Spotlight Figure 6b** NOAA

Chapter 7

Opener NOAA; **Figure 7.2** US Coast Guard; **Figure 7.3a**: NOAA; **Figure 7.3b** Ken Johnson, MBARI; **Figure 7.5** Scripps; **Figure 7.6**

NASA; **Figure 7.7** NOAA/NASA; **Figure 7.9** Sean Chamberlin; **Spotlight Figure 7a**/pg. 121 Johnson Space Center; **Figure 7.12** © Stockbyte/PunchStock; **Figure 7.20** Royalty-Free/CORBIS; **Figure 7.27** Wayne Wurtsbaugh/Utah Sate University; **Figure 7.30** NASA/GSFC

Chapter 8

Opener NOAA; **Figure 8.2a** Peter West/National Science Foundation; **Figure 8.2b** NOAA; **Figures 8.3, 8.8** Sean Chamberlin; **Figure 8.10** © Digital Vision/PunchStock; **Figure 8.20** NASA/GOES; **Figures 8.22a-b** NASA; **Figures 8.23a-b** NOAA; **Figure 8.24** NCEP/NCAR; **Spotlight Figure 8a**/pg. 157 U.S. Navy; **Figure 8.32** NOAA/NCEP; **Figures 8.33a-b** NASA

Chapter 9

Opener Cary Wein; **Figure 9.1** Benjamin Franklin and Timothy Folger's First Printed Chart of the Gulfstream/© Science Vol. 207, 1980, Fig. 1, p. 644; **Figure 9.7** NASA; **Figure 9.8a** NASA; **Figure 9.9b** NOAA; **Figure 9.10b** Courtesy Mark Abbott/NASA; **Figure 9.13** Frank Manaldo/Johns Hopkins Applied Physics Lab, Ocean Remote Sensing Group; **Figure 9.16** NASA/GSFC; **Spotlight Figure 9a**/pg. 184 Tommy Dickey

Chapter 10

Opener Greg Huglin; **Figures 10.1, 10.2** Sean Chamberlin; **Figure 10.3** G. Carl Schoch; **Figure 10.4** Sean Chamberlin and COMET Program at UCAR; **Figure 10.12** NOAA; **Figure 10.14** Scripps; **Figure 10.18** © PhotoLink/Getty Images; **TA Figure 10.2**, top, Royalty-Free/CORBIS; **TA Figure 10.2**, bottom, Sean Chamberlin; **Figures 10.21, 10.23** Sean Chamberlin; **Figure 10.24** NASA; **Spotlight Figure 10.1a** Global Reach of 26 December 2004 Sumatra Tsunami/SCIENCE Vol. 309, 23 September 2005, p. 2045-2048

Chapter 11

Opener Sean Chamberlin; **Figure 11.3** NOAA; **Figures 11a-d** NOAA CO-OPS; **Spotlight Figure 11a**/pg. 225 © Creatas/Punchstock; **Figures 11.16a-b** David Pugh; **Figure 11.18** NOAA; **Spotlight Figure 11b**/pg. 232 NASA

Chapter 12

Opener Sean Chamberlin; **Figure 12.1** Charles Darwin in "A Monograph of the Sub-class Cirripedia with Figures of all the Species," Ray Society, London, 1854/© John van Wyhe 2002-4; **Figure 12.4** Stockbyte/Getty Images; **Figures 12.6a-c** Sean Chamberlin; **Figure 12.6d** © Jeremy Woodhouse/Getty Images; **Figure 12.7a** © Dee Breger; **Figure 12.7b** Oystein Paulsen; **Figure 12.7c, 12.8a** Sean Chamberlin; **Figure 12.8b** © Image 100/PunchStock; **Figure 12.8c** © Georgette

Douwma/Getty Images; **Spotlight Figure 12a**a/pg. 250 © Digital Vision/Getty Images; **Figure 12.11** © image 100/PunchStock; **Figure 12.12a** University of Hawaii/Manoa/Hawaii Ocean Time Series; **Figure 12.13** NOAA; **Figure 12.14** Tommy Dickey; **Figure 12.15** Donna Stern; **Figure 12.17** © 2002, MSU Bioglyphs Project/T. Karson, photographer; **Figure 12.19** NASA; **Figure 12.20** NOAA; **Spotlight Figure 12b**/pg. 260

Chapter 13

Opener © image100/Punchstock; **Figures 13.1, 13.5** Sean Chamberlin; **Figure 13.6** NAS; **Table 13.1 (1)** MHHE/DIL; **Table 13.1 (2)** © Digital Vision; **Table 13.1 (3)** © Don Farrall/Getty Images; **Table 13 1(4)** BW Productions/PhotoLink/Getty Images; **Table 13.1 (5)** NASA; **Table 13.1 (6)** Sean Chamberlin; **Figure 13a**/pg. 282 Field et al. SCIENCE 281, 1998, Fig. 1, p. 238; **Spotlight Figure 13.1** NASA

Chapter 14

Opener © James Gritz/Getty Images; **Spotlight Figure 14a**/pg. 295 © Jeremy Woodhouse/Getty Images; **Figure 14.14** Sean Chamberlin; **Figure 14.19a** Dr. R. Grant Gilmore; **Figure 14.19b** Lance Horn, National Undersea Research Center/University of North Carolina at Wilmington; **Spotlight Figure 14b**/pg. 305 NOAA

Chapter 15

Opener Sean Chamberlin; **Figure 15.1** Ruben Moody; **Figures 15.8a-c; 15.23** Sean Chamberlin

Chapter 16

Opener © Stockbyte/Punchstock; **Figure 16.1** NOAA; **Figure 16.2** © Creatas/Punchstock; **Figure 16.4** Sean Chamberlin

Index

A

abrupt climate change, 6
absolute zero, 114
absorption, of light, 129, 131, 132
absorption coefficients for light in pure
    water, 131
Abyss, 337
abyssal circulation model, 167
abyssal fans, 59, 60
abyssal hills, 60
abyssal plains, 60, 62
abyssal zone, 252
abyssopelagic zone, 252, 253
accessory pigments, 272–273
acetone, 269
acids, 102
Acoustic Doppler Current Profilers
    (ADCPs), 184, 339
acoustics (discipline), 4
acoustic technologies for seafloor mapping,
    53–54, 339
acoustic thermometry, 56
active acoustic technologies, 56, 339
active margins, 58–59
activity model of the scientific method, 5
actualism, 14
adaptations of ocean life, 253–258
    aggregation, 256–257
    bioluminescence, 256, 257–258
    photoadaptation, 253–255
    vertical migration, 252, 255–256
adiabatic changes, 124, 125
Adriatic Sea, 65
Aegean Sea, 65, 240
Aegean Sea Plate, 36
aerosols, as condensation nuclei, 144
aesthenosphere, 43
Africa
    dust storm, satellite image of, 97
    sediment discharge by, 81
African Plate, 36, 65
Age of Fishes, 250
aggregation, 256–257
Agnatha, 251
Agulhas Current, 172, 185
Agulhas Retroflection, 175
aircraft expendable bathythermographs
    (AXBTs), 115
air guns, seafloor mapping with, 53, 54
air-sea interface, 143–145
    clouds, formation of, 144
    defined, 143

evaporation and precipitation, 143–144
    sea ice, role of, 145
    wind stress and turbulence, 144–145
Airy, George Biddell, 57, 192
Alabama, Hurricane Katrina damage,
    209, 210
Alaska, earthquake (1964), 68
Alaska Current, 172, 174
albedo, 45, 145
Alberti, Leon Battista, 140
Alboran Sea, 65
Aleutian Trench, 37
Alfred P. Sloan Foundation, 259, 337
algae
    coralline, 326
    in ice-edge ecosystems, 298
    as phytoplankton, 271
algal reef ecosystem, 325, 326
allometric relationships, 248
altimeters, satellite-based, 232
Altiplano Plate, 36
Alvin (ROV), 63, 338
Amazon River (South America),
    discharge, 315
American Geophysical Union, 35
American Mediterranean, 67
American Samoans, Palolo-cavier feast, 256
*America's Living Oceans: Charting a Course for
    Sea Change,* 10
ammonia
    as a base, 102
    in greenhouse effect, 24
    in primary production, 274
ammonification, 277
ammonium
    in phytoplankton growth, 276
    in upwelling ecosystems, 297
amphidromic points, 228, 229
amphipods, as infauna, 247
Amundsen-Scott South Pole Station
    (Antarctica), 65
Amundsen Sea, 65
Amur Plate, 36
analytical tools, developing, 10–11
Anatolia Plate, 36
anchovies
    below-average temperatures favorable to,
        295, 304, 306
    damage to fishery from El Niño, 155
    as most abundant pelagic fish, 304
    wasp-waist control by, 292–293
Andes Mountains (South America), 37, 38
Andrew, hurricane (1992), 209

anemometers, 140
anglerfishes, 257, 258
angular velocity, 146–147
animals, classification of. *See* classification
anions, 95–96
anoxia, 100, 329
Antarctica
    glaciers, melting and growing of, 231
    ice, thinning of, 133
    melting of ice shelves, 259
    U.S. research stations, establishment
        of, 65
Antarctic-Australian Basin, 65
Antarctic Bottom Water (AABW), 179,
    180, 181
Antarctic Circle, 352
Antarctic Circumpolar Current, 172,
    176, 177
Antarctic Convergence, 181
Antarctic ice sheet, 133
Antarctic Intermediate Water (AAIW),
    180, 181
Antarctic Ocean, mixing of Indian and
    Antarctic Ocean waters, 65
Antarctic Plate, 36
Antarctic Polar Front Zone, 181
Anthropocene Era, 81
    Exploration Activity for, 88–89
anthropogenic carbon dioxide, 102–104
anticyclones, 148
Anvers Island (Antarctica), 65
apex predators, in classical food web, 293
Appalachian Mountains (United States), 60
appendices, 342–352
Aqua, 131, 205, 270, 271
aquaculture, 305
aquatic habitats, human uses, 314
Arabian Plate, 36
Arabian Sea, 100
Archaea (domain), 243, 244, 260
Archean (time period), 41
    cratons, 40
    Snowball Earth hypothesis and, 45
Archimede, 338
Archimedes principle, 57, 123, 231
Arctic Basin, 57, 64
    ice-edge blooms, 299
Arctic Bottom Water (ABW), 180
Arctic Circle, 352
Arctic ice cap
    global warming, effect of, 231
    melting of, 133
Arctic ice floes, expedition to, 168